Benjamin Hossbach   **Christian Lehmhaus**

(德) 本杰明·胡斯巴赫   克里斯蒂安·雷姆豪斯   王晨晖 译

[phase eins].

The Architecture of Competitions 1

建筑竞赛 1

2006 – 2008

辽宁科学技术出版社

Christine and Luis
Theresia, Vera, Judith, and Arne

Thanks for putting up with us.

感谢克莉斯汀、路易斯、特蕾莎、薇
拉、朱蒂斯和阿恩对本书的大力支持

Contents
目录

Benjamin Hossbach, born 1966, and
Christian Lehmhaus, born 1963, are the managing
partners in [phase eins]. They both graduated in
architecture from the Technical University of Berlin.
They are members of the Berlin Chamber of
Architects, the German Institute of Architects (BDA),
the German Association of Consulting Engineers
(VBI), and the German Association for Sustainable
Construction (DGNB).

本杰明·胡斯巴赫（1966年生人）与克里斯蒂安·雷
姆豪斯（1963年生人）是[phase eins].公司的股东。
两人均毕业于柏林技术大学的建筑专业，是柏林建筑
师协会、德国建筑师联盟（BDA）、德国工程咨询协
会（VBI）和德国永续建筑协会(DGNB)的成员。

In 2006, we published a first collection of projects resulting from architectural competitions, from 1995 to 2005. [phase eins]. had the honor to provide a leadership role in these instances. They are all outstanding examples of topical architecture and urban design in Germany and were awarded numerous prizes. This is a clear proof of the positive impact of architectural competitions on the success of construction projects.

Because many of the competitions took place over a rather short period of time, our portfolio has grown to include many exciting projects in less than two years. Thus, to publish a sequel to "The Architecture of Competitions" seems the most obvious thing to do.

As in the first book, the competition projects are presented along with essays on the issues that have concerned us all along, i.e. competitions in general and competition management in particular. The focus of the competition projects and the accompanying essays is on international competitions.

We invited some of our fellow-architects to enrich this volume by giving their impressions and their thoughts about international competitions. All have taken part in numerous competition procedures, either as competitors or as jurors, and they could easily fill a whole book by themselves with accounts of their "Adventures in Competition Land". We extend to them our thanks for their willingness to share their experiences.

Likewise hidden in the wings of true-to-life perspectives, flawless floor layouts and elaborate models are our colleagues' sense of commitment, the high costs they are bound to incur, and the admirable courage of the sponsors who willingly bore those costs in the end. This is why we express our thanks equally to the architects who contributed to this volume with their designs and to our clients, including those whose projects have not made it into the book. The concern of both parties with the quality of the built environment, their ambition to enhance it with projects of their own, forms the basis of the work we do at [phase eins]. It provides us with constant inspiration.

This book goes out to all who – through their interest and involvement in creating architectural excellence – express their concern for transparency in the planning and contract award process, be they architects, specialist planners, academics, private or public officials involved, or developers and principals of building projects: Heroes, all, of the "Construction World"!

Benjamin Hossbach
Christian Lehmhaus

项目的同时也刊登一些国际竞赛和国际竞赛管理问题的短篇文章。

本书邀请了一些曾多次以参赛者或评奖委员会委员身份参与国际竞赛程序的资深建筑师发表他们的看法和见解。十分感谢他们愿意与我们一起分享他们的心得和经验，同时感谢各位同仁出于强烈的责任心提供的大量完美的平面布置图和精心设计，也要感谢主办方最终愿意承担高额成本所表现出来的令人钦佩的勇气。同样我们也要感谢为我们撰稿的建筑师、设计师和我们的委托人，正是他们共同的努力构成我们[phase eins].公司的工作基础，激发我们源源不断的灵感。

本书旨在为那些追求建筑完美的爱好者和实践者服务。无论建筑师、规划专家、学术研究者、相关的个人或官员，还是开发商和建筑项目的委托人，应该说，"建筑行业"的全部精英们均可从本书中获益。

本杰明·胡斯巴赫
克里斯蒂安·雷姆豪斯

我们在2006年出版了第一本建筑竞赛项目设计作品集，收录了1995年至2005年的优秀建筑竞赛项目，[phase eins].公司有幸担任领导角色。获奖项目的设计者所创造的建筑及总体规划均为德国典型建筑和城市设计中的杰出典范，获得了诸多奖项。这十分有力地证明建筑竞赛对于成功建成建筑项目的积极影响。

因为许多竞赛举办时间短，所以我们的作品集中只收录近两年内的优秀的建筑项目。再出版一个"建筑竞赛"的续集尤显必要。和第一本一样，我们在提供国际竞赛

[essays].
[短章]

Statements
观点陈述

Architecture, Taste, and Culture
建筑、品位和文化

Rodolfo Machado, professor in practice of architecture and urban design at Graduate School of Design, Harvard University, and partner in Machado and Silvetti Associates, Boston, USA

Rodolfo Machado，美国哈佛大学设计院建筑学与城市设计应用教授，美国波士顿Machado and Silvetti联合公司的股东。

It has been deeply interesting for me to have participated, as a juror, in two international urban design competitions organised and managed by [phase eins]. First it was the Jabal Khandama competition in Mecca, Saudi Arabia, in 2006 and then the Administration Complex competition in Tripoli, Libya, in 2007.

They were, in a way, similar; I am not just referring to their scale, their dual "by invitation plus open" mode of operation, or the similarities in submission requirements and types of representation: their similarity came from the cultural estrangement of their contexts.

In a few words, what tended to split the body of jurors among the well known lines of "locals and visitors" – as in any good football game – were, above all, matters of architectural taste and the language of images, architectural representation and cultural identity, formal abstraction versus figuration, and such. In other words, the pragmatics of iconography became a central issue. After a few debates it became obvious that the gap between Western architecture and the local vernacular traditions is still real.On the other hand, the power of architecture to represent culture, to stand for – or act as a provider of – a perceived identity, appears to be firmly in place across the board. All of these efforts at communication through architec-ture had in fact a happy ending (or it seemed so to us when the verdicts were announced; one should wait and see what is built before rejoicing). In both cases the best projects won, after much explaining of the reasons for their success, and their advantages of all kinds.

[phase eins]. in their skilled carving of a different professional niche (and/or in finding one more way, other way, to practice architecture and urbanism) was contributing, as much as our professions, to a better world.

我对于参与由[phase eins].公司组织和管理的2006年麦加和2007年的黎波里举办的这两次国际城市设计竞赛，担任评奖委员会委员一直乐此不疲。这两次竞赛不仅在比赛规模、运作模式、参赛要求及设计类型方面有很大的相似之处，在设计内容所反映文化的差异性方面也有很多相同点。

影响评委不同意见的最重要的因素是对建筑的品位、图像的表达方式、建筑的表现手法、文化的特性、严肃抽象还是比喻表达等等。换言之，图解的用语是一个关键的问题。经过多次研讨辩论之后，我们十分清楚地认识到西方建筑与地方民俗传统之间的差距仍然真实地存在。另一方面，建筑对文化的表现力和对同一性的认知力表现都十分恰当。我们认为，在宣布评审结果的时候，这些通过建筑进行的交流确实有了美满的结局。对获奖作品成功之处及其优势进行点评之后，最佳的设计作品胜出。

[phase eins].公司以其专业技能，为我们营造更好的环境，其贡献与我们各位同行一样卓著。

Bottom: Professor Rodolfo Machado (right) and Dr Ahmed M. Shembesh at the jury meeting in Tripoli

下图： Rodolfo Machado教授（右侧）和Ahmed M. Shembesh博士在的黎波里的评奖委员会会议上

Competitions in the East?
在东方的竞赛？

Dr. Sc. Vladimír Slapeta, dean of the faculty of architecture, Brno University of Technology, and professor of history of architecture

Sc. Vladimir Slapeta博士，布尔诺科技大学建筑系主任兼建筑史教授。

Architectural competitions have a long tradition in the Czech Republic, going back to the days of the Habsburg Monarchy. Some worthy of mention designs resulting from competitions have become icons of architectural modernity in our country. These include designs by Josef Gočár for Prague's Old Town Hall (1909) and by Jaromír Krejcar for the Parliament Building on the Letná Plain (1928), as well as Kamil Roškot's entry to the urban-design competition for the new Parliament and Government Complex in Saint Peter's Quarter (1930), and the post-World War Two Collective Residence Project in Litvínov by Eugen Linhart and Václav Hilský (1946).

Since the mid-1980s, the younger generation opposed the restrictions in architectural competitions. In late 1987, the officially recognised Association of Architects was finally forced to launch a completely free architectural competition, the first of its kind after more than 15 years. The tricky task was no less than the extension to the Old Town Hall. It was the eighth competition dealing with an issue that had been occupying Czech architecture for almost a hundred years. As many as 227 architects submitted entries. Even though the decision by the jury, all regime puppets, was more than questionable, the exhibition of the designs attracted immense crowds of visitors and convinced the general public of the need to encourage archi

Bottom Dr. Sc. Vladimír Slapeta (far right) with co-jurors and competitors at the participants' colloquium in Kiev

下图：Sc. Vladimir Slapeta博士（最右侧）与其他评奖委员会委员及参赛者在基辅的专题座谈会上的合影

tectural debate and to juxtapose diverse concepts.

A short time later, in the autumn of 1988, Lower Austria launched a two-stage urban-design and architecture competition for its new seat of government in St. Pölten. The procedure was also open to architects from countries bordering Austria. As a representative of East European countries, I served on the jury of this competition, which sported a field of 166 competitors. The Czech and Slovak entries to this competition fared very well, winning three of a total of nine prizes, and I and my Slovak co-juror gained much important experience, some of which we were able to pass on to architectural circles back home.

The 1990's witenessed the revival of architectural competition.

Initially, these competitions did not run smoothly – no wonder, since Czech and Slovak architects lacked experience with this kind of procedure. After the long break there were only two aged architects who, before World War Two, had acted as jurors, and only a few who had any background in the matter from the days of the Prague Spring in 1968. Thus, a renowned French architect serving on a jury on behalf of the principal was disconcerted when jurors and their alternates took turns appearing, in random fashion. Frequently the work of a juror was not remunerated at all.

Beginning with the establishment of the Czech Chamber of Architects in 1993, competitions took on a more definite format. However, I am sorry to say that in the Czech Republic – probably as in some other East European countries – formalities frequently loomed large. Perhaps this comes from the canonical and doctrinaire belief in regulations and in their efficacy, as a consequence of the deplorable legal situation in these countries under communism. Whereas, a jury should be allowed to interpret rules creatively and magnanimously – after all, nobody can predict what sort of things creative architects will come up with.

And even though today, twelve years after this event, attitudes are more relaxed, in the Czech Republic, more than in any other country, too many competitions have a judicial sequel, some competitors suing the jury or taking the dispute to the Chamber of Architects.

One cause of this shameful state of affairs may be the emergence, within the Chamber of Architects, of a class of officials that see themselves more as bureaucrats than as architects. Supported and encouraged by the attorneys employed by the Chamber, for whom formalities *per se* take precedence over content and concept, they unfortunately do not realise that excessive formalism in such matters can discredit the institution of the architecture competition as a whole, nationally and internationally. Competitions have always been the salt in the soup of architectural discussion. Without them, we would be all much poorer – neither the city of Brasília nor the Sidney Opera, the Berlin Philharmonic, the Grande Arche of La Défense, or Berlin's Potsdamer Platz would have seen the light of the day without the architectural competition. From this point of view, I find it a matter of utmost importance that henceforth competitions, not only on the national level but also on the global stage, be under the guidance of a competent, specialised firm of architects such as [phase eins]. This practice will help to set new standards, raise existing ones, and further develop the culture of architectural competitions across borders – also towards the East!

建筑竞赛在捷克共和国历史悠久，可以追溯到哈普斯堡君主制时期。从竞赛中脱颖而出的一些设计师如今已经成为我国建筑的现代主义杰出代表。

自从80年代中期以来,年轻的设计师们一直反对建筑竞赛中的诸多限制。1987年末,受到官方认可的建筑师协会最终不得不举办一次15多年来首次完全自由的建筑竞赛，多达227名建筑师报名参赛。尽管评奖委员会成员都是政体的傀儡，其获奖结果的裁定令人质疑，但是设计作品的展示吸引了无数的参观者，使公众相信了鼓励建筑学讨论和不同建筑观念并存是非常必要的。

不久以后，在1988年的秋季，下奥地利州举办了一次城市设计和建筑竞赛，以争取其在圣波尔顿政府中的新的一席之地。这次竞赛也对奥地利邻国的建筑师开放，我作为东欧国家的一名代表，担任评奖委员会委员。捷克和斯洛伐克参赛者在这次竞赛中表现出色，拿到全部的九大奖项中的三个，而我本人也获得了许多重要经验。

90年代出现了建筑竞赛的复兴。最初，这些竞赛运作得并不顺利——这一点不足为奇，因为捷克和斯洛伐克的建筑师缺乏竞赛程序方面的经验。当时评奖委员会委员的工作根本没有任何的酬金。1993年，捷克建筑师协会成立以后，建筑竞赛才有了更加明确的格式。然而令人遗憾的是，在捷克共和国，也许就像在其他的东欧国家一样，拘泥形式问题非常严重。可是，我们是应该允许评委创造性地解释规则的，毕竟无人可以预测到富有创意的建筑师会提出什么样的设计。

尽管在12年后的今天人们的态度更加放松随意，但是在捷克共和国有很多的竞赛在举办之后都有司法程序。这是因为一些参赛者，或者起诉评奖委员会，或者提交争议由建筑师协会仲裁。这种情况可能是因为在建筑师学会中出现了一群认为自己更是政府官员的人，他们在协会受雇律师的支持下，认为形式重于内容和观点，因而没有意识到在建筑竞赛这类事情上过分形式主义可以在国家范围内、甚至在世界范围内败坏竞赛机构的名声。一直以来竞赛都增加了建筑讨论的趣味性，如果没有竞赛，我们的思想就会变得贫瘠，也就不会有像巴西利亚城、悉尼大剧院、柏林交响乐团、拉戴芬斯区的新凯旋门和柏林波茨坦广场这些伟大的建筑出现。从这一点来看，我发现无论是国家级别的竞赛，还是世界级别的竞赛，有像[phase eins].公司这样有能力的专业性公司的指导是非常重要的。这种作法会有助于确立新标准、提高现有标准、进一步发展跨国界建筑竞赛的文化——也包括东方！

Architectural Competitions
建筑竞赛

The most established figure on the Jordanian architecture scene, Jafar Tukan, runs a thriving practice in Amman. Mr. Tukan is active in professional organisations for architecture, engineering, protection of the historic built environment, and the fine arts in Jordan and Lebanon.

Jafar Tukan，约旦建筑界的最著名人士，在安曼从业，生意红火。Tukan先生十分积极地参与在约旦和黎巴嫩的建筑、工程、有历史意义的环境保护及美术等专业组织的活动。

In the Middle East until only a few years ago, awareness of organizing competitions was still unfamiliar. The idea of the prospective client was, in most cases, that a competition is like window shopping for a commodity. Clients where not aware of the cultural as well as the economic implications that the creative process of designing a building involves. They were not also aware of the significance of the evaluation process and criteria.

For all the above reasons I took the decision not to participate in any competition where the international rules of competitions are not strictly applied, implemented, and followed through.

A few years ago I was invited by [phase eins]. to be a member of the jury for a competition they were organising for a project in Abu Dhabi. That experience was very impressive to me. It took care of all the anxieties that used to discourage me from participating in competitions. There, you find a team dedicated to the organising process from day one, meticulously preparing a detailed and thorough brief, assigning fair compensation for all competitors, selecting the jury members very meticulously, defining very clear milestones for meetings, questions and answers, and submissions, with complete transparency.
In this way [phase eins]. represented to me the full solution to the chronic problem of competitions in the

Middle East and, I am sure, in many other places in the world. To me, they were the first office that specialises in organising architectural competitions as I have not heard of such specialised offices before. A very important feature of their service is that, because of their professionalism, organisation, and transparency they could attract the top names in the field of architecture to offer their services to [phase eins]. clients.

在中东，竞赛的组织安排一直是不广为人知的，这种现象一直持续到几年以前才有所好转。在大多数情况下，有潜力的委托人认为竞赛就像是某种商品的浏览橱窗一样，而没有意识到设计建筑这个创造性的过程也包括了文化以及经济影响，更没有意识到评估过程和评估标准的重要意义。

由于上述这些原因，我选择不去参加那些不严格应用、贯彻和坚持国际竞赛规则的竞赛。几年前，我受[phase eins].公司之邀，为阿布扎比一个项目举办的一次竞赛担任评奖委员会委员。那次经历对我来说非常难忘，抵消了过去阻碍我参加竞赛的种种焦虑。因为这个团队在组织过程中从始至终一直兢兢业业，精心准备详尽全面的任务说明，向全部参赛者支付合理酬金，认真选择评奖委员会委员，对会议的重要事件、问题答案及设计作品均作出明确的规定……并且完全公开、透明。

因此，在我看来，[phase eins].公司代表了中东及世界许多其他地方的有关竞赛长期存在的弊病的完全解决方案。[phase eins].是我所听说过的第一家专门组织建筑竞赛的公司，他们的专业性、组织性和透明性足以吸引建筑领域顶级人物的加入，这是他们服务的一个重要特色。

Bottom Jafar Tukan (center right) and Gulzar Haider, chair of the jury, in Amman

下图： Jafar Tukan（右中）和评奖委员会主席 Gulzar Haider在安曼

A new Administration Complex in Tripoli
的黎波里的新行政办公建筑区

Professor Peter Zlonicky, architect and urbanist, visiting professor at the universities of Venice, Trento, Zurich, and Vienna; honorary member of the Leibniz Institute for Regional Development and Structural Planning, Berlin

Peter Zlonicky教授，建筑师兼城市规划专家，威尼斯大学、特兰托大学、苏黎世大学和维也纳大学的客座教授，柏林莱布尼兹地区发展和结构规划协会的荣誉会员。

The first large-scale urban-design and architecture competition in Libya: With no previous experience in this area, people here still expect internationally competitive standards and results that will reflect this. Accordingly, the sponsor shows quite some interest in the process, the architecture firms being invited and in the jury.

The competition scenario meets the expectations of our Libyan colleagues and matches their level of interest.

- A top-level competition management prepares the competition brief, proposes potential competitors and structures the procedure clearly and transparently in all its essential stages.
- An international field of high-performance firms with experience in similar projects takes on the task *in situ*.
- From the moment the competition entries are opened, equally experienced jurors establish a true team spirit, advise the sponsor, and hold discussions with all the competitors.

For most of the invitees this was their first time working in Libya. Accordingly, several questions were asked of the Libyan sponsor:

- Should the architectural messages reflect the state's self-image? The answer to this question is remarkable for its relaxed self-assuredness: We do not need any representational architecture, we need no dominant motif for the city; we want the architectural quality to communicate the specific character of the place.
- Is the new administrative complex to stand on its own or be an integrated part of the city of Tripoli? Answer: It would be best to strike a proper balance between the specificity of the site and its structural integration with the rest of the city. However, the competitors should use their own judgment in deciding.
- Are there trends in Libyan architecture, traditional elements the new concept should embrace? Answer: The sponsor rather expects designs that can set an international example. It would be great to achieve an icon representing the new Libya. But there are also the traditional design aspects to consider, of light and shadow, courtyards and gardens and materials suitable for a good quality of life in the Libyan climate.
- Are there planning laws, are there standards to be complied with? Answer: Sure, there are quite a number of regulations governing planning and construction in Libya, but here it is a matter of giving international perspectives that are most pertinent.

The yield: There is a clear overall winner: the entry by Léon Wohlhage Wernik Architekten. An emblematic sign within the cityscape, embedded among olive groves. A

Bottom Professor Peter Zlonicky
at the jury meeting in Tripoli

下图：Peter　Zlonicky教授在的黎波里
的评奖委员会会议上

beautiful park framed by congress buildings, mosques, and the secretariats. An architecture that translates traditional elements into novel forms. The unanimous decision in favor of this entry reflects the outcome of a common learning process. While, to begin with, the jurors have different preferences, they ultimately settle on the criterion of a special message for the project, one that would inform the self-image of Libyan governance. With their high-quality design Léon Wohlhage Wernik Architekten fulfill these expectations.

What remains? Best-practice competition procedures are planted on Libyan soil. All parties expressed appreciation for the competence with which the competition was managed. The jury's work was well received – by a host having no previous standard to judge by and whose focus quickly turned to a speedy realisation of the project.

What permanent impressions will be left? – Meticulous preparation and close cooperation between the sponsor and the competition managers provided an optimum basis for the work of all who took part: the impartiality, openness and intellectual curiosity of both Libyan and international members of the jury, the readiness to listen, to clarify positions through deliberations, the sponsor's respectful attitude towards the contributions of the competitors, and wonderful hospitality.

这是利比亚首次大规模城市设计和建筑的竞赛。虽然在这一领域没有经验，我们仍然期待能有国际性的竞赛标准和成果。与此同时，主办方对竞赛过程、受邀建筑公司及选择评奖委员会委员都表现出极大的兴趣。这次竞赛的方案符合我们利比亚同事的期望和利益。

一家世界顶级的竞赛管理公司准备竞赛任务说明，推荐参赛者，公开透明地明确所有必要的组织程序环节。
选择国际范围内的一些有进行类似项目实际经验的高效能的建筑事务所。

从竞赛开始之日，由同样经验丰富的评奖委员会树立起一个真正的团队精神，对主办方提供建议，与所有参赛者进行讨论。
这次竞赛是大多数的受邀参赛者在利比亚的第一次。与此同时，向参赛者提问关于利比亚主办方的几个问题：

问：建筑作品表现主题应不应该反映国家的自我形象？
答：（该问题的答案因其轻松的自信而引人注目）我们不需要任何有代表性的建筑，也不需要表现城市占主导地位主题的作品。我们需要的是能够表现地方特色的优秀的建筑作品。

问：新行政办公建筑区是单独存在还是作为的黎波里整体的一部分存在？
答：最好是能够在表现自身特征和与城市其他建筑整合两个方面找到一个适当的平衡点。然而，参赛者应该自己进行决断。

问：在利比亚，建筑新概念应不应该包含传统因素的趋势？
答：主办方更愿意设计师的作品能够国际化。如果能够设计出一个代表新利比亚的地标性建筑固然很好，但是也应该考虑到传统设计方面的因素，比如光线和阴影，乡村、花园和其他适合表现利比亚生活质量的材料的运用。

问：有没有规划方面的法律？有没有需要遵守的标准？
答：当然，在利比亚有很多规划和建筑法规，但是这次竞赛的目的是提供非常恰当的国际视角。

竞赛的成果：十分明显，这次竞赛的大赢家是Leon Wohlhage Wernik建筑事务所。他们的作品用新颖的形式诠释了传统的要素，评奖委员会一致决定其获胜。开始评委的意见不同，但是最终确定了这个能够反映利比亚政府管理形象项目的评定标准，这是共同学习进步的结果。Leon Wohlhage Wernik建筑事务所用高质量的设计满足了他们的期望。

竞赛的影响：在利比亚开始有了最佳的竞赛程序，参与各方对竞赛管理人员的能力赞赏有加，评委会虽然没有既定的评价标准，但是也受到了好评。

持久的影响：主办方和竞赛管理公司的认真准备和密切合作：程序公平、公开，评委求知欲强、善于倾听然后深思熟虑作出决定，主办方对竞赛者作品的尊重和欢迎，为参与各方的工作提供了最宜的基础。

International Competitions

国际竞赛

International Competitions
国际竞赛

Architectural competitions thrive on communication between those involved. An international competition implies the encounter of different cultures that adds excitement, complexity, and the chance to learn from each other to the procedure.

国际竞赛参与各方之间的活跃沟通，增加不同文化交流的刺激性和复杂性，为相互学习竞赛程序提供机会。

What do the competitions documented in the present volume have in common? Let us look at the competitions for the concert-theater venue in Amman, the Darat King Abdullah II for Culture and Arts, Jabal Khandama and Jabal Omar developments in Mecca, Eco Bay in Abu Dhabi, the Administration Complex in Tripoli: Apart from the fact that all these projects are highly complex and located in Arab countries, what they have in common is that their principals made exacting demands as to the quality of planning, in each instance they wanted something special and they were willing to do everything required to achieve this goal.

Besides being willing, they were open-minded and determined to win over foreign professionals to take part. They wanted them to contribute their know-how and services as competing planners, and to act, as needed, as neutral jurors, experts, moderators, and competition manager. Consequently, all the relevant procedures had an international character.

Along with the similarities, there were also, however, marked differences, which did not always originate from the projects themselves. They had to do with the actors involved and the local specifics. Thus, it may be rewarding to take a closer look at both the general and specific skills international competitions call for, and the special issues they raise.

Today, projects are being carried out routinely across the world in the same way that the economy as a whole communicates and acts on a global scale. Thus, what was out of the ordinary for our clients was not the issue of employing foreign consultants but that like many German clients it was their first time working under the rules of a well-structured competition procedure which by virtue of its international character had that extra dimension.

In many European countries, architectural competitions are a tried and tested instrument used both for planning and awarding planning contracts. But even in the public sector in Germany (traditionally a stronghold of architectural competition) making competitions mandatory is being more and more questioned. On the one hand this is due to the fact that, in general, public authorities are retreating from their role as principals, and on the other it is a response to what seems to be an increasing formalisation of European public procurement law. Frequently, the Europe-wide announced competition is being replaced by a negotiated procedure as per VOF (allegedly because it is easier to control), or by a partial or complete private funding approach that steers clear of the public procurement law. However, in the private sector an opposing trend can be observed, not least at the inter-

Bottom Jurors Manfredi Anello, Faisal Al-Banani, Donald Bates, and Jafar Tukan visiting the competition site in Tripoli

下图：评奖委员会委员们参观的黎波里的竞赛现场

Top Jury colloquium in Tripoli
上图：在的黎波里的评奖委员会专题座谈会

national level. Here, principals do not simply submit to mandated procurement guidelines like in Europe; instead, if they are convinced by the strength of strategic arguments, they deliberately choose the competition as an instrument of structured design optimisation and procurement of planning services. This makes the competition procedure more significant and offers the prospect of it becoming an international standard.

The question how a well-ordered competition procedure with its fundamental principles of fairness, equality of opportunities, project assessment by an independent jury of architects, transparency, etc. can be established as a norm at the international level is answered in the same way as when the question is asked at the national level. The common point is a basic acceptance of the principle of "performance through competition." While this concept has long gone unchallenged in sports and in many other social and economic contexts, it is not taken for granted in the cultural setting (which is where, we emphatically believe, architecture and urban design belong), and this urgently needs to be understood.

The necessity of establishing this connection between competition and performance is felt most poignantly in countries where this concept has not been generally accepted as yet. With regard to competition as a procurement instrument, it is undeniable that clearly structured, transparent procurement procedures are

less widespread in many countries (particularly non-European ones) than in Germany, where it is taken, if not as a matter of course, then as a desirable goal.

Add to this the importance of a discourse-based approach to preparing and making planning decisions. This is generally accepted neither in Germany nor in other democracies, and there are even more remarkable instances of projects carried out in countries still in the initial stages of democratisation or guided by quite different societal models. The lesson of the discourse-based approach is that even in such societies competition can be successfully utilised as a planning and procurement tool.

What counts are not just societal structures, but people. The principals must be open to novel approaches, possess an interest in maintaining quality, respect the performance of the experts and be willing to make a commensurate commitment of their own. On their part, the architects in turn must possess a high degree of openness to the demands of the client, be ready to take unconventional approaches when required and refrain from imposing upon their clients' solutions that are outside their experience.

The disregard of these general conditions may result in a competition becoming a purely academic exercise without any usable outcome. Soon, such an exercise will be replaced by one where the principal directly commissions an architectural firm that he is already familiar with. Or else the client will resort to concurrent and multiple commissioning in order to obtain a range of choices, an inefficient and expensive method.

Regardless of how carefully and responsibly a principal proceeds with such a course of action, it still constitutes a risk for both sides, one that can be avoided once the opportunities provided by an orderly competition are more fully known.

In our experience, what was decisive for the success of the projects described here was that we were proactive in encouraging the principals to get involved in the process, and that we "localised" the procedures by including colleagues from host country, thus avoiding the imposition of a ready-made catalogue of actions, a strategy just all too typical of the Euro-centric attitude. There may have been a time when the idea of providing "development aid" to some parts of the world by importing a prefabricated system of rules, competitors and jurors may have been justified, but in this day and age, when in almost any country we can find highly educated counterparts to work with, it requires a different approach if the competition concept is to be successfully "imported".

The conclusions to be drawn from running international competitions can be reintroduced into current discussions in Germany and Europe about reorganising the competition industry, in particular with regard to such aspects as anonymity, the principal's obligation to honor the promise that the competition winner will be commissioned, copyright issues, and the question of how these aspects can best be handled in procedural terms so as to satisfy the needs of all the parties involved. We will return to these issues with the aid of concrete examples.

Irrespective of prevailing general political or economic conditions, we can say that optimum results cannot be achieved without an intelligent, open attitude on the part of the decision makers. Every knowledge-able property developer or building department head is aware of the fact that optimum general conditions must be obtained from the earliest stages of a develop-ment project, and that the chances for optimisation in terms of cost-saving or quality enhancement decreases as the project progresses. This is true for every country in the world, regardless of the project's location. This makes it all the more incomprehensible that prin-cipals frequently do not regard making adequate budgetary provisions for this initial project phase as something essential. The examples described in this book, however, provide convincing proof that it is possible to provide a tailor-made competition procedure that respects budgetary constraints for any project.

The crucial steps for project optimisation must be taken as early as the pre-design phase, when deci-sions on the proposed location, concept, and program are being made. It is during this "phase one" that it is

essential to prepare and make the fundamental choices determining the project for best results. To this end, all the experts and stakeholders involved in the planning process should meet at a "round table". And quite often the economic and political conditions guaran-teeing a project's success have still to be created, both internally, i.e. on the part of the principal, and exter-nally at the level of civic administrations, financiers, and the general public. With international competi-tions this is no less true than with German ones.

To demonstrate successful use of "the planning tool" (the name we use to describe archi-tectural competition), it is best to take a closer look at some concrete examples.

Jabal Khandama Development Project in Mecca, Saudi Arabia

At the end of 2005, Jabal Khandama Development Co., a subsidiary of the Saudi Arabian Fakieh Group, commissioned us to prepare and conduct an urban-de-sign and architectural project competition for a devel-opment area of some 60 hectares, adjacent to Mecca's Grand Mosque. Residential projects, hotels, and supply facilities for some 100,000 pilgrims and residents were to be realised on this site. In addition, optimal perspec-tives of the Grand Mosque and in particular the Kaaba were to be provided, and a traffic concept was to be developed that guaranteed optimum accessibility to this central part of the City of Mecca. Owing to the task's complexity, the procedure was run in two stages, the first of which produced a great variety of possible solutions without unduly stressing the resources of the participating firms or burdening the budget with a collateral increase in architectural fees. Also, this type of procedure allows one to optimise, based on the outcome of stage one, the tasks set for stage two.

An international field of ten architecture and planning firms took on the challenge, and in September 2006 an equally international jury, in conjunc-tion with the developer and the High Commis-sion of the City of Mecca, chose the entry by the French-Lebanese consortium Atelier Lion archi-tectes urbanistes (Paris), Dagher, Hanna & Partners and Abdul-Halim Jabr (both from Beirut).

The experiences gained from the competition, the technical discussions during the preparatory stage and the colloquia as well as the wide acceptance of the first-placed proposal show clearly the importance of making the proper procedural choice and its positive impact on the quality of the planning process.

Administration Complex in Tripoli, Libya

In autumn 2006 we were asked to coordinate an international architectural competition for the new governmental administration complex in Tripoli, the capital of Libya. On a 240-hectare area south of the present city, some 30 edifices were to be planned, among them the parliament building, various ministries and a hotel. The jury, composed (as with the Saudi Arabian project) of nationals and non-nationals, met in May 2007, and the result was announced in June 2007 within the framework of a UN-Habitat conference. The winner was the Berlin-based firm Léon Wohlhage Wernik Architekten.

With this procedure, we succeeded once more in establishing sound principles for preparing and conducting competitions in a country where, at first sight at least, one would not have expected a propensity to accept such principles.

It is certainly true that there exist at present certain conditions in Libya that favor the opening-up of the national market to the global marketplace. With this comes a huge demand for international know-how and services. Unlike in Saudi Arabia, our client was a public entity, and it was remarkable to see how matter-of-factly they dealt with the many steps of the procedure, with technical and content issues as well as formal and process aspects. Also, the communicative competence and the personal commitment of our Libyan counterparts were much closer to the European attitude toward these issues than was the case in Saudi Arabia. This facilitated cooperation in many respects .

Darat King Abdullah II in Amman, Jordan

In July 2007, the Greater Amman Municipality commissioned [phase eins]. to coordinate a competition for the concert and theater building "Darat King Abdullah II" in Amman. The Royal House preferred

Left Impressions from the competition procedures for the Jabal Khandama Development and the Jabal Omar Development projects in Mecca

左图：来自麦加项目竞赛程序的印象

this course of action to commissioning a stellar international architecture firm directly. Of all the international competition projects presented here, this seems the one most firmly rooted in the native soil and most likely to be implemented. Surely this has to do with the economic situation of a country that is poor in raw materials and obliged to utilise its resources as carefully as any European country.

If we were to indicate criteria that allow one to predict the success of a project, we would point, for example, to the promptness and precision of the answers given by the principals to questions asked by external advisors, to their linguistic competence and the regularity of their attendance at meetings. In all these categories, our counterparts in Amman score highest. This does not denigrate in any way the personal competence of the sponsors of other projects described here. Despite difficult general conditions, the Amman lineup raises high hopes for the future of this project.

The decision to award two equal first prizes confirms this assessment as it reflects the client's generosity and responsible management in economising financial resources.

Conclusion

So, what are the differences between competitions conducted outside of Germany, and in the Middle East or North Africa in particular, as compared to similarly ambitious competitions here back home? In order to be able to give an answer, to point out differences and similarities, one should pose the question with respect to each of the different parties involved in a procedure: How do the motives, goals and methods of the clients differ (if at all)? Is there a difference in terms of what is expected of competitors and jurors, or of what the competitors themselves expect? And, last but not least, what is expected of the competition manager?

To open the discussion of possibly different motives, goals, and methods in different countries (Germany as local example, the Middle East and North Africa as international ones), we will address the question: What motivates a developer or indeed any client to launch a competition in a situation not governed by the constraints of contract-award regulations? The answer from international examples does not markedly differ from the answers one gets in Germany: What is sought is the surplus value added to a project by architectural excellence, meticulous preparation, fine-tuned content elements, etc. – all of them advantages to be obtained in both scenarios. A certain difference may be due to the fact that international expertise may introduce novel aspects to the project; that involving international expertise is then offered as proof of neutrality in the preparation and conduct of the procedure.

The expectations competitors face regarding the number of plans, diagrams, and models to be delivered and the degree of detail required from visualisations will, as a rule, be higher outside of Germany. This may be because non-German decision makers, compared to their counterparts in Germany, ask for more detailed and more easily readable visual representations as a basis for their decisions. As a rule, they do not accept the degree of abstraction currently accepted in Germany. This attitude will persist at least as long as such demands are complied with. In addition, the required amount of technical and structural detailing, of representation of services, circulation and logistics concepts is considerably higher than in Germany.

The demands put on the manager of an international competition are radically different from those put on the coordinator of a German regional or national competition. In particular, as the client barely has any preconceived ideas about the standard and conduct of the procedure, the competition manager can – and must – question the reasonableness of each of the usual steps. However, as a result one can say that many sub-processes of run-of-the-mill competition procedures, in terms of structure, are transferable to international competitions as well. In the end it is precisely the presentation of these ready-made components during the acquisition stage that often tips the balance in favor of a commission. There are a number of fundamental differences that spring specifically from the international character of the competition and affect every single one of its aspects: language problems, cross-cultural differences in understanding certain concepts, differences in standardisation and metric systems, and the sheer size of logistical problems. Also, as a rule, competition briefs must be prepared in several languages, meetings be moderated in a language that often is not the coordinator's mother tongue, and travel must be managed (including visa procurement) so that all those involved are at the right place at the right time. Reviewing all the projects coordinated by us over the recent years, a trend can definitely be seen, not only in the Arab countries, but also in Eastern Europe and Central Asia.

To begin with, the projects presented here are definitely strokes of luck, and as yet nobody can say if these extraordinary events will be able to serve as models

for "normal" projects. All the same, it is remarkable that in all these cases either a second competition was commissioned after the conclusion of the first one or there is a serious interest in doing so. This, too, confirms that our approach is reasonable, successful and acceptable to principals and the other parties involved, be they competitors, jurors, or politicians.

To conclude, one can say, in view of the examples presented, that architecture and planning competitions are efficient instruments of planning and commissioning on the international level. Their structural differences, compared to similarly ambitious German ones, are smaller than initially expected. Their genuine differences are less due to their being conducted outside of Germany and more to the fact

Top View of Tripoli's Old Town and harbour
上图：的黎波里老城区和海港的鸟瞰图

that they are competitions on an international scale and, importantly, transcend local borders. Any differences as to motives, goals, and methods of the principals can be found by comparing public clients to sponsors from the private sector, but even this method does not yield a clear-cut result in the countries in question, nor, for that matter, in various others.

And how will the clients from Saudi Arabia, Libya and Jordan assess the issues here discussed?

Apparently, the procedural principles that are important in this context, namely equity, transparency, the binding force of the commission promise, judging by an independent, competent jury, and careful preparation of the procedure (and, thus, of the project) are well appreciated by those clients who have gained plenty of experience with what can happen (and what, as a rule, actually happens) when these same principles are disregarded.

And this kind of experience, together with the knowledge that German engineering amounts to more than just DIN standards and technical advantage, but that it also means adherence to schedules, a fair amount of dependability and incorruptibility, and providing a well-structured framework – all of these things are perceived as virtues which the international public associates with Germany and German ways. These are the foundations on which to run successful competitions overseas.

Top Scenes from the participants' and jury colloquia and the jury meeting during the Administration Complex competition in Tripoli

上图： 在的黎波里行政办公建筑区项目期间，参赛者和评奖委员会委员举行的专题座谈会及评奖委员会会议的场景

Top Jury meeting in Amman
(front LTR: M. Schindhelm, O. Maani, D. Staples,
J. Bolles-Wilson, L. Attel, K. Kada)
上图：在安曼的评奖委员会会议

目前所举办过的有文件证明的竞赛的共性是什么？除了参与竞赛的全部项目都在阿拉伯选址、建筑结构复杂之外，他们的共性还包括项目委托人对规划质量的要求明确，喜欢项目与众不同并为此不遗余力。此外，他们思想开放，决心争取外国专家加入，希望他们或作为竞争者来贡献自己的技术和服务，或根据需要作为中立评奖委员会委员、专家、监考员、竞赛管理人员做事。因此，所有相关程序都有国际性的特点。

然而，国际竞赛也有许多明显的不同。这些不同不总是因项目本身而产生，而是与项目的参与者和当地的特殊性有关。因此，更深入了解国际竞赛要求的一般和专门技能及其相关特殊性的问题是非常有益的。

欧洲范围内的竞赛正被一种协商程序、或者是一种摆脱政府采购控制的私人部分出资或全额出资的方法所代替。然而，在私人范围内，人们可以发现一种完全相反的趋势，项目委托人慎重地把竞赛看成是结构设计最优化和获取规划服务的一种工具。这使得竞赛程序更为重

要，甚至使其国际标准成为可能。如何把以光明正大、机会平等、建筑师评奖委员会独立进行项目评估、公开透明等为基本原则的井然有序的竞赛程序确立为国际水准的建筑竞赛规范？其回答与确立国家竞赛规范的方法相同，即要首先认同在运动界和社会经济领域一直占绝对优势地位的"通过竞争表现"这一原则。建立竞赛与表现的联系，在至今仍未普遍接受这一观念的国家尤显必要。用一种对话交流形式来进行准备和作出规划决定的重要性更加强调了这一点。即便在民主国家和一些仍处于民主初级阶段的国家或不同社会结构的国家里，竞赛作为规则和实现规划目标的工具也可以被成功地加以运用。

真正重要的不仅仅是社会结构，而是社会中的人。项目委托人必须乐于接受新方法，对永续的质量怀有兴趣，尊重专家的工作，愿意履行自己相应的义务。建筑师必须非常乐于接受客户的要求，也应乐于采用非常规的方法，禁止用客户不了解的方法来欺骗他们。轻视这些一般条件可能会使竞赛成为一种纯粹的学术实践，这样就

不会产生任何实际的成果。

根据我们的经验，这些项目之所以成功的决定性因素是我们比较鼓励项目委托人参与到竞赛程序中来，通过吸收主办国建筑师使竞赛增添地方色彩，从而避免一成不变的程式化模式。在当今这个时代，任何一个国家的建筑师都非常有水平，要想成功引进竞争观念，就需要使用不同的方法。通过举办历次国际建筑竞赛我们总结出来的经验，在讨论建筑竞赛行业重组问题时，可以被重新引入，尤其是关于匿名、项目委托人践行其诺言使竞赛获奖者可以获得委托的义务、版权问题以及为满足参与各方需求在程序上如何对这些问题作出最佳处理等问题。可以说，不考虑政治、经济这样大的条件，决策者没有理性的、开放的态度就无法取得最好的结果。有知识的开发商或建筑商都很清楚最佳条件必须从开发项目的初期阶段获得，随着项目的进行，节约成本、加强质量方面的最佳条件会逐渐降低。不管项目的选址如何，这一点适用于任何国家。因此，项目委托人通常更不理解在工程初期作出充分预算规定的必要性。本书列出的项目实例充分说明提供一个重视预算限制的"量体裁衣"的竞赛程序的可能性。

制定项目优化方案的关键步骤必须早在设计之前，即选址、构想和计划的阶段进行。为了达到这一目的，参与规划的专家和股东应该召开"圆桌会议"。很多情况下，保证项目成功实施的政治、经济条件仍然可以从项目内部的项目委托人和项目外部的民政、财政和大众中努力获得。

为了证明建筑竞赛这一规划工具的成功运用，我们最好更加仔细地研究一下下面具体的实例。

沙特麦加的Jabal Khandama开发项目：

2005年末，Jabal Khandama开发公司委托我们为邻近麦加大清真寺的一块60公顷的地段准备并举办一次城市设计和建筑项目竞赛。由于任务复杂，竞赛程序分为两个阶段。第一阶段不过分强调参与竞赛的建筑事务所的实际资源，也不因为建筑费用间接增长来限制预算。然后在第一阶段成果的基础上设定第二阶段的任务。全世界范围内10家建筑规划事务所接受挑战，参加竞赛。2006年9月评奖委员会同开发商及麦加最高委员会一起选出了获胜者。

此次竞赛获得的经验、在准备阶段进行的技术性讨论和专题座谈会以及对首次提议的广泛认可清楚地表明制定适当竞赛程序的重要性及其对于规划过程和质量的积极影响。

利比亚的黎波里的行政办公建筑区：

2006年秋季，我们应邀为的黎波里的新行政办公建筑区协调一次国际性的建筑竞赛，计划在城南一片240公顷的区域兴建大约30座大厦，其中有议会大厦、各部委办公楼和一家酒店。2007年5月召开评奖委员会，6月按照联合国人居署的框架规定，宣布了评选结果。获胜者是柏林的Leon Wohlhage Wohlhage Wernik Architekten建筑事务所。

在这次竞赛程序中，我们又一次成功地在一个一开始很难接受这些规范的国家里确立了准备和举办竞赛的健全的规范。现在利比亚具备有利于国内市场向全球开放的条件。相应的，对国际技术和服务的需求也非常巨大。与沙特不同，我利比亚的项目委托人是一个公开实体，他们处理程序步骤、技术和内容以及形式和进程问题时都本着实事求是的态度。此外，利比亚同行的沟通能力和个人承诺与欧洲对这些问题的态度更为接近。这些在很多方面都有利于竞赛参与各方的合作。

约旦阿曼的Darat King Abdullah II音乐大剧院：

2007年7月，The Greater Amman Municipalityw公司委托[phase eins].公司为阿曼的音乐大剧院Darat King Abdullah II组织协调一次竞赛。王室更愿意直接委托一家国际性的大建筑事务所来做。阿曼原材料匮乏，必须像欧洲国家一样慎用资源。

如果要指出预示项目成功的标准的话，我们会说，比如，

是项目委托书对项目外部的咨询机构所提出问题的迅速和精确回答,是项目委托书的语言能力,是项目委托人出席会议的定期性。尽管综合条件艰难,但是安曼阵容提高了项目前景的厚望。项目委托人肯定了评奖委员会的决定,授予两个一等奖,反映其在经济财政资源方面的慷慨和责任管理。

结论:

所以,与在德国本土举办的同样抱负不凡的竞赛相比,在德国以外其他地方举办的竞赛和单在中东或北非举办的竞赛之间有什么不同?为了回答这一问题,提出异同点,我们应该对竞赛程序的参与各方提出下面的问题:委托人的动机、目标和方法有什么不同?委托人对参赛者和评奖委员会委员,或者参赛者对自己的期望有不同之处吗?最后,但也是重要的一点,对竞赛管理人员有什么样的期望?

为了公开讨论不同国家可能存在的不同动机、目标和方法,我们提出下列问题:开发商或任何委托人在不受合约判授规则的限制的情况下推动竞赛举办的动机是什么?从国际竞赛

实例中得出的答案与在德国我们得出的答案并无明显不同:两者追求的都是通过建筑的卓越、精心的准备和优质的内容元素等来追加项目的剩余价值。某一分歧的产生可能是因为国际性技术可能把一些新颖之处引入到项目中来,也可能是因为所涉及的国际性技术在程序准备和实施过程中只作为一种中立的证明出现。

德国以外的地方在规划方案、图表和提供设计作品的数量和形象化要求的详细程度方面,对竞赛者的期望要更高一些,这可能是因为外国决策者与他们的德国同行相比要求更详细、其形象表现要求更易懂,这样他们才能作出决策。他们通常不接受现在德国普遍认同的抽象程度。只要同意项目决策者的这些要求,这种态度就会继续存在下去。此外,要求的技术含量、结构细部特征的含量、室内设施网描绘的含量、发行和物流观念的含量也比德国的高得多。

国际竞赛对竞赛管理人员的要求与德国地区或国家级竞

赛对项目协调人员的要求截然不同。尤其是委托人对于竞赛程序的标准和管理没有任何成见，所以竞赛管理人员能够、也必须对程序的常规步骤的合理性提出质疑。因此，一般化的竞赛程序的许多关于结构的子流程也可以转而用于国际竞赛中去。最后，在项目采集阶段准确地提供这些现成的要素经常起着决定性的作用。有许多主要的分歧是因为竞赛的国际性质产生的，竞赛的方方面面都受到影响：语言问题、理解一些观念时跨文化的分歧、标准化和公制的差别、物流问题。此外，通常竞赛任务说明书必须用几种语言准备，会议必须用一种通常不是项目协调人员的母语来调节并且安排交通安排事宜,以使参与竞赛的各方在规定的时间到达规定的地点（包括购买护照）。

重新审视我们近年来参与协调的全部项目，我们肯定会看到一种趋势，这种趋势不仅在阿拉伯国家存在，而且在东欧和中亚的国家存在。

首先，这里所提供的建筑项目肯定是偶然的，到目前为止，我们不能说他们代表"正常"项目。但是他们证实了我们的方法是合理的、成功的，得到了项目委托人和其他不论是竞赛者、评奖委员会委员、还是政客等项目涉及到的所有参与各方的认可。

总之，鉴于这些实例，我们可以说，建筑和规划竞赛是国际平台上建筑一种规划和委托的工具。与类似规模的德国竞赛相比，他们在结构上的差异比最初所期望的小得多。这种差异的产生更多地是因为竞赛的国际规模大大超越地方的界限。通过比较公共项目委托人和私人项目委托人，我们可以发现委托人的动机、目标和方法，但是这种方法没有在这些国家里产生明确的结果。那么，沙特、利比亚和约旦的项目委托人们会怎样评价这里讨论的这些问题呢？十分明显，公平、公开、委托承诺的约束力、评奖委员会的独立称职、精心准备这些程序等竞赛程序原则深受那些委托人的欢迎，因为他们对如不遵守这些原则会造成的后果有充分的经验。有了这种经验，了解德国工程不仅仅意味着德国标准和技术优势，而且德国工程也坚持按照预定计划、相当的可信和清廉性、提供结构良好的框架——这些都是国际公认的德国及德国方式的优点，也是成功举办国际竞赛的基础。

Left Visit to the competition site,
participants' colloquium, Eco Bay Project in Abu Dhabi
左图：参观竞赛现场、参赛者的讨论会和阿布扎比的生态湾项目

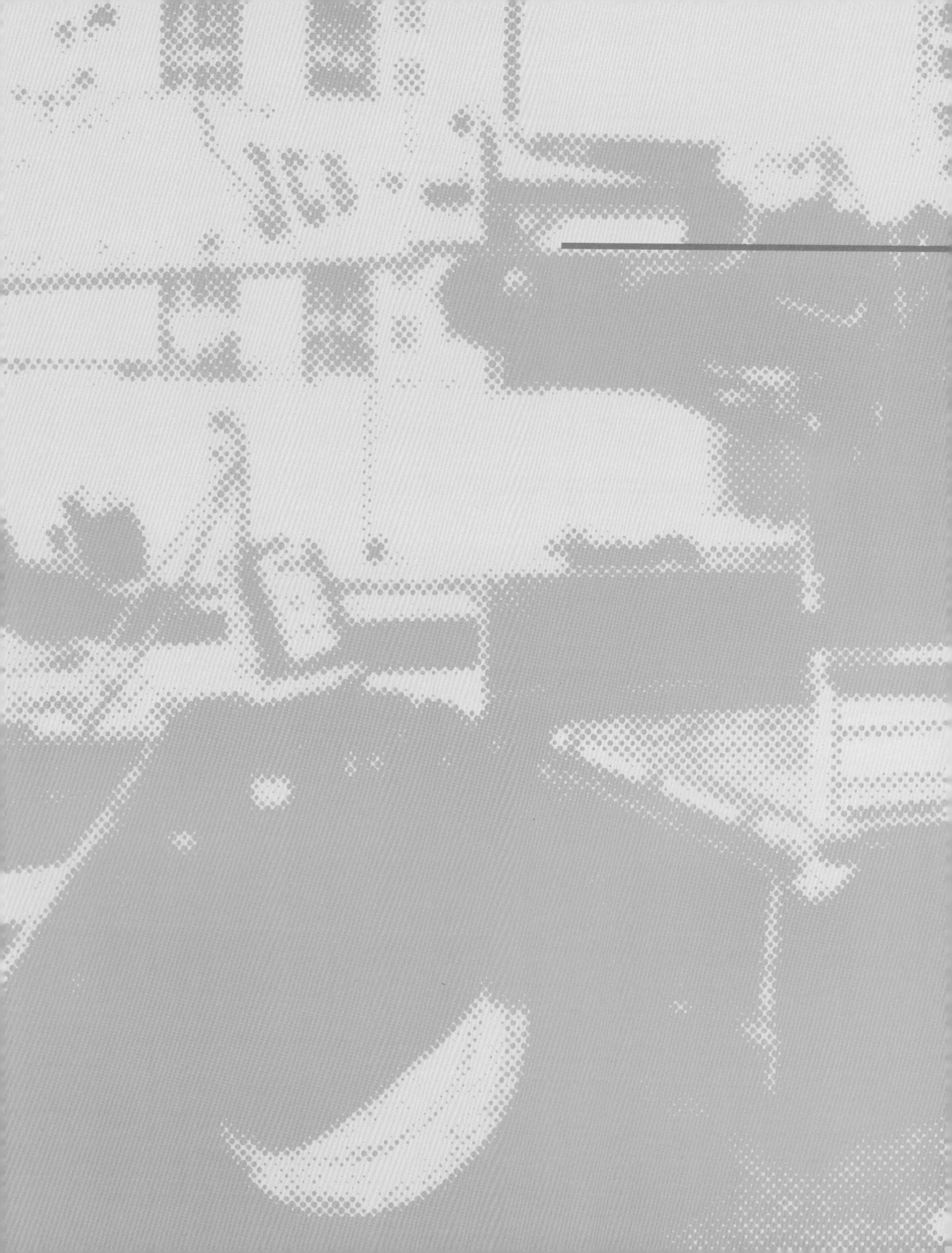

Choosing the Competition Procedure

竞赛程序的选择

Choosing the Competition Procedure
竞赛程序的选择

Who is eligible to take part? How much time does the procedure require? What are the costs? Who are the ones who make the decisions? These are some of the many questions raised during the delicate process of deciding what kind of procedure to use in a competition.

谁有资格参加竞赛程序？需要多长时间？费用多少？决策者是谁？这些问题都是在决定竞赛程序的这个微妙过程期间提出的众多问题中的一些问题。

Of similar importance for the success of a team working on a project is the choice of the proper procedure for selecting the architect. The first thing is to clearly state that there should indeed be a choice, as even this fact is frequently overlooked. Various standard procedures exist, each with its own "controls" for fine-tuning purposes, thus making the number of available procedural forms huge or, at times, unmanageable. There is no single form that can be applied to every project, nor is there only one procedure applicable to a given project.

The decision on the procedural form must be made at the very beginning of a project. It is all the more important to prepare the decision calmly and with the support of competent consultants. Any kind of procedure is a highly abstract entity, miles away from the hands-on experience of lively jury discussions and the chance to view actual designs.

Consultants take on a huge responsibility; they alone (even in the best of cases) are familiar with available options, can weigh them against each other, explain them to the sponsor, and defend their choice. In assessing the advantages and disadvantages of the different procedural modes, dogmatic arguments should be avoided. The important thing is to make sure that a competition is the correct

option in the first place. Choosing the most appropriate approach, rather than a quest for the "ideal" competition procedure is the overall goal.

There is little awareness of how important it is to assess the total value a competition adds to a project; consider how much a competition speeds up the subsequent planning and approval processes. How much more efficient is a design obtained from a transparent and instructive selection process than one obtained in other ways? And how much more acceptable is such a process in the eyes of the general public? Such questions can rarely be answered with hard figures, but they are pertinent still, as long as the process is well coordinated and all those involved act as one.

After the political strategic feat of convincing the client that it is basically desirable to have a competition, comes the much simpler task of detailing the technical, legal and financially strategic subtleties of the various procedures and describing the available "fine-tuning controls" mentioned before. The first issue to address is on what legal basis the competition is to be held. Are there comprehensive national laws or equivalent regulations (in a given instance) for conducting competitions, and is the procedure perhaps part of a public contract-award procedure? Could it be an international competition based on relevant UNESCO principles and UIA guidelines? In all the pertinent rulebooks is an almost universal consensus as to the definition of the fundamental principles governing all competitions. These are principles first evolved when the earliest competitions were held in the 19th century: "equal opportunities for all competitors", "adequate remuneration", "unequivocal language describing the commissioning criteria", "independence and competence on the part of the jury", and "precise formulation of tasks". The potential for customising the procedure resides in these tantalising questions: What means should be employed to attain these goals? How may one use, in a responsible fashion and without violating basic principles, the interpretive leeway contained in recommended instruments and pathways? Who will be eligible to participate? Architects only? Or engineers as well? How will competitors be found, by public announcement or by an internal selection procedure? How many jurors are needed to advise the principal in the decision process? Is the procedure to be conducted anonymously, or does it make sense to permit, during the period of competition itself, more immediate

Bottom Jury meeting during the competition for the ThyssenKrupp Quarter, Essen (LTR: Dr. E. D. Schulz, chair of the board ThyssenKrupp AG; Dr. M. Grimm, CEO ThyssenKrupp Real Estate; Kaspar Kraemer, president BDA; B. Hossbach)

下图：德国埃森的蒂森克虏伯办公建筑物举办竞赛期间所进行的评奖委员会会议

dialogue involving the jury, the competitors, and the expert consultants? What is the available time span, and might the procedure be split into two stages so as to yield, initially, a wider choice of ideas? What fees will be paid to the competitors, what prizes will be awarded, and what deliverables must be submitted? What are the rules governing the authors' copyright? A final but not inconsiderable point: It is crucial to answer all these questions unequivocally, before the start of the competition proper and to instruct all the parties involved accordingly, otherwise misconceptions could spoil the project from the start.

Once we admit the major importance of the procedural choice for a project's success, the crucial issue within this process, in our experience, is quite a different one. Here, too, the images conjured up by the metaphors of matchmaking, role casting, or headhunting have their equivalent in the reality of the competition manager's work: It involves a meticulous and comprehensive description of the design task. Only if the task is described with total clarity, consistently and realistically can the architects (the potential soul-mates, the actors and the headhunted) find proper answers and provide the designs required. In other words: Only then can the client and the architect find each other.

为建筑师选择适当程序对于项目团队的成功是非常重要的。首先要明确地声明选择确实必须存在。每一种不同标准的程序都有自己的优质"操作控制器",因此可用的程序形式众多,或者是有时难以管理。不存在形式适用于所有项目的情况,也不存在只有一种程序适用某一项目的情况。

在项目最开始的时候必须作出程序形式的决策。心平气和地准备和争取能干的咨询专家的支持更加重要。任何程序都是一种高度抽象的,远非像评委在辩论会上举手表决和浏览设计作品那么简单。

咨询专家的责任重大,他们自身熟悉可能得到的选择项,能够权衡并向主办方进行解释,为其选择进行辩护。在评估不同程序模式的优劣时,辩论应该避免教条主义。重要的事情是首先确认竞赛是正确的选择,竞赛的总体目标是选择最适合的竞赛程序,而不是寻求理想的竞争程序。

很少有人能意识到评估竞赛附加到项目的全部价值的重要性。想想竞赛对推动后续规划和批准过程的作用有多大,经过透明性强、指导性强的选择,竞赛程序所得到的设计比用其他方法得到的设计更有效率。这种过程在大众看来更易接受的程度为多少?这些问题很难用生硬难懂的数字来回答,但是只要过程协调顺畅、各要素统一和谐,他们仍是恰当的。

在努力说服委托人应该要竞赛程序以后,紧接着要完成一些更为简单的任务:细化各种程序技术、法律和金融上关键的微妙的事项,研究前面提到的优质"操作控制器"的可行性。第一个问题是关于举办竞赛的法律基础问题。有没有举办竞赛的综合的国家法或相应的规定?竞赛程序是不是公众授予合同程序中的一个部分?这个竞赛会不会是以联合国教科文相关规则和国际建协准则为基础的国际性的竞赛?所有的适当的规则手册对于控制所有竞赛的基本原则的定义都有一个普遍的共识。这些原则是从19世纪早期竞赛中演变而来的,即"所有竞赛者机会均等"、"适当的报酬"、"用清楚明白的语言委托标准"、"评奖委员会独立称职"、"任务说明的明确表达"。

按照委托人的具体要求制定程序潜力取决于下列问题:应该使用什么方法来达到这些目标?在不违反基本原则的情况下,可以怎样负责地运用建议方法和途径中包含的解释的灵活性?谁有资格参加竞赛?只是建筑师吗,还是也有工程师?怎样找到竞赛者,是通过公示还是通过内部选择?在决策过程中需要多少评奖委员会委员给委托人出谋划策?竞赛是匿名进行,还是应该允许评奖委员会委员、竞赛者和咨询专家进行更为直接的对话?可以获得多长的时间间隔?有没有可能把竞赛程序分成两个阶段,以便在项目之初就产生更多观点可供选择?要付给竞赛者什么费用,要颁发什么奖项,必须递交什么设计作品?控制作者版权的规则是什么?最后,但并非不重要的一点,是在竞赛开始前明确回答这些问题并相应指导参与各方是非常关键的,否则错误的想法从一开始就会破坏整个项目。

依我们的经验来看,一旦我们承认选择竞赛程序对于项目成功的重要性,这一过程中的关键问题就变得非常不同。这里被婚姻经纪人、演员经纪人或猎头公司这些比喻想象出来的形象在竞赛管理人员的实际工作中都可以找到相应的等价物:包括细心、详细地说明设计任务。只有任务得到彻底、明确的规定,建筑师才能真正恰当地回答并提供要求的设计作品。换言之,只有那个时候,委托人和建筑师才能真正互相了解。

Export of Competition Management Services

竞赛管理服务的出口

Export of Competition Management Services
竞赛程序的选择

In these times of globalisation, not only do goods travel around the world, but so do legions of engineers, consultants, and all sorts of specialists. As architecture these days has become a global issue, it is only logical that the know-how required at the start of a construction project also crosses borders.

在全球化时代，不仅货物在全世界流通，工程师、咨询专家和各种专家团队也可以走遍世界。因为当今建筑已经成为一个全球性的问题，在建筑项目初期阶段所要求的技术也要跨越国界，这是符合逻辑的。

In presenting to the general public more of the many competitions it has coordinated, [phase eins]. provides a welcome opportunity for expounding some thoughts on the role of the competition manager. It advances the Architectural Competition model as an institution as well, in what amounts to a reimportation of a previously successfully exported system. Within the framework of competitions conducted in North Africa and the Middle East, [phase eins]. has had repeated success in establishing essential procedural principles of European-style competitions elsewhere. And it has done so in countries where one would have expected "unethical" practices or competitions for "the best idea" without the promise of the winner being commissioned. There were no procedures in place whereby sponsors accept and assert as matters of fact such principles as anonymity, equality of opportunities, the promise of a contract award and equitable remuneration. [phase eins].'s success is reflected not only by the excellence of the competition outcomes and the positive feedback given by principals, competitors, and jurors but also by the subsequent use in those countries of the competition approach. The exportation of the competition approach is in effect an export of the "competition management" service profile, and one should not underestimate the opportunities it offers the architect who has specialised in the start phase, in getting projects off the ground.

Traditionally, many competitions are coordinated by sponsors who frequently are public authorities, or by architects who perform this service among others. In some instances even the Chamber of Architects takes on this role. Accordingly, it is often reduced merely to preliminary examination, or becomes a passive role during crucial phases of the procedure such as jury meetings! Thus, instead of creating a fruitful "round-table atmosphere" where the parties who (as the sponsor's representatives or experts in the field) have accompanied the project over a long period share know-how, the initiative is often abandoned to the resident "Godfather", i.e. to whoever is the most senior person involved. When one envisions the competition coordinator (i.e. the competition manager) as the person in charge who in his or her capacity as the "phase-one architect" is available to support the project and its sponsor during the critical initial stages, the opportunities this role presents become obvious. It provides a technically competent moderator for important meetings and above all a consultant versed in the crucial areas of task definition, the choice of procedure, the appointment of jurors, and of course the preliminary examination. This does not challenge the role of the jurors but strengthens it, as the parties involved can defend their positions more clearly once the competition manager is in place and doing his job.

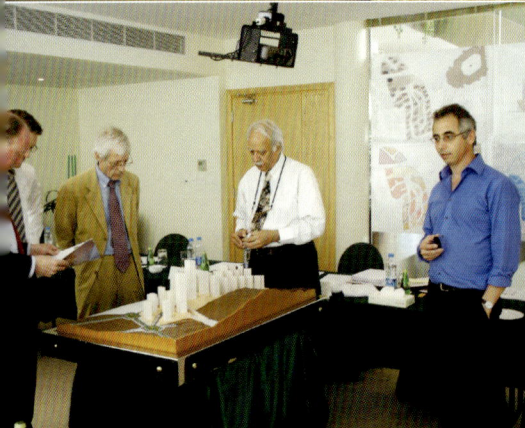

The recognition of the competition manager's role and active support for it will not only benefit architects seeking to excel in this field, but enhance the quality and reputation of competitions in general. As the importance of the start phase to a project is not yet generally recognised, many projects here in Germany continue to suffer budgetary shortfalls, so that a completion under optimal conditions is out of the question. A clear, unequivocal description in institutional terms of the services to be rendered and a definition of minimum standards and qualifications is what the competition model under a competition manager provides and this would constitute a welcome boost for both the success of projects at home as well for an export market for competition management services abroad.

[phase eins].公司向公众提供其组织协调过的许多竞赛实例，欢迎就竞赛管理人员角色的作用陈述不同的看法。作为一家专业机构，它在已经十分成功的出口体系的再输入过程中提高建筑竞赛的模式水平。[phase eins].公司在北非和中东举办的建筑竞赛框架下，已经多次成功地在欧洲以外地区确立了欧洲方式竞赛的基本程序原则。在期望"缺乏职业道德的作法"或"最佳点子"竞赛的同时，而不承诺获胜建筑公司授以委托的竞赛的国家里也非常成功。在主办方认为匿名、机会平等、承诺授予合同和适当报酬为事实问题的地方没有适当的竞赛程序。

[phase eins].公司的成功不仅表现在竞赛产生的优秀设计作品、项目委托人、竞赛者和评奖委员会的积极反馈上，也表现在那些国家后续对该竞赛方法的采用上。竞赛方法的出口实际上是"竞赛管理"服务资料的出口，我们不应该低估其为建筑师提供的机会，因为他们在项目初期和顺利开始阶段都十分专业。

从传统意义上来讲，很多竞赛都是同主办方（通常为公共权力机构）或由服务的建筑师组织协调的。在一些情况下，甚至建筑师协会担负此任。所以，竞赛经常被简化为初试，或者是在评奖委员会会议这样的竞赛程序中的关键阶段时表现被动消极。因此，竞赛并没有创造一种

Left Scenes from the competition procedures in Saudi Arabia, Libya, Jordan, and South Africa
左图：在沙特、利比亚、约旦和南非的竞赛程序的场景

长时间参与项目各方分享技术的富有成效的"圆桌会议"的环境，而是把主动权只留给常驻的"教父"，即最资深人士。当我们把竞赛组织协调者想象成在初期关键阶段作为"第一阶段建筑师"，可以对技术项目及主办方负责的时候，这个角色所带来的机会非常明显。它为重要会议，尤其是在任务定义、程序选择、评奖委员会委员任命和初试这些重要方面进行的咨询提供有力技术的主持人。这并没有动摇，相反还加强评奖委员会的权威性，因为一旦竞赛管理人员就位开始工作，参与各方就可以更清楚地为自己的立场进行辩护。

对竞赛管理人员角色及其对竞赛的积极支持的认可，不仅可以使在这一领域寻求卓越的建筑师受益，而且也增强了整个竞赛的质量和声誉。由于项目初始阶段的重要性仍未被人广泛认识，许多在德国本土的项目继续遭遇预算短缺。这样一来，最优条件下的竞赛是没有可能出现的。

用机构术语对所要提供服务的明确描述和对最低标准和资格的定义是在竞赛经理人管理下的竞赛模式所提供的结果，这将有力地推动国内项目及国外竞赛管理服务出口市场的成功。

The Architectural Competition Model in the Corporate World

企业时代的建筑竞赛模式

The Architectural Competition Model in the Corporate World

企业时代的建筑竞赛模式

Corporate architecture is intrinsic to a viable communication strategy, for many companies. Interestingly enough, the methods employed for choosing the best architect and the optimum design end up reflecting the development of corporate philosophy and a company's rootedness in society.

对于很多公司来讲，公司建筑本质是一种切实可行的沟通策略。十分有趣的是，用来选择最佳建筑师和最优设计的方法最终却反映出公司价值体系的发展和公司在社会上牢不可破的地位。

Bottom　BMW World in Munich (COOP HIMMELB(L)AU)

下图：慕尼黑的BMW世界

adidas, BMW, Mercedes, Porsche, Telekom, ThyssenKrupp, and Volkswagen — in recent years a large array of major German corporations have taken on important new construction projects which ended up finding their way into architecture magazines. Many smaller and medium-sized enterprises have followed suit. The frequency of such projects may be merely coincidental, or they may reflect several years of capital expenditure backlog finally catching up. At any rate, what it actually means is that many of these projects are the outcome of formal architectural competitions which is a new phenomenon. Before now, competitions (as opposed to referral procedures and multiple assignments on the same subjects) were primarily a planning instrument used by public authorities. Private initiatives usually greeted the idea of conducting a competition with a distrustful frown or, at best, a condescending smile: This way of going about things was regarded as outmoded, inflexible, cumbersome, and expensive. So, has there been a change of mind? And, if this is the case, what is the reason? What opportunities does it open up for architects?

All the projects sponsored by the firms mentioned above belong to this class of high-profile building initiatives, with strategic importance for the owner: corporate headquarters, company museums, in-house training centers, manufacturing buildings, sales outlets, or complexes combining several of these functions. What they have in common is that the owners constructed them themselves. And they share this feature with numerous public projects. A closer look reveals that even the functions of these company buildings are comparable to those of typical public buildings, like city halls, museums, educational institutions, and even churches. What matters in both cases is architectural quality, the projection of company values and goals in conjunction with an enhancement of the public space and a well-structured or (at best) consensually legitimised contract award process.

Competition procedures for adidas in Herzogenaurach (adidas World of Sports, adidas Factory Outlet and Adi Dassler Brand Center), BMW in Munich (BMW World), Leipzig's BMW Factory, and ThyssenKrupp in Essen (ThyssenKrupp Quarter) were coordinated by [phase eins].

As a matter of course for such projects, architects should take on the role, among others, of a competition manager. Apart from coordinating organisational activities and the tasks performed within the framework of the preliminary examination of competition entries, the architect provides standard architectural services in the sense of HOAI, in particular by formulating the competition task. Some of these services can be categorised as belonging to Service Phase one, while others, not listed there, more or less constitute a "Service Phase zero". If so stipulated, the competition manager can be brought into the process as early as the project development stage, e.g. when the site is being selected, when strategic goals are being revised and integrated, or when the building program is being developed and all these activities have to be coordinated. Services such as these, while not covered by HOAI, are among those for which architects are ideally qualified and should not be left to project controllers, quite independently of whether or not a project is being developed within the framework of a competition (cf. Susanne Knittel-Ammerschuber, "Erfolgsfaktor Architektur – strategisches Bauen für Unternehmen", 2006). Projects run as competitions are particularly suited for the inclusion of a "Service Phase zero", as it is the competition manager's assignment, not the competitors', to structure the task, ask the right questions, and to give it a precise formulation. It is this — the formulation of the task — that is the first line of communication within the project, and holds the key to the project's success or failure.

In this process, finding the proper tone, a manner of speaking intelligibly to both the principal and the competitors, is a key challenge. Clients must be able

to recognise vocabulary they are familiar with from their corporate roles and brand-value manuals, while architects should not feel that what has been handed to them is an advertising brochure rather than a competition brief. As the next step, it must be decided what demands the architectural plan projected must meet. Again, public administration bodies (who are more often than not the third partner involved) have a language and specific concerns of their own, both of which must be addressed so as to ensure maximum support of the project from that side as well. In all this, the cardinal rule should be transparent relationships between all participants in the process.

Every company has its Corporate Identity Manual, which, however, has rather limited value as an information source for task formulation. As a rule it usually provides little outside the areas of letterhead design, placement of the company logo and choice of headline colors, all of which have some bearing on designing the competition brief but on not much else. Of much higher information value but rarely spelled out are the firm's corporate culture and brand values: this is where strategic goals for positioning and overall presentation of the company and its brands are formulated. Indeed even here the focus is mainly on products rather than on the company's corporate architecture, a more universal, long-term feature more deeply rooted in its organisational structure and reflecting its greater degree of social commitment. Therefore, a thorough and intensive dialogue with the client about the way the company staff (and CEOs in particular) think and act becomes essential.

With all the projects mentioned, the final competition brief documents included information on the company's history and its organisational structure as well as extracts from documents defining the corporate image and the creative presentation of the company. After all, the two parties should get to know each other. Thus, the areas listed above are considered of equal importance to the "classic concerns", the standard technical issues such as data on the location of the project site, existing vegetation, accessibility, or the composition of the subsoil.

Crucially, what it comes down to is whether the project's architectural design specifications can be derived from the company's stated corporate philosophy, and whether this is at all a sensible or desirable course of action, within the context of the competition task.

The practice of continuing to clarify all aspects of the task throughout the competition period was crucial to the success of these high-complexity projects. A process of approximation, it is at the same time a learning process on the part of the principal and should not be suspended over the course of the competition. In this regard a two-stage competition is particularly appropriate, possibly with a cooperative second stage where the latter allows a more focused specification of the task, if need be through direct discussions on the basis of preliminary designs.

In each case, architectural competition was instrumental to success: It enabled us to coordinate the planning process with the public authorities, to secure acceptance by the general public, and to do so in a timely fashion — in meetings that lasted several days! This involved a remarkable commitment on the part of such public officials as mayors and government ministers to achieve the successful completion of what were in effect private-sector projects. At the same time, the process raised the level of discourse during colloquia and jury meetings to a new high, among the competition board-members and architects involved (who naturally were no less committed).

It is not surprising that this development is not an automatic one: Not every Saul becomes Paul. At the same time as the projects that were decided through competition were launched, other equally important projects undertaken in some cases by the same companies and also by a number of others were launched as well, but through direct commissioning or unregulated and less transparent procedures. As a result many, many buildings are still being constructed that will not attain major importance for these companies, their staff or for the general public. The power of decision in this matter lies, as always, in the hands of the principals and their representatives. As is so often the case in construction, what really matters is the awareness of how important the culture of the building process is, an indispensable element of which is a culture of planning. In this context it is all the more important to point to those companies and their respective decision-makers at the executive level who have taken on this responsibility and helped to launch the kinds of major projects that were then realised through the competition approach. And by the same token the more these methods, which have been so successful, are explained and promoted in a balanced way, so much more will a competition approach to producing corporate architecture enhance the quality of the same.

adidas、BMW、梅塞德斯—奔驰、保时捷、电信、蒂森克虏伯和德国大众——近年来，一大批德国大公司已经开始进行重要的新建筑项目，这些项目都被收录到建筑杂志中。许多小一些的公司和中型企业也纷纷效仿。项目这样频繁的出现可能只是巧合。不管怎样，许多项目都是正规建筑竞赛这一新现象的成果。在此之前，竞赛主要是公共权力机构运用的一种规划工具。这种作法已经被看成是过时的、不灵活的、麻烦的和价格昂贵的。所以，观念是否发生了变化？如果变了，为什么？它为建筑师们开创了什么样的机会？

所有上述提及建筑公司主办的项目属于知名度高的建筑的表率，对业主来说具有非常重要的意义：公司总部，公司博物馆，室内训练中心，生产加工建筑物，品牌直销店，或者管理建筑区。他们的共同之处是由所有者自己建造的。无数公共项目也有这个特点。如果我们更仔细一些，我们就会发现这些公司大楼的功能性堪比那些典型的市政厅、博物馆、教育机构，甚至是教堂一类的公共建筑。公司大楼也好，公共建筑也罢，重要都是建筑质量、公司价值观和与增强公共空间和结构良好的正统化相结合的授予合同过程的体现。

奥拉赫的Adidas、慕尼黑的BMW、莱比锡的BMW工厂和埃森的蒂森克虏伯这些项目的竞赛程序都是由[phase eins].公司组织协调的。这类项目的建筑师自然应在其他同行中间充当起竞赛管理人员的角色。除了协调组织结构方面活动的任务、执行初赛框架内的任务以外，建筑师按照HOAI的意义，特别通过规定竞赛任务提供标准的建筑服务。竞赛管理人员就会在项目开发阶段，即在选址、修订和整合战略目标，或者是开发建筑计划、协调全部活动的阶段尽早参与进来。制定建筑任务、提出正确问题并给出明确规定是竞赛管理人员的任务，而不是竞赛者的任务——这是项目沟通的第一界限，对项目成功与否至关重要。

在这个过程中，找到一个对委托人和竞赛者都适合的会话方式非常关键。委托人必须能认可他们熟悉的法人角色和品牌价值手册中的词汇，建筑师也不应该觉得他们得到的只是广告宣传册，而是竞赛的任务说明。第一步，必须决定规划的建筑计划必须满足什么样的需要。此外，公共管理机构有语言方面和他们自己的具体问题，为了确保该方也最大程度地支持项目，这两种问题也必须要说明。所有这些问题中，基本规则应该是过程中参与各方关系的公平透明化。

每家公司都有自己的公司形象手则，然而它对于任务规定的信息来源的价值相当有限。它通常除了能提供与设计竞赛任务说明有些关系的抬头设计、公司标识的放置、抬头色彩选择方面的信息以外，对竞赛的帮助甚少。公司文化和品牌价值相比更有参考价值，却很少进行详细说明。实际上，这里强调的重点主要是产品而不是公司建筑，一个更广泛、更长期的深深植根在组织结构中反映其社会承诺的特色。因此，与委托人就公司员工的思维和行为方式问题展开深入全面的对话是非常重要的。

上述提及的全部项目，最终的竞赛任务说明都包括了公司历史、组织结构以及从文件中节选出来的公司形象的定义和创造性的展示。毕竟，双方应该彼此了解。因此，上述所列的问题对于建筑"关注传统"，即项目选址、现状植被、可及性或下层土构成方面的数据这样的标准性的技术问题，同样重要。

接下来关键的一步是项目的建筑设计规格说明能否从公司规定的公司哲学中获得，这是否是一个在竞赛任务中的明智的或可取的行动方案。

在整个竞赛期间阐明任务的各个方面的做法，对于这些高度复杂的项目的成功非常关键。接近完美的过程同时也是一个委托人学习的过程，不应该被在竞赛进行的过程中暂停下来。在这一点上，竞赛分两个阶段是特别恰当的，在第二阶段如有需要，应该通过初次设计基础上进行直接的讨论来对任务进行更为详细的规定。

在每一个实例中，建筑竞赛对于项目成功都是非常有帮助的：它让我们能与公共权力机构协调规划过程。赢得大众认可，并且是以一种非常及时的方式——仅开几天的会议而已！这包括获得市长、政府部长一类政府官员对实际上是私人性项目的成功完成所作出的承诺。同时，这个过程把专题座谈会和评奖委员会会议的谈话水平提高到一个新的高度。

这种发展不是单独进行的，发展结果也不尽相同。同时，在决定通过竞赛推出项目的时候，其他一些同样重要的项目有时也由同样的公司和一些其他公司进行，但是却是通过直接委托或不够规范、透明的程序进行的。因此，很多正在建设中的建筑物无法为这些公司、公司员工或大众获取重要地位，因为这类事情的决定权总是掌握在委托人及其代表的手中。在建造过程中真正重要的事情经常是对建筑程序文化重要性的认识，这种文化是规划文化中一个不可分割的要素。因此，向那些公司及其那些负责帮助进行当时通过竞赛方法通过的主要项目的管理层决策者指明该点更为重要。同样，这些已经成功的方法解释得越均衡参与各方利益，用竞赛的方法产生的公司建筑质量越高。

Top BMW World and BMW Tower in Munich

上图：慕尼黑的BMW世界和BMW高楼

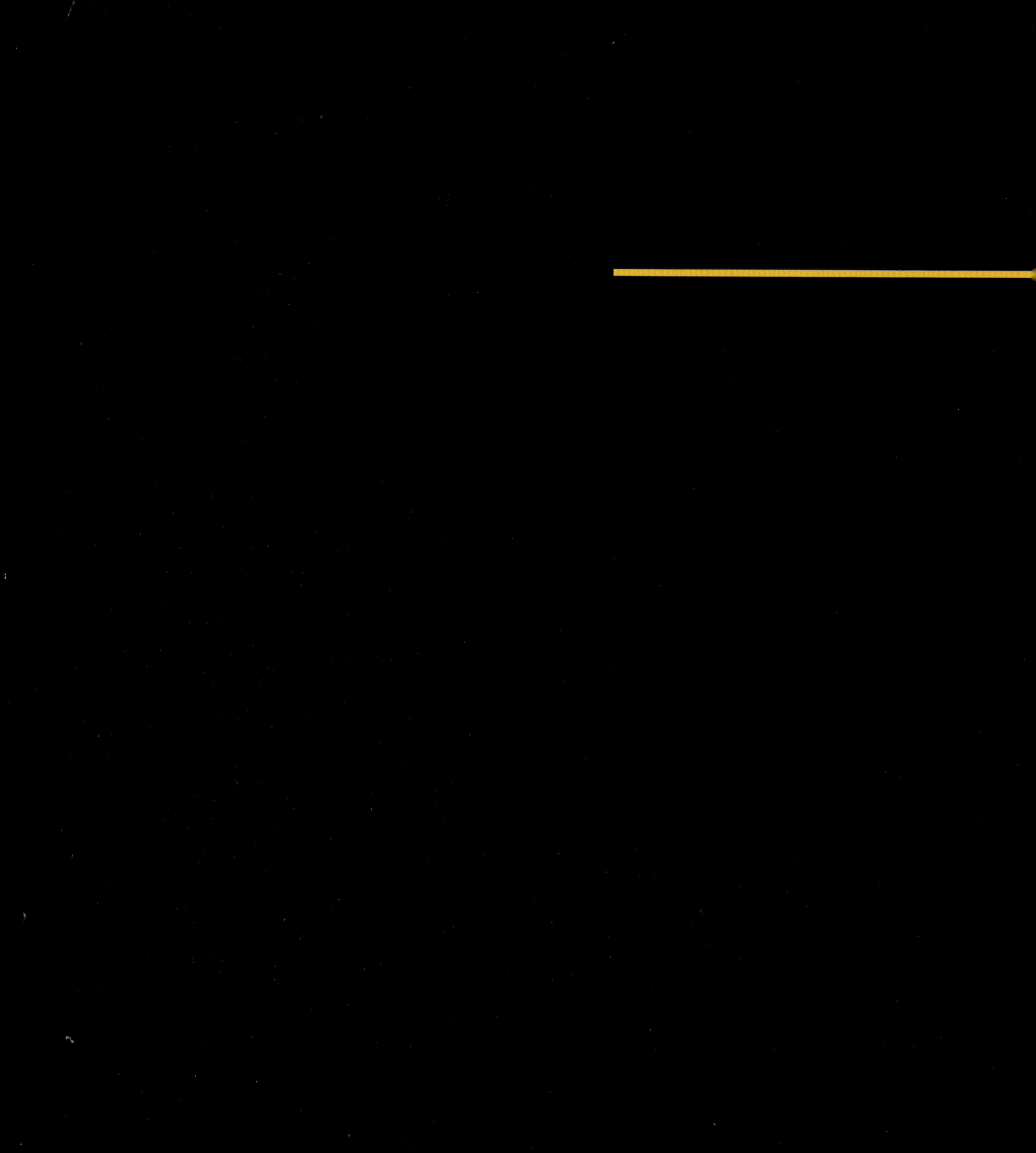

[projects].
[设计项目]

Selected Projects
精选设计项目

This section illustrates, with drawings selected from among those entered, 16 competitions coordinated by [phase eins].

这一部分详细介绍由[phase eins].公司组织协调的15次竞赛程序中精选出来的规划设计作品。

Eco Bay Project
Sorouh Real Estate, Abu Dhabi
2006

生态湾项目
阿布扎比Sorouh房地产公司
2006

Khandama Development Project
Jabal Khandama Development Co., Mecca
2006

Khandama开发项目
麦加Jabal Khandama开发公司
2006

Jabal Omar Development Project
Architecture Competition
Jabal Omar Development Co., Mecca
2006

Jabal Omar开发项目
建筑竞赛
麦加Jabal Omar开发公司
2006

Jabal Omar Development Project
Engineering Competition
Jabal Omar Development Co., Mecca
2006

Jabal Omar开发项目
工程竞赛
麦加Jabal Omar开发公司
2006

Swiss Pavilion at the World EXPO 2010, Shanghai
Swiss Confederation
2006-2007

2010年中国上海世博会的瑞士展馆
瑞士建筑师 联合会
2006—2007

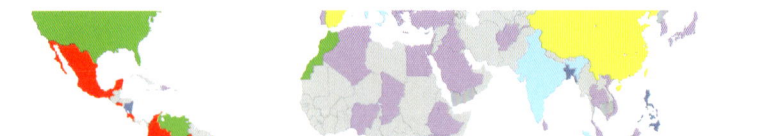

Holcim Awards 2006
Holcim Foundation
2006

霍尔森可持续建筑大奖2006
霍尔森建筑师联合会
2006

ThyssenKrupp Quarter, Essen
ThyssenKrupp AG
2006

德国埃森的蒂森克虏伯办公建筑
埃森克虏伯股份公司
2006

Museum Folkwang
City of Essen
2006-2007

Folkwang博物馆
德国埃森
2006—2007

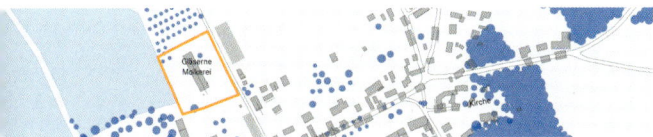

Glass Dairy, Münchehofe
Gläserne Molkerei GmbH
2006–2007

玻璃奶品厂
Glaserne Molkerel GmbH
2006—2007

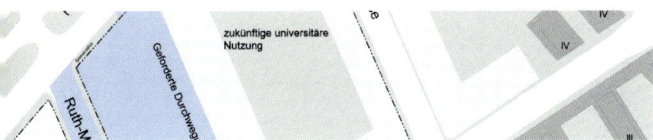

Riedberg Campus Commons, Frankfurt/Main
Ministry of Higher Education, Research, and the Arts
Federal State of Hessen
2006

Riedberg校园公用地，德国美因河畔
法兰克福埃森联邦州高等教育、研究
和艺术部
2006

New KfW Building at the Senckenberganlage,
Frankfurt/Main
Kreditanstalt für Wiederaufbau (KfW)
2005–2006

新KfW大楼，Senckenberganlage，
德国美因河畔法兰克福
KFW公司
2005—2006

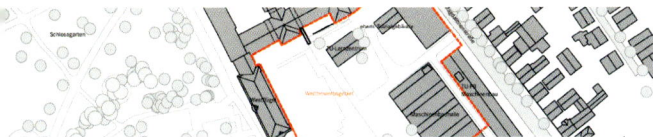

University and State Library, Darmstadt
TU Darmstadt
2005

达姆施塔特大学和州立图书馆，德国
达姆施塔特TU公司
2005

Southbank Projekt, Stellenbosch
Spier Holdings
2006

南部海岸项目，南非斯伦博斯
Spier股份公司
2006

Administration Complex, Tripoli
ODAC
2006–2007

行政办公建筑区，利比亚的黎波里
ODAC公司
2006—2007

Tower at Suk Al Thalath Al Gadeem, Tripoli
Libya Africa Investment Portfolio (LAP)
2007–2008

Suk Al Thalath Al Gadeem高楼，
利比亚的黎波里
利比亚非洲投资组合公司
2007—2008

Darat King Abdullah II for Culture and Arts, Amman
Greater Amman Municipality (GAM)
2007–2008

Darat King Abdullah II，约旦阿曼
大阿曼市政公司
2007—2008

Benthem Crou
BRT Bothe

Lab Architecture St

O

KSP Engel un

wel
Richter Teherani

Eco Bay Project in Abu Dhabi
阿布扎比生态湾

dio

tner & Ortner Baukunst

Snøhetta
T. R. Hamzah & Yeang
Zimmermann

Eco Bay Project Abu Dhabi
生态湾 阿布扎比

Restricted project competition preceded by an application procedure
限制严格的建筑项目竞赛，开始前有申请程序。

Resort Hotel

Competition Site

block 53-2-2
block 53-2-1
block 53-3
block 52-2

collector street B
collector street A
collector street C
loop road

Khor Al Baghal

Upper Village

Central Business District

loop road

Central Park

Entertainment District

Private Development

Marina

Marina Village

Lower Village

Qarm

substation and district cooling

substation and district cooling

marina canal

collector street

地点 Abu Dhabi 时间 02/2006-04/2006 主办方 Sorouh Real Estate PJSC represented through Mounir D. Haidar   人数 7 competitors   面积 134.6 ha   竞赛费用 150,000 USD   专业评奖委员会 Xaveer de Geyter, Brussles; Prof. Rainer Mertes, Stuttgart; Jafar Tukan, Amman   其他专业评奖委员会 Dominique Lyon, Paris   专家评奖委员会 Mounir D. Haidar, Sorouh Real Estate, chief executive officer, Abu Dhabi; Faris Suhail Al Yabhouni, Sorouh Real Estate, vice chairman, Abu Dhabi   其他专家评奖委员会 Adnan Sahli, Sorouh Real Estate, project's director, Abu Dhabi; Dr. Moustafa Baghdadi, Sorouh Real Estate, senior urban planner, Abu Dhabi

The "Abu Dhabi 2030 Urban Structure Framework Plan" is to lay the basis for future growth and development of tourism in the capital of the United Arab Emirates, up to the year 2030. The point is for Abu Dhabi to experience the boom enjoyed by Dubai, but without repeating its mistakes in urban design. A key aspect of the plan is maximum conservation of natural resources. One of the large development areas is Reem Island. Of three projects located on Reem Island, one is the 140-hectare "Al-Shams Abu Dhabi", developed by Sorouh Real Estate under the patronage of Sheikh Mohammed bin Zayed Al Nahyan. Its master plan envisages several theme focused residential developments connected by canals and, in concept, evoking an environment-savvy lifestyle. The area will accommodate tourist attractions, as well as top end residential use. A two-block 37,000 square metres site in the northwest portion of the master plan (prepared by RNL International) was chosen for the competition. A detailed spatial-use-plan (focusing on overall objectives and marketed under the brand name "Eco Bay") was overlaid upon the two-block area. One half of the approx. 200,000 square metres gross floor area is slated for residential use, under the motto "Healthy and Green Living"; the other half will be hotel space, offices, a health clinic, a "Green Nutrition Center", a swimming pool complex, and a shopping mall. The space-use-plan included a small mosque. A block cooling plant was to be worked into the design as well.

"阿布扎比2030城市结构框架规划"的目的是为沙特首都旅游业到2030年的未来发展打下基础,旨在使阿布扎比经历迪拜的繁荣,重复在城市设计上所曾犯下的错误。规划的重要方面是最大限度地保护自然资源。最大的开发地区之一是里姆岛,包括三个项目,其中一个是由Sorouh地产公司在Sheikh Mohammed bin Zayed Al Nahyan资助下开发的"阿布扎比的阿尔沙姆斯"。其总体规划的设想强调由运河连结居住环境的发展,在观念上引发一种了解环境的生活方式。这个项目提供酒店住宿及高端住宅区。位于总平面西北部分的一个两区3.7万平方米的设计被选参赛。在此基础上有一个详细的空间利用规划,约2万平方米的建筑总面积秉承"健康绿色生活"信条,铺上石板用作住宅,另一半将建起酒店、办公楼、健康诊所、"绿色营养中心"、游泳池建筑区和商场。空间使用规划的设计中也包括了一个小型清真寺和一个蓄热式供冷设备。

Aerial view　鸟瞰图

Competition site　竞赛地点

Competition site　竞赛地点

View of the town　城镇景色

1

4

7

2

5

3

6

Qualified participants
合格的参赛者

1
1st prize　一等奖
BRT Bothe Richter Teherani,
Hamburg

2
1st prize　一等奖
T. R. Hamzah & Yeang Sdn. Bhd.,
London

3
3rd prize　三等奖
Ortner & Ortner Baukunst, Vienna

4
Further participant　其他参赛者
Snøhetta, Oslo

5
Further participant　其他参赛者
KSP Engel und Zimmermann
Architekten, Frankfurt/Main

6
Furher participant　其他参赛者
Benthem Crouwel, Amsterdam

7
Further participant　其他参赛者
Lab Architecture Studio, Melbourne

BRT Bothe Richter Teherani Hamburg
BRT Bothe Richter Teherani建筑事务所 汉堡

"It reads as an urban ensemble of smoothly integrated volumes that transform the site with its restrictions to a unique and unseen development."

"城市的整体规划把几个体积平滑地融为一体，将原来的局限打破，变成一个独特的开发区域。"

作者 Hadi Teherani, Jens Bothe, Kai Richter   合作伙伴 Joachim Landwehr, Heike Hillebrand, Sven Breuer, Bernd Muzey, Benjamin Holsten   建筑系在校生 Kai Bämfer, Simone Gemeim, Jana Hagenacker, Arne Kersten, Arne Lösekamm,Tabea Müller Wüsten
专家 landscape planning: Breimann & Brumm, Hamburg BRT Engineering GmBH

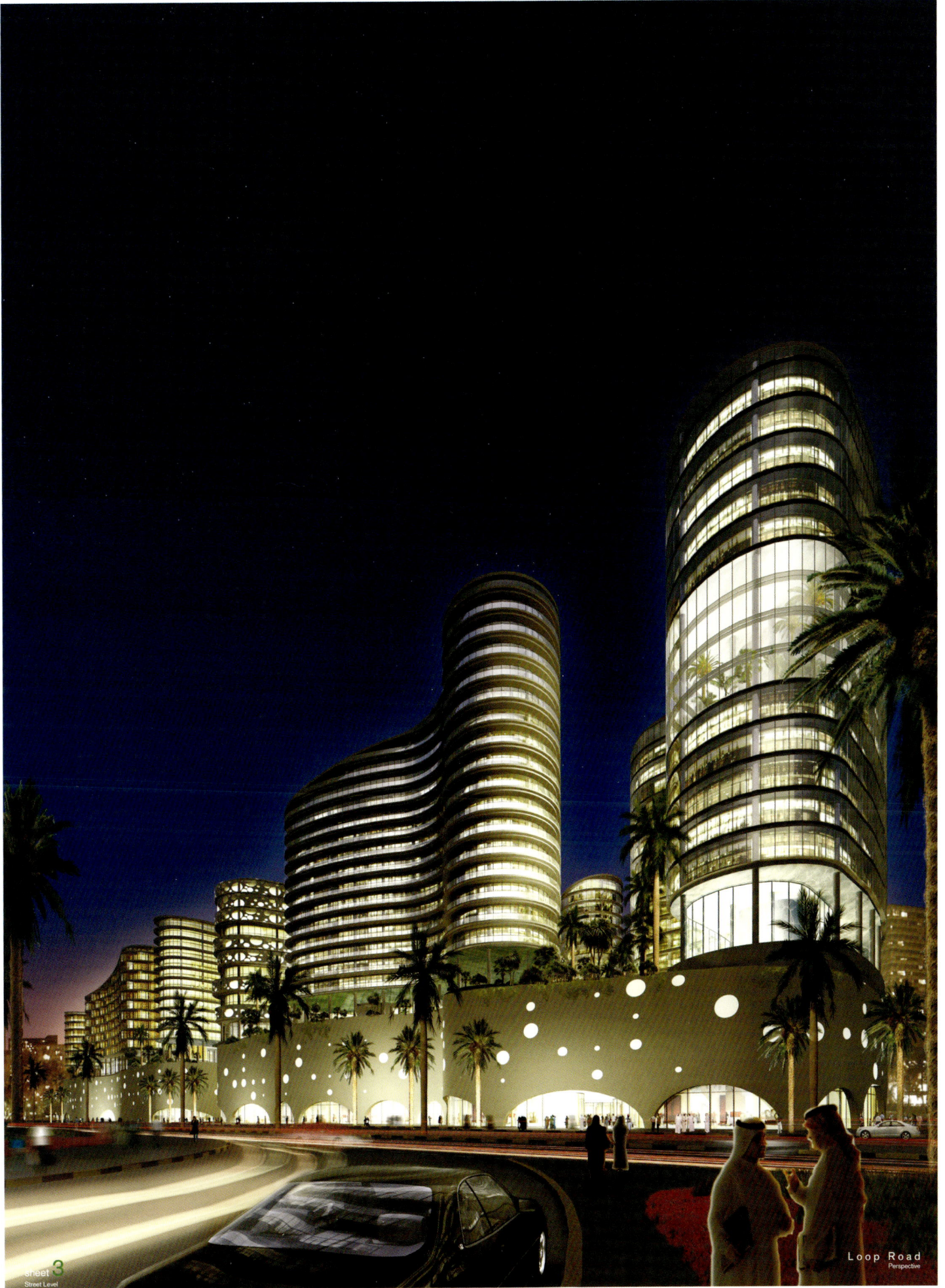

Eco Bay Project, Abu Dhabi

Loop Road
Perspective

Sheet 3
Street Level

Mangrove Business Park
Perspective

mangroove park city
Eco Bay Project, Abu Dhabi

Powertower
Plan View scale 1:500

Residential
Plan View scale 1:500

Residential
Plan View scale 1:500

Residential
Plan View scale 1:500

Residential
Plan View scale 1:500

Apartment Hotel
Plan View scale 1:500

Eco Clinic
Plan View scale 1:500

Buisiness Appartments
Plan View scale 1:500

Tenant Offices
Plan View scale 1:500

Mangrove Towers
Residential | Power Tower | Eco Clinic | Office | Apartment Hotel scale 1:500

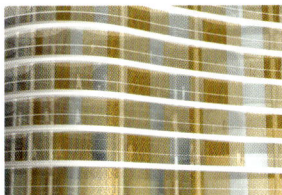

Studio
3 Bed + Room
1 Bed + Room
1 Bed + Room
2 Bed + Room
1 Bed + Room
1 Bed + Room
2 Bed + Room
2 Bed + Room

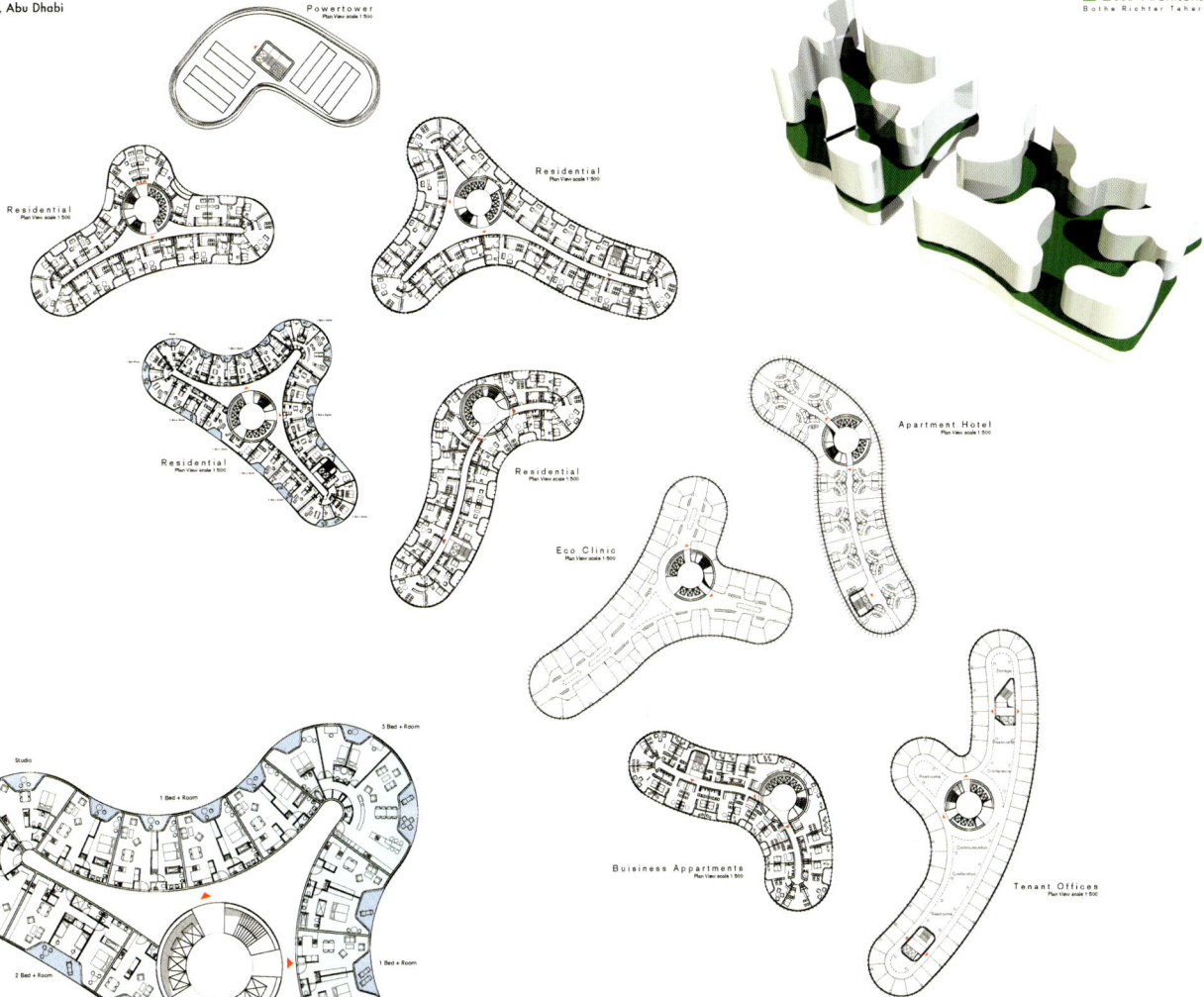

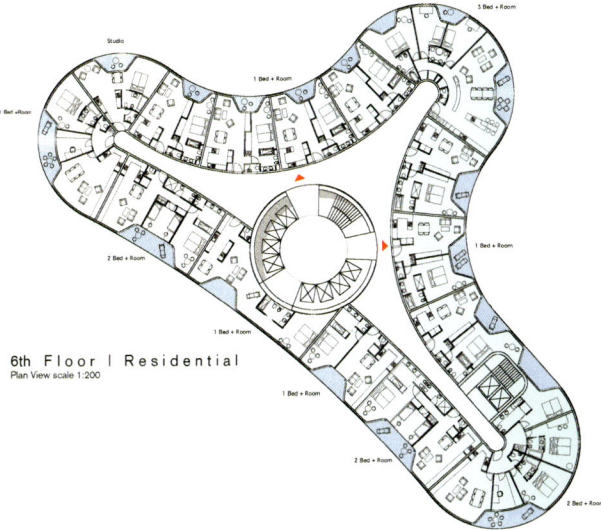

6th Floor | Residential
Plan View scale 1:200

Samples Facade
Individual Loggia | Sunshading

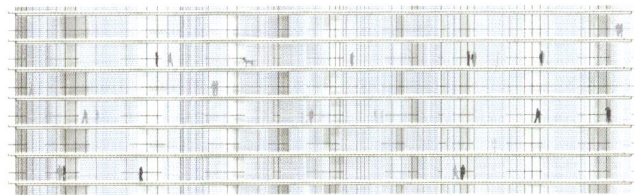

+ 47.05 M
+ 43.75 M
+ 40.45 M
+ 37.15 M
+ 34.40 M
+ 30.55 M
+ 27.25 M

6th Floor | Residential
Exemplary Elevation scale 1:200

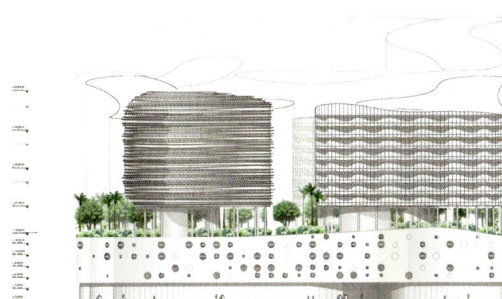

North-West
Elevation scale 1:500

sheet 7
Mangrove Towers

Powertower

Residential
Plan View scale 1:500

Residential
Plan View scale 1:500

Residential
Plan View scale 1:500

Residential
Plan View scale 1:500

Spa
Plan View scale 1:500

Green Nutrition Center
Plan View scale 1:500

Buisness Appartments
Plan View scale 1:500

Buisness Center
Plan View scale 1:500

Women's Fitness

Locker rooms

Locker rooms

Men's Fitness

1st Tower Floor | Spa
Plan View scale 1:200

Mangrove Residential Park | Mangrove Business Park
Elevated Landscape scale 1:500

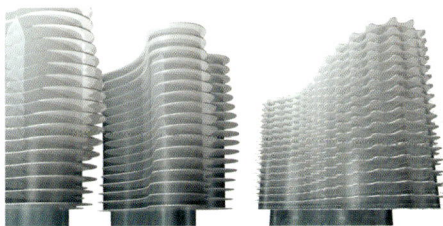

Samples Facade
Balcony

Mangrove Business & Residential Park
Landscape Elements

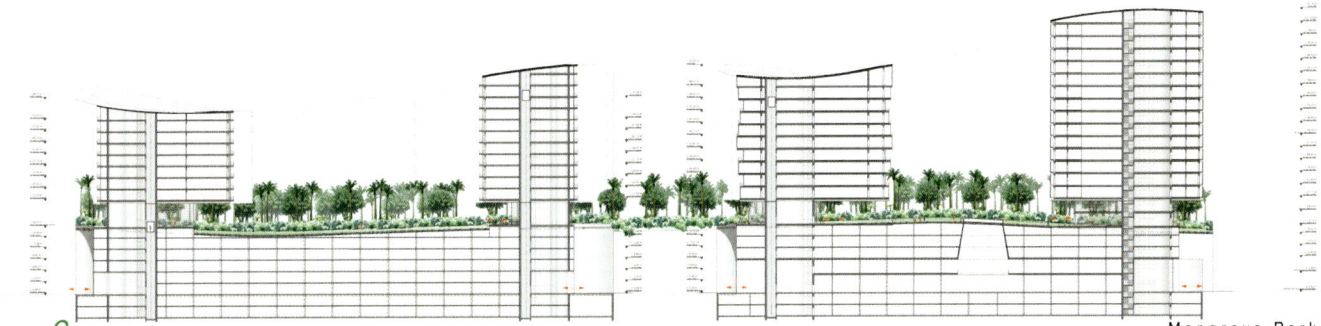

sheet *6*
Mangrove Park

Mangrove Park
Longitudinal Section scale 1:500

T. R. Hamzah & Yeang Sdn. Bhd. London
T. R. Hamzah & Yeang Sdn. Bhd.建筑事务所 伦敦

"... a semi-enclosed and passively cooled pedestrian street and courtyard atrium around which the various programs and buildings are organised"

"……半封闭、遮阴的人行街道和院落商街周围的建筑物整齐有序。"

作者 Dr. Kenneth Yeang   合作伙伴及建筑系在校生 Andy Chong, Joyce Foo Hui-Sing, Haytham Danoon, Virginia Nolan, Ainin Nurul, Abdul Qadir Abas, Mohd. Khairi, Mohd Som Bin Mohamad, Lawrence Lek, Rifa'at Ahmad

T. R. Hamzah & Yeang Sdn. Bhd., London

Tower 5 (29 storey)

Tower 4 (26 storey)

Tower 3 (20 storey)

Tower 2 (16 storey)

Tower 1 (13 storey)

Eco Bay Project, Abu Dhabi

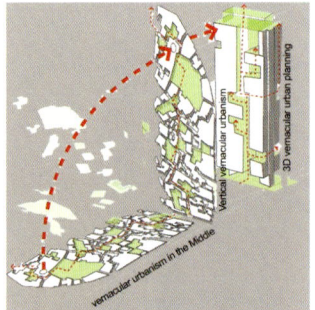

Vertical vernacular urbanism

Concept Diagram

Continuous Green

Green Atrium

Concept
3D showing continuous
green screen and atrium

strong current

Phasing Diagram

1. Shopping Centre (East Wing)
2. Car Park
3. Residence & Hotel
4. Residence 1
5. Residence 2
6. Shopping Centre (West Wing)
7. Office Tower
8. Office Tower 2
   - Eco Clinic
   - Nutricious Centre
   - Business Centre

Khor Al Baghal

Circulation Diagram

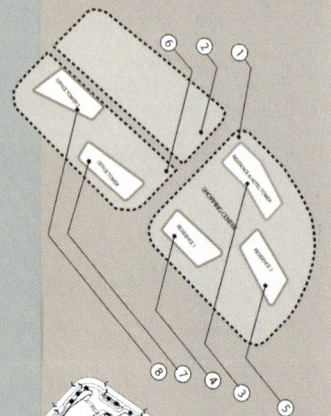

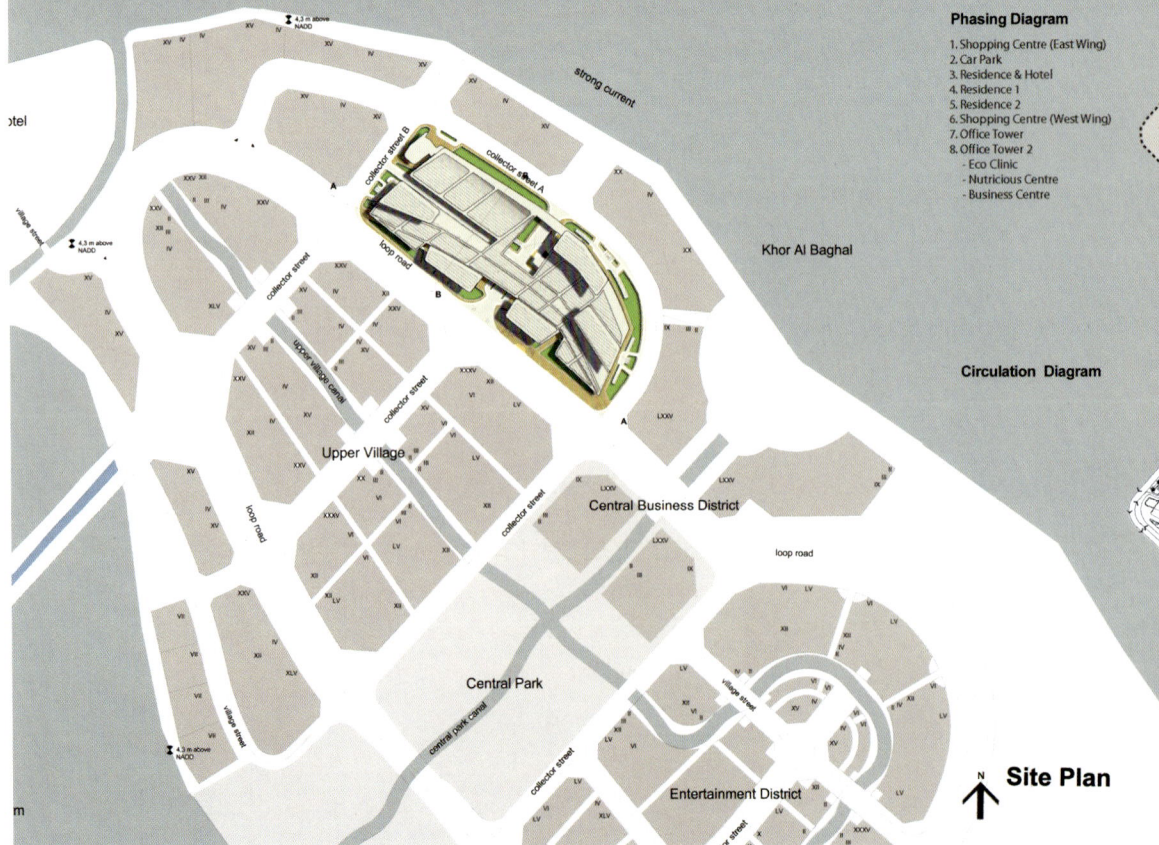

Upper Village

Central Business District

loop road

Central Park

Entertainment District

N

Site Plan

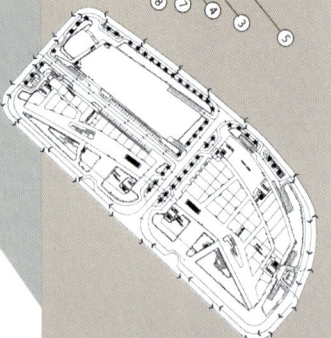

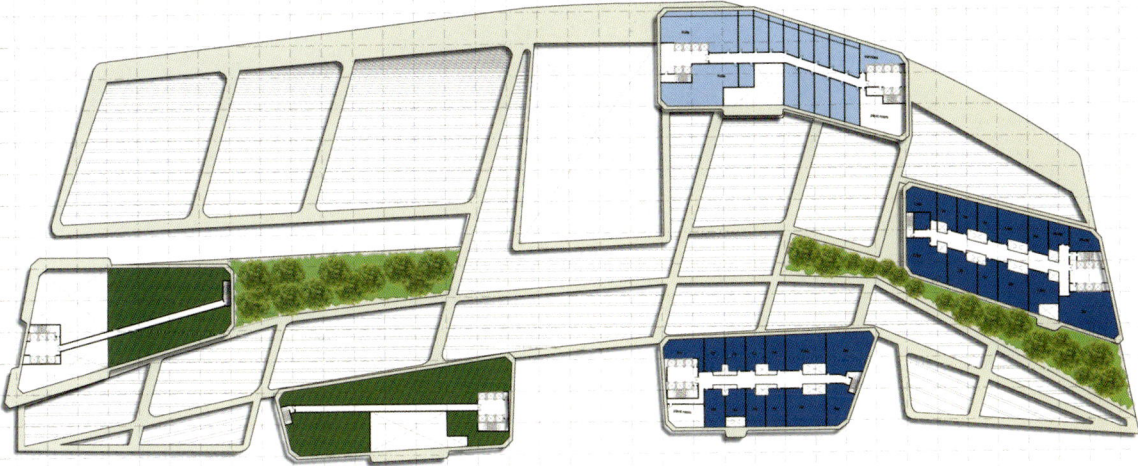

Level 14
Scale 1:500

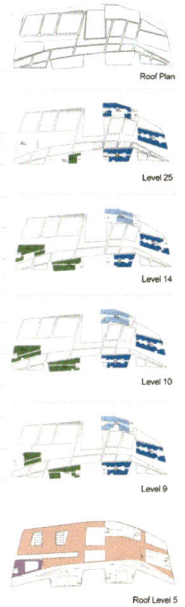

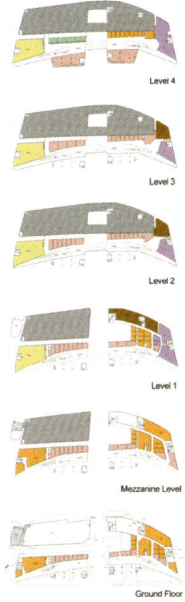

Level 1
Scale 1:500

Ground Floor Level
Scale 1:500

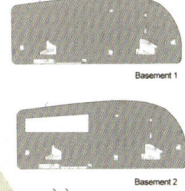

Roof Plan

Level 25

Level 14

Level 10

Level 9

Roof Level 5

Level 4

Level 3

Level 2

Level 1

Mezzanine Level

Ground Floor

Basement 1

Basement 2

Spatial Program
Legend
Residential
Hotel Apartments
Tenant offices
Business center
Eco Clinic
Retail
Green Nutrition Center
Eco Court and Eco Street
Spa
Eco Knowledge Center
Mosque
Cooling Substation
Parking Spaces
Circulation Areas

Ortner & Ortner Baukunst Vienna
Ortner & Ortner Baukunst 维也纳

"The outer shell is covered with a fine mesh that gives all elements a coherent appearance – a 'network of cells' with circular openings for entrants or special functions."

"外壳由一层细网格覆盖，使整体元素和谐——带圆形开口的'单元网状物'供入户或特殊功能之用。"

作者 Prof. Laurids Ortner, Prof. Manfred Ortner   合作伙伴 Helena Feldmann, Sebastian Wiswedel, Philipp Kapteina, Christian Kosnar, Meriton Redza, Tobias Mauritz   自由建筑师 Christian Heuchel; structure: Krebs und Kiefer, Kalsruhe, Prof. Dan Constantinesen   专家 building services: Integ Consulting Engineers, Berlin; Prof. Georg Mayer; landscape planning: Neumann und Grusenburger, Berlin; Thomas Grusenburger; traffic aspects: ambrosius blanke, Bochum; Dr. Phillip Ambrosius; animation and renderings: Archimation, Berlin; façades and metal construction: Wolfgang Priedemann, Berlin

Eco Bay Project, Abu Dhabi

Nightview Loop Road

Interior quality of office spaces

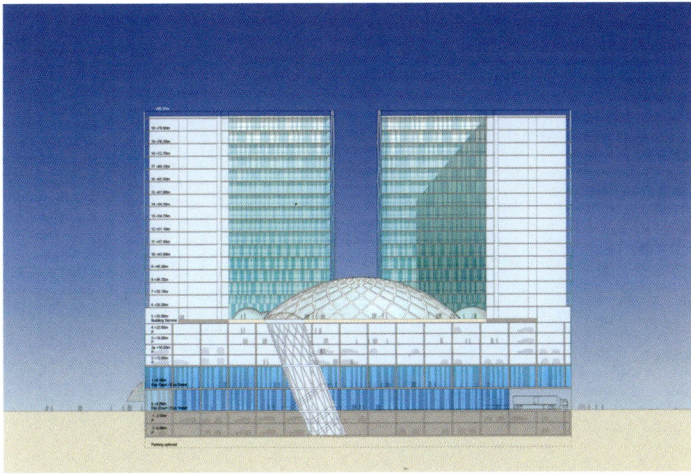

Transversal Section 1:500

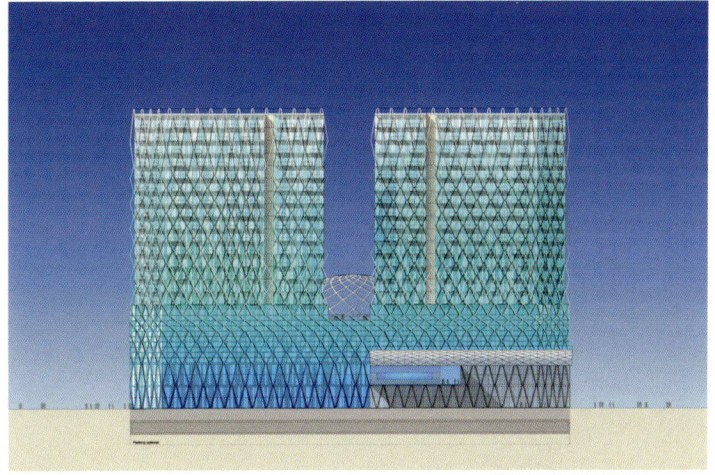

Elevation Collector Street A 1:500

Tower Level
Residential
Offices
Hotel

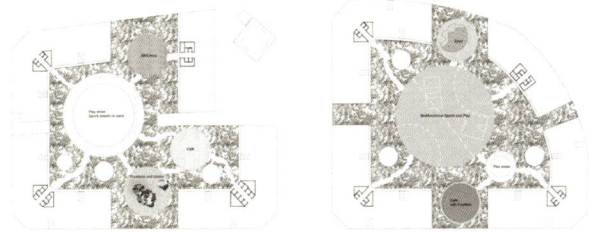

Roofgarden
Building Services

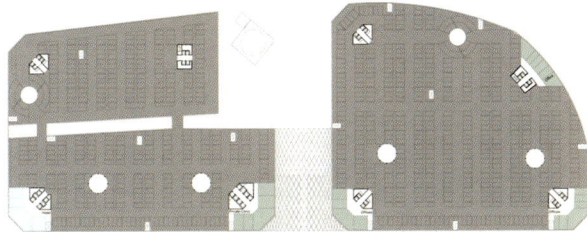

Podium Level 3 - 4
Parking
918 Parking Spaces

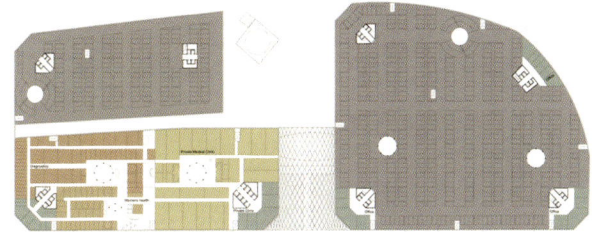

Podium Level 2
Eco Clinic
Parking
697 Parking Spaces

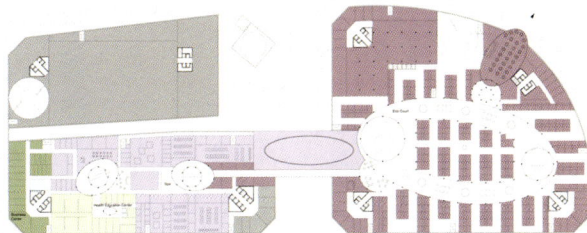

Podium Level 1
Eco Court
Spa

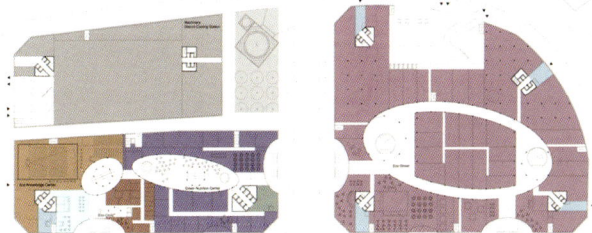

Entrance Level
Eco Street
Green Nutrition Center
Eco Knowledge Center

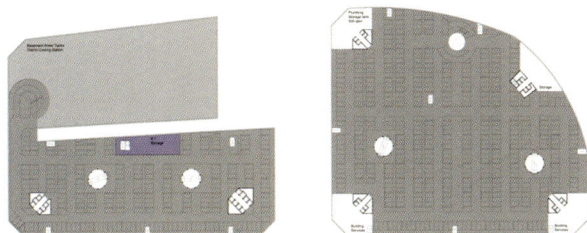

Underground Level -1
Parking
Building Services
Storage
698 Parking Spaces

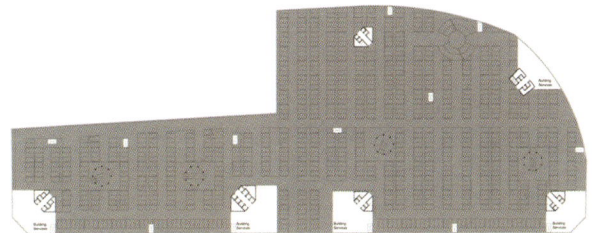

Underground Level -2
Parking
Building Services
Buidling Services
831 Parking Spaces

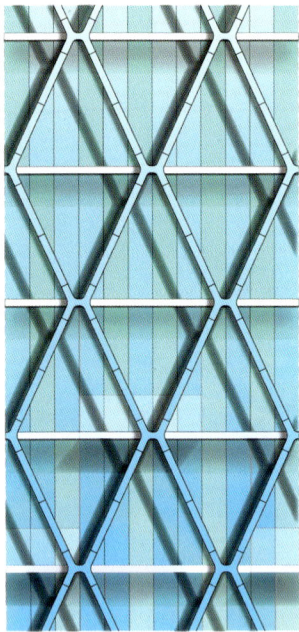

Elevation Loop Road 1:50

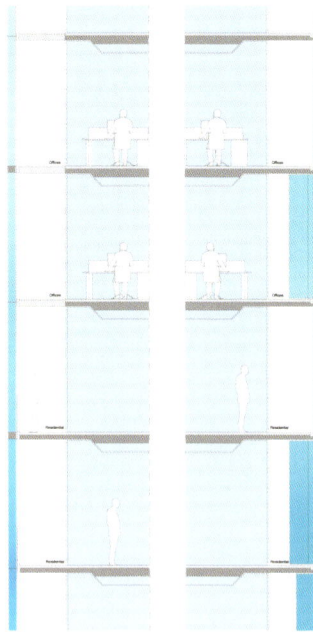

Section Loop Road 1:50 Tower

Section Roof Garden 1:50 Tower

Elevation Roof Garden 1:50 / Residential

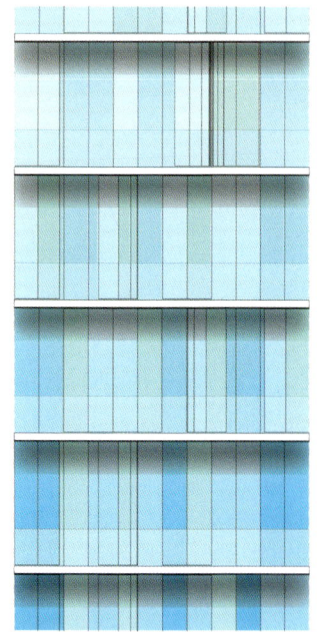

Elevation Roof Garden 1:50 / Offices

Plan View Loop Road 1:50 / Residential

Plan View Roof Garden 1:50 / Residential

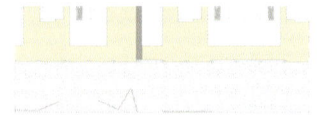

Plan View Roof Garden 1:50 / Offices

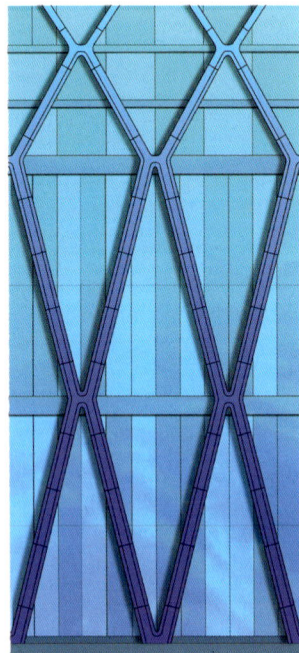

Elevation Loop Road 1:50 Podium

Section Loop Road 1:50 Podium

Sketch Tower of Light

Hotel Apartements 1:200

Typ A small   32 sqm UA
Typ B large   48 sqm UA

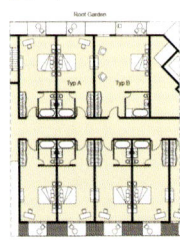

Tenant Offices 1:200

Typ A standard tenants   83 sqm UA
Typ B micro tenants   45 sqm UA
Typ C macro tenants   29 sqm UA

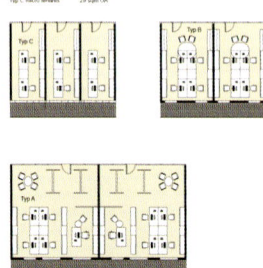

Residential 1:200

Typ A studio   32 sqm UA   Typ D bedrooms   114 sqm UA
Typ B 1 bedroom   50 sqm UA   Typ E bedrooms - room 120 sqm UA
Typ C 1 bedroom - room 79 sqm UA   Typ F bedrooms   130 sqm UA

Roof Garden / Sphere

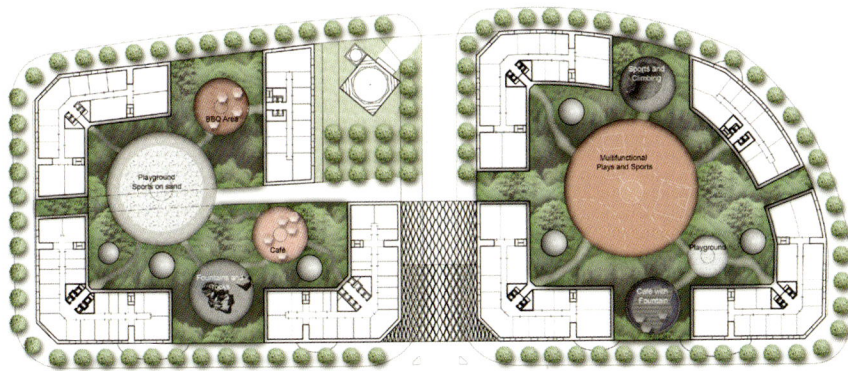

Roof Garden

fine mesh as an icon of nature

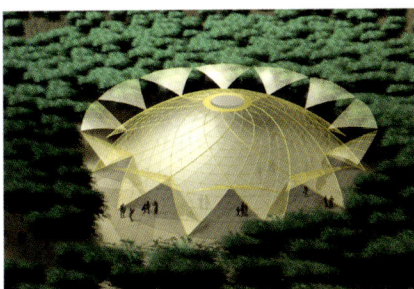

Sphere closed during the summer

Sphere opened during the winter

Bamboo field

Snøhetta Oslo
Snøhetta 建筑事务所 奥斯陆

"An architecture that is responsive to the demands of site and climate whilst also reflecting the aspirations of modern urban living."

"一个建筑物不仅反映出环境和气候的要求，同时也反映出现代城市生活所激发出来的灵感。"

作者 Robert Greenwood, Ibrahim El Hayawan　合作伙伴及建筑系在校生 Andreas Heier, Bartek Milewski, Ibrahim El Hayawan, Joshua Teas, Marianne Saetre, Robert Greenwood　专家 environmental design: BDSP Partnership Ltd., London, Klaus Bode

CONCEPTS AND STRUCTURES

Our goal for this proposal has been to create a foremost example of modern urban architecture. We propose an architecture that is responsive to the demands of site and climate whilst also reflecting the aspirations of modern urban living. The result is a unique vision for a new urbanity combining commercial and residential programme elements to form a vibrant urban neighbourhood. Rooted in it's cultural tradition but providing new arenas for social interaction within a modern urban context. This proposal places emphasis on creating an appropriate and attractive residential quarter that is woven together with an energetic and busy commercial and retail hub. Our aim has at all times been to adhere to simple and robust principles of sustainability. Nature itself has been our inspiration in finding the basic ordering structures for the proposal; waves in the sand, ripples in the water and the opening segments of a palm leaf. These natural structures are all examples of form that is organic in it's true sense. We propose an urban structure that is not an imitation or a representation of nature but one that is truly organic and natural in it's underlying form. Our proposal is open ended and flexible, allowing for growth, change and unlimited variation. The modern urban reality demands a structure that can respond to the complexity and the contradictions that define urban living. The concept must allow for an ever changing and evolving programme, responsive to particular market demands and users aspirations. The proposal envisages flexibility in phasing, allowing the different plots and areas to be developed in a manner that is responsive to the market.

HANGING GARDENS

These natural elements are the primary visual and ordering concept in our proposal. They provide a contextual setting for healthy and green living. The gardens form green strands that are woven across the whole site, forming a visual and contextual background for the residential development. These elements are fully integrated into the sustainability concept for the project, providing a shaded and comfortable microclimate within the development. A sense of intimacy and the presence of nature is a vital ingredient to a successful residential environment. The traditional courtyard garden has for centuries provided the context for domestic living, a shaded and cool oasis around which the residential spaces can be ordered and grouped. To create this large-scale modern urban setting we have chosen to transform the courtyard into vertical gardens. These green walls and the courts they describe provide the opportunity for all the apartments to enjoy a natural setting associated with the qualities of intimacy and privacy

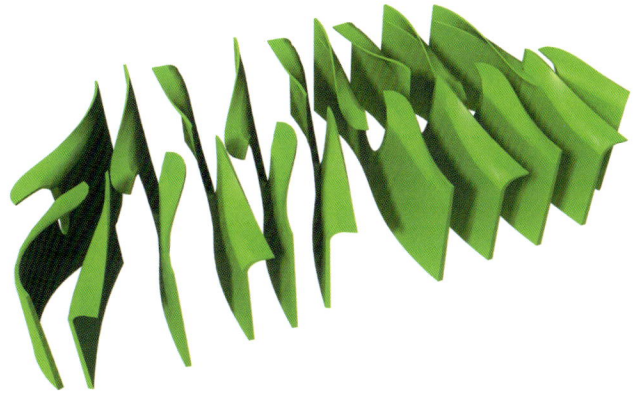

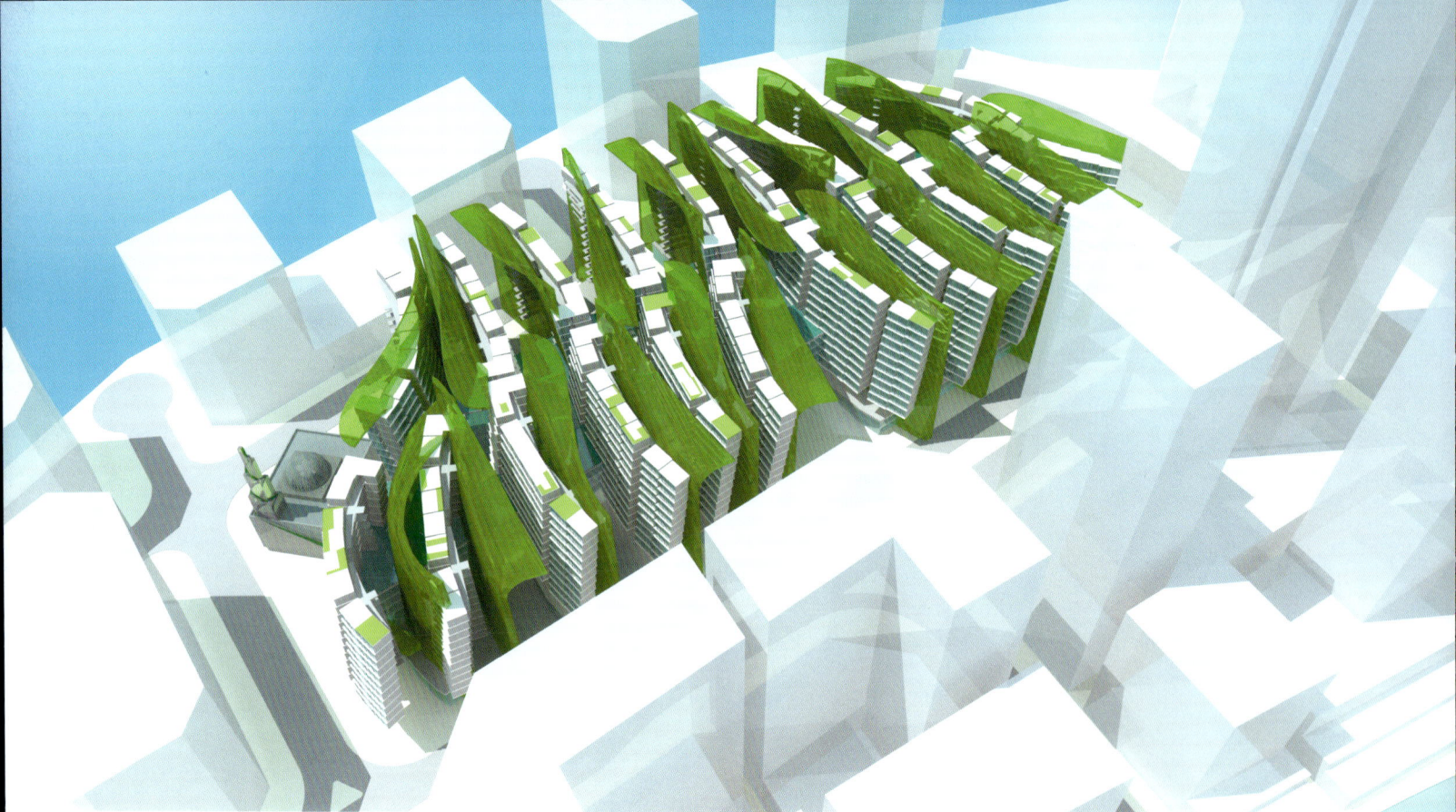

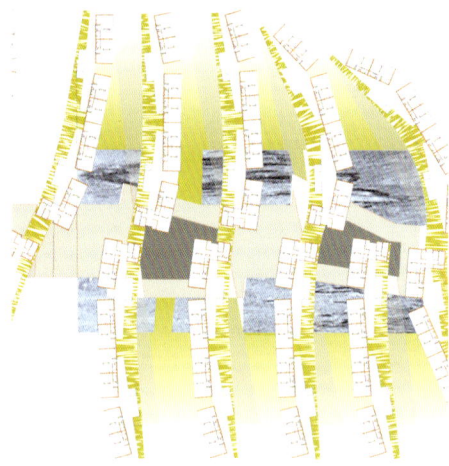

RESIDENTIAL   APARTMENTS

Both in terms of programme area and more importantly architectural expression, the apartments form the dominant element in our proposal. We have chosen to give this portion of the urban plan a predominantly residential character, in contrast to the neighbouring commercial developments to the east. Whilst the commercial will strive upwards towards the vertical, residential and cultural programme tends towards the horizontal, appreciating aspects of intimacy and context. We have chosen to arrange the apartments horizontally across the whole site, connecting the two plots with a unifying structure. The apartments form curvilinear strands, stretching across the site from north to south. Each strand of apartments is associated with it's own vertical hanging garden. All elements of vertical access and servicing are integrated into the vertical garden. The apartments are arranged in groups of 4 units, each sharing a common entrance patio associated with a dedicated elevator connecting to the parking. The entrance patios create a village atmosphere, the context for an intimate collective environment set within the garden elements. Upon entering from the patio the apartments are planned for efficient modern dwelling, offering a high degree of intimacy and privacy. All the rooms enjoy unobstructed views across the shaded exterior court to the neighbouring garden. We have placed the emphasis on providing 1-bedroom apartments in the range of 45 –65 m2. Larger units are located at the ends of the rows forming a spine long the middle of the development where the orientation of the units alternates from the northern to the southern side of the garden element.

HOTEL APARTMENTS – The hotel apartments are integrated into the residential structure, enjoying the same qualities as the apartments. Located to the northeastern sector of the site, the Hotel has it's own entrance and lobby from the Eco Street, with direct elevator connection to the parking below.

OFFICES LEVELS 4 - 7

ECO STREET LEVELS 1 - 2

OTHER PODIUM FUNCTIONS LEVELS 0 - 3

TENNANT OFFICES – The tenant offices are organised as a spine running along the centre of the site on levels 4-7. The spaces are split up into varying sizes, offering the opportunities for both smaller and large enterprises. The office spaces will have the opportunity for entrances and lobbies accessed from the Eco Street. Dedicated elevators will connect all of the office levels directly to the Eco Street and to the Parking below. In planning the office accommodation we have put the priority on creating an attractive and stimulating office environment. The office spaces are sufficiently deep to allow for full flexibility in planning at the same time as giving all the work spaces views out to the exterior garden courts or to the landscaped courtyards sunk through the Eco Street along the central spine.

BUSINESS CENTRE – Centrally located on a corner site at street level this function will be an important element activating the street and connecting to the larger urban development. The Business centre will enjoy both street access and a direct connection to the Eco Street and offices above.

ECO CLINIC & GREEN NUTRITION CENTRE – We propose a prime location for the Clinic and Nutrition centres adjoining the Loop Road. These functions are planned over 4 levels, vertically connecting the Loop Road with the Eco Street. The street level entrance and lobby allow the functions to be orientated both inwards to this development and outwards to the larger urban community.

SPA – Similar to the Clinic and Nutrition centre, the Spa is conceived as a connecting function. Planned over 3 levels, the main entrance will be located on the northern collector street, with an upper Eco Street entrance on level 2. As with the clinic this will give the opportunity to link towards the larger community.

ECO COURT AND STREET – The Eco street and court are proposed as the main connecting elements, spanning the length of the site and connecting all of the functions. This is the heart of the project and social forum that can make this a truly mix use development, combining residential, retail and commercial functions in a holistic concept. In its form the Eco Street and Court are inspired from the historical Soukh, providing a market place where people can meet, eat relax and socialise. This continuous space spans across the collector street in the form of a bridge and is punctuated by 4 naturally ventilated and landscaped entrance courts. At both the northern and southern end the Eco Street opens down to street level, integrating the mosque and knowledge centre with this market function. The plan form allows for retail outlets of varying size and form giving the possibility for a multiplicity and complexity of use.

MOSQUE – We have chosen to give this important public function a prominent street level location. Situated on the northwestern corner of the site the mosque will be an important public element connecting out to the surrounding urban community. Conceived as a separate element independent of the main structure, the mosque can adopt a more formal architectural language responding to the cultural and historical context.

ECO KNOWLEDGE CENTRE – Similar to the Mosque, the Knowledge centre is planned as a separate element marking the end of the Eco Street and connecting to street level. This prominent location offers the opportunity to establish a cultural landmark within the larger urban context.

PARKING – The parking facilities are distributed on the 2 basement levels and the central area of the ground floor plan. This distribution allows for access to both plots from the central collector street C and eradicates the need for any parking garage facades towards the public streets. The ground floor level parking is located directly under the Eco Street and the office spine above, this will be the ideal location for parking serving these functions. This parking level is punctuated by the landscaped courtyards in the Eco Street, thus allowing daylight to penetrate into the garage and forming the access point up into the Eco Street. The lower parking levels will primarily serve the apartments, with direct elevator connections to the entrance patios above.

SECTION 1:200

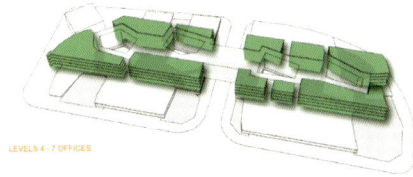

LEVELS 4 - 7 OFFICES

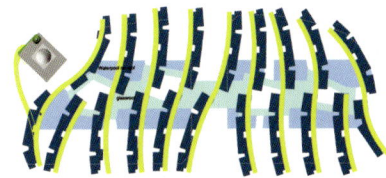

LEVEL 8-14

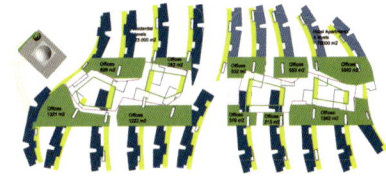

LEVELS 4 - 7

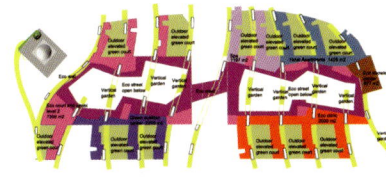

LEVEL 3

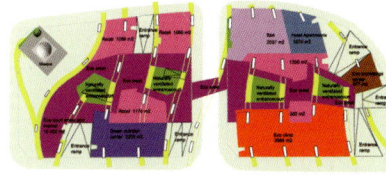

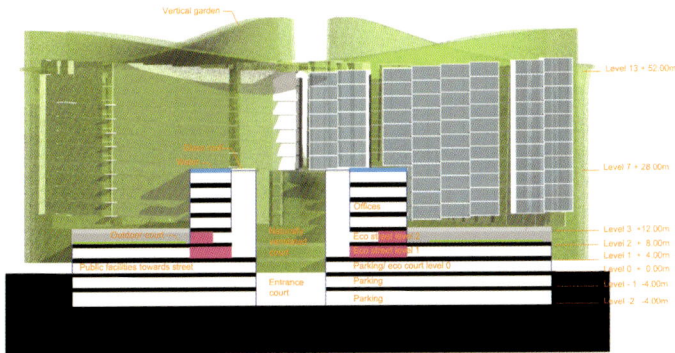

SECTION 1:500

ELEVATION 1:500

SECTION
AREAL DIAGRAM

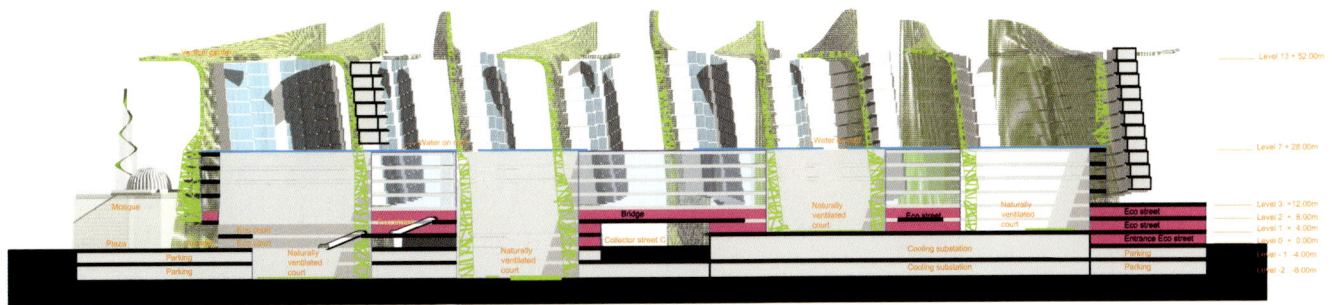

SECTION 1:500

KSP Engel und Zimmermann Architekten Frankfurt/Main
KSP Engel und Zimmermann 建筑事务所 德国美因河畔法兰克福

"The leading goal was to interpret the lively qualities of an old oriental city and create a unique object bay applying advanced techniques to traditional forms." "A vertical Medina."

"主要目标是反映出一座东方古城的充满活力的特色，把高科技运用到传统形式中，从而创造出一个独特的生态湾。" "垂直的麦地那。"

作者 Jürgen Engel, Michael Zimmermann   合作伙伴及建筑系在校生 Ala Ramdan, representative and director, Middle East; Gregor Gutscher, Thomas von Girsewald, Antonio Vultaggio, Ramona Becker, Sylvia Grüning   专家 structural engineers: Schlaich, Bergermann und Partner GbR, Stuttgart; Sven Plieninger; mechanical services: IPB Berchtold, Sarnen; Peter Berchtold; landscape architecture: WES & Partner Landschaftarchitekten, Hamburg; Wolfgang Betz

KSP Engel und Zimmermann Architekten, Frankfurt/Main

Eco Bay Project, Abu Dhabi

Engel und Zimmermann
Architekten

Sorouh

Eco Bay Project Abu Dhabi

2 | 4 | 6 | 8 |

KSP Engel und Zimmermann

elevation collector street c . scale 1:200

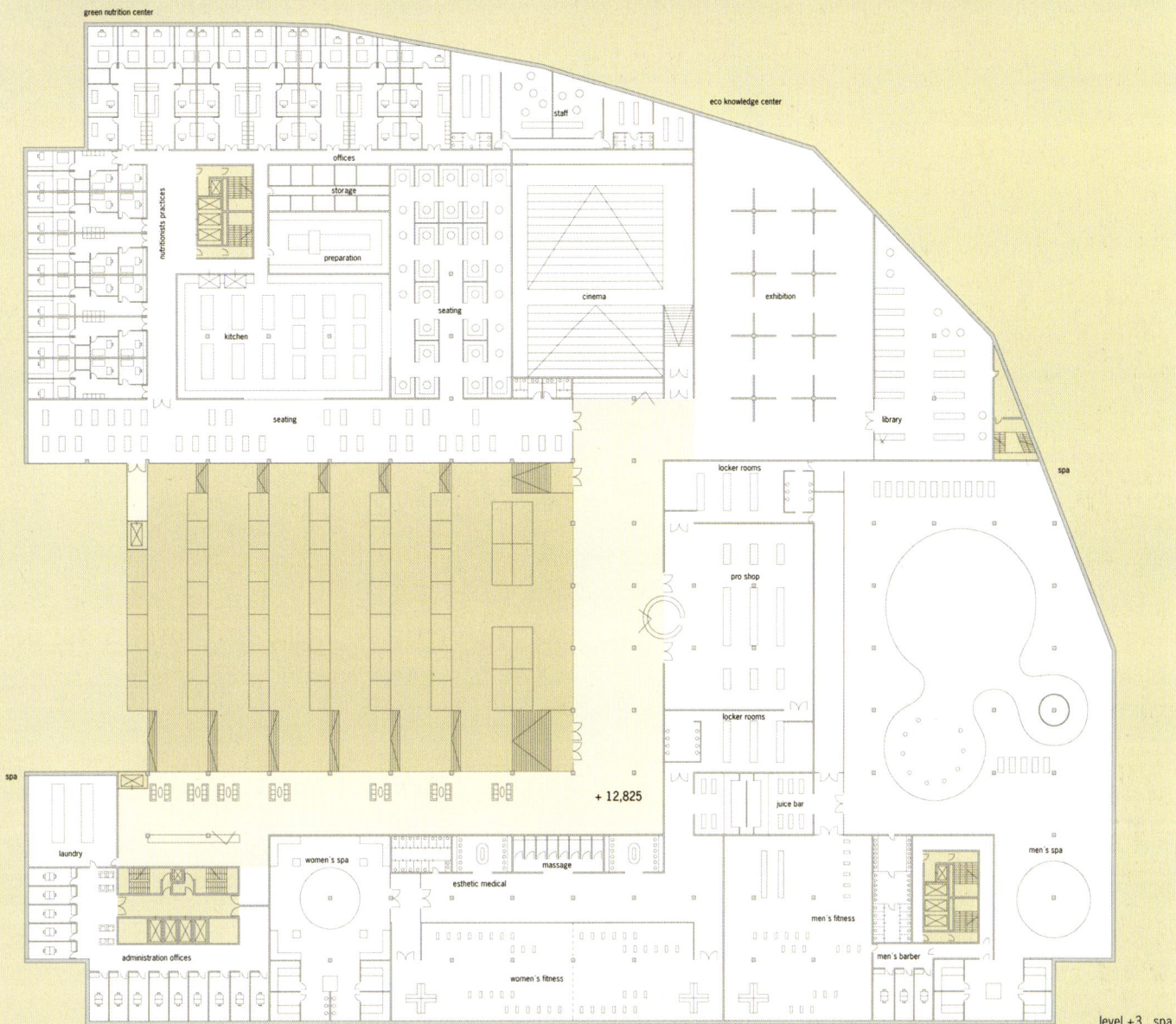

green nutrition center

eco knowledge center

staff

offices

storage

nutritionists practices

preparation

seating

cinema

exhibition

kitchen

seating

library

seating

locker rooms

spa

pro shop

+ 12,825

locker rooms

juice bar

men's spa

spa

laundry

women's spa

massage

esthetic medical

men's fitness

men's barber

administration offices

women's fitness

level +3 . spa

women's lounge

void

void

reception

children creche

men's lounge

technic

womans health

conference

diagnostics

diagnostics

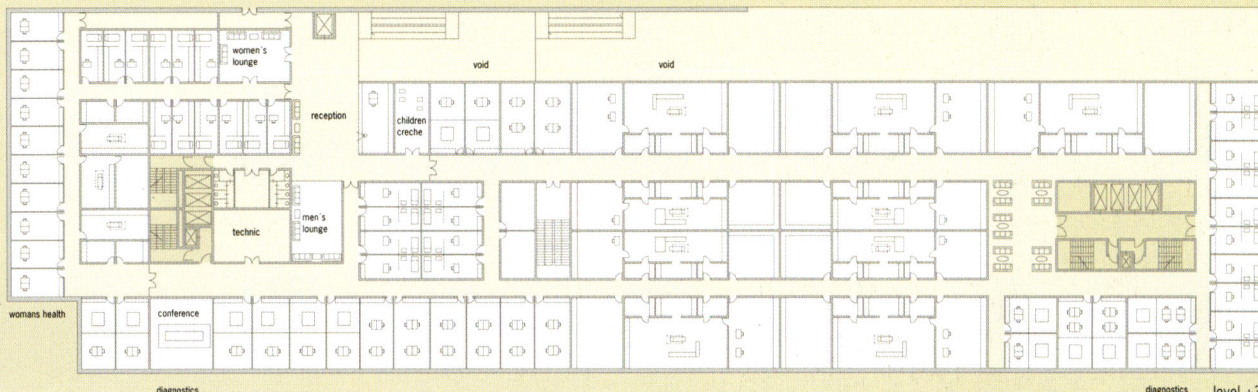

level +3 . eco clinic

Benthem Crouwel Amsterdam
Benthem Crouwel建筑事务所 阿姆斯特丹

"An introverted yet strikingly recognisable design."
"Eco bay forms an indoor urban square that adds to the spatial diversity of the peninsula."

"含蓄而又特别引人注目的设计。"
"生态湾在半岛上形成一个内部城市广场，增加了其空间结构的复杂性。"

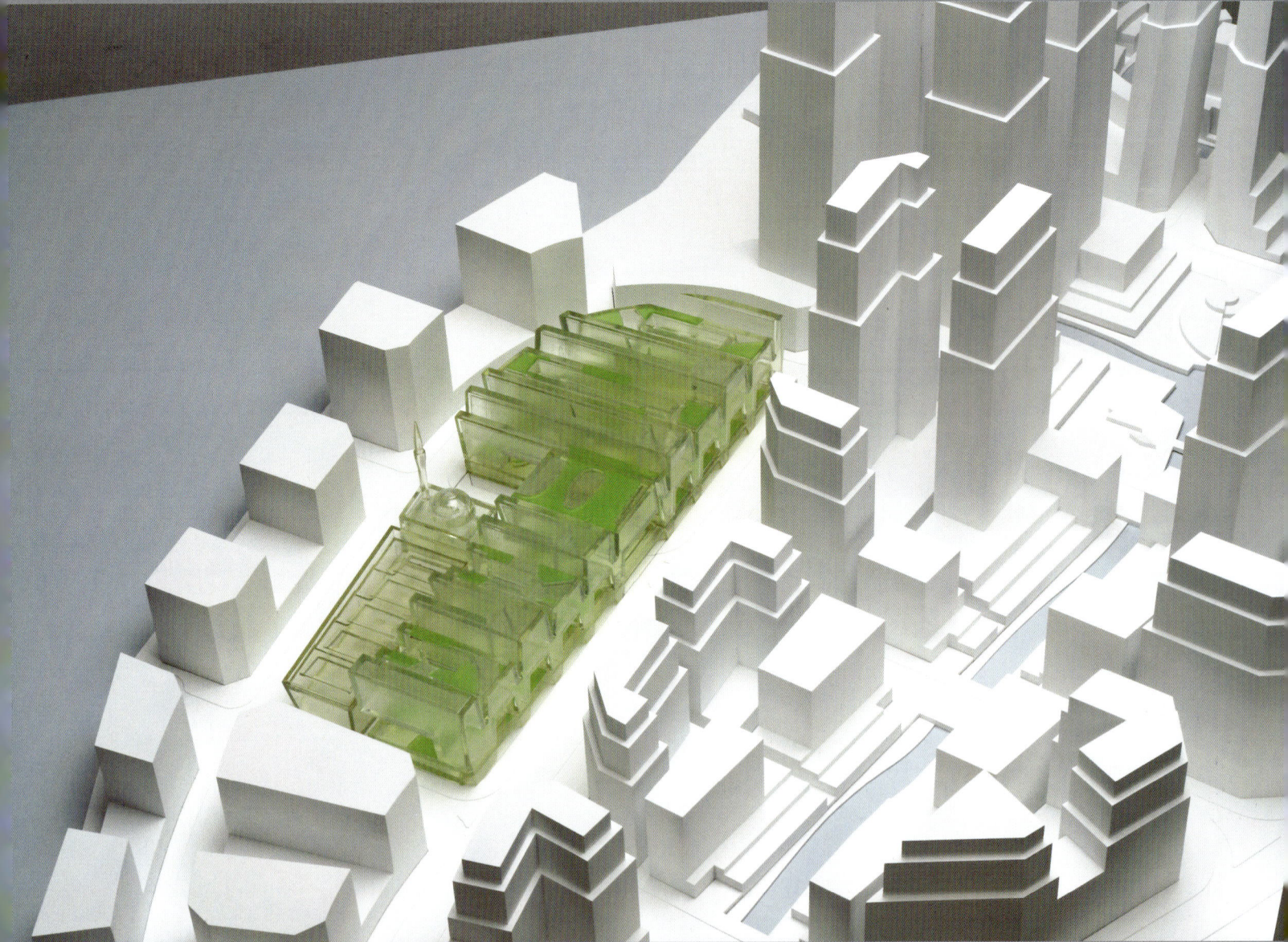

作者 Marcel Blom, Joop Paul, Bart Brands, Paul Van Bergen　合作伙伴及建筑系在校生 Benthen Crouwel Architekten BV bna, Anja Blechen, Boudewijn Bakker, Marleen van Driel, Michael Jaggoe, Jeroen Jonk, Mijke de Kok, Markwin Margaretha, Jelle van der Neut, Evert Pronk, Ronno Stegeman, Jan Torringa, Marten Wassmann, Jos Wesselmann, Guido de Witt Arup bv Munir Al-Hashimi, Jules Misere, Jaap Wiederhoff　专家 Karres en Brands Landschapsarchitecten BV, Lieneke van Kampen, Uta Krauset

approach on loop road

volumes, boulevards, worlds

Benthem Crouwel, Amsterdam

Eco Bay Project, Abu Dhabi

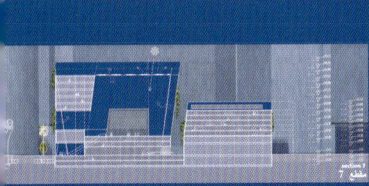

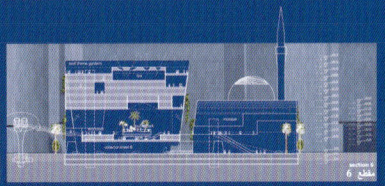

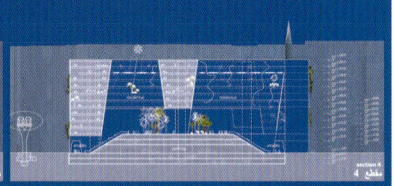

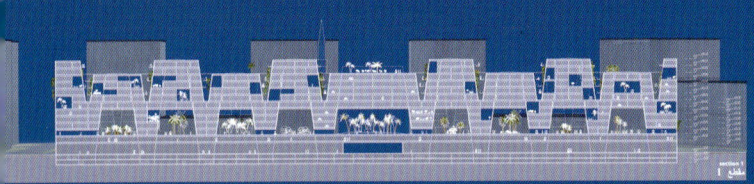

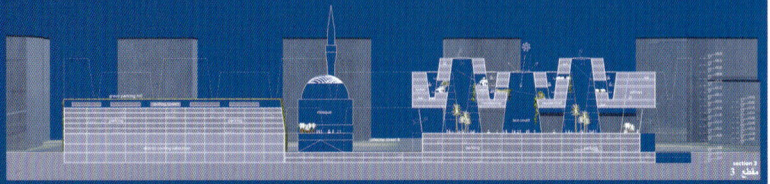

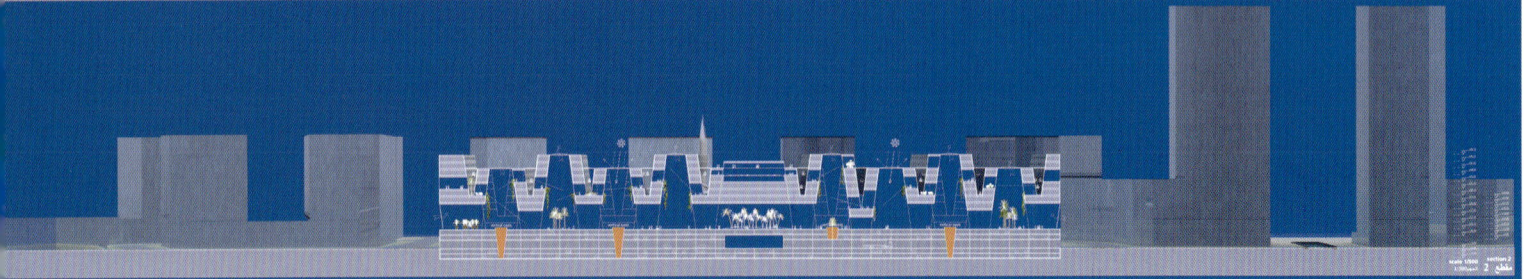

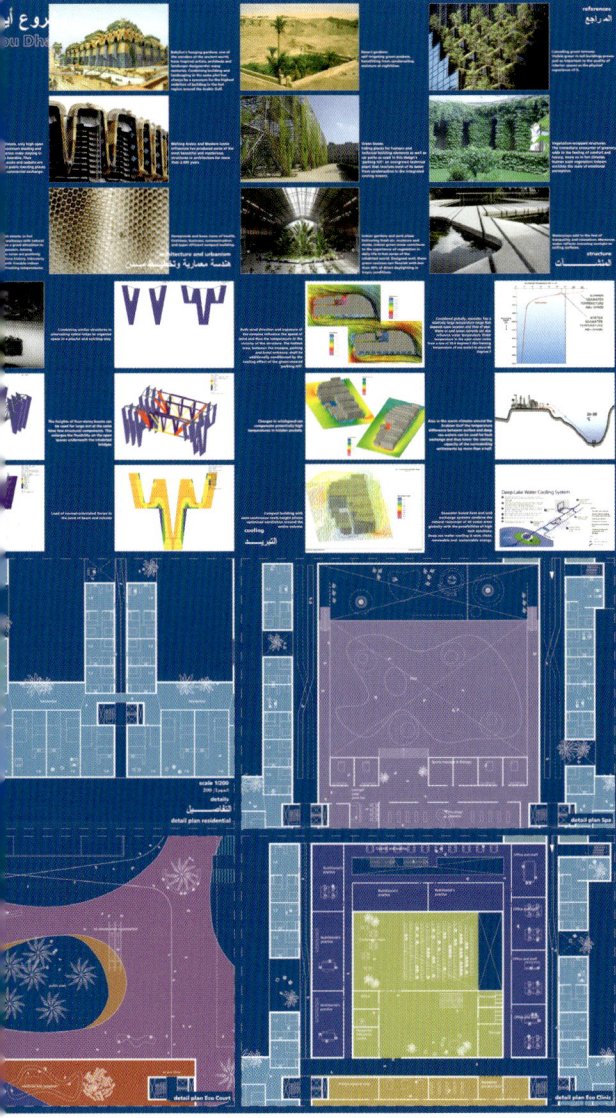

باي

Lab Architecture Studio Melbourne
Lab 建筑事务所 墨尔本

"The focus here is on 'relationships' – wether spatial, operational, or environmental [...] it offers a complex matrix of lapping facilities, intertwined experiences, and mutually supportive aspirations."

"焦点在于空间的、运作的，或者环境的'关系'上。它提供了重叠的设施、缠结在一起的经验和互相支持的志向的一个复杂的母体。"

作者 Donald Bates, Peter Davidson, David Racz 合作伙伴及建筑系在校生 Ben Milbourne, Julian Canterbury, Alvin Lee, Wayne Sanderson, Chuin Wee Lin, Will Hosikian, Daniel Wolkenberg, Hendy Wijaya, Nicole Yean Ling Tang, Winnie Ha, Monique Brady, Sophie Drying 专家 structural engineer: Peter Bowtell; traffic engineer: David Shrimpton; services engineer: Russell Jessop; structural engineer: Carolina Bartram; environmental engineer: Patrick Bellew

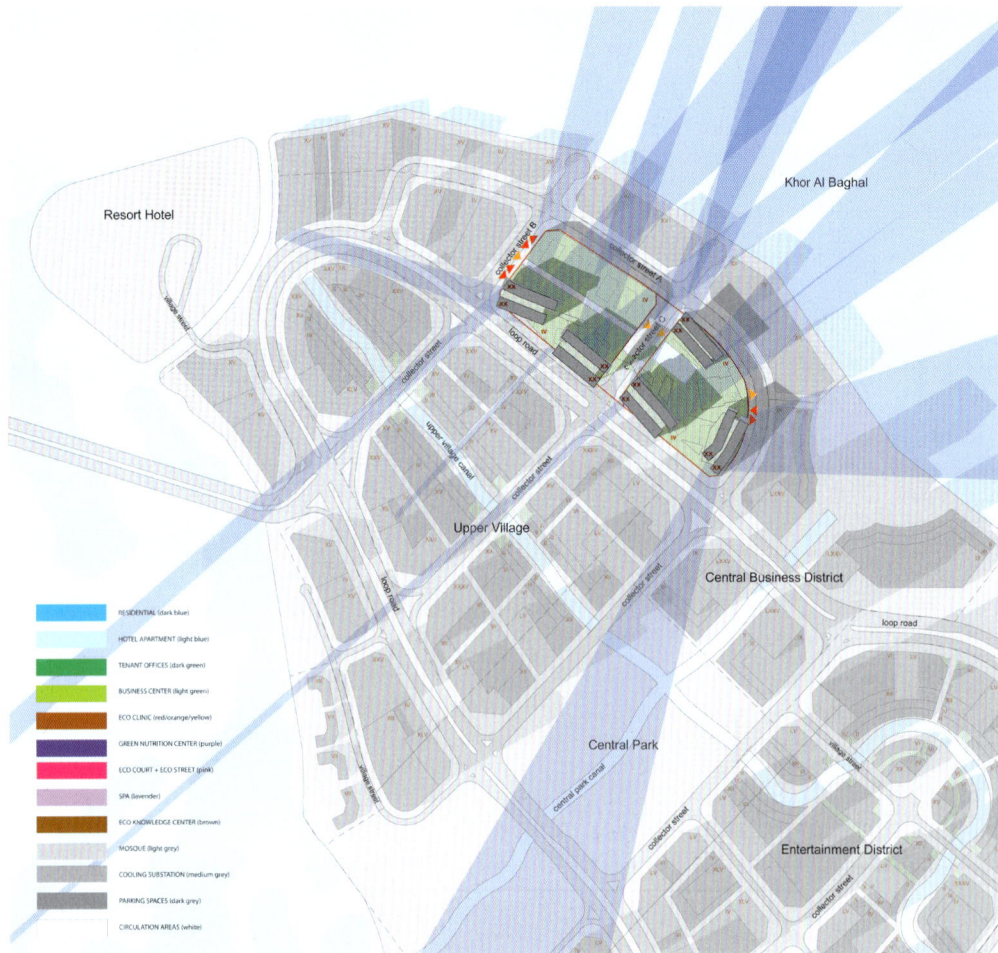

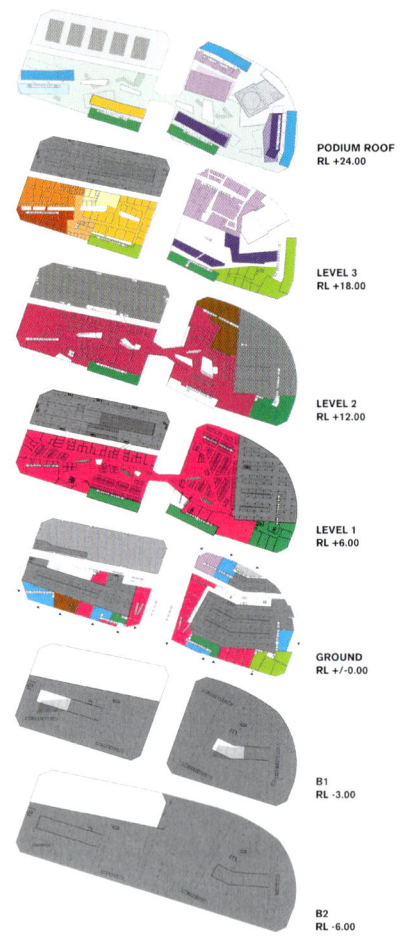

Resort Hotel

Khor Al Baghal

Upper Village

Central Business District

Central Park

loop road

Entertainment District

RESIDENTIAL (dark blue)
HOTEL APARTMENT (light blue)
TENANT OFFICES (dark green)
BUSINESS CENTER (light green)
ECO CLINIC (red/orange/yellow)
GREEN NUTRITION CENTER (purple)
ECO COURT + ECO STREET (pink)
SPA (lavender)
ECO KNOWLEDGE CENTER (brown)
MOSQUE (light grey)
COOLING SUBSTATION (medium grey)
PARKING SPACES (dark grey)
CIRCULATION AREAS (white)

PODIUM ROOF
RL +24.00

LEVEL 3
RL +18.00

LEVEL 2
RL +12.00

LEVEL 1
RL +6.00

GROUND
RL +/-0.00

B1
RL -3.00

B2
RL -6.00

ECO BAY PROJECT
SHAMS ABU DHABI

SITE CONTEXT

SCALE 1:2000 @ A1   MECCA   NORTH

LAB architecture studio   1

project no
28.03.06

SOUTH WEST AERIAL VIEW

ENVIRONMENTAL TERRACE

DAY SPA POOL

RESTAURANT TERRACE

DISTRICT COOLING PLANT COOLING TOWERS

NORTH WEST AERIAL VIEW

RL 12.00
LEVEL 2

ALUMINIUM EDGE STRIP

DOUBLE GLAZED GLASS
PLANK SYSTEM

INTERNAL GLARE PROTEC-
TION SCREEN

RL 6.00
LEVEL 1

EXPOSED FLOOR SLAB

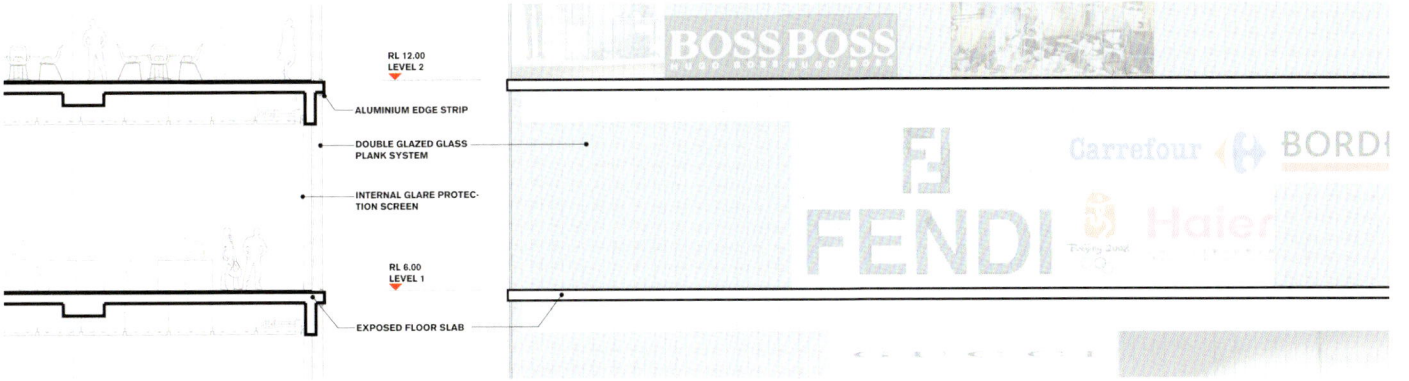

PODIUM DETAIL SECTION

RETAIL FACADE DETAIL ELEVATION

TYPICAL MALL ORGANISATION

limmited range of retail tenancy sizes and configurations. limited range of retail types and other functions. linear/limited connections between zones

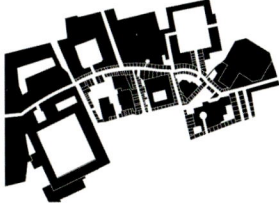

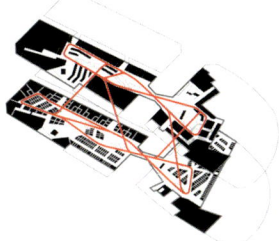

SECTION OF ISFAHAN SOUKH

very broad range of tenancy sizes and configurations. broad range of retail types and other functionsinear. linear/limited connections between zones

ECO-COURT ORGANISATION

very broad range of tenancy sizes and configurations. very broad range of retail types and other functions. network of connections between zones, encouraging retail synerg and diversity of retail experience

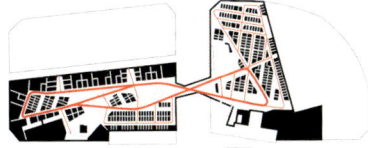

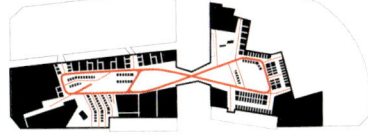

ECO-COURT CIRCULATION

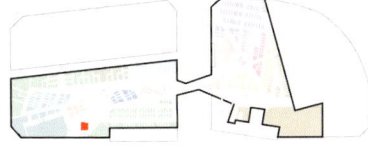

ECO-COUR RETAIL ZONEING

GARDEN VOIDS

DAY SPA

MOSQUE

ECO TERRACE

RETAIL LEVEL 2

RETAIL LEVEL 1

GROUND LEVEL ENTRY

ECO-COURT PERSPECTIVE

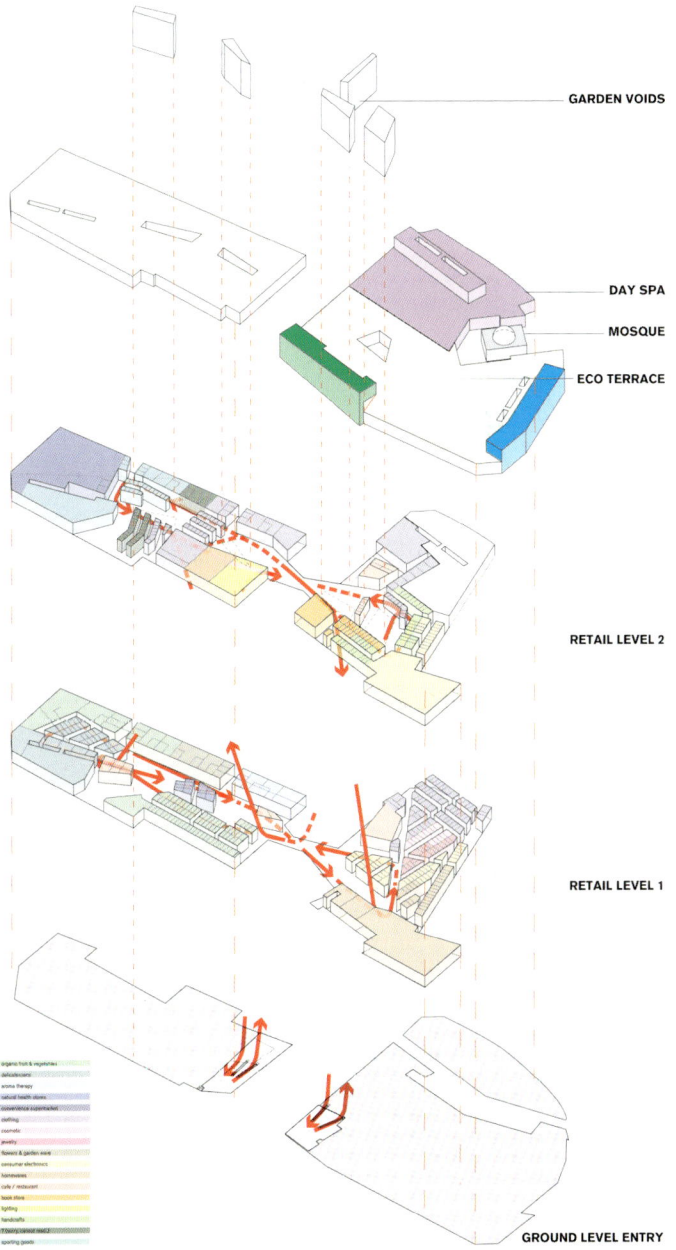

ECO-COURT PROGRAM RELATIONSHIPS

ECO-COURT SECTIONAL PERSPECTIVE

Ateliers Lion a

DB Architecture

Hijjas Kastu

ACXT – IDOM Group

rchitectes urbanistes

Khandama Development Project in Mecca

麦加Khandama开发项目

Associates

Jafar Tukan

Khandama Development Project **Mecca**
Khandama开发项目 麦加

Restricted two-stage project competition for architects and city planners preceded by an application procedure
限制严格的建筑师和城市规划人员的项目竞赛，分为两个阶段。竞赛开始前有申请程序。

地点 Mecca   时间 05/2006–09/2006   主办方 Jabal Khandama Development Co. represented through
Sheihk Abdul Rahman Abdul Qader Fakieh   人数 Stage 1: 10 competitors (thereof 9 who submitted); Stage 2: 5 competitors
面积 60 ha   竞赛费用 195,000 USD   专业评奖委员会 Prof. Dr. Abdulrahman Alangari, Riyadh; Prof. Nezar AlSayyad, PhD, Berkeley; Prof. Dr.
Max Bächer, Darmstadt; Mohammad Saeed Farisi, PhD, Jeddah; Prof. Dr. Gulzar Haider, Lahore, Pakistan;
Dr. Kayvan Karimi, London; Prof. Rodolfo Machado, Boston; Prof. Heinz Nagler, Cottbus; Dr. Mahmud Bodo Rasch, Stuttgart
and Jeddah/Madina   评奖委员会助理 Prof. Heinz Nagler, Cottbus; Dr. Mahmud Rasch, Stuttgart

Mecca in the Hejaz, a region of western Saudi Arabia, is the birthplace of the prophet Muhammad and, beside Medina, the holiest place of Islam. It is enclosed by the mountain ranges. Since time immemorial, the Grand Mosque with the Kaaba has been at its center. In recent years, the Saudi government has made enormous efforts to cope with the soaring number of visitors. Huge development projects have been initiated with a view to channeling the flow of pilgrims, making the Hajj safer, and providing accommodation and food for the visitors. Other planning objectives were to coordinate the further development of Mecca's inner city in response to the increase in regional population. They included enhancing the inhabitants' quality of life, and putting to better use valuable properties that offer a view of the Grand Mosque or are situated close by. The competition site includes the Jabal Khandama, a 65-hectare development zone immediately east of the Grand Mosque with a very demanding topography, as its name implies (jabal being the Arabic word for mountain). The stipulated 2.4 million square metres gross floor area is to accommodate hotels, residential and retail use, as well as public facilities, and generous public-prayer spaces in sight of the Kaaba. The principals attached great importance to a high-performance access system and a consistent design approach that responds sensibly to climatic conditions, to the history of the site, and to its cultural and religious significance.

麦加位于沙特阿拉伯西部，是先知穆罕默德的出生地，这旁边的麦迪那是伊斯兰教的圣地。麦加周围群山环绕，很久以来大清真寺一直坐落其中。近年来，沙特阿拉伯政府为了满足数量剧增的旅游者的需要，下大力气建设大型开发项目，来为旅游者提供食宿。同时，也制定其他一些规划来协调麦加内城的未来发展，包括增强麦加居民的生活质量、更好发挥大清真寺及周围景观的优势特点等。竞赛地点包括Jabal Khandama，毗临大清真寺东面的一块65公顷的开发区，协议规定240万平方米的建筑总面积将被用作酒店、住宅和商店以及其他公共设施和宽敞的公众祈祷处。项目委托人非常重视高性能的门禁系统和适应当地气候条件、历史地位和文化宗教重要性的可持续性的设计方法。

Aerial view　鸟瞰图

Competition site　　竞赛地点

Competition site　　竞赛地点

The Great Mosque　大清真寺

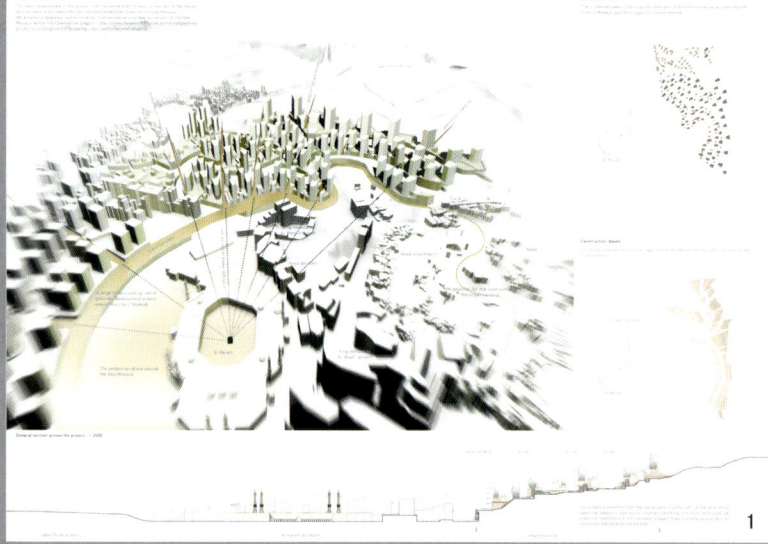

GUIDING IDEAS AND PHILOSOPHY OF THE PROJECT   SET A

1

URBAN CHARACTER

THE MAIN ARTERY
THE VERTICAL CIRCULATION
THE OLD FABRIC RECREATED
THE SOCIAL FLOW

4

1st Ring Road

Al-Shamiyyah

Shubaikah

1st Ring Road

NEW LINK ROAD CONNECTING THE BUS
TERMINAL AND ZONE A

NEW BUS TERMINAL & MAIN CAR PARKING
"TRANSFORIUM"

CERTAIN BUILDINGS IN ZONE A ARE REPOSITIONED IN
ORDER TO CREATE SMOOTH TRANSITION THROUGH
NEW DEVELOPMENT AREAS

THE MAIN PROMENADE THE PALM FOREST

THE ROAD NETWORK OF ZONE A IS
MODIFIED TO REALIZE FULL INTEGRATION
WITH THE OVERALL TRANSPORTATION
SYSTEM

7

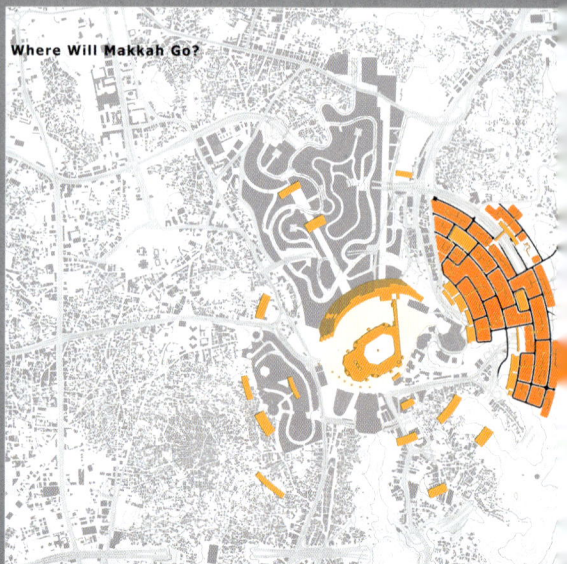

Where Will Makkah Go?

THE PRAYER PLATFORM
THE UPPER PLAZA
THE BREATHING LUNG
AL KIBLA

THE PRAYER PLATFORM
THE COMMERCIAL ZONE
THE SOCIAL HUB
THE SHADED RETREAT

THE NEW MARKET PLACE
THE BUILT MOUNTAIN
THE OPEN FABRIC
THE INTERACTIVE
ENVIRONMENT

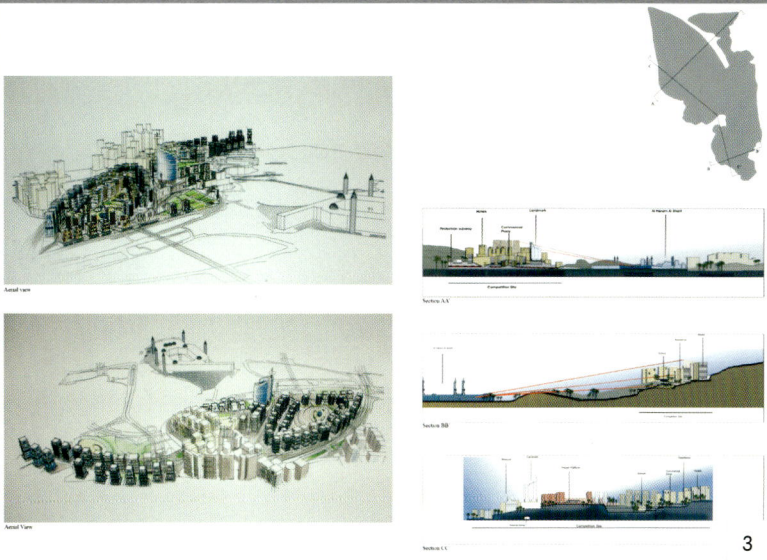

Qualified participants
of the 1st stage
第一阶段合格的参赛者

1
Ateliers Lion architectes urbanistes,
Paris, France

2
Consolidated Consultants
Engineering & Environment –
Jafar Tukan Architects, Amman,
Jordan

3
Dar Al Handasah Nazih Taleb &
Partners, Beirut, Lebanon

4
ACXT – IDOM Group, Bilbao, Spain

5
Hijjas Kasturi Associates Sdn.,
Kuala Lumpur

6
Llewelyn Davies Yeang, London, UK

7
DB Architecture & Consulting,
Istanbul, Turkey

8
Albert Speer und Partner,
Frankfurt/Main, Germany

9
Saucier + Perrotte Architects,
Montreal, Canada

1

2

4

5

3

Qualified participants of the 2nd stage
第二阶段合格的参赛者

1
1st prize　一等奖
Ateliers Lion architectes urbanistes,
Paris, France

2
2nd prize　二等奖
Consolidated Consultants
Engineering & Environment –
Jafar Tukan Architects, Amman,
Jordan

3
Honorary mention　荣誉奖
ACXT – IDOM Group, Bilbao, Spain

4
Honorary mention　荣誉奖
Hijjas Kasturi Associates Sdn.,
Kuala Lumpur

5
Second round　第二轮
DB Architecture & Consulting,
Istanbul, Turkey

Ateliers Lion architectes urbanistes Paris
Ateliers Lion建筑城市规划事务所 巴黎

"The basic requirement is to maximise direct visual connection to the Haram [and] to create efficient and safe pedestrian links to the Holy Mosque."

"基本要求是最大限度地建立与内室的直接视觉联系，创造便捷安全的通向大清真寺的行人通道。"

作者 Yves Lion, Fadhallah Dagher, Abdul Halim Jabr   合作伙伴及建筑系在校生 Ateliers Lion architectes urbanistes: Claire Piguet, David Jolly, Sajin Lee, Maroun Lahoud, Triet Le Minh, Moritz Krüger Dagher, Hanna & Partners: Fouad Hanna, Joseph Rustom, Christian Zahr, Pierre Khoury, Jamil Khoury, Maya Rafieh   自由建筑师 modelmaking: Maquetas y Projectos de Architectura, Barcelona; images: Auralab, Paris
专家 transportation planning: Citec, Philipp Gasser, Geneva

Aerial perspective

The basic requirement of the project is to maximize direct visual connection to the Haram. Also primary is to create efficient and safe pedestrian links to the Holy Mosque.
We propose a spacious «urban landing» that serves as a further expansion of the Holy Mosque within the Khandamah project. It also curves beyond the limits of the competition project to accomplish the following urban and fonctional aspects.

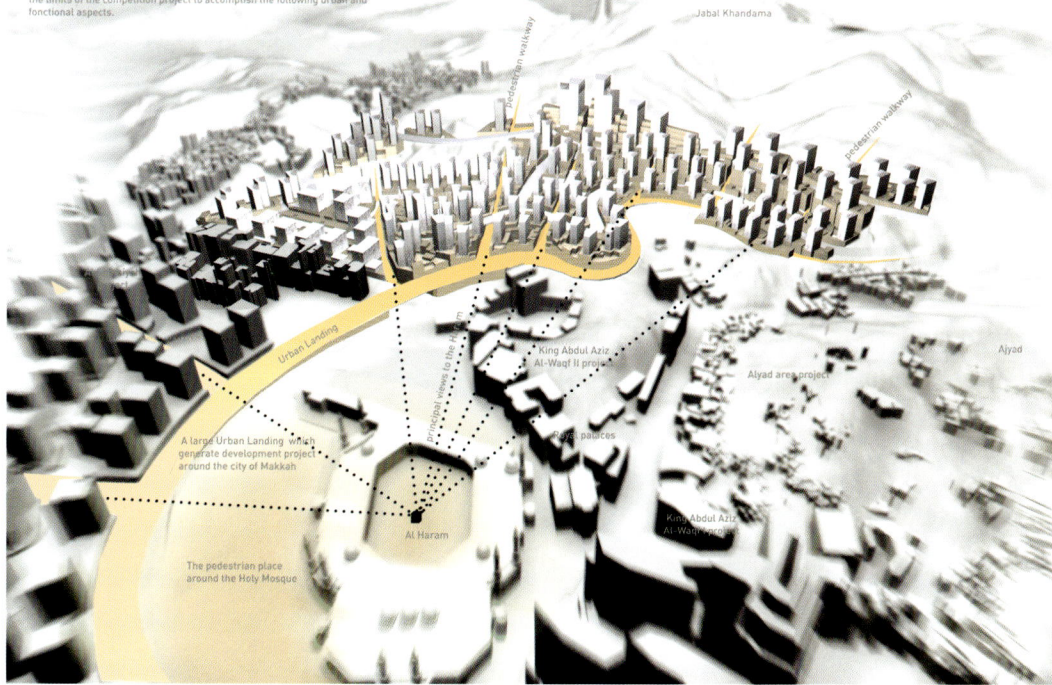

Construction levels

The clustered towers offer a gentle diversity of built form that equally blends with historic Makkah and the rugged mountain beyond.

Construction bases

The bases rationalize the urban agglomeration and anchor the public domain of the development.

Jabal Khandama

pedestrian walkway

pedestrian walkway

Ajyad

Urban Landing

King Abdul Aziz
Al-Waqf II project

Ajyad area project

principal views to the Haram

Royal palaces

King Abdul Aziz
Al-Waqf II project

A large Urban Landing which generate development project around the city of Makkah

Al Haram

The pedestrian place around the Holy Mosque

General section accross the project 1:2000

urban landing   street   street   street

Our proposal benefits from the topography to better utilize the land, through a unifying base that adapts to site levels. It contains parking,

Vehicular traffic flow during off season

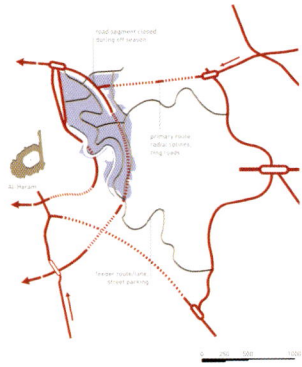

Off season bus circulation

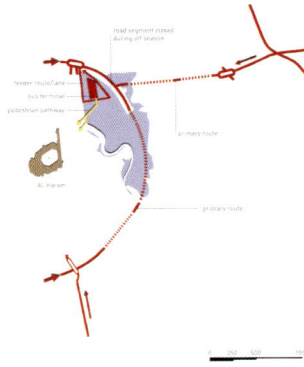

Network row during peak hours or seasons

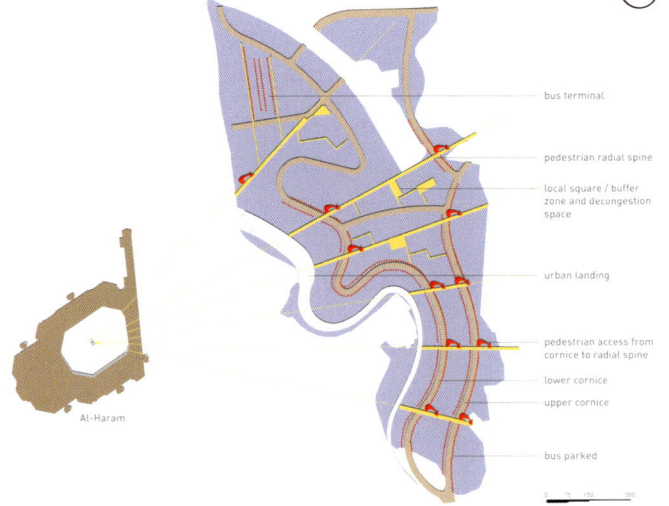

bus terminal

pedestrian radial spine

local square / buffer zone and decongestion space

urban landing

pedestrian access from cornice to radial spine

lower cornice

upper cornice

bus parked

Al-Haram

Vehicular traffic flow during HAJJ season

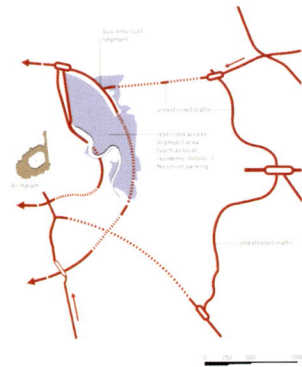

HAJJ season bus circulation

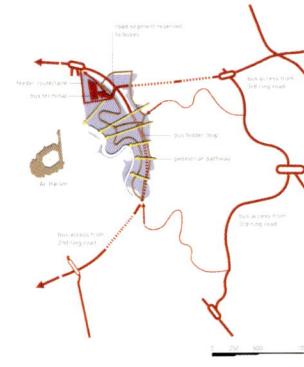

Separation between pedestrian radial spines and vehicular network

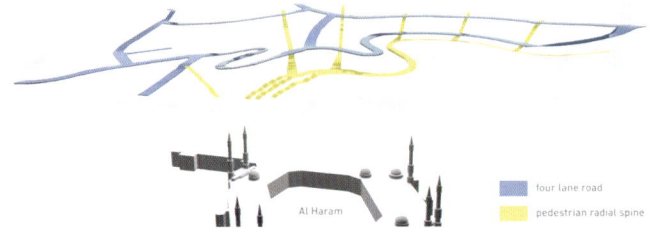

Al Haram

four lane road
pedestrian radial spine

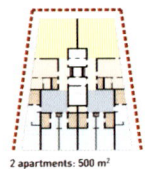

2 apartments: 500 m²

1 apartment: 600 m²

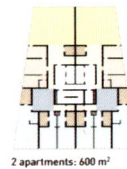

2 apartments: 600 m²

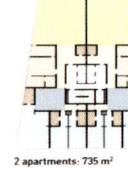

2 apartments: 735 m²

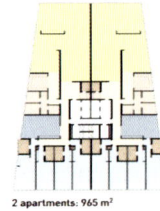

2 apartments: 965 m²

2 apartments: 500 m²

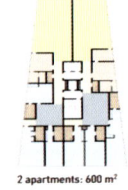

2 apartments: 600 m²

2 apartments: 735 m²

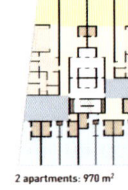

2 apartments: 970 m²

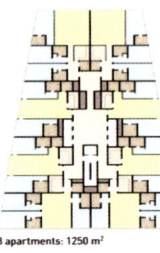

8 apartments: 1250 m²

APARTMENTS

- - - - façade with view towards the Holy Mosque

reception / dining

lobby / corridor

family room

bedroom

bath / toilet

services

shaft

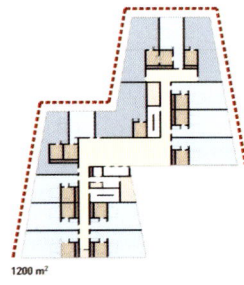

1200 m²

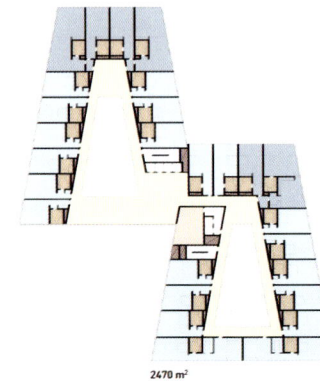

2470 m²

HOTELS

- - - - façade with view towards the Holy Mosque

typical guest room

suite

bath / toilet

lobby / corridor

service

shaft

2nd ring road

314

336

348

354

four-lane-road

367

2nd ring road

391

387

Al Shamiyah project

pedestrian walkway

302

314

314

314

320

338

335

340

345

362

355

330

332

331

365

pedestrian walkway

333

347

386

holy place

urban landing

336

four-lane-road

356

pedestrian walkway

346

349

343

390

King Abdul Aziz
Al-Waqf II project

royal palaces

urban landing

340

four-lane-road

357

pedestrian walkway

362

320

pedestrian walkway

lower cornice

365

upper cornice

330

367

369

Al Haram

pedestrian walkway

400

1st ring road

366

372

Ajyad hospital

410

King Abdul Aziz
Al-Waqf I project

four-lane-road

117

Khandama Development Project, Mecca

prayer room

level 30

level 25

level 20

high-rise hotel

level 15

minaret

mosque entrance

prayer room

high-rise hotel

level 15

high-rise hotel

level 10

commercial services

drop zone

upper cornice
four-lane road

public facilities

hotel entrance

hotel facilities

public stairs

level 10

level 5

lower cornice

retail

public facilities

urban landing

pedestrian walkway

public stairs and elevators

hotel

apartements

urban landing

existing cliff

existing tunnel

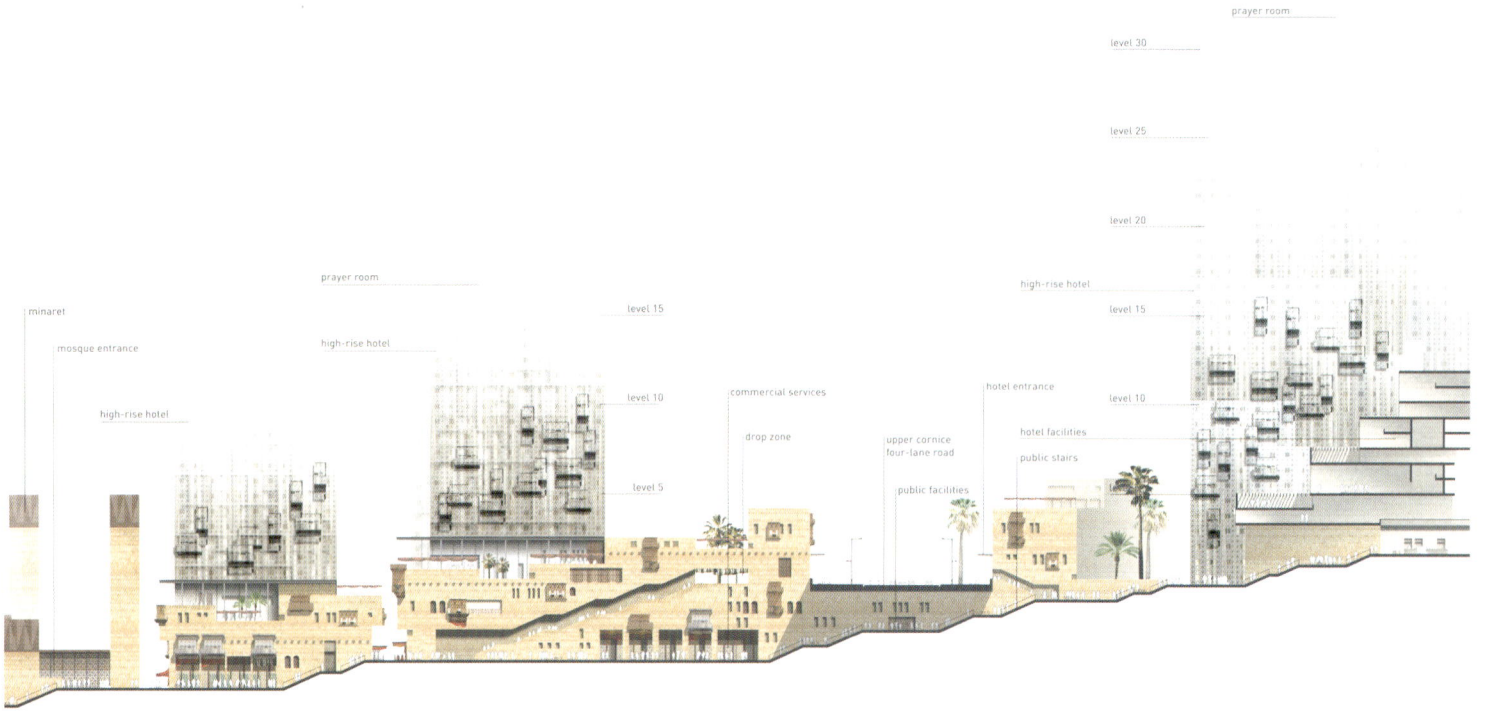

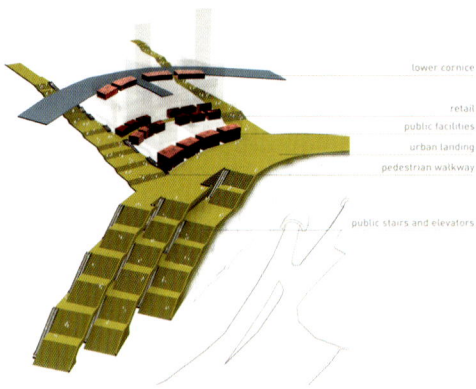

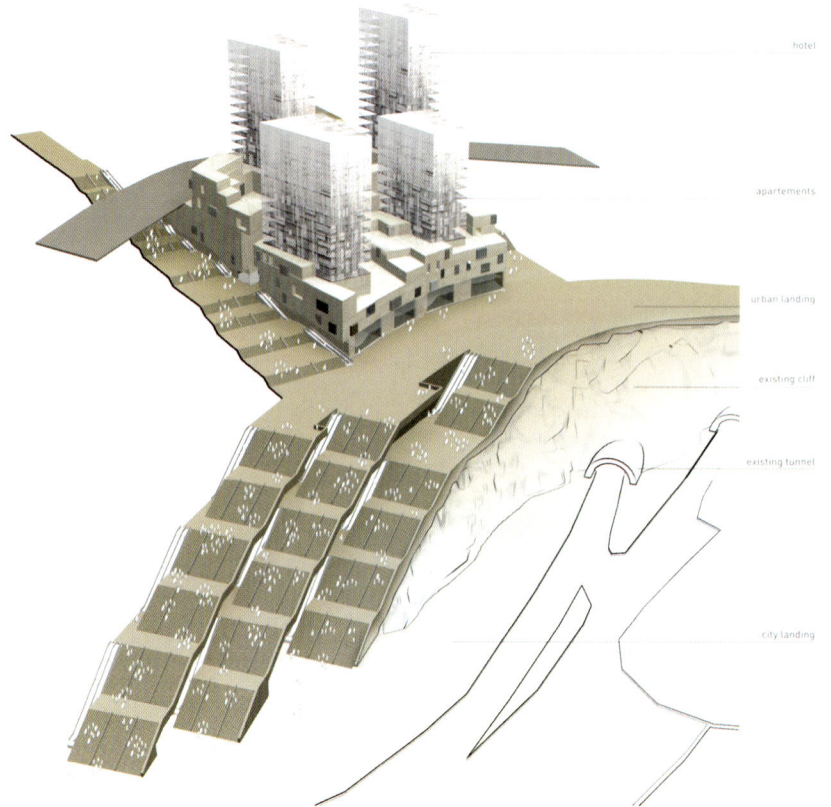

3D model of inner organization

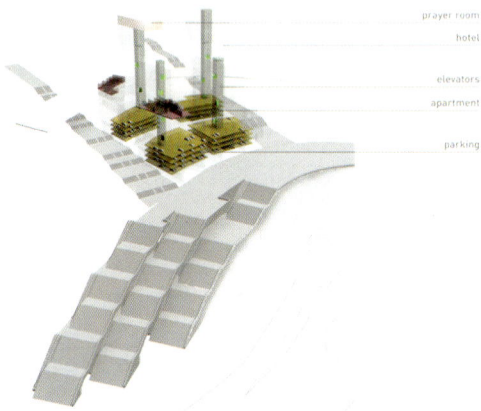

prayer room

hotel

elevators

apartment

parking

city landing

Khandama Development Project, Mecca

Consolidated Consultants Engineering & Environment Jafar Tukan Architects Amman

工程环境综合咨询公司——Jafar Tukan建筑事务所 安曼

"Dealing with the loss of public space and public life and addressing major environmental degradation [by] the urban mountain oasis."

"营造城市山野绿洲，来解决公共空间和公共生活缺失和主要环境恶化的问题。"

作者 Jafar Tukan, Rami Daher, Yasser Darwish   合作伙伴 Consolidated Consultants Engineering & Environment - Jafar Tukan Architects: Saba Innab, Jumana Hamadani, Andaleeb Al-Bezreh, Ahmad Seyam Turath, Amer Al-Jokhader, Maiss Razem, Bilal Sami, Siba Tawalbeh
自由建筑师 consulting architect: Mohammad Abbas, Amman; architectural presentation: Tamer Shaqwara/Quattro Design, Amman
专家 (Consolidated Consultants Engineering & Environment) traffic engineer: Issam Bilbesi; road engineers: Hisham Salhi, Rania Arnaoot

Rock Garden | Walkalators to main Arteries | Stairs & Escalators for lower Haram Plaza to main Arteries | Looking to Haram

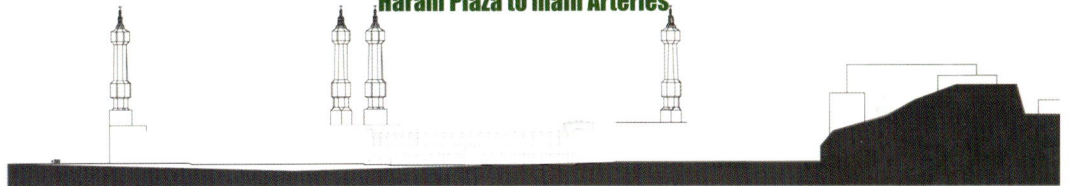

Looking to Haram

Royal Palaces

BUS TERMINAL AND PLAZA CONNECTING TO HARAM.

CUT OUT AREA NO.1 WRAPPING UP AROUND THE MOUNTAIN "BUILDING ,CORRIDOR METAPHORE" (HOTELS,PRAYER HALLS,TERRACES,COMMERCIAL).

NATURAL ROCK PLAZA WITH ENVIROMENTALLY RESPONSIVE TOWERS.

RESIDENTIAL COURTYAL COMPLEX.

MAIN PEDESTRIAN ARTERY WITH WALKELATORS,STAIRS, ESCALSTORS AND SHADED AREAS.

PLAN #17

SECTION 3_3

RESIDENTIAL COURTYAL COMPLEX.

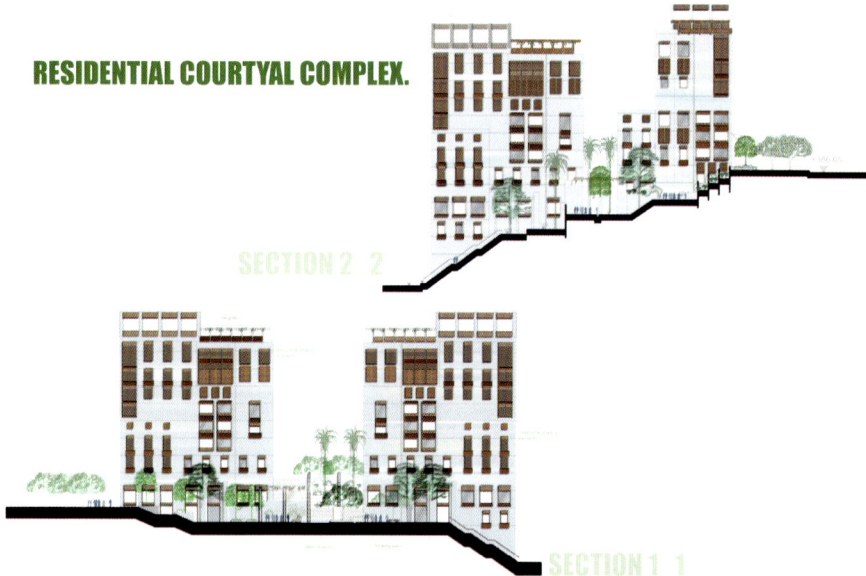

SECTION 2_2

SECTION 1_1

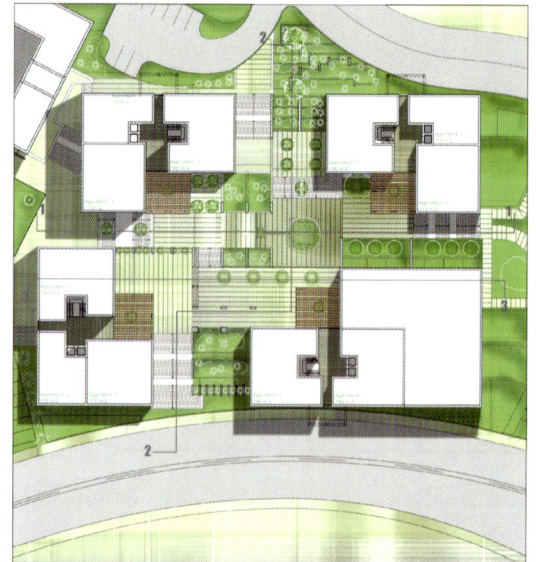

090306

Green Pedestrian Arteries and Prayer Platforms

Pedestrian public arteries (Green Pedestrian) (that cross the contours) (commercial, hotels, restaurants) supported by and connected to Platforms (prayer platforms) encompassing residential and other social services such as schools, condominiums, mosques, and residential neighborhoods.

The Green Arteries will provide a perfect chance to move people upward through the use of pedestrian walkways, escalators, travelators and other methods and techniques.

The arteries and the platforms will provide informal commercial areas for pilgrims and travelers.

These Platforms and Arteries will provide maximum visual accessibility to the Kab'ah (viewing the Kab'ah is considered an act of worship).

Pilgrims and Informal Trading during Hajj Season

(Hotel, Commercial, Prayer Hall)
Building Corridor Metaphore
(Wrapping around the mountain)

Shaded Areas
(Arteries Looking to Haram)

PLAN #15

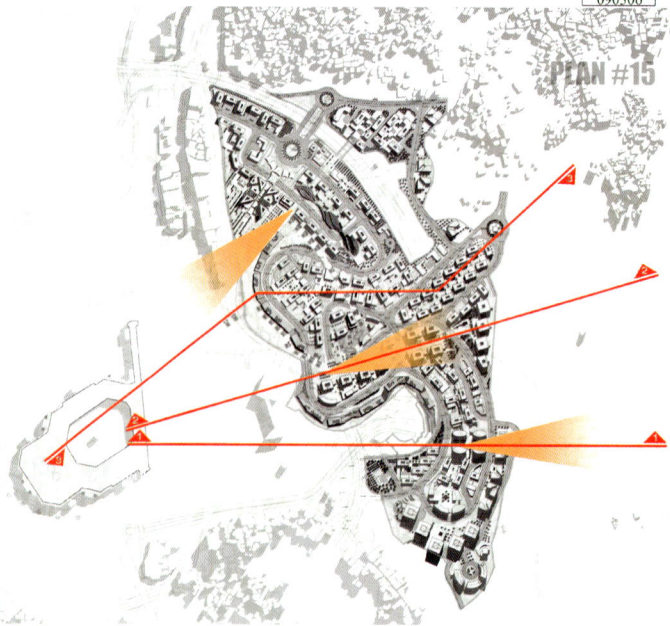

Residential

Commercial - Hotels opening on to Arteries

Commercial - Residencial

(Hotel, Residencial, Prayer Hall)
Building Corridor Metaphore
(Wrapping around the mountain)

Open Plaza Connecting with Haram

Building with partial view to Haram

Building with full view to Haram

Upper floor with full view to Haram

PEDESTRIAN TRAFFIC SYSTEM

Al-kaa'ba

090306

SCALE 1:2000

ACXT – IDOM Group Bilbao
ACXT——IDOM集团 西班牙毕尔巴鄂

"Caressing the rocky and desertic topography that Muhammad experienced with the intention to respect the landscape to a maximum."

"拥抱穆罕默德经历过的岩石和沙漠地貌，最大限度地尊重景观。"

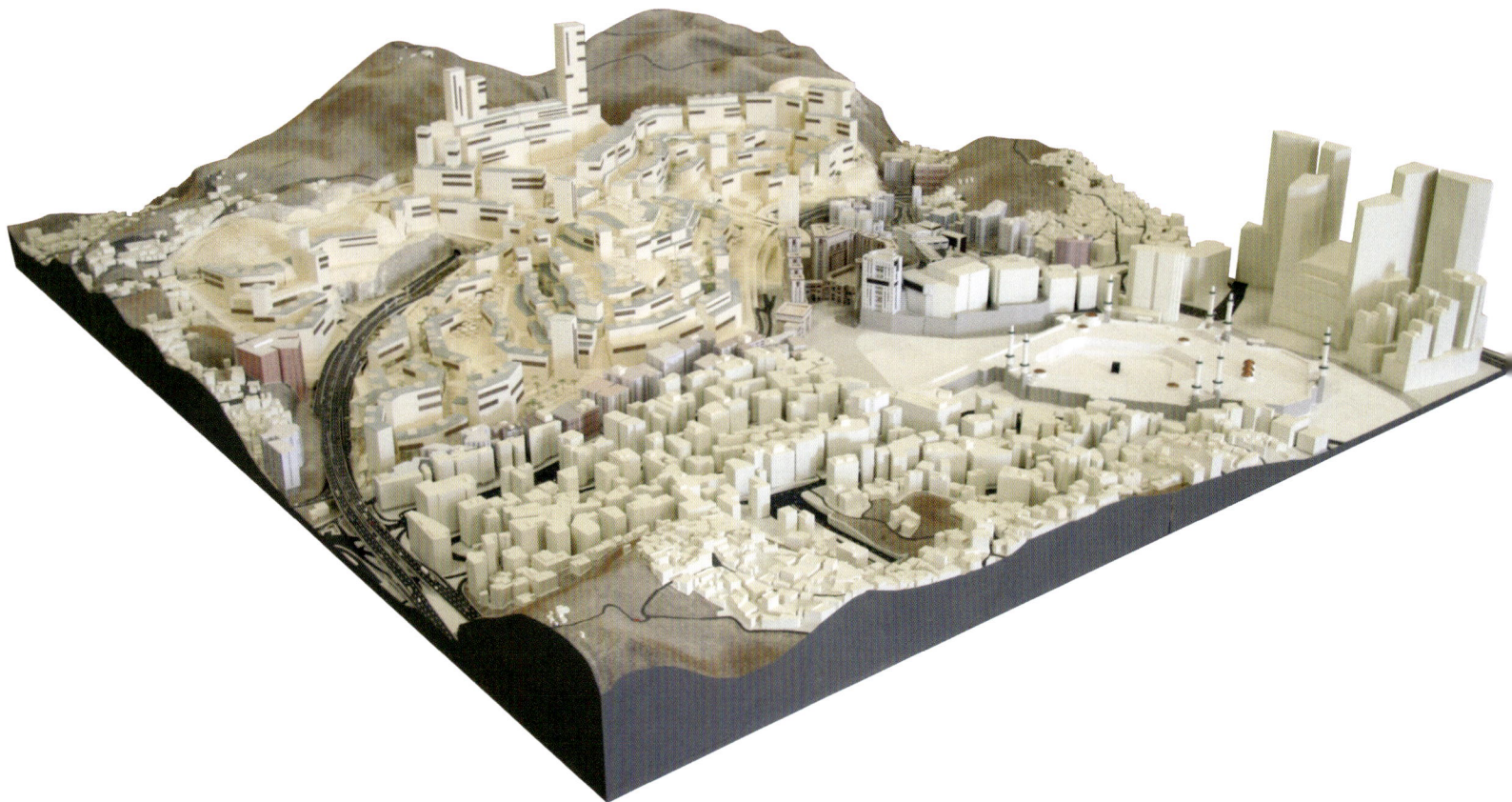

作著 Juan Pablo Puy, Juan Carlos Valerio, Khalid Memhood   合作伙伴 Marc Potard, Aurora Macias, David Moreira, Pilar Pardo, Leyre de Lecea, Ricardo Figueirido, Leire Alonso   **Freelancers** Manuel Leira, Diego Urunuela   专家 civil engineers: Miles Shephard, Raul Coleto, José Ramon Ruiz, Ignacio Blanco, Javier Aja our proposal is the synthesis of the morphology of a traditional islamic city and the singular skyline of the city of makkah.

schematic view of a traditional islamic city

schematic view of the proposed morphology

skyline

view from southern plaza

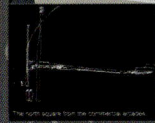

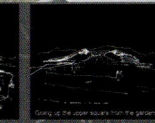

The north square from the commercial arcades.   The north square from the commercial arcades.   The north square seen from the upper square.   The north square seen from a hotel room.   Going up the upper square from the gardens.

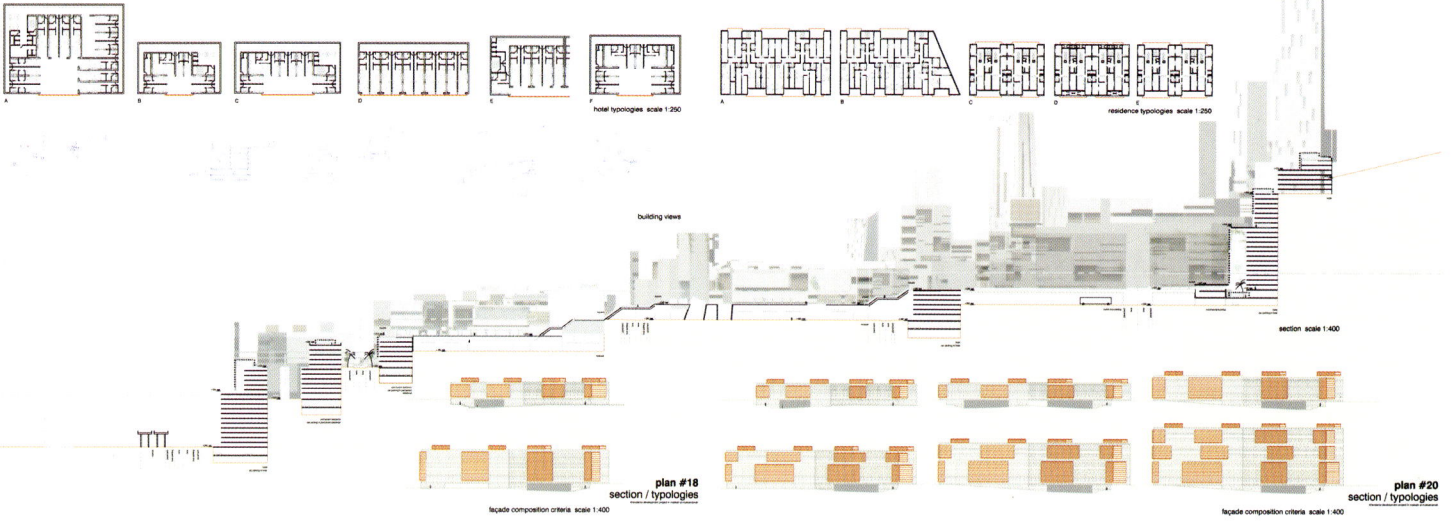

hotel typologies scale 1:250

residence typologies scale 1:250

building views

section scale 1:400

plan #18
section / typologies

plan #20
section / typologies

façade composition criteria scale 1:400

façade composition criteria scale 1:400

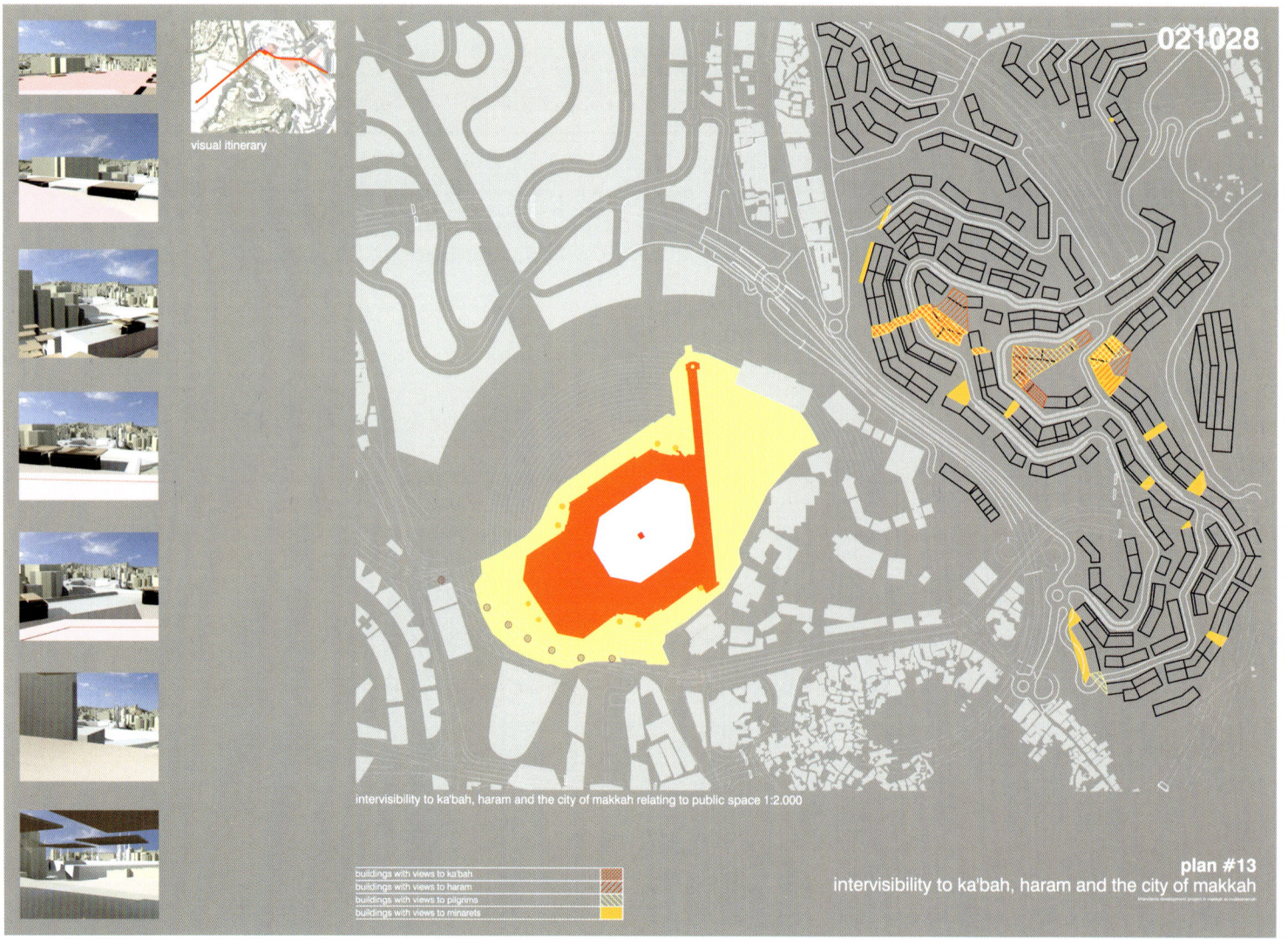

021028

visual itinerary

intervisibility to ka'bah, haram and the city of makkah relating to public space 1:2.000

buildings with views to ka'bah
buildings with views to haram
buildings with views to pilgrims
buildings with views to minarets

plan #13
intervisibility to ka'bah, haram and the city of makkah

plan #3
site plan
khandama development project in makkah al-mukkarramah

021028

Hijjas Kasturi Associates Sdn. Kuala Lumpur
马来西亚吉隆坡Hijjas Kasturi联合事务所 吉隆坡

"The focus of this urban plan is on the creation of an atmosphere of immersive spirituality."
"这个城市规划的重点是创造一种冥想的氛围。"

作者 Hijjas Kasturi, Azmilhram Md. Isa (Perunding Trafik), Faisol B. Hussain (T & T Konsult)   合作伙伴 Azaiddy Abdullah, Kris Budiarto, Jamaliah Misri, Mohammad akmal Mohammad, Mohamad Rifa'l, Mohamad Adini, Nazirull Safry Paijo, Ng Ee Reei, Laziah Ahmad Desa, Mohamad Qasifullah Jaslenda   专家 Environmental Analysis Group (EAG), Department of Building Technology and Engineering, Kulliyyah of Architecture and Environmental Design, International Islamic University, Malaysia

Hijjas Kasturi Associates Sdn., Kuala Lumpur

129

Khandama Development Project, Mecca

THE UPPER EDGE: FIVE "PILLARS"

The urban plan for Jabal Khandama suggest a formation of the **Upper Edge** of the site by placing taller structures near the mountain's ridge. Five iconic towers are amongst these structures, which trace the concentric circle of the Al-Haram, enclosing and containing the Holy Sanctum. The imposing height of the five towers (201.6 meters) are determined by the highest peak amongst the five principal mountains which encircle the valley of Al-Haram i.e Sh'ab Ammer at +620.00 meters from sea level. The datum line established for project shall not exceed this level.

The five towers, due to their relative position to Al-Haram, will appear imposing while at the same time give the city emphasis and significance as it "crowns" the mountain. On spiritual level, the five towers allude to the five rukn or pillars of Islam. The religion fully inscribe the five pillars as the foundation that must be practiced by all Muslims. At night, the towers will look like five beacons, all of which delineates toward the Holy Ka'aba, further emphasising the significance of the icon. The apex of these towers are angled towards the

centre of Al-Haram circle, suggesting humility, respect and even "dialogue" between the towers with the Ka'aba; between the mundane or profane with sacred; between modernity and extreme antiquity. For those willing to contemplate, these five towers; the crown and their allusions to the rukn, are more than just mundane hotel towers.

datum line of Tower 4 +604.3m
datum line of Tower 3 +516.9m
datum line of Tower 5 +529.2m
datum line of Tower 2 +505.9m

datum line of Masjidil Haram +381.8m

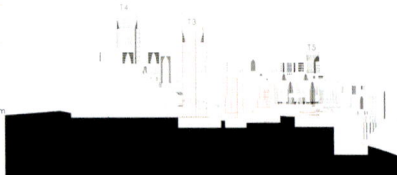

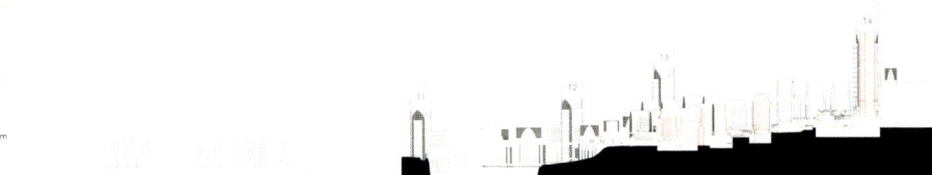

SECTION 1   JABAL KHANDAMA

datum line of Tower 4 +604.3m
datum line of Tower 3 +516.9m

datum line of Tower 5 +529.2m

datum line of Masjidil Haram +381.8m

SECTION 2   JABAL KHANDAMA

datum line of Tower 4 +604.3m

datum line of Tower 3 +516.9m

datum line of Tower 2 +505.9m
datum line of Tower 1 +486.3m

datum line of Masjidil Haram +381.8m

SECTION 3   JABAL KHANDAMA

datum line of Tower 4 +604.3m
datum line of Tower 3 +516.9m
datum line of Tower 5 +590.2m

datum line of Tower 2 +505.9m
datum line of Tower 1 +486.3m

datum line of Masjidil Haram +381.8m

SECTION 4   JABAL KHANDAMA

2nd Ring Road

Al Masjed Al Haram

Ghazza

open space/buffer/
landscaping area
building plot
M   mousallah
T   tower
PO   podium
BH   building height
BT   bus terminal

10

SCALE 1 : 1000

DB Architecture & Consulting Istanbul
DB建筑咨询事务所 土耳其伊斯坦布尔

"The project attempts to introduce a contemporary model for dealing with the identity problem of Islamic cities today."
"这个项目旨在为解决当今伊斯兰教城市的特性问题提供一种建筑模式。"

作者 Derman Bunyamin, Cem Ilhan   合作伙伴及建筑系在校生 Dick Hökenek, Murat Yüksel, Serhat Yavuz, Aysegül Akgül,
Güven Cimenoglu, Orcun Köken, Rüstem Cakmakli

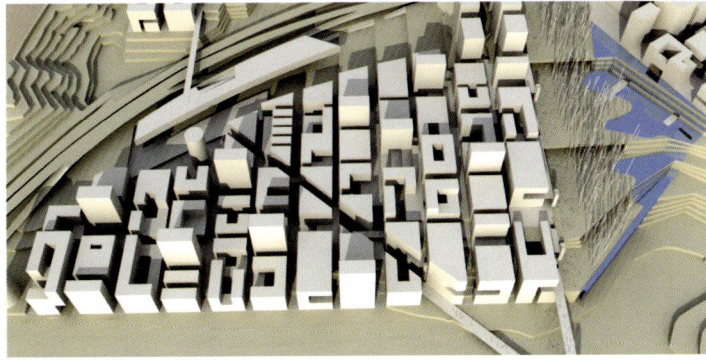

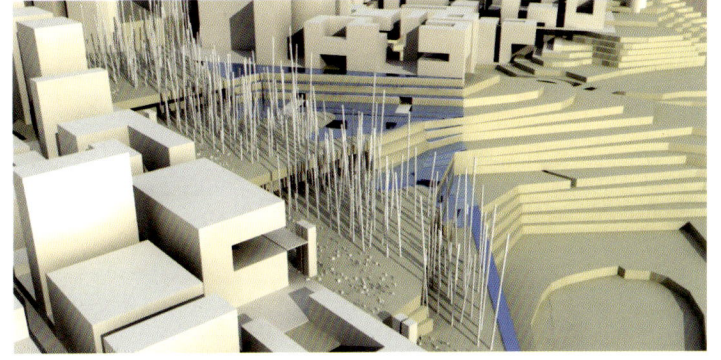

397766

SECTION A-A

HOTEL TEMPORARY RESIDENCE TEMPORARY RESIDENCE PERMANENT RESIDENCE

HOTEL HOTEL TEMPORARY RESIDENCE HOTEL TEMPORARY RESIDENCE PERMANENT RESIDENCE PERMANENT RESIDENCE

HOTEL TEMPORARY RESIDENCE TEMPORARY RESIDENCE PERMANENT RESIDENCE

A-A

A-A

CUT-OUT PLAN
1:250

CUT-OUT PLAN

SET E

LEVEL +36.00

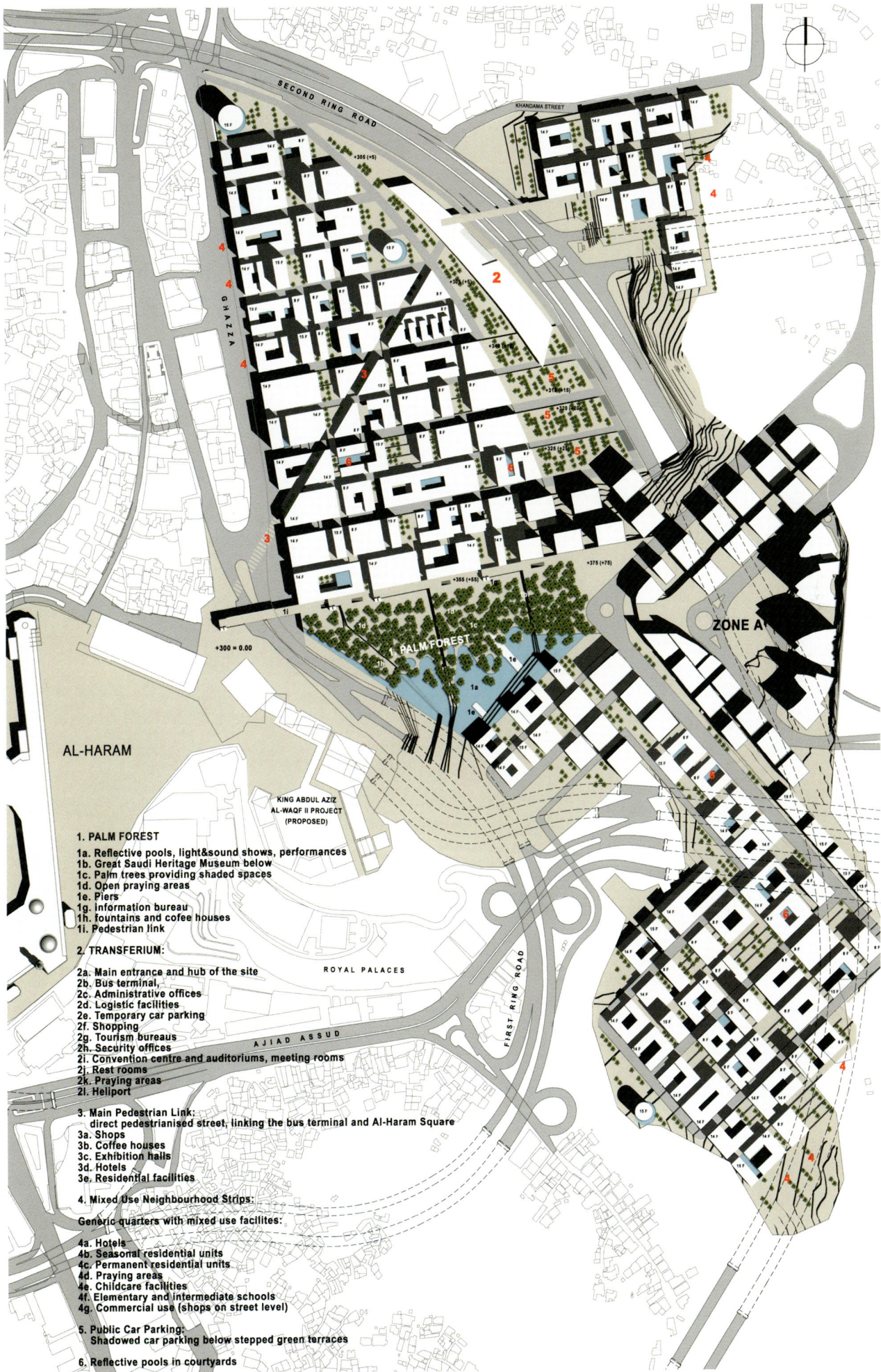

SECOND RING ROAD

KHANDAMA STREET

GHAZZA

ZONE A

PALM FOREST

AL-HARAM

KING ABDUL AZIZ
AL-WAQF II PROJECT
(PROPOSED)

ROYAL PALACES

FIRST RING ROAD

AJIAD ASSUD

+300 = 0.00

1. PALM FOREST

1a. Reflective pools, light&sound shows, performances
1b. Great Saudi Heritage Museum below
1c. Palm trees providing shaded spaces
1d. Open praying areas
1e. Piers
1g. information bureau
1h. fountains and cofee houses
1i. Pedestrian link

2. TRANSFERIUM:

2a. Main entrance and hub of the site
2b. Bus terminal,
2c. Administrative offices
2d. Logistic facilities
2e. Temporary car parking
2f. Shopping
2g. Tourism bureaus
2h. Security offices
2i. Convention centre and auditoriums, meeting rooms
2j. Rest rooms
2k. Praying areas
2l. Heliport

3. Main Pedestrian Link:
direct pedestrianised street, linking the bus terminal and Al-Haram Square
3a. Shops
3b. Coffee houses
3c. Exhibition halls
3d. Hotels
3e. Residential facilities

4. Mixed Use Neighbourhood Strips:

Generic quarters with mixed use facilites:

4a. Hotels
4b. Seasonal residential units
4c. Permanent residential units
4d. Praying areas
4e. Childcare facilities
4f. Elementary and intermediate schools
4g. Commercial use (shops on street level)

5. Public Car Parking:
Shadowed car parking below stepped green terraces

6. Reflective pools in courtyards

Moriyama & Teshim

Jabal Omar Development Project in Mecca: Architecture Competition

麦加Jabal Omar开发项目：建筑竞赛

Gerber Architekten

a Architects

Jabal Omar Development Project Mecca
Jabal Omar开发项目 麦加

Restricted project competition for architects preceded by an application procedure
限制严格的建筑师项目竞赛，开始前有申请程序。

地点 Mecca 时间 05/2007-06/2007 主办方 Jabal Omar Development Co. 人数 2 competitors
面积 23 ha 竞赛费用 200,000 USD 专业评奖委员会 Prof. Dr. Gulzar Haider, Lahore; Kaspar Kraemer, Cologne;
Prof. Yves Lion, Paris; Prof. Rodolfo Machado, Boston; Jafar Tukan, Amman 其他专业评奖委员会 Abdul-Halim Jabr, Beirut

The architectural competition for the Jabal Omar project had as its central focus the planning of a large inner-city development in the Holy City of Mecca, west of the Grand Mosque. Planning constraints were similar to those encountered on the Jabal Khandama project: spatial proximity to the Kaaba, Mecca's dramatic changes in contour, exacting climatic conditions, and enormous pressure for development from the real estate market. The aim of the competition was to design, within the confines of an existing masterplan, three buildings of exemplary importance to the overall project while respecting the intrinsic structures and underlying principles of an Arabian city. Planning was based on a master plan by the Paris firm Ateliers Lion which had been selected in a previous competition. Ateliers Lion solved crucial issues of access to the area, and the layout of lots and building typology (plinths and point blocks). The competition included, on Lot A, an approx. 140 metres high Twin Tower high-rise ensemble, allowing approx. 250,000 square metres of gross floor area for hotel and retail uses. On two smaller lots, the program asked for comparatively less tall buildings (approx. 40,000 and 15,000 square metres, respectively) for hotel, retail and residential uses.

Jabal Omar开发项目的建筑竞赛重点强调对位于大清真寺以西的麦加圣城中一大片内城开发的规划。对规划的系统规定参数与Jabal Khandama项目的类似；与圣堂在空间上的临近、确切的气候条件、来自房地产市场开发的巨大压力。举办竞赛的目的是在现有总体规划的框架之内，在重视固有结构、强调阿拉伯城市道义原则的同时，设计三个有代表性的重点建筑物。规划以在前一次竞赛里选出的巴黎Ateliers Lion建筑城市规划事务所提供的总平面图为基础，该建筑师事务所解决了门禁、场地布局和建筑类型等关键的问题。竞赛地点包括在A区约140米高的双峰塔式高层建筑的总效果、25万平方米建筑总面积用作酒店和商店。在两个小一些的场地上，规划要求相对不太高的建筑物来作酒店、商店和住宅。开发整个区域地标性的建筑决非普通的问题。

Aerial view　鸟瞰图

Competition site　竞赛地点

Competition site　竞赛地点

The Great Mosque　大清真寺

139

Jabal Omar Development Project, Mecca

1

2

1
1st prize　一等奖
Moriyama & Teshima Architects,
Toronto

2
2nd prize　二等奖
Gerber Architekten, Dortmund

Jabal Omar Development Project, Mecca

Moriyama & Teshima Architects Toronto
Moriyama & Teshima建筑事务所 多伦多

"We set our sights on reproducing and upgrading the views to the Ka'bah, and on creating the kind of spatial awareness that orients one's attention on the Ka'bah."

"我们着眼于重建和提高对天房的认识，也着眼于创造出把人们注意力引向天房的空间感知。"

作者 Moriyama & Teshima, Toronto/Ottawa/Ryadh 合作伙伴及建筑系在校生 Gord Doherty, Shawn Geddes, Eric Klaver, Aisha Marsh, Tara McCarthy, George Meng, Ajon Moriyama, Jason Philippe, Carol Phillips, Elias Saoud, Kathryn Seymour, George Stockton, Cheiwei Tai, Drew Wensley, Scott Wiseman, Steve Culver, Will Klassen, Tom Ngo, Ryan Trinidade 自由建筑师 modelmaking: OHM Industrial Designers; Islamic cultural advisor: Syed Nouman Ashraf; elevator technology: Dover/ThyssenKrupp Elevator 专家 mechanical & electrical engineers: Buro Happold; structural & building envelope engineers: Halcrow Yolles

Jabal Omar Development Project, Mecca

+301.00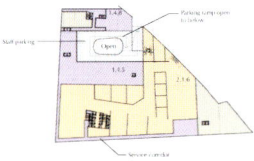

+304.00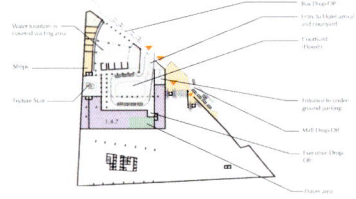

+308.00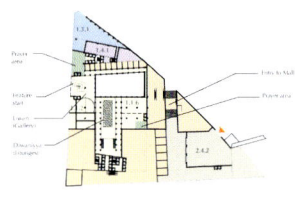

+311.50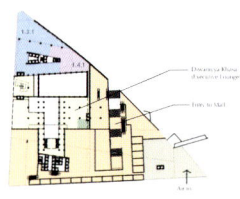

DECORATIVE STONE BALLAST

LOWER PARKING GARAGE

WATERFALL

MIHRAB

LIMESTONE PAVING

MIHRAB

FABRIC CANOPY

INFINITY EDGE WATERFALL

WATERFALL

MIHRAB

WATERFALL

WATER RILL

MIHRAB

Water from Tower A

Water polishing garden sub-surface flow plant and aggregate stone matrix in cell basin

Recycled water

Potable water

Central septic system

Roof Top Musallah

View to Masjid Al Haram

Seperate flow

Water feature

Irrigation

Pump

Conditioning 1

Catch Reservoir

Conditioning 3

Conditioning 2

Villa terrace +333.00

Water fall

Skylight

Water polishing garden sub-surface flow plant and aggregate stone matrix in cell basin

Roof Top terrace +330.00

Water feature

Seperate flow

Irrigation

Terrace at Banquet / Conference level +323.00

Water feature

Water feature

Courtyard +304.00

Open to Hotel parking below

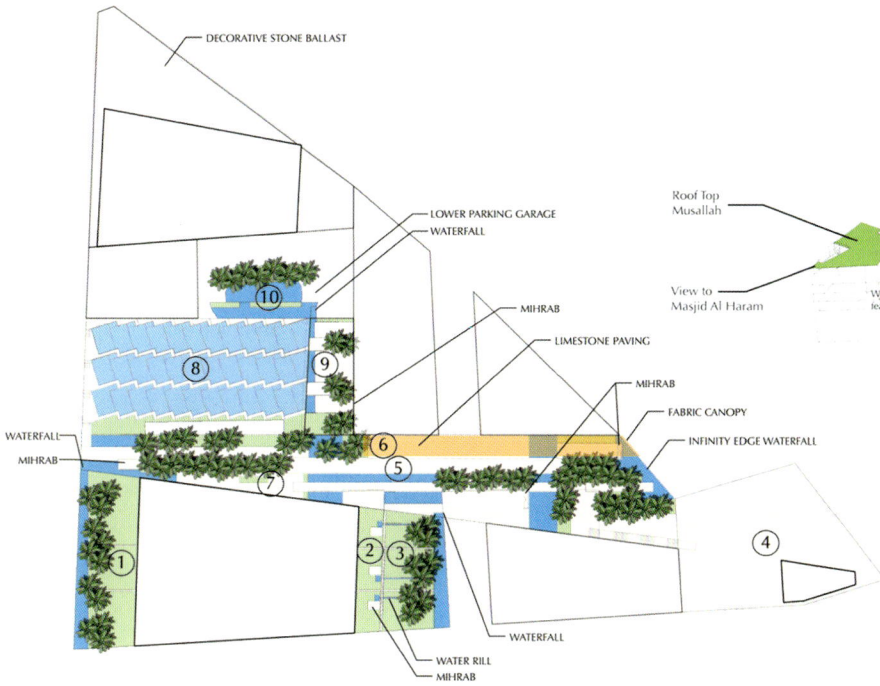

SITE PLAN - PLOT A : ENTRANCE LEVEL

1:500

1 PRIVATE VILLA TERRACE - 333m
2 PRIVATE VILLA LOWER TERRACE - 333m
3 PRIVATE VILLA UPPER TERRACE - 336.5m
4 MUSALLAH ROOF TERRACE - 340m
5 PUBLIC TERRACE - 330m
6 EXECUTIVE HOTEL ENTRANCE - 330m
7 VILLA AND HOTEL ENTRANCE - 330m
8 WATER POLISHING GARDEN - 326.5m
9 CONFERENCE ROOM TERRACES - 323m
10 HOTEL ARRIVAL COURTYARD - 304m

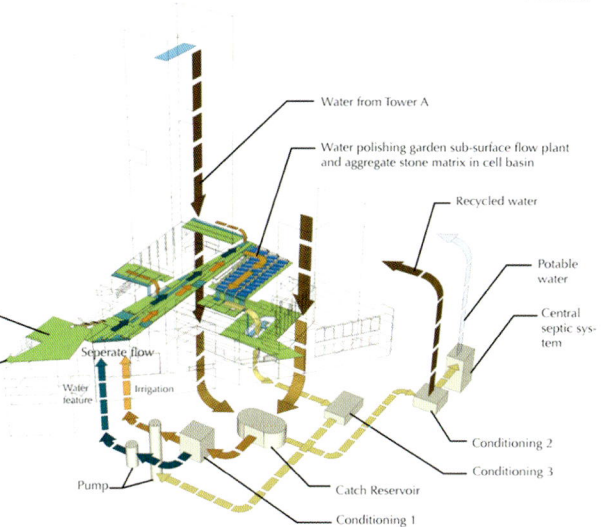

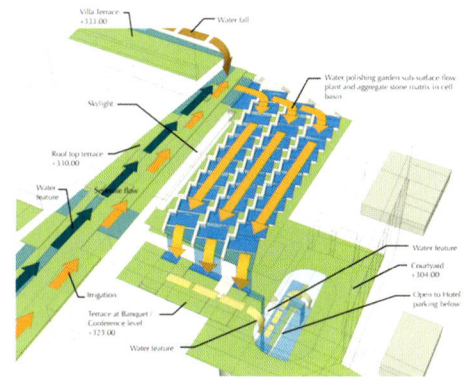

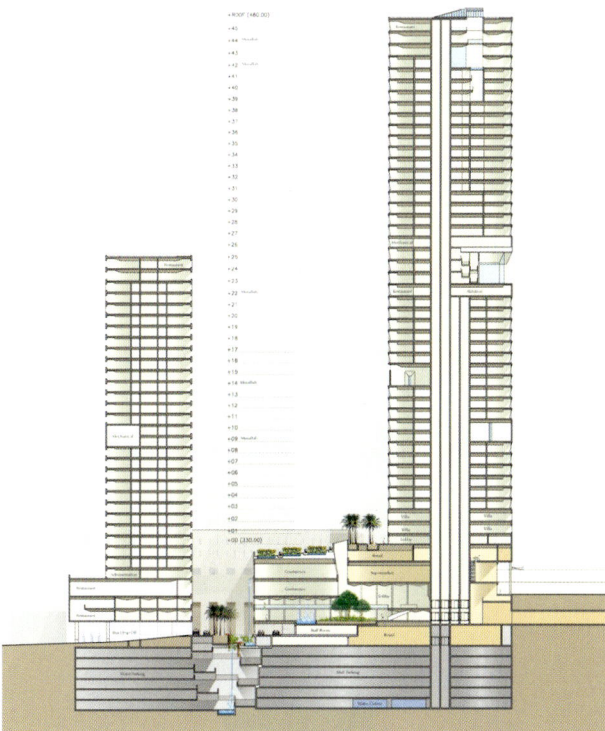

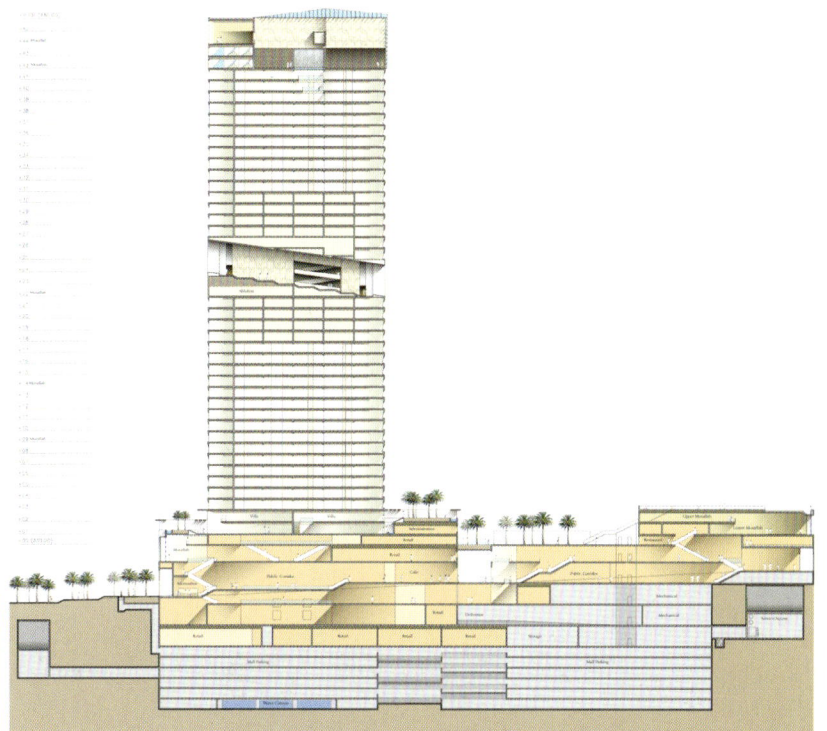

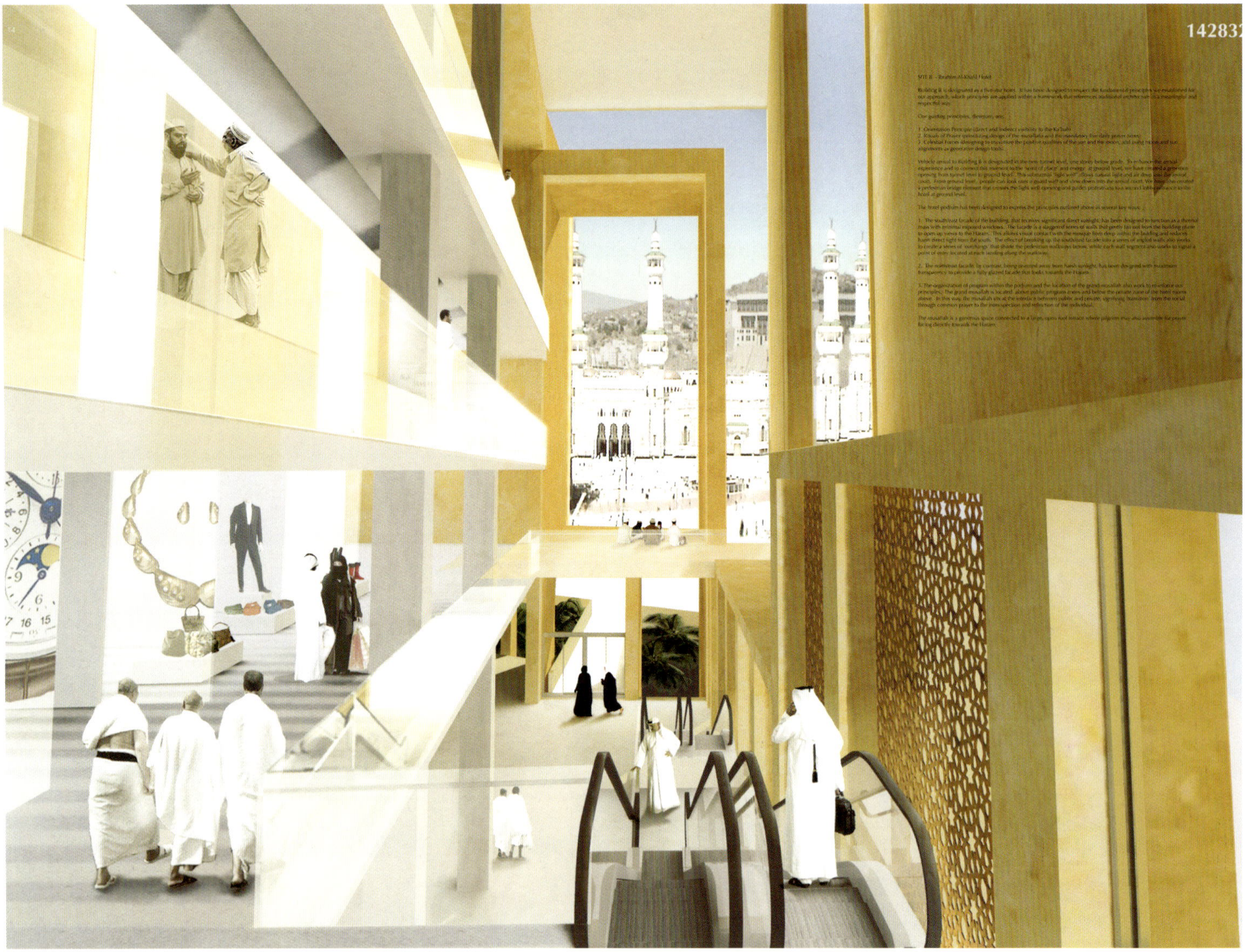

DHUHIR (EARLY AFTERNOON) MUSALLAH

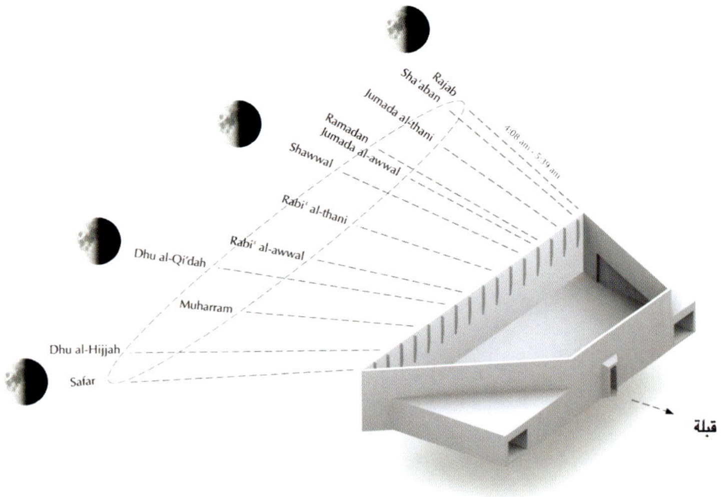

Rajab
Sha'aban
Jumada al-thani
Ramadan
Jumada al-awwal
Shawwal
Rabi' al-thani
Rabi' al-awwal
Dhu al-Qi'dah
Muharram
Dhu al-Hijjah
Safar

4:08 am – 5:19 am

FAJR (PREDAWN) MUSALLAH

قبلة

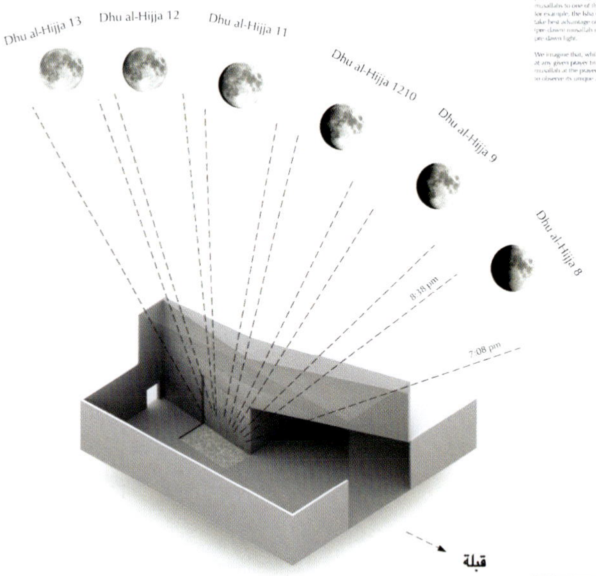

Dhu al-Hijja 13   Dhu al-Hijja 12   Dhu al-Hijja 11
Dhu al-Hijja 1210
Dhu al-Hijja 9
Dhu al-Hijja 8

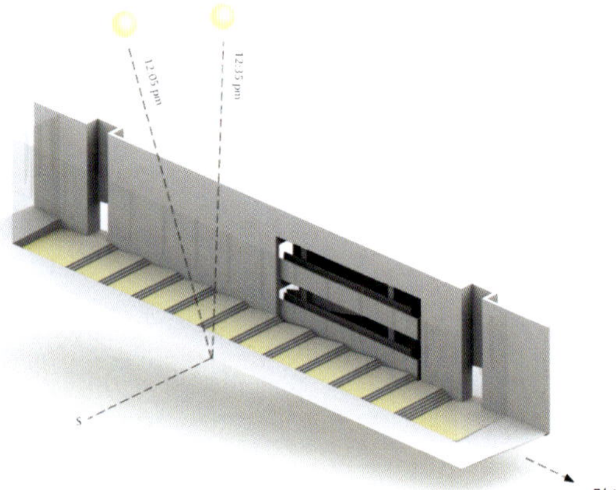

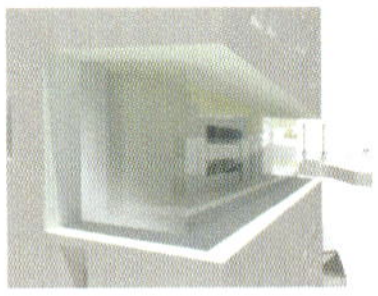

8:18 pm

7:08 pm

قبلة

ISHA (EVENING) MUSALLAH

12:03 pm   12:35 pm

DHUHIR (EARLY AFTERNOON) MUSALLAH

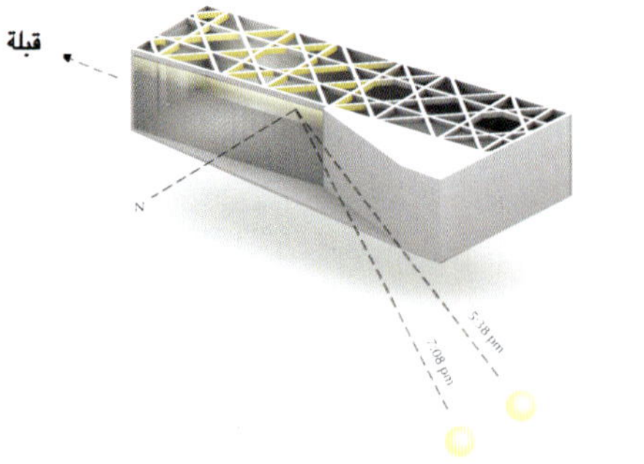

قبلة

قبلة

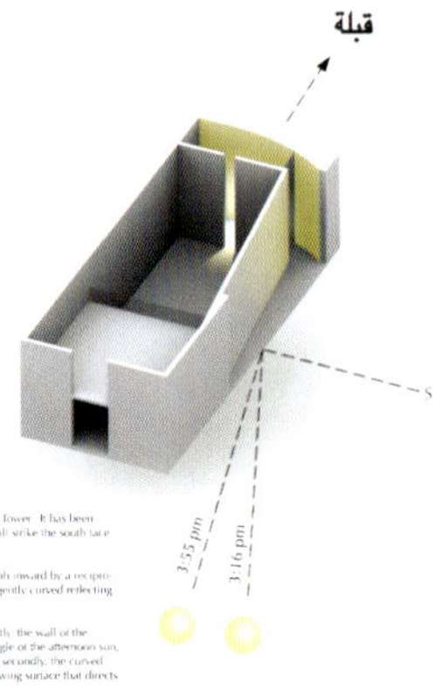

5:18 pm

7:08 pm

MAGHRIB (SUNSET) MUSALLAH

قبلة

3:55 pm
3:16 pm

'ASR (LATE AFTERNOON) MUSALLAH

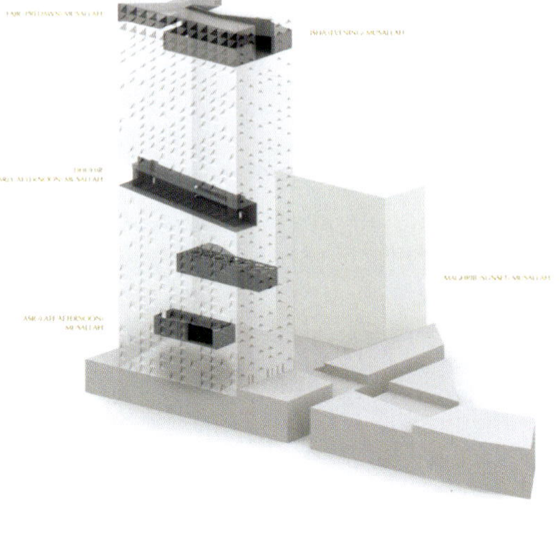

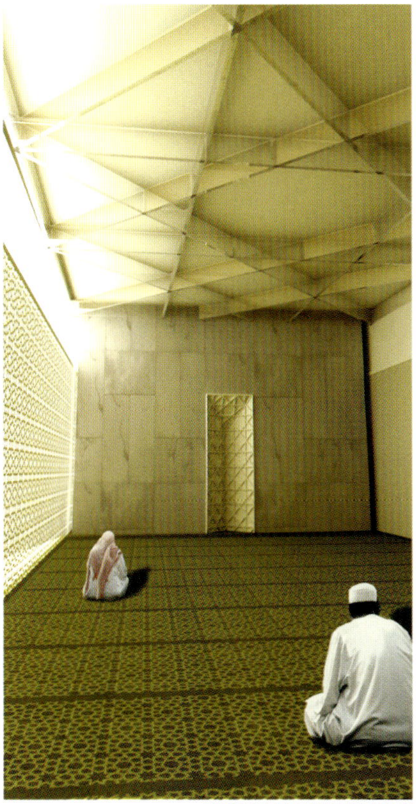

MAGHRIB (SUNSET) MUSALLAH

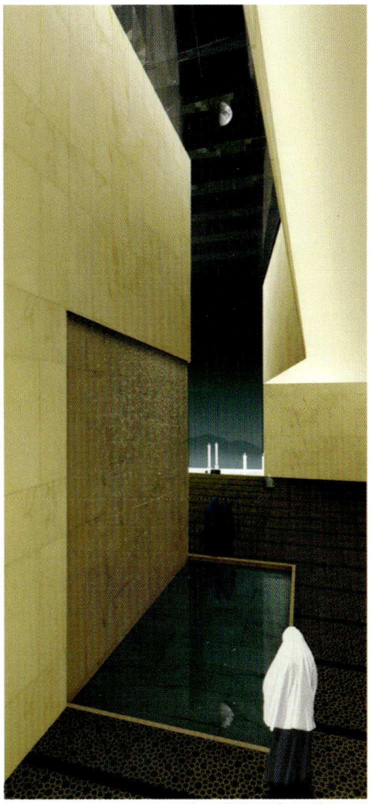

ISHA (EVENING) MUSALLAH

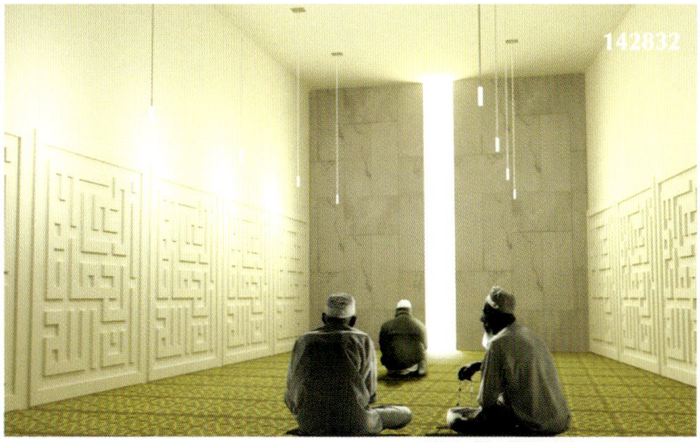

'ASR (LATE AFTERNOON) MUSALLAH

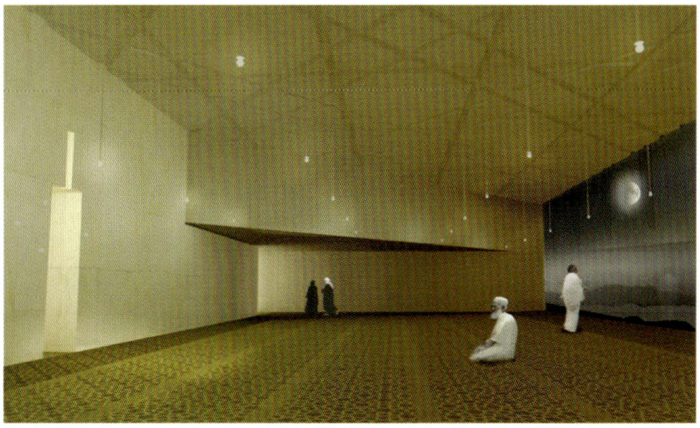

FAJR (PRE-DAWN) MUSALLAH

142832

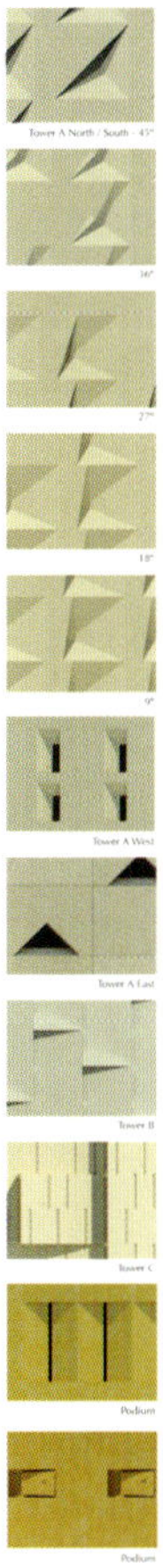

149

ذو الحجة 8   ذو الحجة 9   ذو الحجة 10   ذو الحجة 11   ذو الحجة 12   ذو الحجة 13

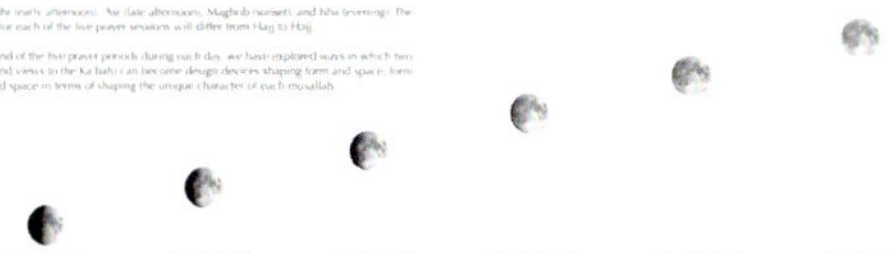

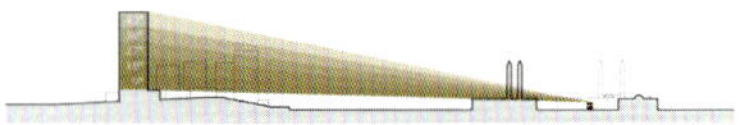

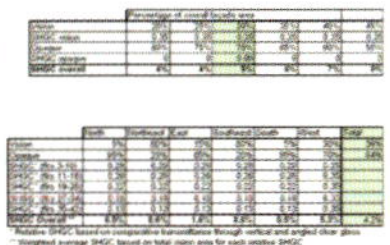

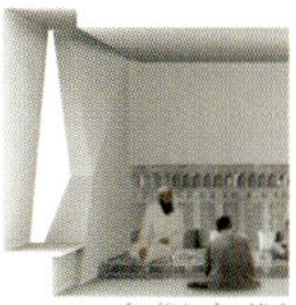

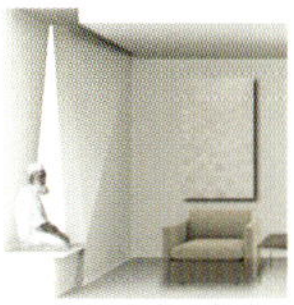

CONCEPT

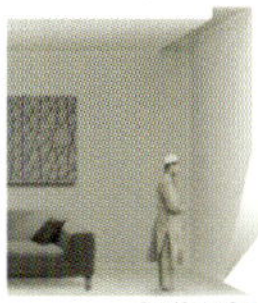

STRUCTURAL SYSTEM

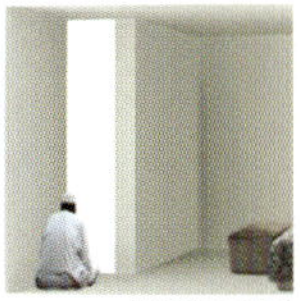

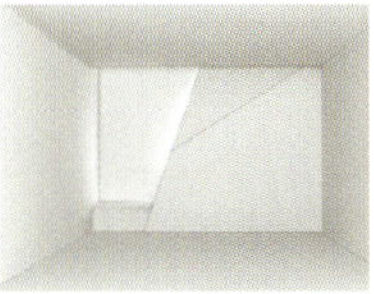

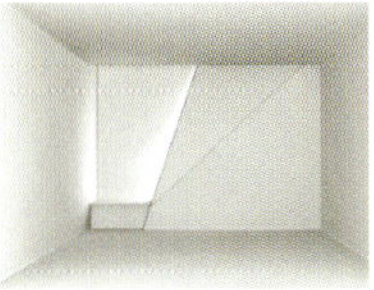

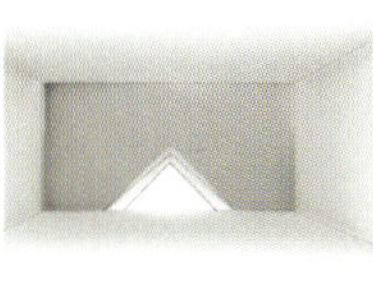

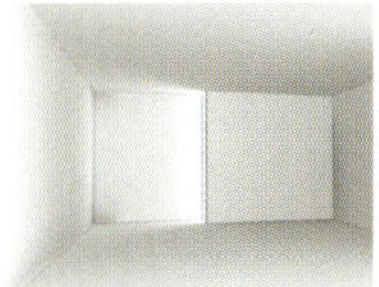

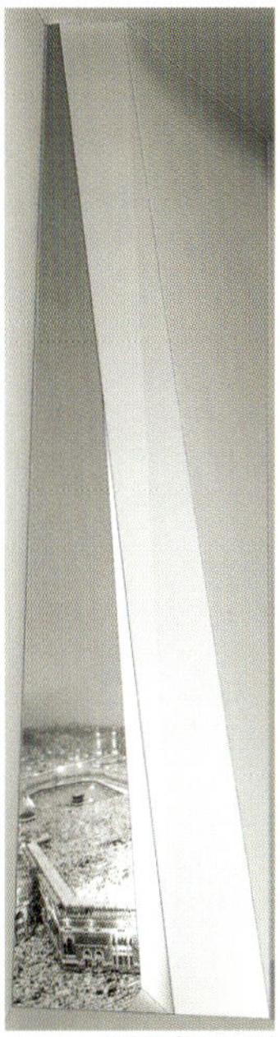

STRUCTURAL

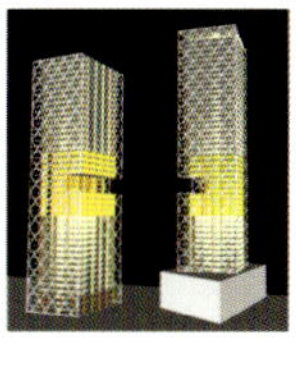

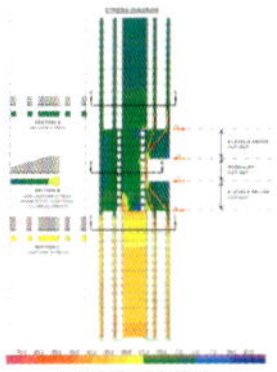

THERMAL LOADING

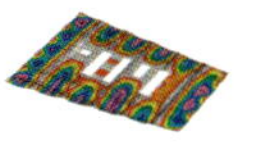

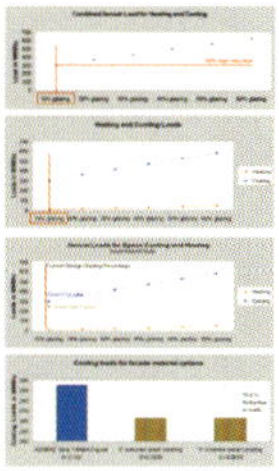

THERMAL MASS

NATURAL VENTILATION

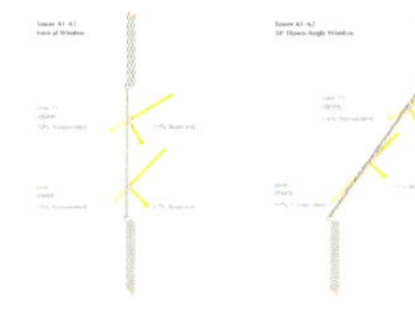

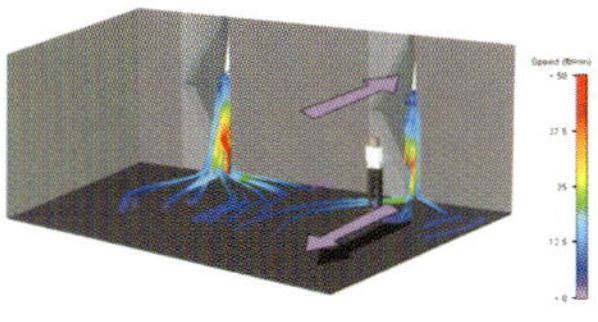

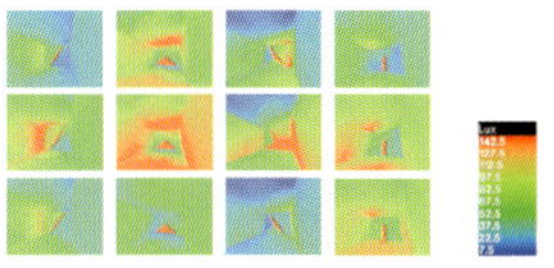

DAYLIGHT

Gerber Architekten Dortmund
Gerber建筑事务所 德国多特蒙德

"The large city blocks form a dense urban pattern with narrow streets, in keeping with the region's climatic conditions. While the plinths appear rather grounded, the high-rises turn into elements of the sky, with which they seem to fuse."

"段轨床基座似乎非常接近地面，高层建筑融入天空，大城市街区形成了与气候条件相协调的街道狭窄的密集的城市式样。"

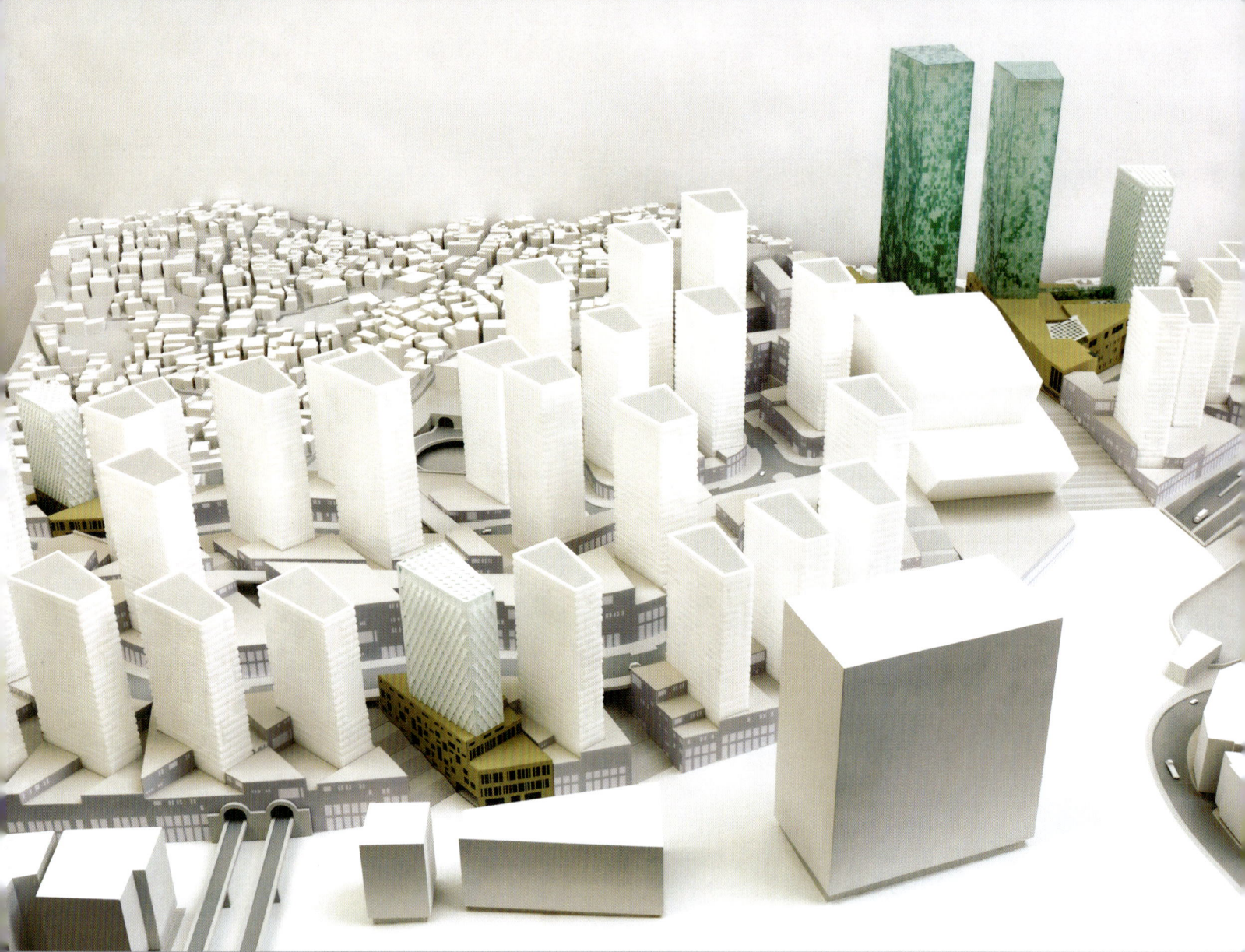

作者 Prof. Eckhard Gerber   合作伙伴 Siegbert Hennecke, Nicole Juchems, Jan Kallert, Sandra Kroll, Stefan Lemke, Thomas Lücking, Nathalie Ponamarev, Florian Rist, Caroline Ting   专家 mechanical & electrical engineers: Peter Moesle, DS-Plan AG; façade engineer: Mr. Barf; structural & building envelope engineer: Florian Förster

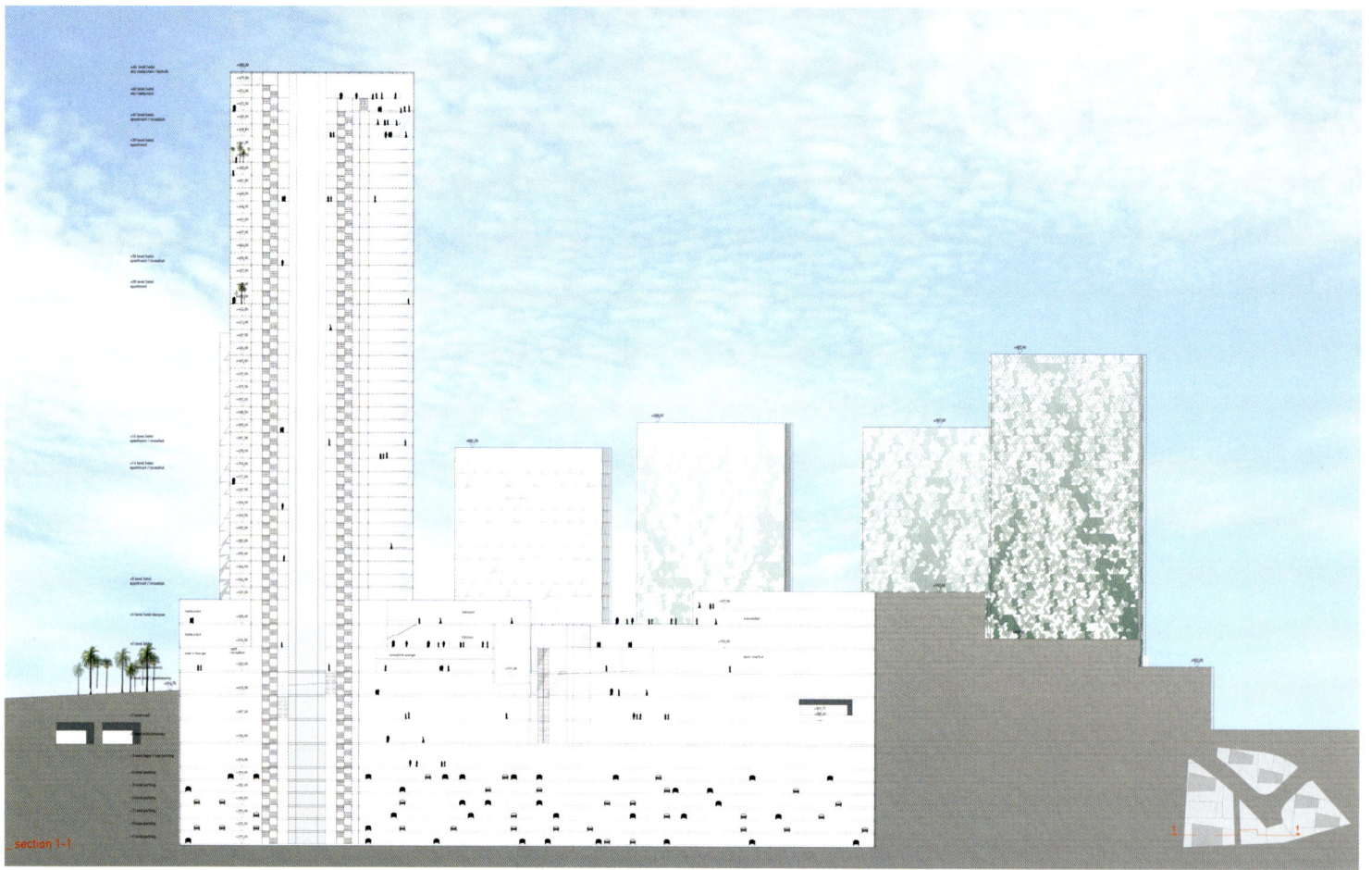
section 1-1

section 2-2

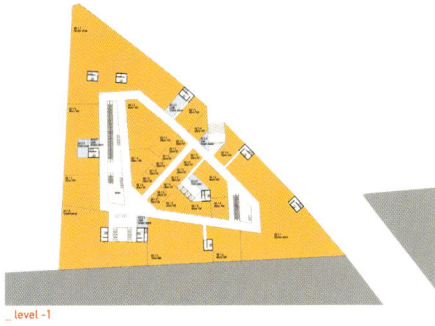
_ level -1

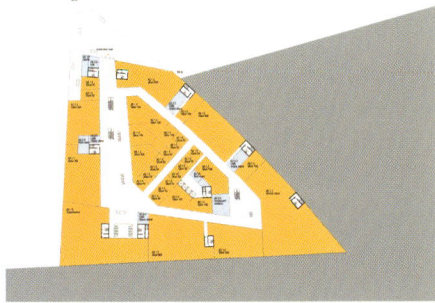
_ level +4

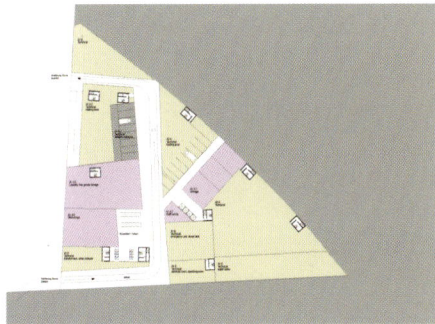
_ level +9

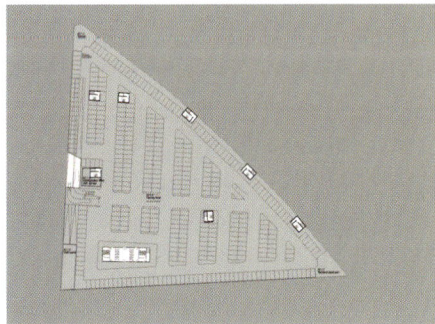
_ level -2

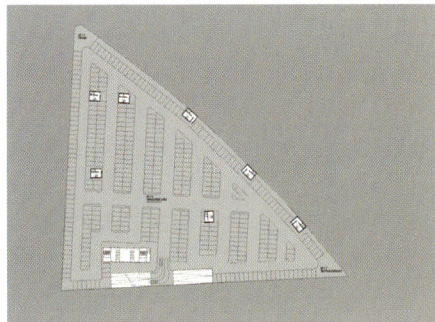
_ level +3

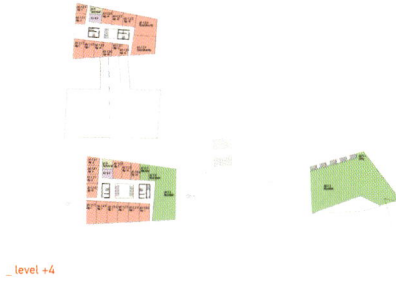
_ level +8

_ level -3

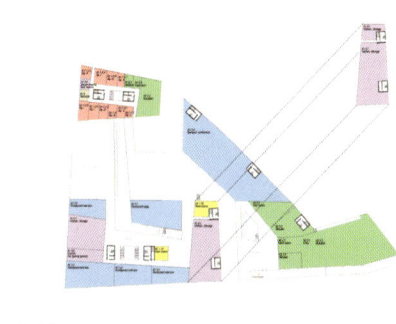
_ level +2

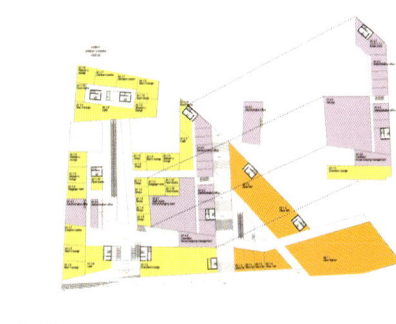
_ level +7

_ level -4

_ level +1

_ level +6

_ level -6

_ level 0

_ level +5

_ elevation seen from stairs

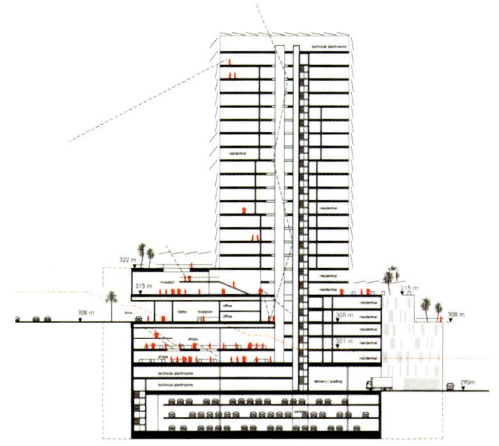
_ longitudinal section

_ elevation seen from road e

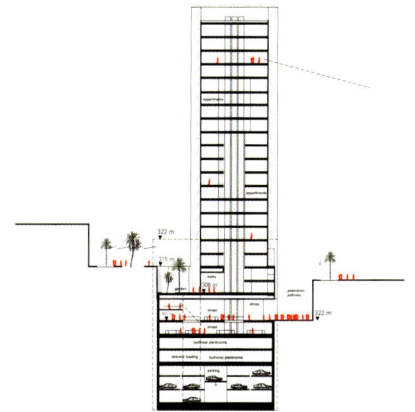
_ transversal section

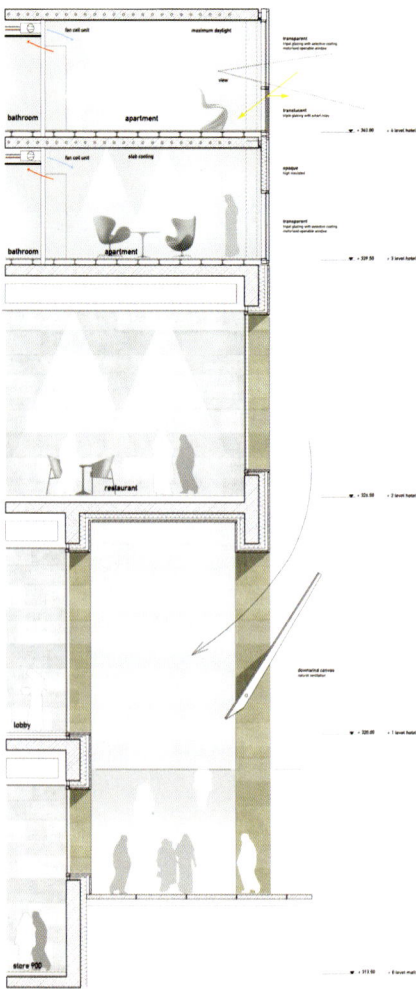

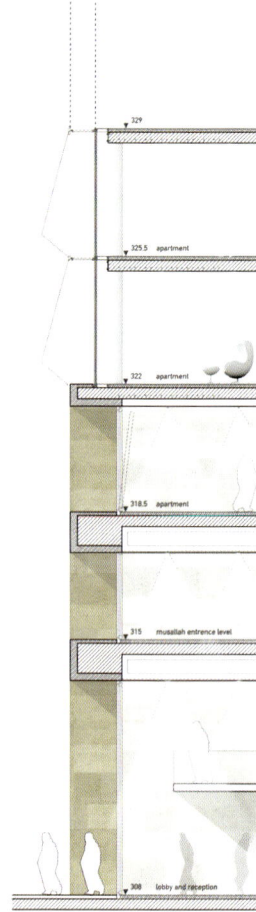

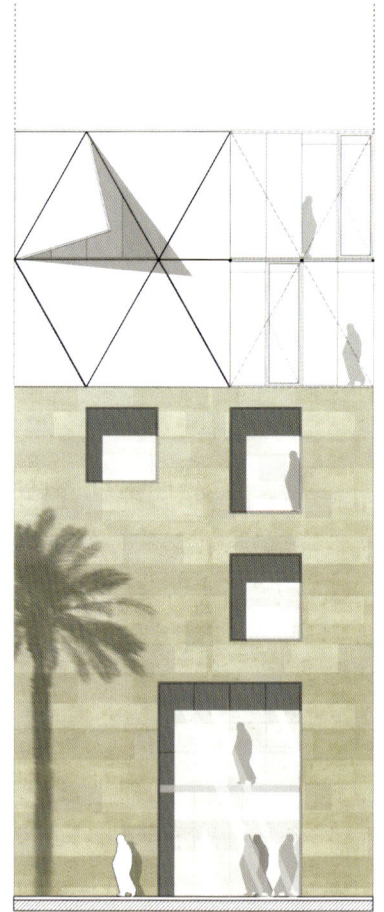

_ detail section of the facade plot a __ 1:50

_ detail elevation of the facade plot a __ 1:50

_ detail section of the facade plot c __ 1:50

_ detail elevation of the facade plot c __ 1:50

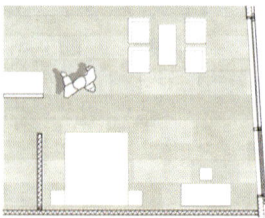

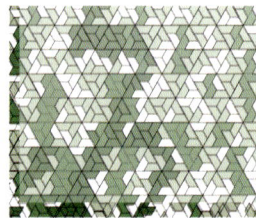

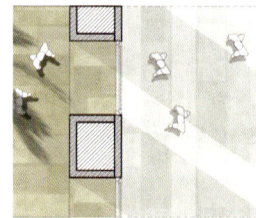

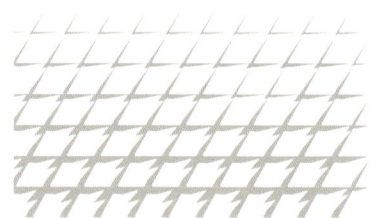

_ detail floorplan apartment plot a __ 1:50

_ facade twin tower north

_ detail floor plan +308.00 plot c __ 1:50

_ facade tower plot c

_ materials

_ translucent

_ opac

_ local natural stone

_ canvas

_ environmental engineering

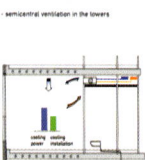

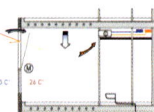

_ semicentral ventilation in the towers

- small energy consumption
- optimal and efficient natural ventilation with motorized openings
- high heat insulation good daylight use
- accumulation of condensation water is avoided by intelligent motor control. window can only opened at reasonable phases (e.g. during the

- high cooling capacities with small cooling peak loads

- high comfort during night hours (sleeping, air drought is avoided)
- noise is avoided, no fans are in operation during most of the sleeping time

_facade details twin tower north

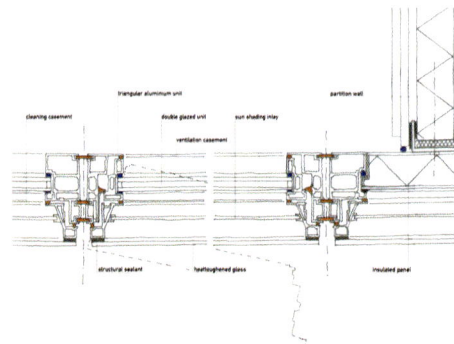

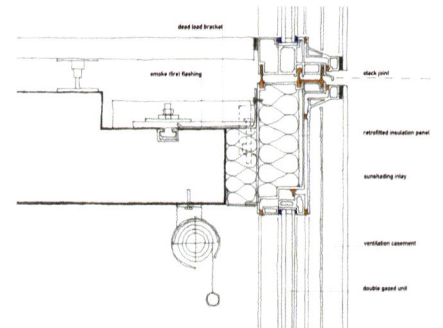

FACADE GRID DERIVED FROM TRADITIONAL PATTERNS

Traditional Islamic patterns usually show strong three-, four- or six-fold symmetry.

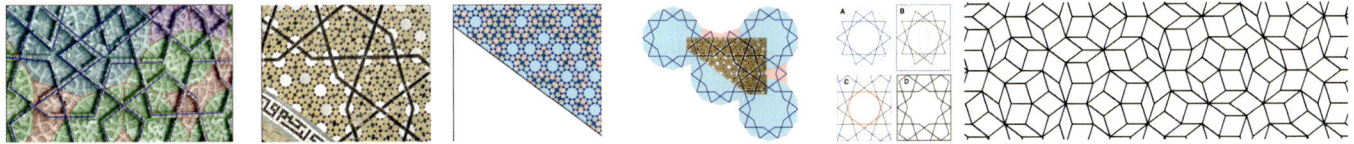

About 500 years ago Islamic architects developed five- and ten-fold symmetry patterns, closely related to the aperiodic patterns discovered by Penrose in the 1960ies.

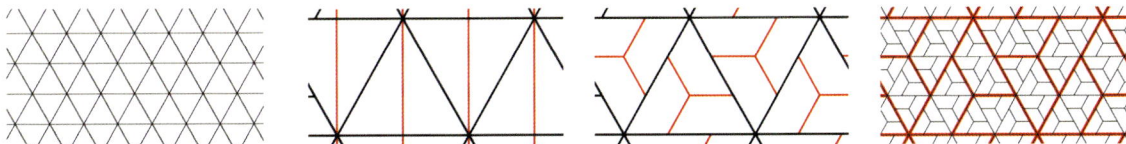

The facade tiling is based on a simple regular triangular tiling which is subdivided in several ways to create different periodic patterns as shown here. Just as many traditional patterns this tiling does contain itself at a smaller scale.

Based on the proposed grids various facades are possible. The large glass facades shows pseudo random elements that enable great flexibility and link to aperiodic patterns an the complexity of medieval patterns without copying them. In addition to that avoiding regular patterns creates a camouflage effect, which gives a nice uniformity the whole plot.

Buro H

von Ge

Jabal Omar Development Project in Mecca: Engineering Competition

麦加Omar开发项目：工程竞赛

Jabal Omar Development Project **Mecca**
Jabal Omar开发项目：工程竞赛　麦加

Restricted project competition for engineers preceded by an application procedure
限制严格的工程师项目竞赛，开始前有申请程序。

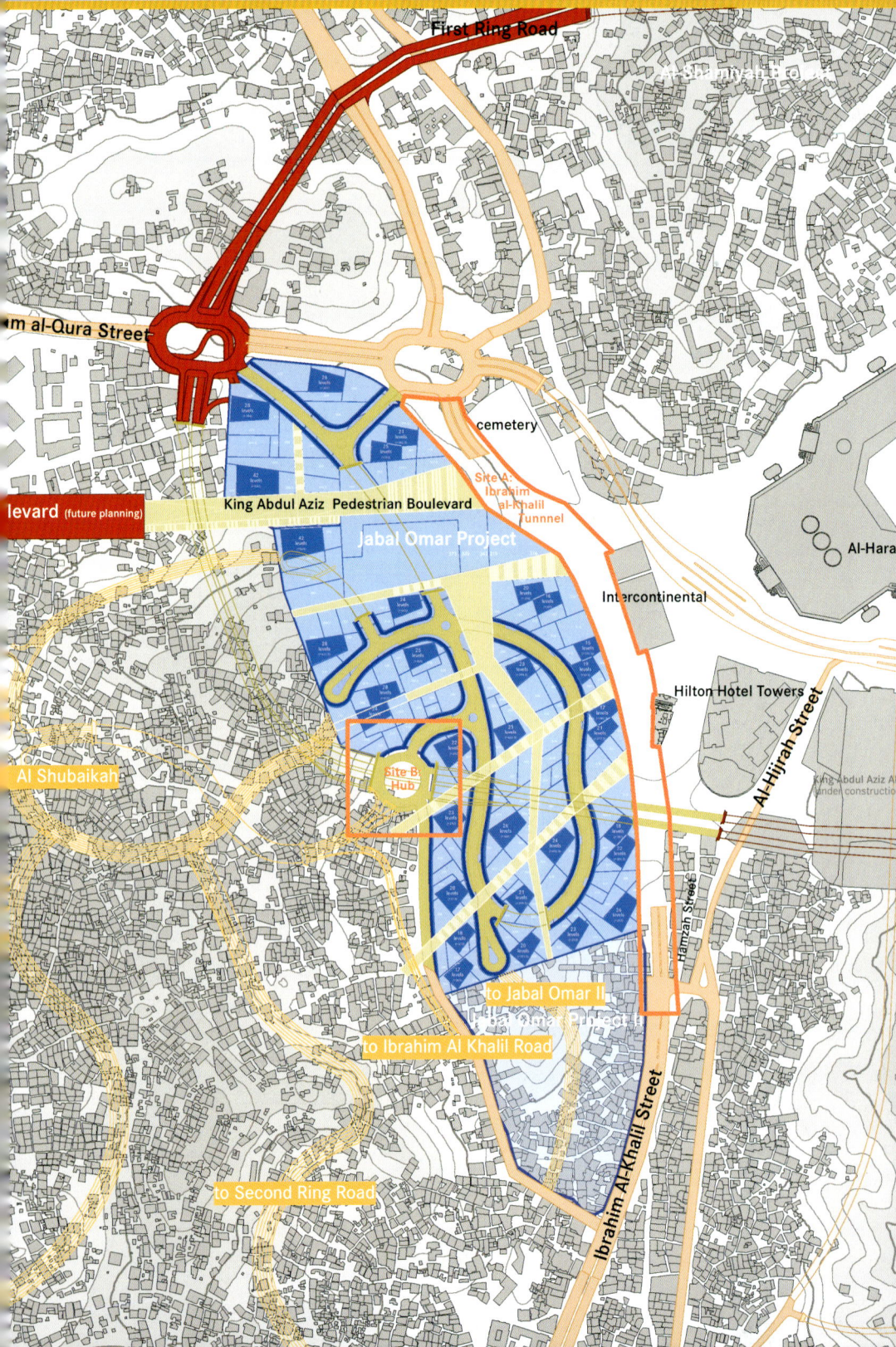

地点 Mecca
时间 04/2007–03/2008
主办方 Jabal Omar Development Co.
人数 2
面积 approx. 23 ha
竞赛费用 200,000 USD
专业评奖委员会 Ralf Baumann;
Schüßler Plan, Berlin;
Heinrich Burchard, Berlin;
Prof. Dr. Jürgen Ruth, Weimar;
Dr. Dieter Tetzner,
DMT engineers, Leipzig;
Rembert Wösthoff, Dorsch Consult Traffic
and Infrastructure GmbH, München

This competition featured the same inner-city development area west of the Grand Mosque as the Jabal Omar competition, and both were conducted together. Its brief was the design of two traffic-bearing structures and involved an interdisciplinary mix of infrastructure and architectural issues. Here as well the point of departure was the master plan by the Paris firm Ateliers Lion. Using traffic forecasts and adjusting for the city's difficult topography, the competition established a grid for access routes to the new quarter. The point of the competition was the technical and artistic development of two key infrastructure projects within the broader Jabal Omar concept: the Ibrahim Al-Khalil Tunnel and the Jabal Omar Hub. The found area above ground will extend the square of the Grand Mosque, so that it borders the Jabal Omar project's first row of buildings. In turn, the Hub will be an intricate, multi-faceted structure distributing traffic in the southwest. It will connect Jabal Omar and other developments along the way to the innermost ring of Mecca's speedway network. The connector to the ring will be a road running in a tunnel beneath the project area. Due to the complexity and social significance of these structures, the competitors were asked to improve their submitted designs after the time allotted to the task had elapsed: This became the basis for the jury's ultimate decision.

此次竞赛与Jabal Omar开发项目竞赛一样，都是对大清真寺以西内城的开发。两次竞赛同时进行。任务是设计两层交通结构，涉及基础设施相互之间的混合和建筑方面的问题。由巴黎Ateliers Lion建筑城市规划事务所提供的总平面图也是同样的出发点。利用运输量预测，对城市的艰难的地貌相应进行调整，竞赛确立一个进入新的建筑物当中的地标。竞赛的目的是大Jabal Omar理念中两个关键基础设施在技术和艺术上的开发：Ibrahim Al–Khalil隧道和Jabal Omar中心。隧道的地上部分使大清真寺广场得以延伸到与Jabal Omar项目前几排的建筑物相接。反过来，中心成为一个错综复杂的多层建筑结构，分流西南部的交通，把Jabal Omar和其他开发项目与麦加高速公路网的最内环连接在一起。
由于这两大建筑结构的复杂性和社会重要性，要求参赛者在完成分配任务之后对所递交的设计作出改善，改善之后的设计是评奖委员会最终评奖的基础。

Aerial view　鸟瞰图

Competition site　竞赛地点

Competition site　竞赛地点

Competition site (in the foreground the Great Mosque)
竞赛地点（大清真寺为前景）

161

Jabal Omar Development Project, Mecca

Buro Happold London/Berlin
Buro Happold建筑事务所 伦敦/柏林

"The architectural design of the above ground structures takes its inspiration from several elements within the landscape and surrounding urban topography."

"建筑物的地上结构的设计取自景观和周围城市地貌当中几个元素所激发的灵感。"

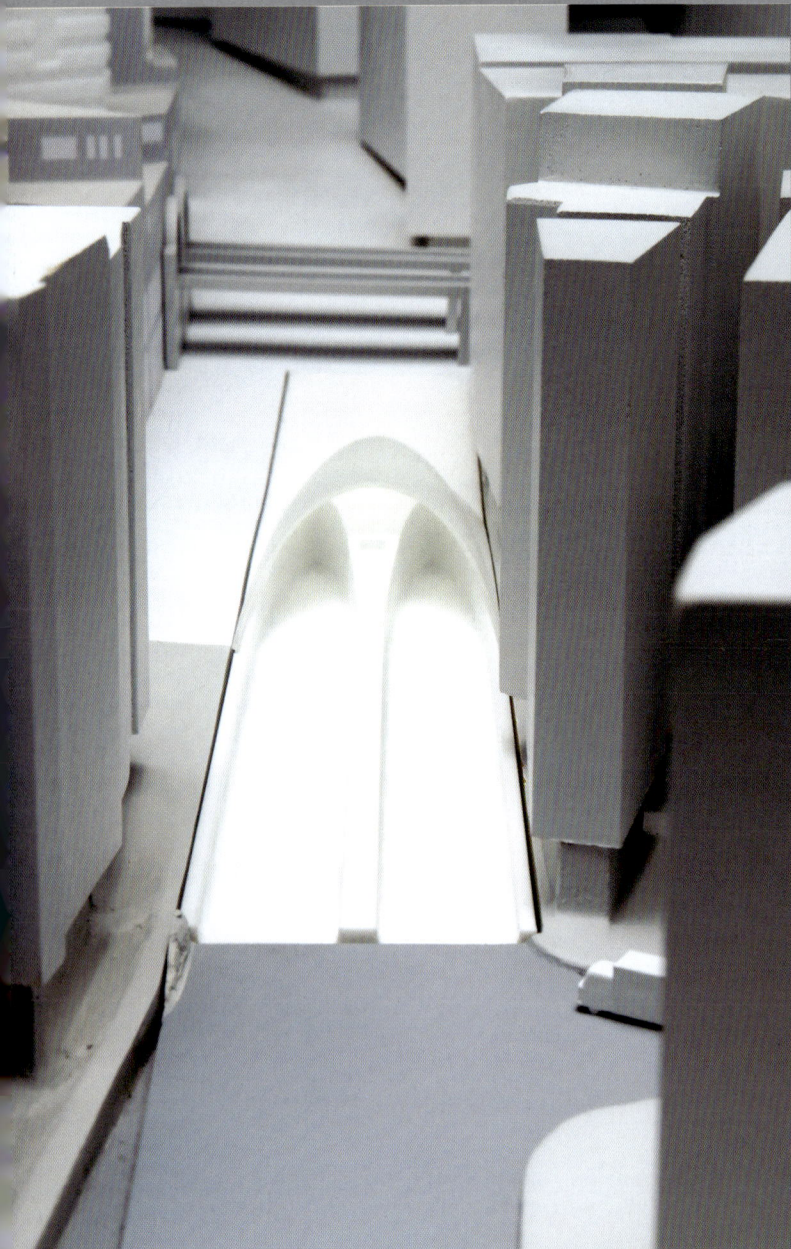

作者 Buro Happold Ltd., London/Berlin; Paul Rogers, Mark Taylor in cooperation with Baseler und Hoffmann, Essingen and Hakes Associates, London/UK

Visualisation Hub

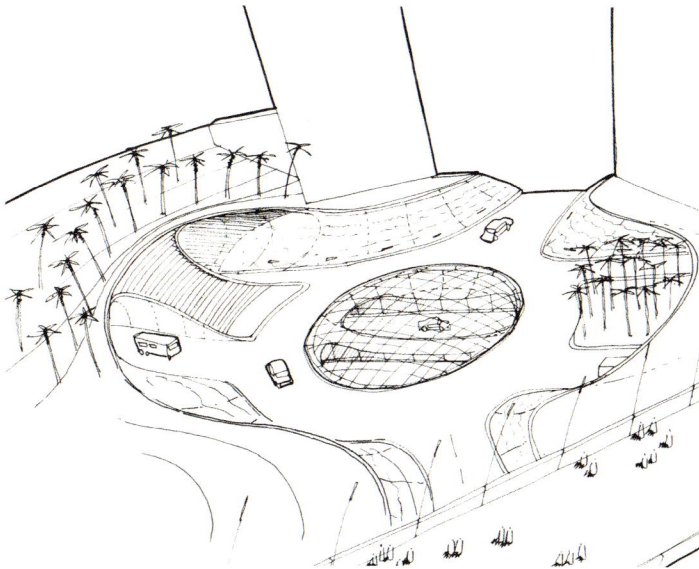

Hub Design Concept

In order to ensure continuity to the Jabal Omar site masterplan and infrastructure design the concept for the hub utilises the same design concepts as for the tunnel portals. When these concepts are applied to the site for the hub this allows for the design development of a circular grid shell roof light respectively shading design which spans over the centre of the hub.

This provides the opportunity to bring controlled levels of shaded daylight down into the lower ring road tunnel also creating a clearly identifiable feature within the tunnel, marking the hub as the important connection to the Jabal Omar Project. The space between on / off ramps is landscaped using progressive geometries generated by the grid shell roof structure.

163

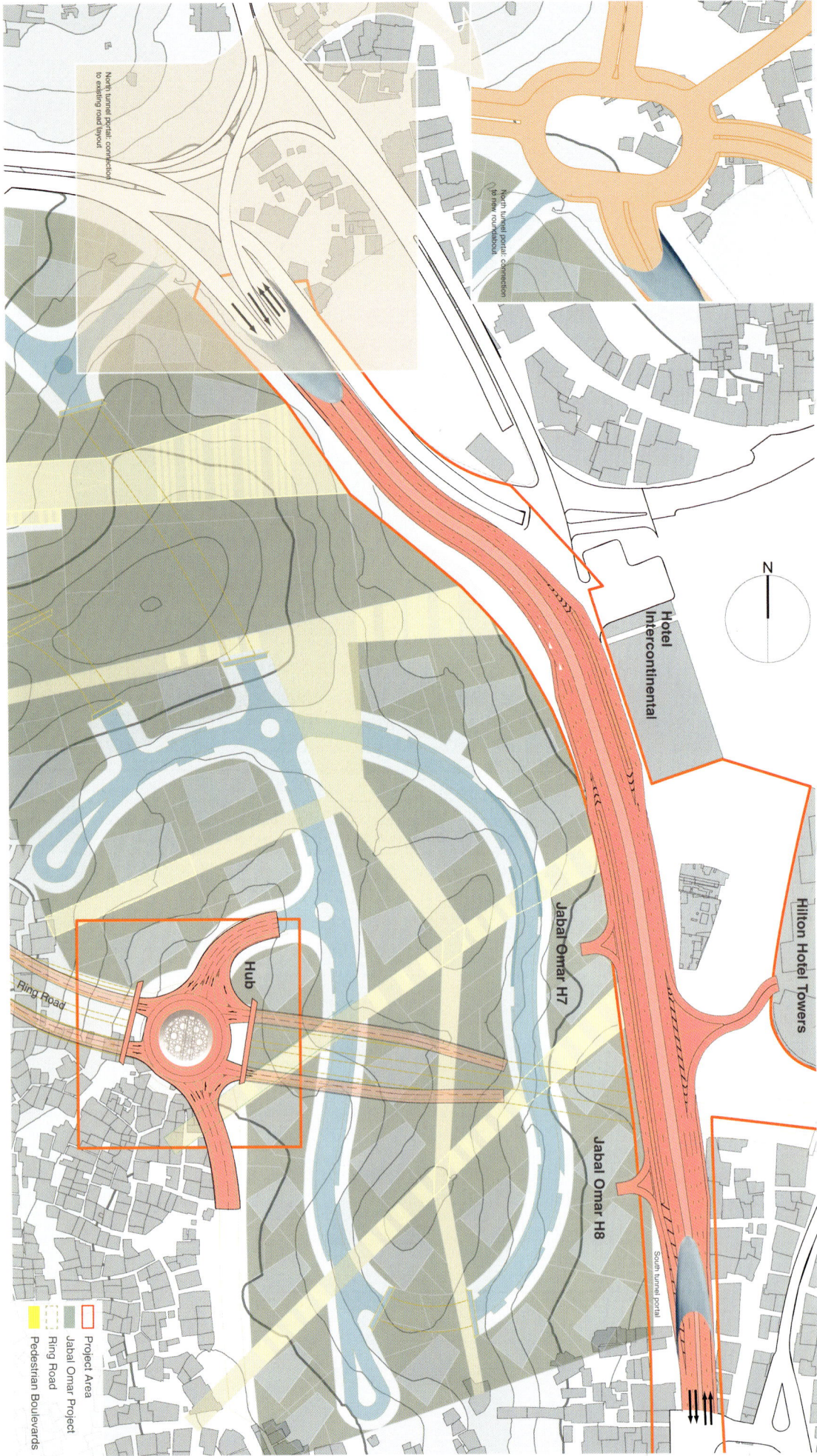

North tunnel portal: connection to existing road layout

North tunnel portal: connection to new roundabout

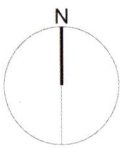

N

Hotel Intercontinental

Hilton Hotel Towers

Jabal Omar H7

Jabal Omar H8

Ring Road

Hub

South tunnel portal

Project Area
Jabal Omar Project
Ring Road
Pedestrian Boulevards

Buro Happold, London/Berlin

Tunnel Portal Design Concept

The tunnel portal is conceived conceptually to progressively ease down in scale from the wide valleys and hills of the surrounding landscape of the Makkah region, through to the dense urban grid and into the new Jabal Omar Tunnel. The architectural design of the above ground structures takes its inspiration from several elements within the landscape and surrounding urban topography. These include dunes, the wadior oasis and traditional screens or shading devices. These are arranged so that the tunnel portals are brought to ground level.

The tunnel portals are surrounded by safety barriers (height approximately 140 cm) to avoid people climbing down to the road level.

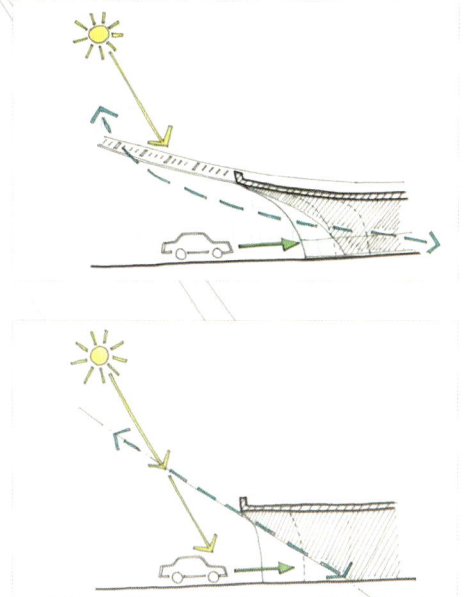

Jabal Omar Development Project, Mecca

Tunnel Sections

Cross sections (not in scale)

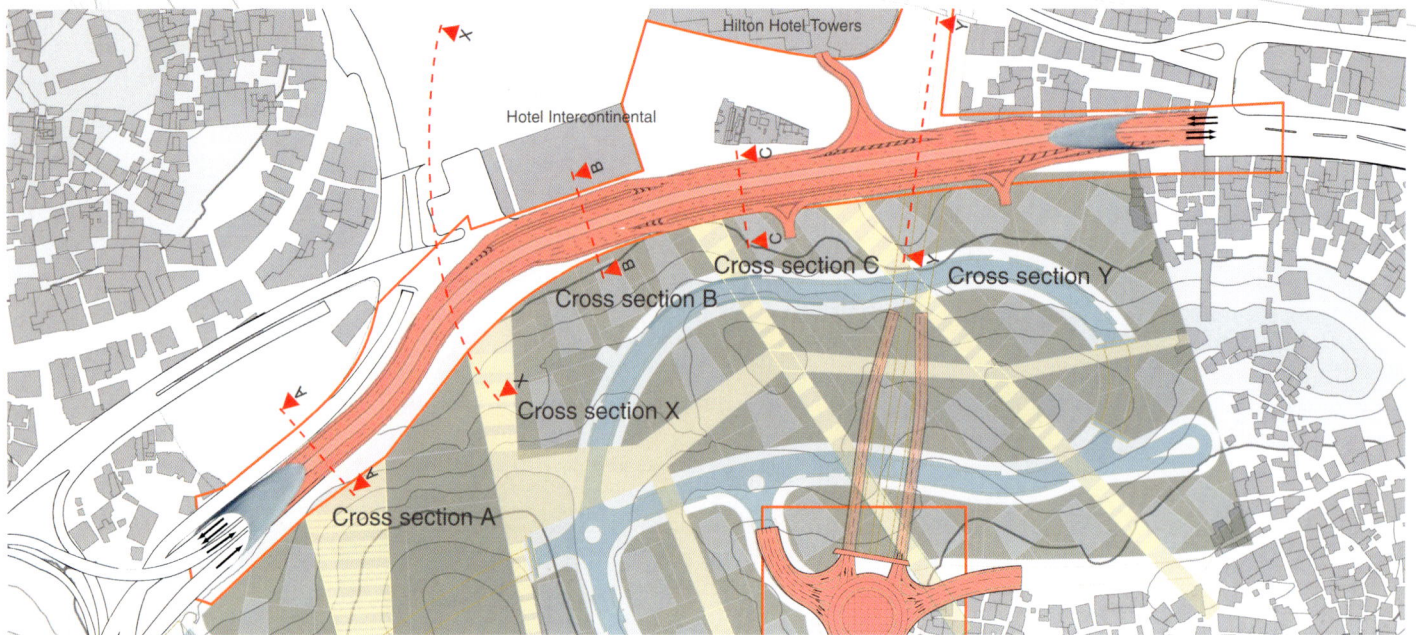

Hilton Hotel Towers

Hotel Intercontinental

Cross section C

Cross section Y

Cross section B

Cross section X

Cross section A

Cross section X – X in relation to neighboring buildings (not in scale)

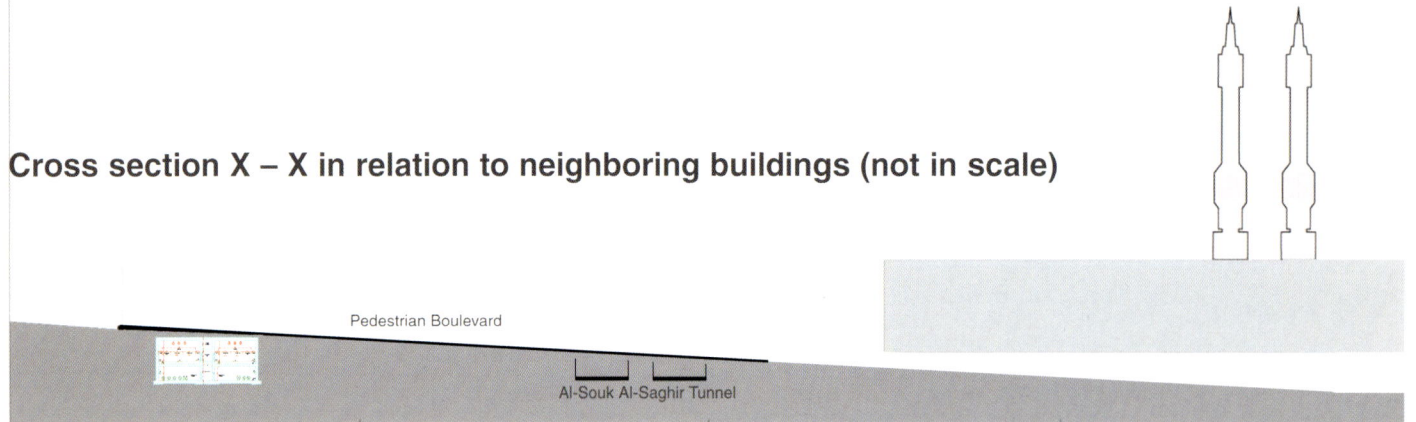

Pedestrian Boulevard

Al-Souk Al-Saghir Tunnel

Longitudinal section in relation to neighboring buildings (not in scale)

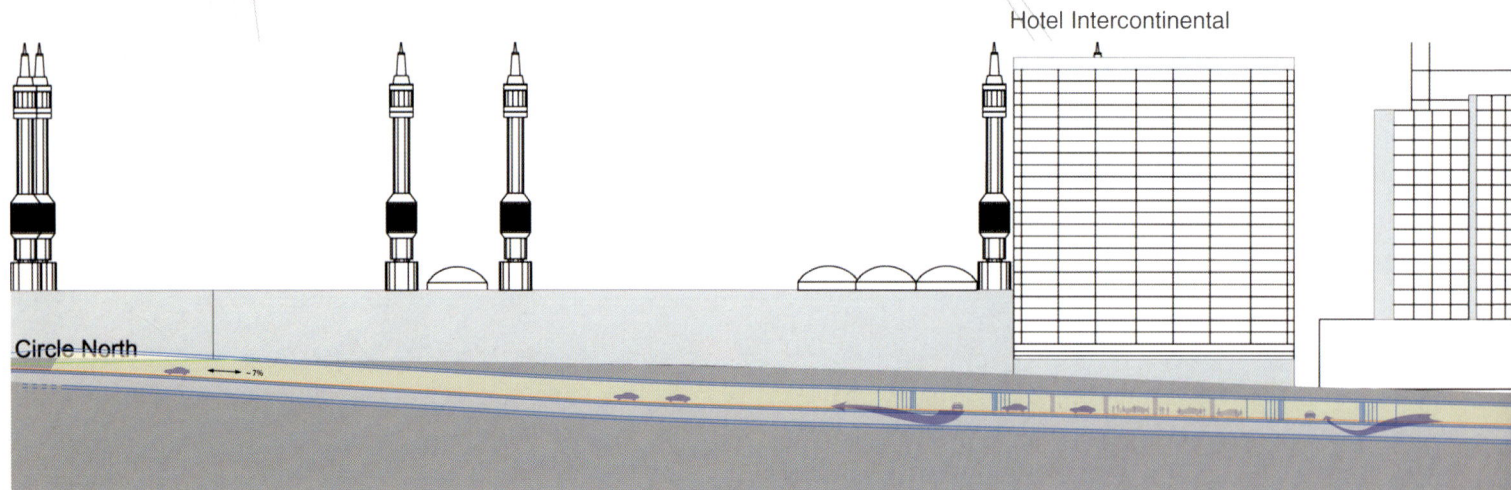

Hotel Intercontinental

Circle North

Cross section A (near northern portal)

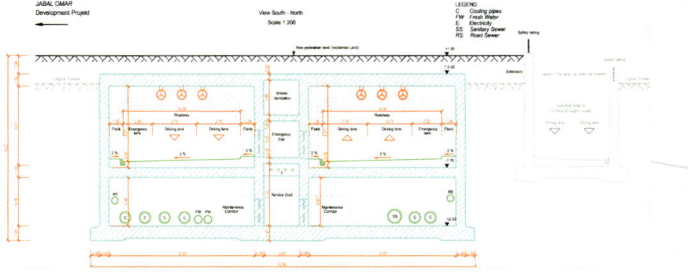

JABAL OMAR
Development Project

View South - North
Scale 1:200

LEGEND
C   Cooling pipes
FW   Fresh Water
E   Electricity
SS   Sanitary Sewer
RS   Road Sewer

Cross section B (access to Hotel Intercontinental)

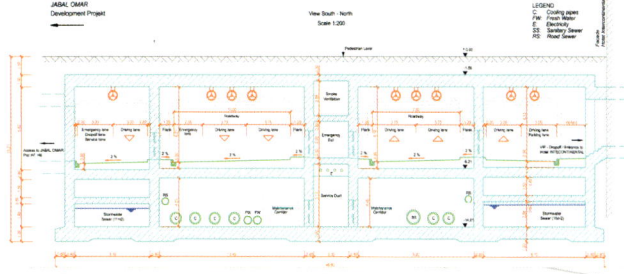

JABAL OMAR
Development Project

View South - North
Scale 1:200

LEGEND
C   Cooling pipes
FW   Fresh Water
E   Electricity
SS   Sanitary Sewer
RS   Road Sewer

Cross section C

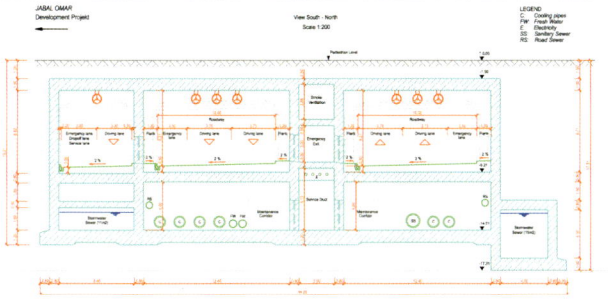

JABAL OMAR
Development Project

View South - North
Scale 1:200

LEGEND
C   Cooling pipes
FW   Fresh Water
E   Electricity
SS   Sanitary Sewer
RS   Road Sewer

Cross section Y – Y

Jabal Omar Project

Hilton Hotel Towers

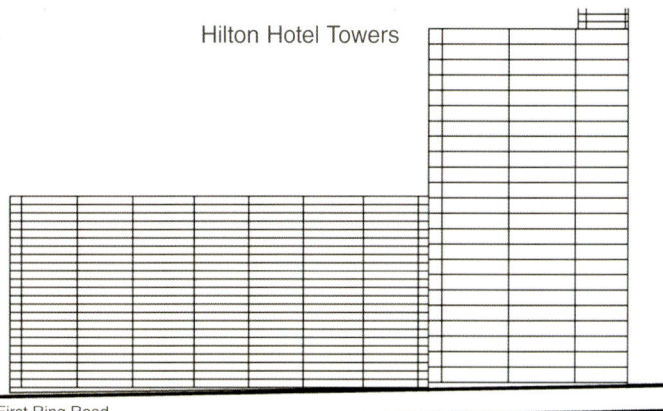

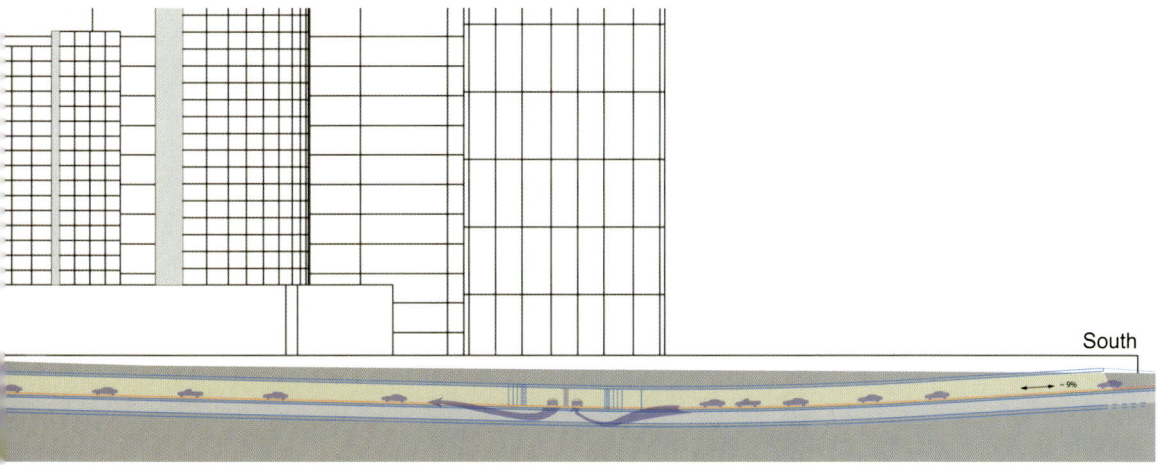

First Ring Road

First Ring Road

Hilton Hotel Towers

South

Safety concept

Schema information flow :   processed by control unit ⟶   Reflex control ⟶

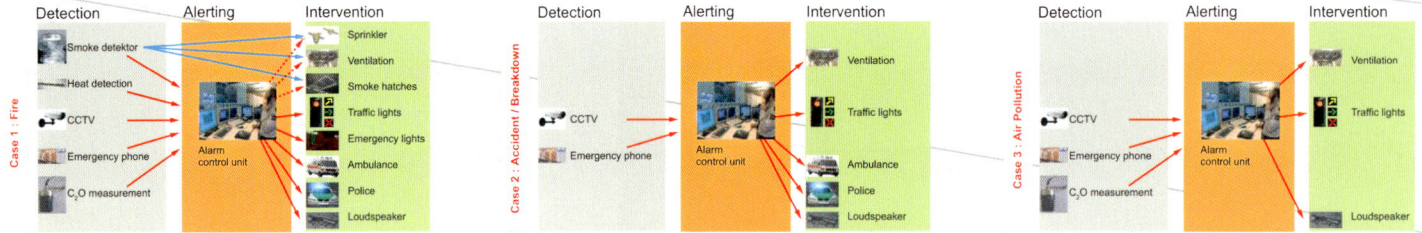

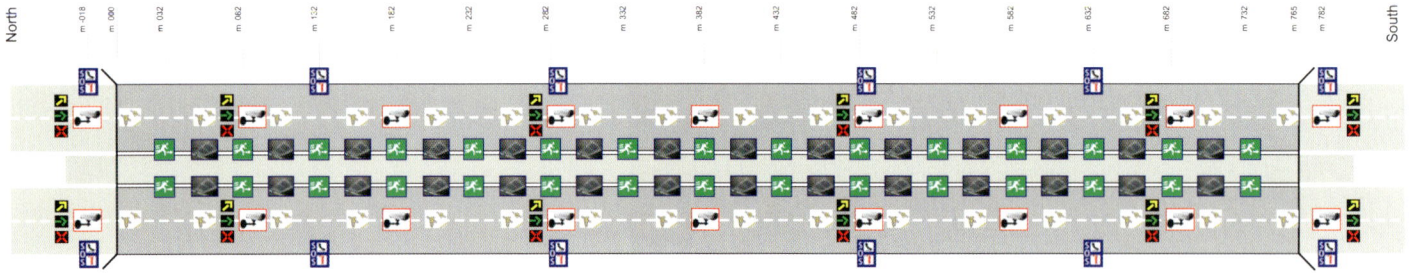

Rescue

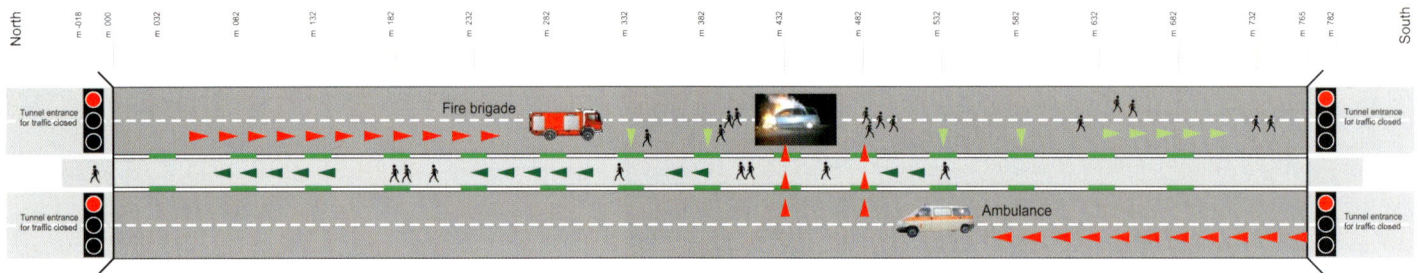

Longitudinal section through emergency exit and smoke ventilation (schematic)

Ventilation system

Lighting

VIP underground drop off

VIP «Drop Off» – Hotel Arrival Concept

The below ground drop off to the Intercontinental Hotel leads directly off the main tunnel. In order to minimise the transfer of road noise and to create a tranquil arrival area the interior walls of the VIP drop off area have been designed to be a series of water features. The entrance to the Intercontinental Hotel is protected by a glazed screen with an automatic sliding door, which enables the lobby to remain quiet, climatically controlled and free from traffic noise and vehicle fumes. As the VIP drop off is adjacent to the foundations of the Intercontinental Hotel the openings are punched through to – 1 basement level to create the internal VIP reception area for the Intercontinental. A grand sweeping staircase and glass lift take the VIPs either directly to their suites or up to the main reception.

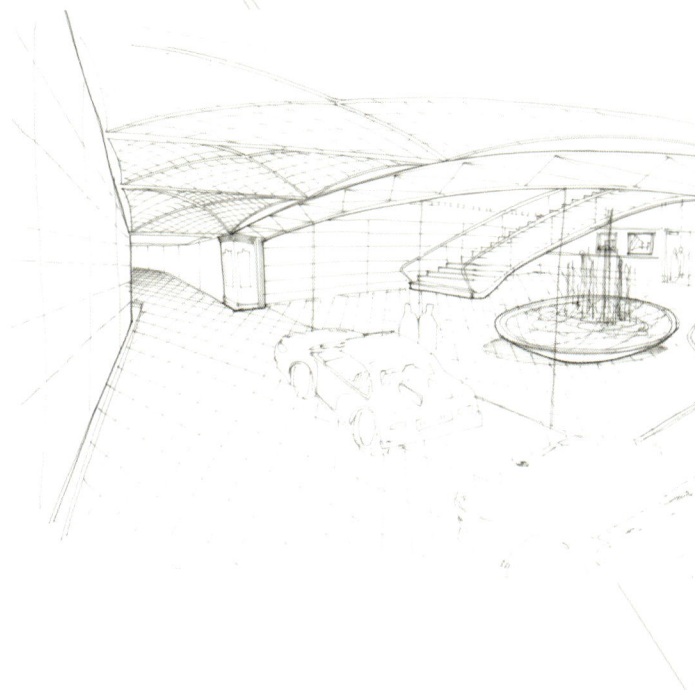

Buro Happold, London/Berlin

Jabal Omar Development Project, Mecca

von Gerkan, Marg und Partner Architekten Hambur
Von Gerkan, Mary und Partner建筑事务所 汉堡

"The local tradition leads to geometry. Geometry was applied on the functional requirements of the structure. The detail and colour of the structure is determined by the characteristic of the chosen material concrete. The expressive power of the structure is enhanced by a well-directed illumination."

"当地传统使几何学在建筑结构的功能要求上得到运用。建筑结构的细部特征和色彩取决于所选具体材料的特点。照明良好也增强了建筑结构的表现力。"

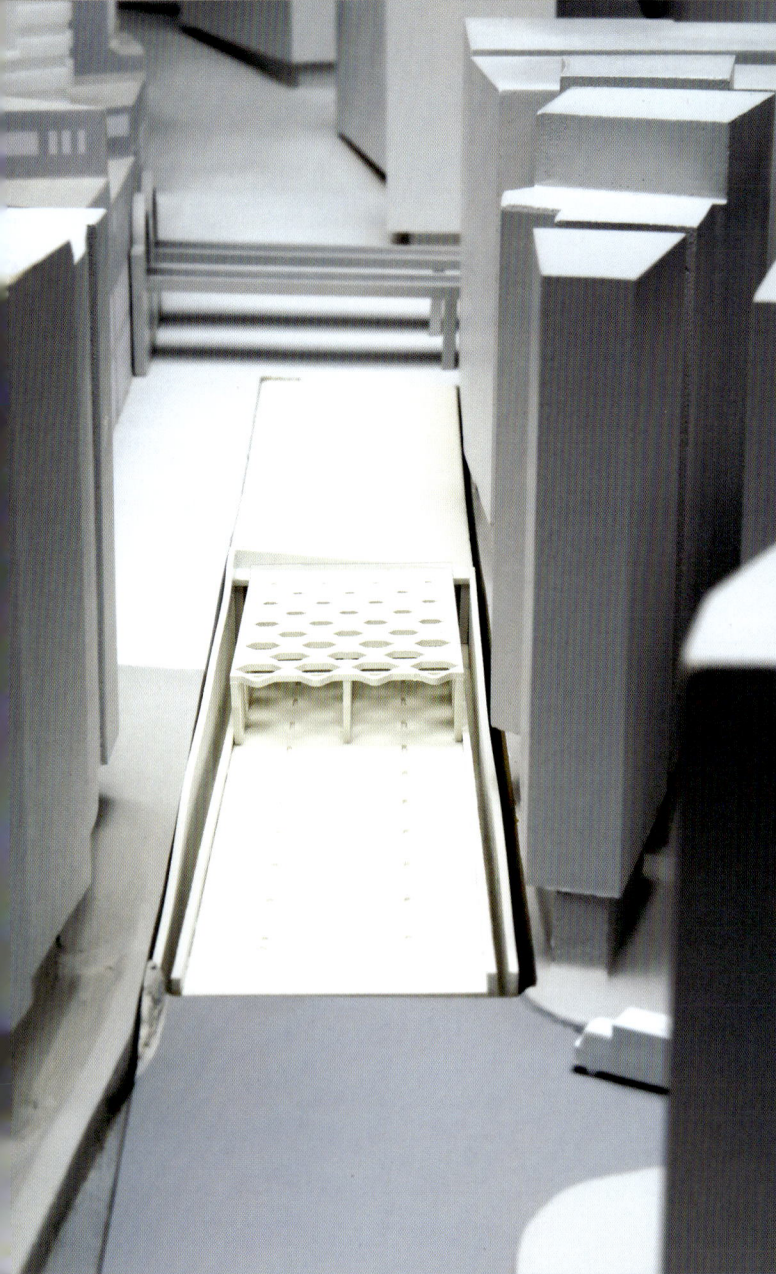

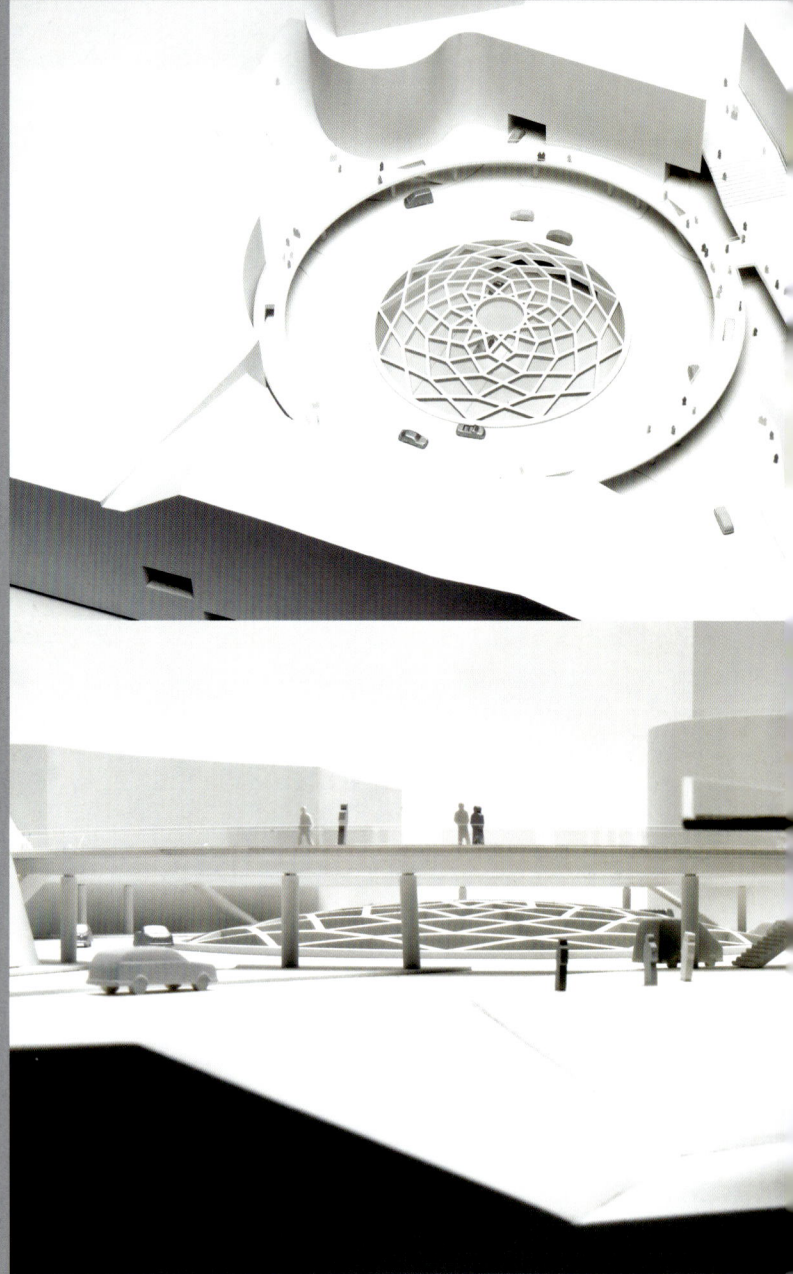

作者 von Gerkan, Marg und Partner, Jürgen Hillmer, Hamburg   专家 traffic engineers: Dr. Brenner Ingenieurgesellschaft, Aalen; structural analysis: Werner Sobek Ingenieure, Frankfurt

von Gerkan, Marg und Partner Architekten, Hamburg

Jabal Omar Development Project, Mecca

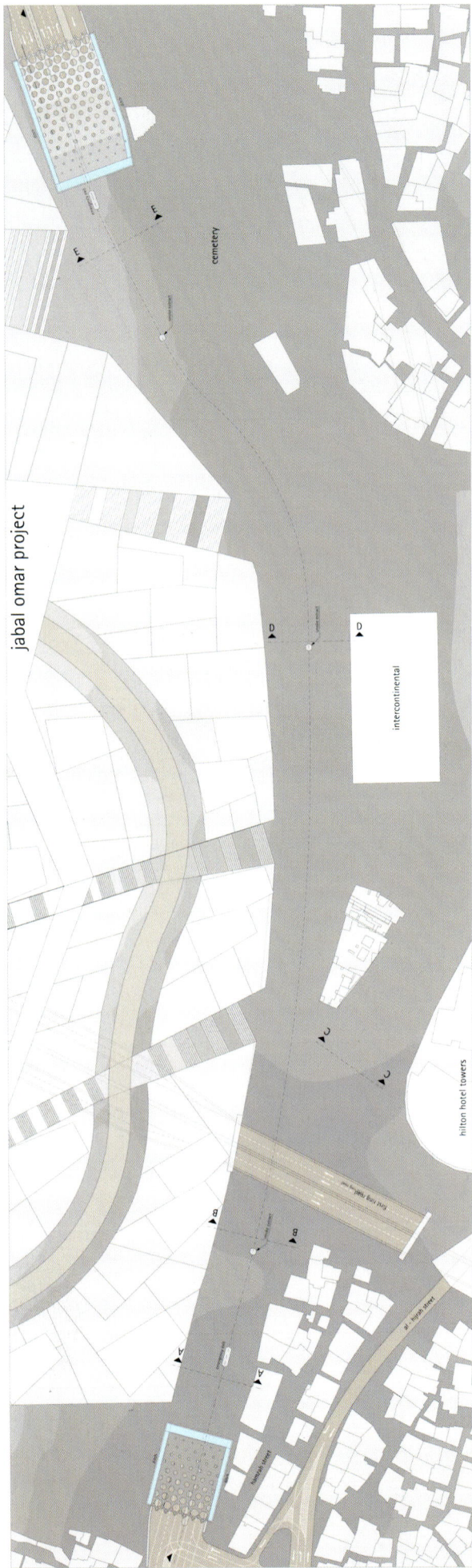

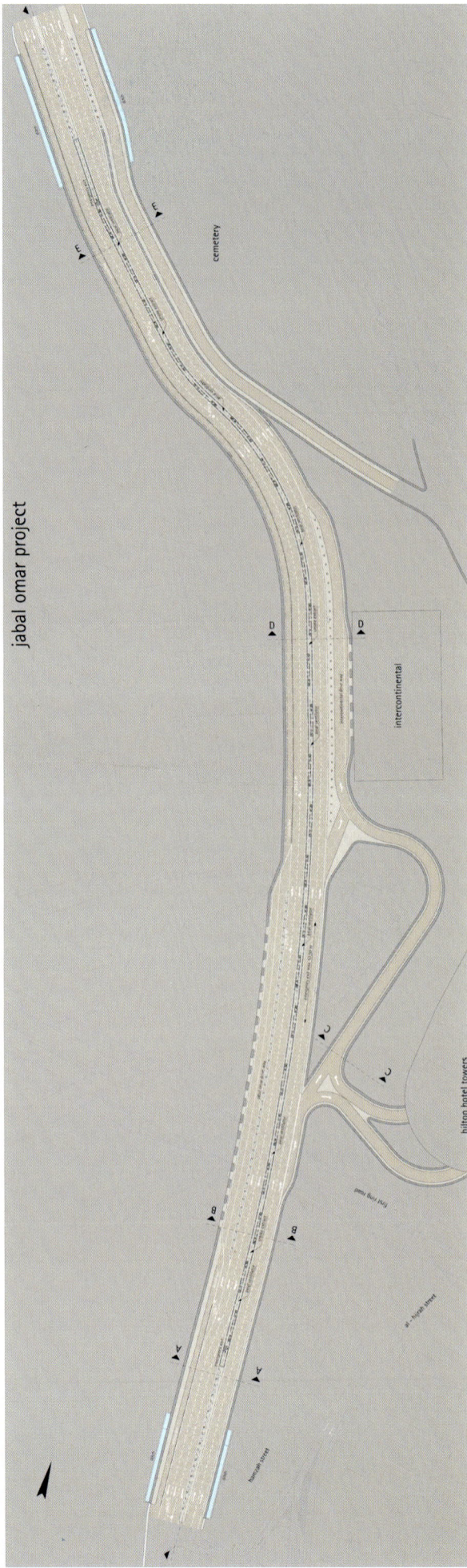

jabal omar project

cemetery

intercontinental

hilton hotel towers

jabal omar project

cemetery

intercontinental

hilton hotel towers

Jabal Omar Development Project, Mecca

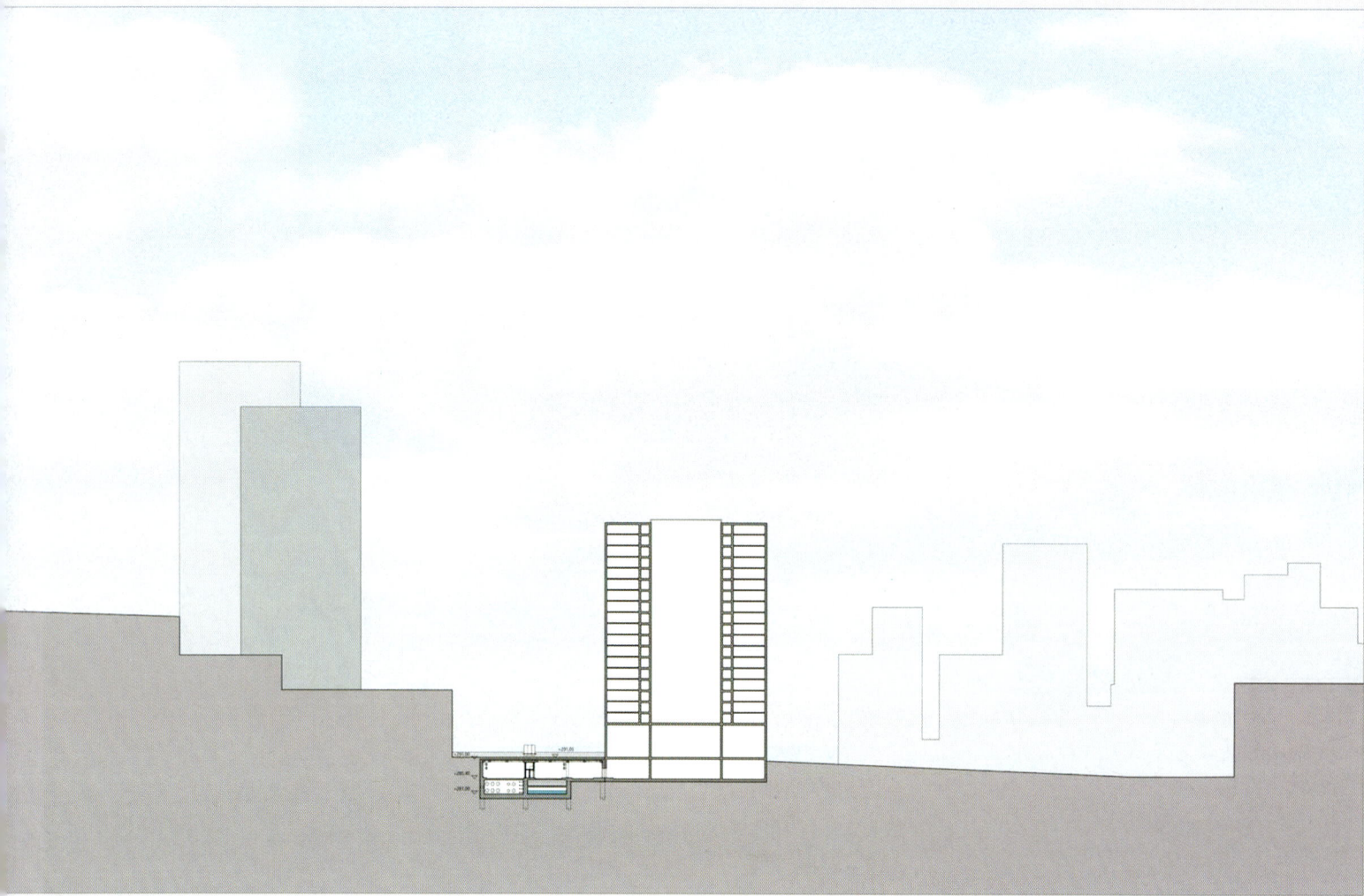
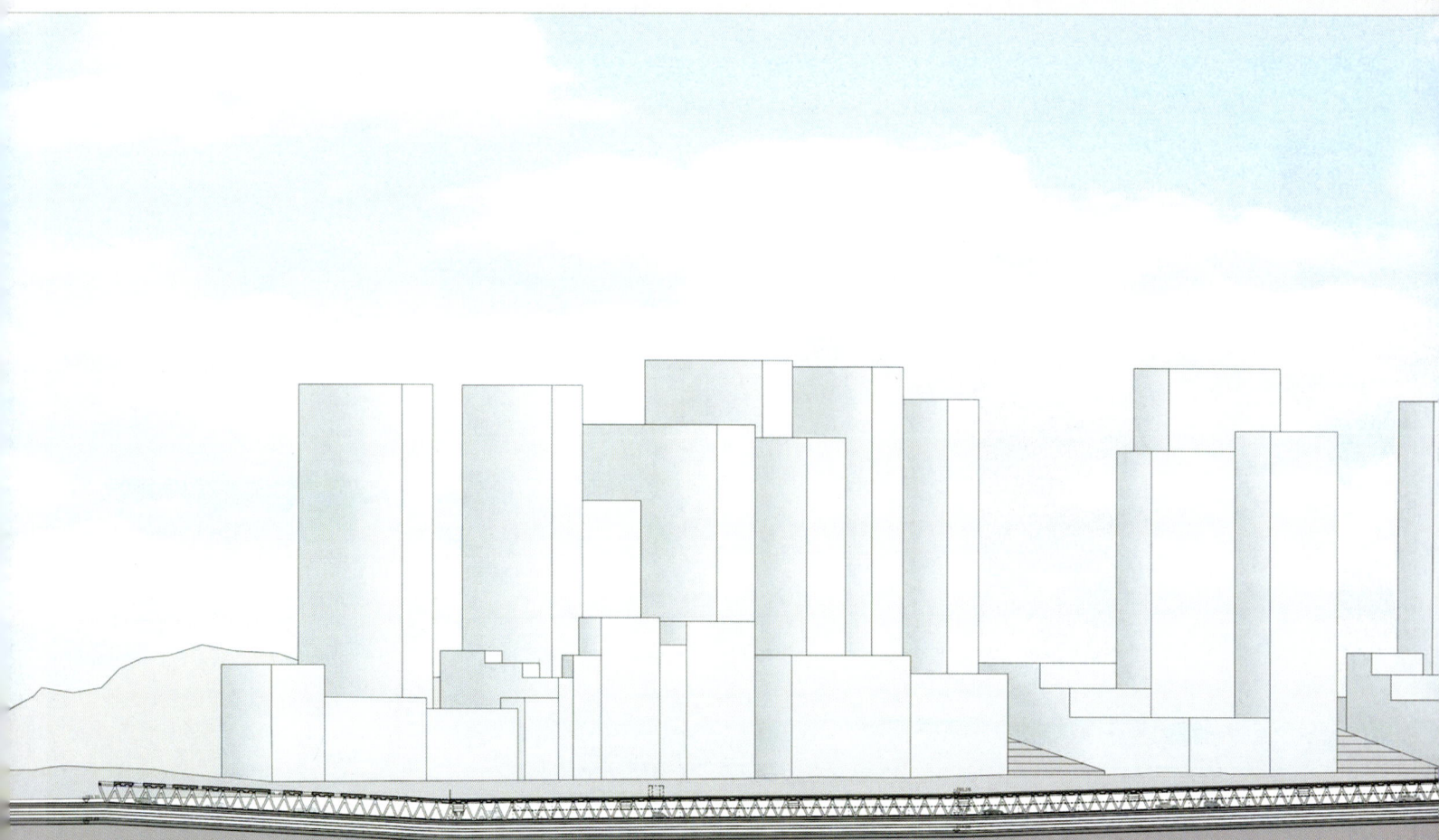

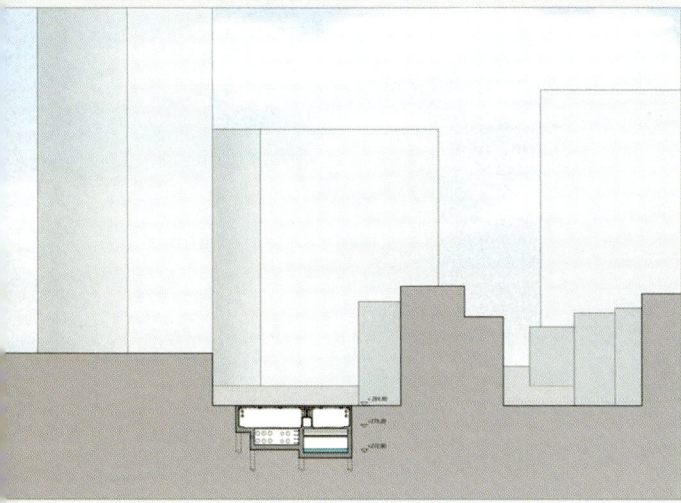
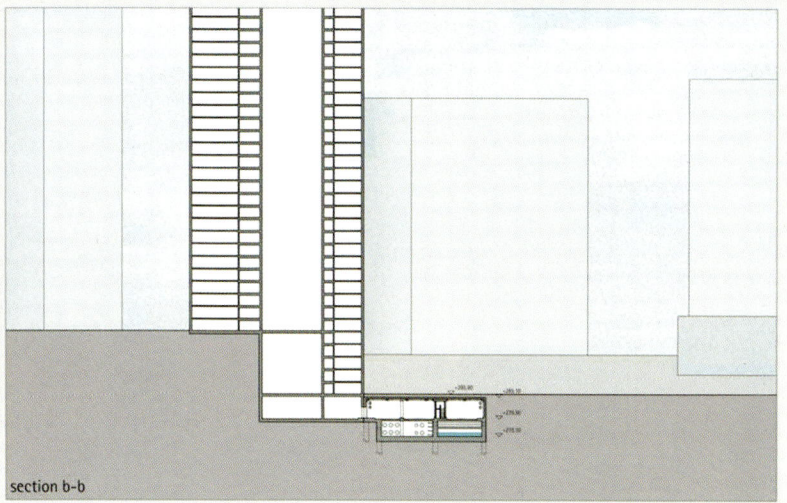

section b-b

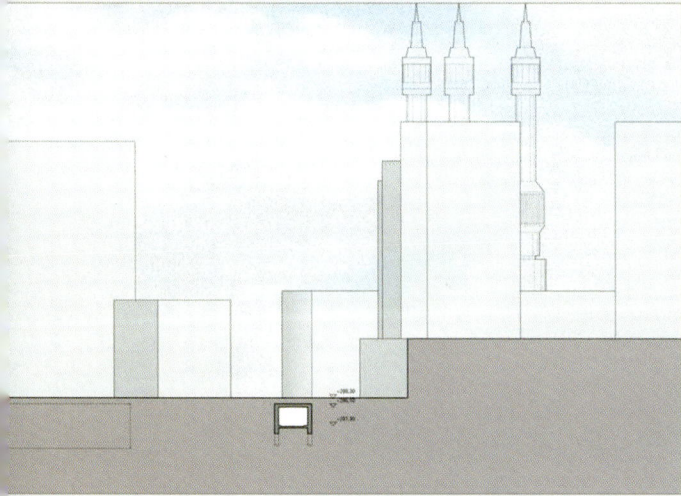
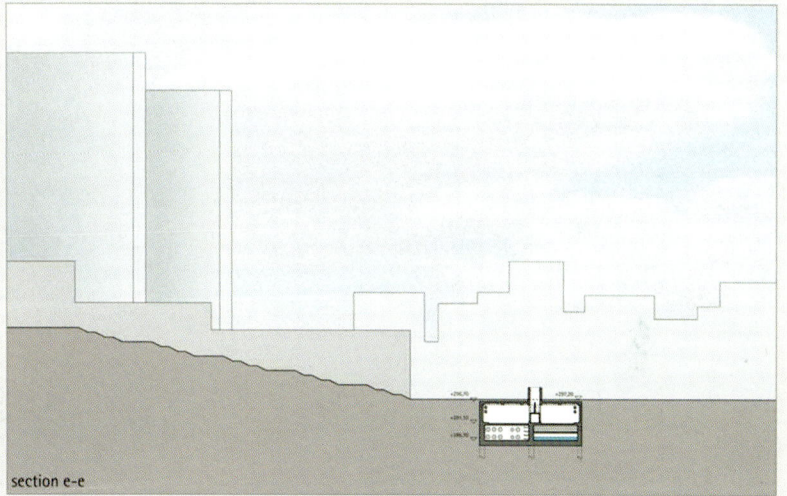

section e-e

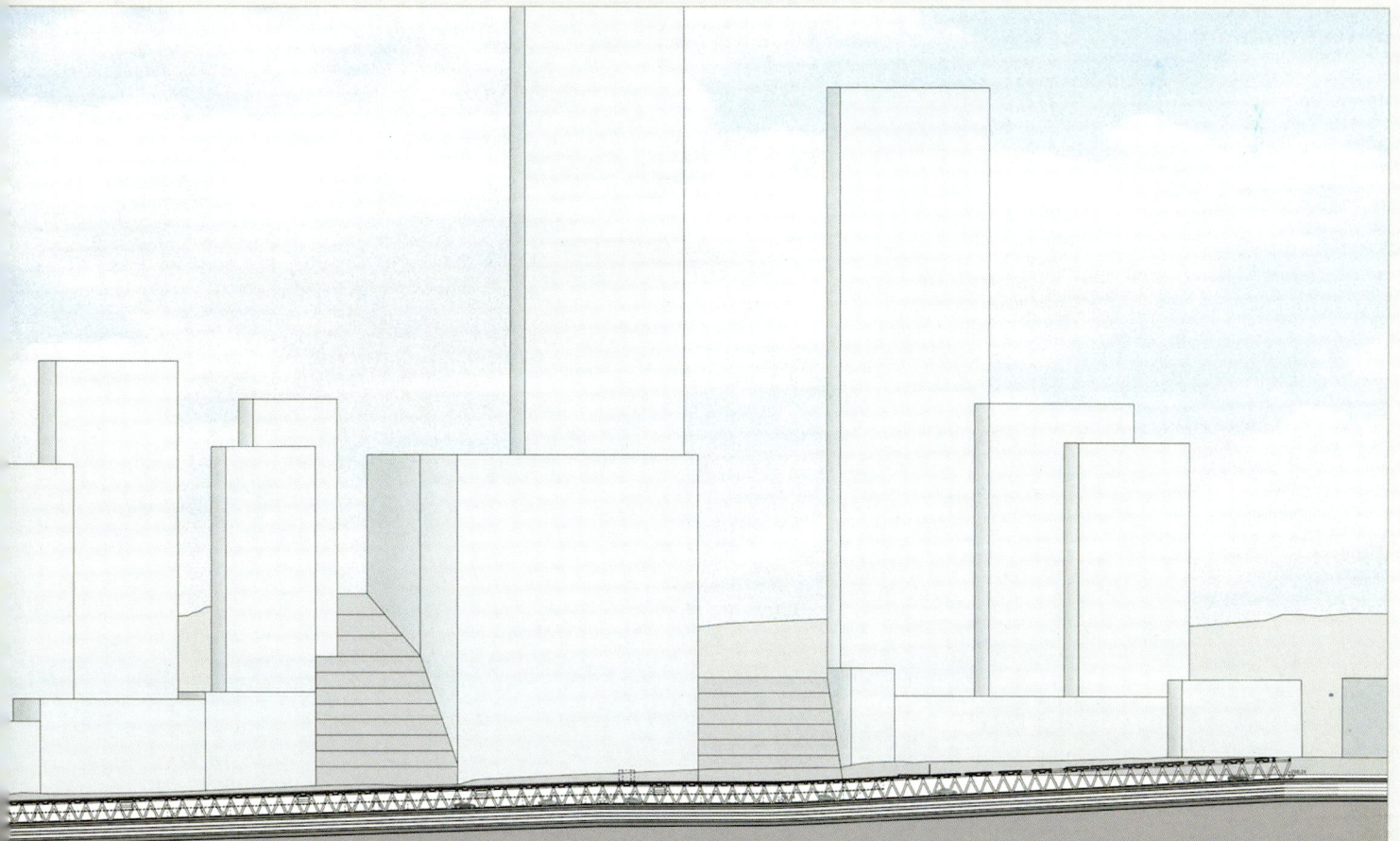

von Gerkan, Marg und Partner Architekten, Hamburg

Jabal Omar Development Project, Mecca

Architectural Concept

The area of the projected tunnel is situated west of the Holy Mosque, the Haram. This unique neighbourhood is answered with a proposal keeping innovation and continuity in balance.

In order to achieve reasonable harmony with functional requirements and the urban structure of Makkah as a whole, the chosen guiding principle is based on structure as source for a unique expression. The state-of-the-art construction technique is applied for an geometrical interpretation of traditional architectural elements.

The local tradition lead to geometry. Geometry was applied on the functional requirements of the structure. The detail and colour of the structure is determined by the characteristic of the chosen material concrete. The expressive power of the structure is enhanced by a well directed illumination.

The Hub is based on the same design principles. Daylight can pass the open structure in the centre and mark the entrance to the Jabal Omar Development along the First Ring Road. The pedestrian level refers to the hub as raised ring. The structure of the pedestrian level is designed modest in order to create a hierarchy between the ornament in the centre and the ring.

Traffic

- Tunnel

The projected tunnel runs in the course of Ibrahim Al-Khalil Street in the north-south-direction. The existing ground rises in the northbound direction with a height difference of 15 m at approx. 800 m length. The western boundary of the tunnel is given by future development of hotel complexes; the eastern border is set by the existing buildings of Hotel Intercontinental and Hilton Hotel Towers as well as a cemetery. In addition, relevant constraint points are four intended pedestrian paths, crossing Ibrahim Al-Khalil Street and leading to Al-Haram Al-Sharif Mosque as its main access lines from the west.

According to the task description, two hotel drop-offs, both at level -1, were to be considered for the existing hotel facilities east of the tunnel (Hotel Intercontinental) (E.25). To get there by driving southbound, a u-turn is facilitated at the southern intersection following the tunnel ramps in order to enter the northbound sections.

In the southbound tunnel sections, subgrade hotel drop-offs for the Jabal Omar plots H7 and H8 are provided.

For the main tunnel roadways, widths of 7.50 m (2 3.75 m lane width) were employed, plus 0.25 m hard shoulder (for markings) on both sides. An additional service lane of 3.00 m is provided in the southbound tunnel sections (according to the minutes of the 2007-06-05 Participants' Workshop). Emergency walkways are provided right hand of the roadway.

The connection to the existing road network is provided north of Ibrahim Al-Khalil Tunnel by a signal-controlled intersection. This necessitates the removal of the existing flyover bridge which not can be connected to the intersection system. Operation of the existing tunnel can be maintained by integrating its western access ramp in the approach of the new tunnel. The eastern exit ramp is connected to the proposed intersection.

In the south, a design proposal with two offset signalized intersections is shown in the plan sheet. In the same way as for the northern intersection, type, dimensions and shape of that intersection was also taken from the working plan. Here, the western part of the double-intersections is intended to be the main intersection with the Ibrahim Al-Khalil Street as the major traffic stream. The eastern partial intersection is lower ranked.

It has to be explicitly pointed out that no capacity analysis was conducted for neither of the connecting intersections north and south of the Ibrahim Al-Khalil Tunnel as no traffic data have been provided in the competition documents. Likewise, the given layout from the working plan of the competition documents was basically adapted without further optimization, as optimization measures would require valid traffic data.

Therefore, a successful operation and smooth traffic flow under real traffic conditions can not be guaranteed. Rather than that, the fear exists that the intersections might not be able to handle peak traffic volumes and thus traffic might possibly collapse.

Proposed traffic routing concept during the construction of the Ibrahim al-Khalil tunnel

From the traffic engineering perspective a subdivision into 4 building stages is required. As a continuous availability of 4 through-lanes in (2 in each direction) is demanded, the only possible working solution is providing these lanes west of the construction site in the Jabal Omar area. Solid soil and steep rising terrain will necessitate a considerable earthworks effort. However, from the side of the loadbearing structure of the tunnel, it seems not possible to divide the construction in longitudinal stages where tunnel sections are built one after another.

Although the usage of the Jabal Omar plot for the temporary road will be elaborate, the benefit is that the traffic routing can be maintained over the whole construction stage, so there is no need to subdivide the traffic routing concept into different stages. In the same manner, the uncomplex routing concept does not need further graphical explanations.

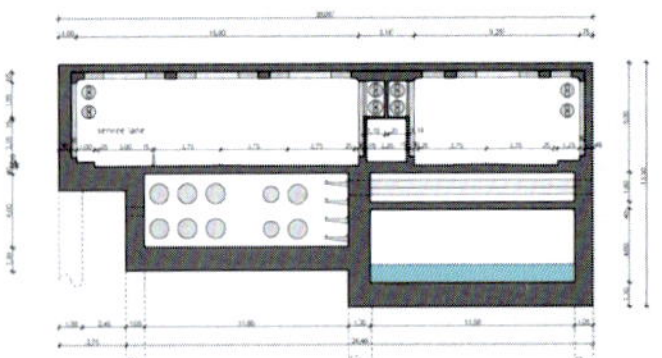

Section A-A

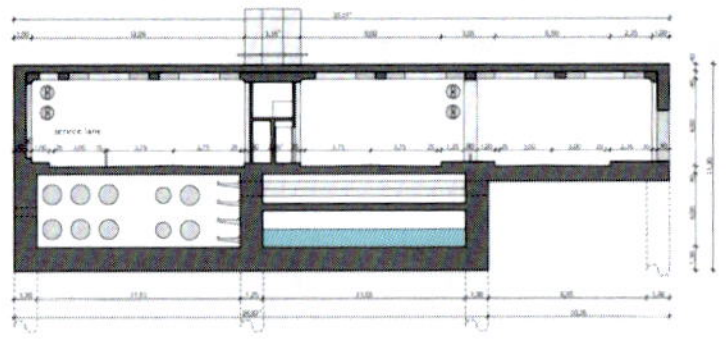

Section D-D

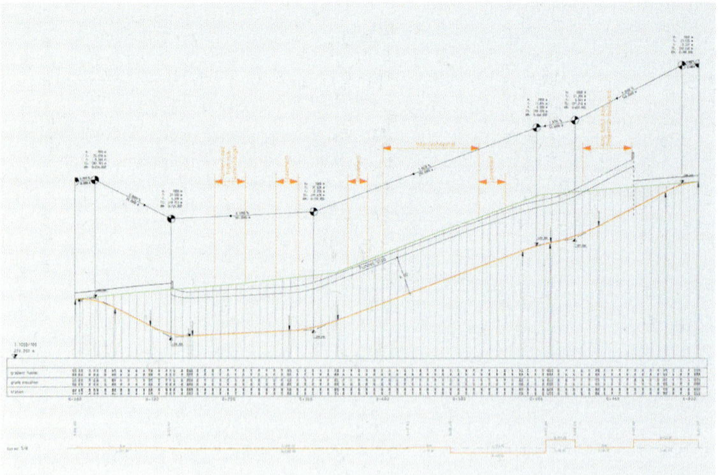

Layout Plan

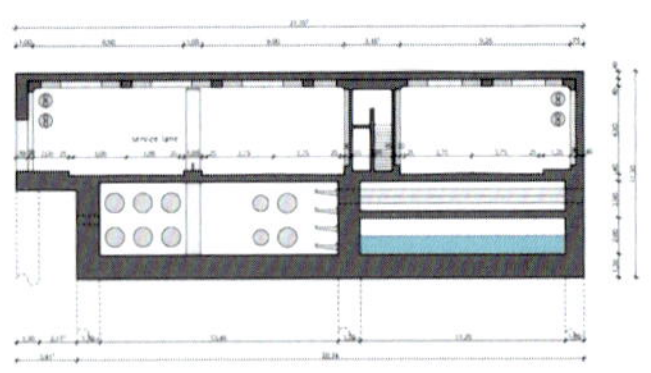

Section B-B

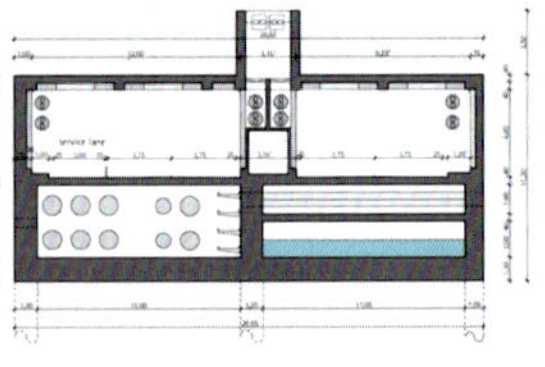

Section E-E

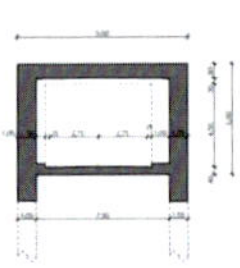

Section C-C

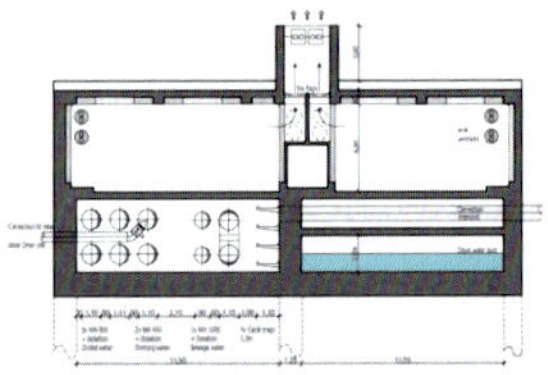

technical section

scale 1:200

Structure

- Tunnel

Following arguments make the top-to-bottom construction method for Al Khalil tunnel reasonable:

Geometry

The tunnel is a 2 by 2 lane tunnel close below the surface. It is not a crossing of a mountain or a hill but a lower of the existing road level to make area on top of the road mainly crossable. Above stated will be the elimination criterion for the tunnel boring and the NATM method.

Excavation

Excavation needs space beside the tunnel to provide workspace and scarping soil. That means additional excavation volumes. If required scarping angles could not be reached more advanced methods of retaining soil are required as there are anchoring or anchoring combined with shotcreting. Another aspect could be keeping off groundwater.

Limitation of construction site

Using top-to-bottom method requires little site area and provide a finished surface soon before finishing the complete tunnel. Space could be used for example for traffic management during tunnel construction .

These arguments led to the top-to-bottom construction method for Al Khalil tunnel

The steps for construction are:

1) Set the pile walls on east and west boundary of the tunnel. Depending on detailed groundwater conditions the piles could be just touching or for water protection they should be overlapping.

2) 2) Set the piles for the intermediate slab support separating the lane directions. These piles could be arranged with spacing, for example approx. 5m, and will later be incorporated into the intermediate wall.

3) Prepare the slab support, that could be for example a pile cap beam.

4) Construct the prestressed tunnel slab. The slab will be used for carrying vertical loads and horizontal loads due to stabilize the tunnel walls against soil loading. The main beams could be prefabricated or in-situ concrete. Due to geometrical conditions in-situ construction will be preferable.

5) Excavating tunnel below the slab.

6) Construct utility channels below the road lane

7) Construct the slab on the utility channels which will be the later road level

8) Cover the raw pile wall with 25cm concrete wall

9) Shape the intermediate wall between the piles either as inclined columns like at the ramps or as closed concrete walls

Under certain conditions as there are single sided excavating of the pile wall for the depth of tunnel plus utility channel there could be required temporary wall to wall bracing during construction the utility channel.

The dimensions for the main elements tunnel are:

Piles Æ80 cm

Slab 30cm

Beams width/depth 50/50cm under slab

Columns width/depth 60/80cm

Building Services Engineering - Tunnel

Technical System

The media supply for the districts, which will be opened up by the tunnel building, is planned according to the general offer. There are plans for six pipes, diameter 800 mm, for cooling water, two pipes, diameter 600 mm, for drinking water and a waste water pipe, diameter 1000 mm. On top of that there are four cable paths for the supply with electricity (medium and low voltage) and telecommunications.

The piping will be done in a separate media tunnel below the automotive tunnel. It is expected that the density of installation will be higher on the Jabal Omar Projectes side. For this reason the media tunnel will be built on the west side of the tunnel. The stormwater-tunnel is on the tunnel's east side. As the stormwater-tunnel is supersized due to the installation height in the media tunnel, it is possible to cross it for connecting objects on the east side of the tunnel. The remainder of the cross sectional area still fits the demands in the general offer.

Ventilation engineering

The projected ventilation equipment inside the tunnel has to fulfill 3 main objectives:

- Reducing CO2 percentage inside the air in side the tunnel to permissible rates
- Reducing the obscuration caused by soot in the exhaust fumes and tire abrasions
- Transport fire gases out of the tunnel in case of a fire

Four pairs of axial fans are used in each traffic direction for the usual ventilation of the tunnel building. The fans are positioned in regular intervals and transport the air in the direction of the tunnel. The regular distance between the fan pairs is approx. 100 m. By doing this, it is possible to produce a homogeneous perfusion according to the acknowledged rules of technology. The fans are reversible, so they are able to transport the air with or against the traffic direction. The construction of the axial fans will be done to obey the appropriate building regulations regarding the maximum concentration of CO and the obscuration caused by soot and tire abrasions.

It is acceptable for tunnels of this kind and length to transport fire gases by the pre-mentioned axial ventilators to the end of the tunnel and blow them into environment. The smoke has to be conducted a long way through the tunnel, which contradicts the high security demands given in the general offer. To ensure maximum security for the tunnel users there are extra chimney flues (regular distance approx. 200 m), that lead the fire gases directly to the outside. The chimney flue, which is the closest to the fire, will be activated and opened and the axial fans transport the fire gases to the selected chimney from both sides. So it is ensured, that only certain parts of the tunnel will be affected by fire gases.

See explanatory text:

Lighting system

Emergency-lighting and escape-route marking

Communication systems

Fire alarm system

Fire-fighting systems

Interactions of the safety systems

Power supply

Electrical systems in the services tunnel

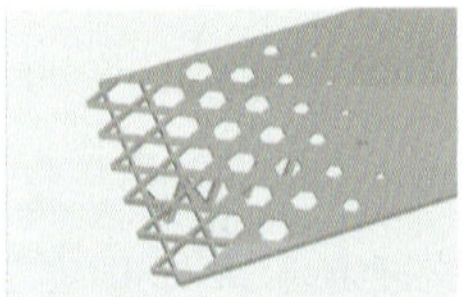

Descent from open roof to closed roof structure

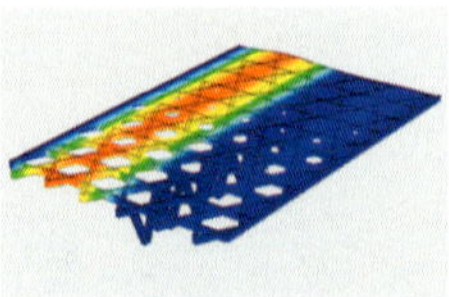

Deflection due to prestress in main beams

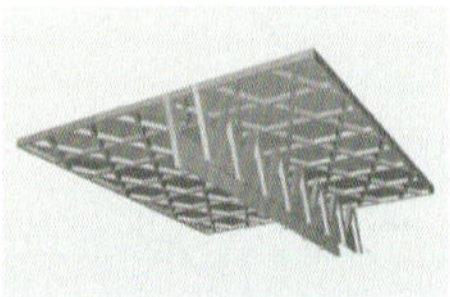

View to inner tunnel with safety tunnel and ventilation channel above

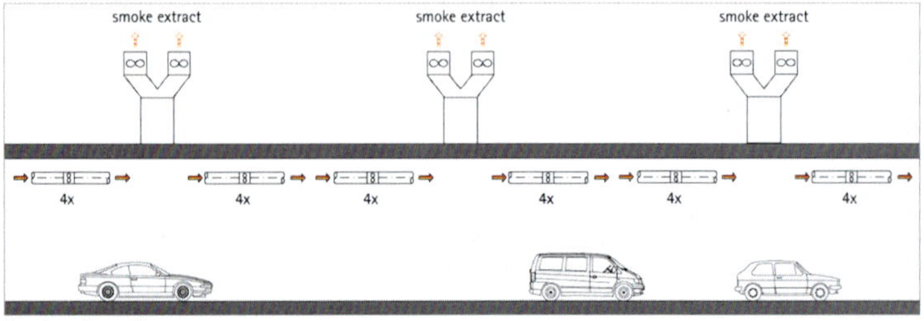
smoke extract
smoke extract
smoke extract
4x
4x
4x
4x
4x
4x

Normal operation condition

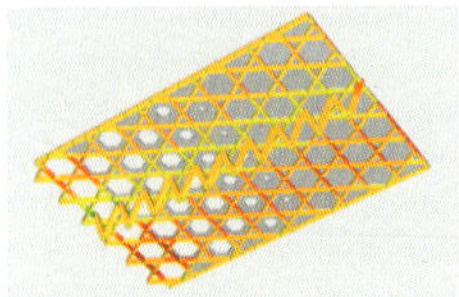

Bending moments in beams

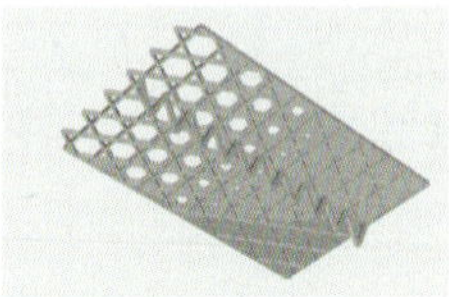

Descent from open roof to closed roof structure

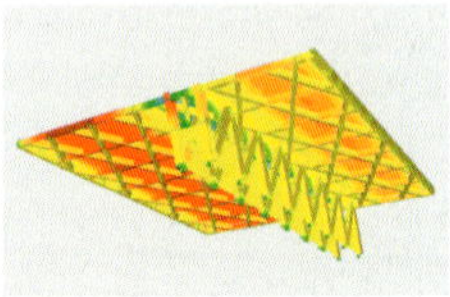

Bending moments in inner tunne

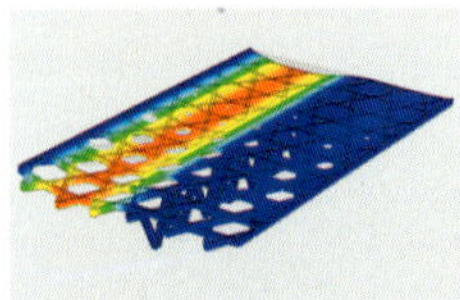

Deflection due to vertical loading

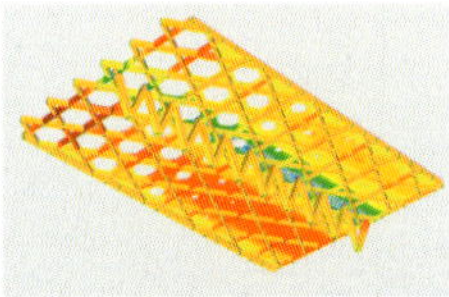

Bending moments in slab and beam

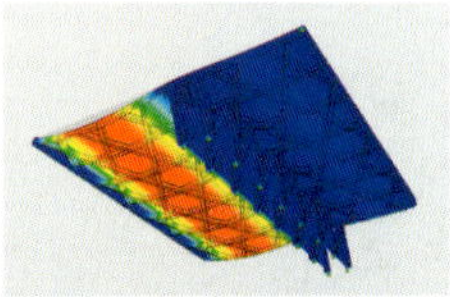

Deflection due to vertical loading in inner tunnel

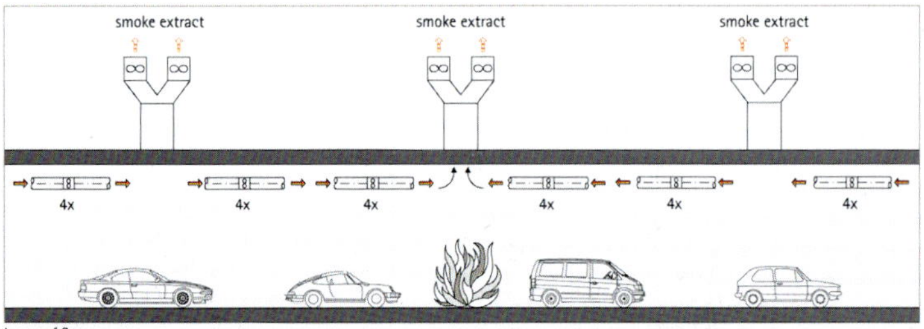
smoke extract
smoke extract
smoke extract
4x
4x
4x
4x
4x
4x

In case of fire

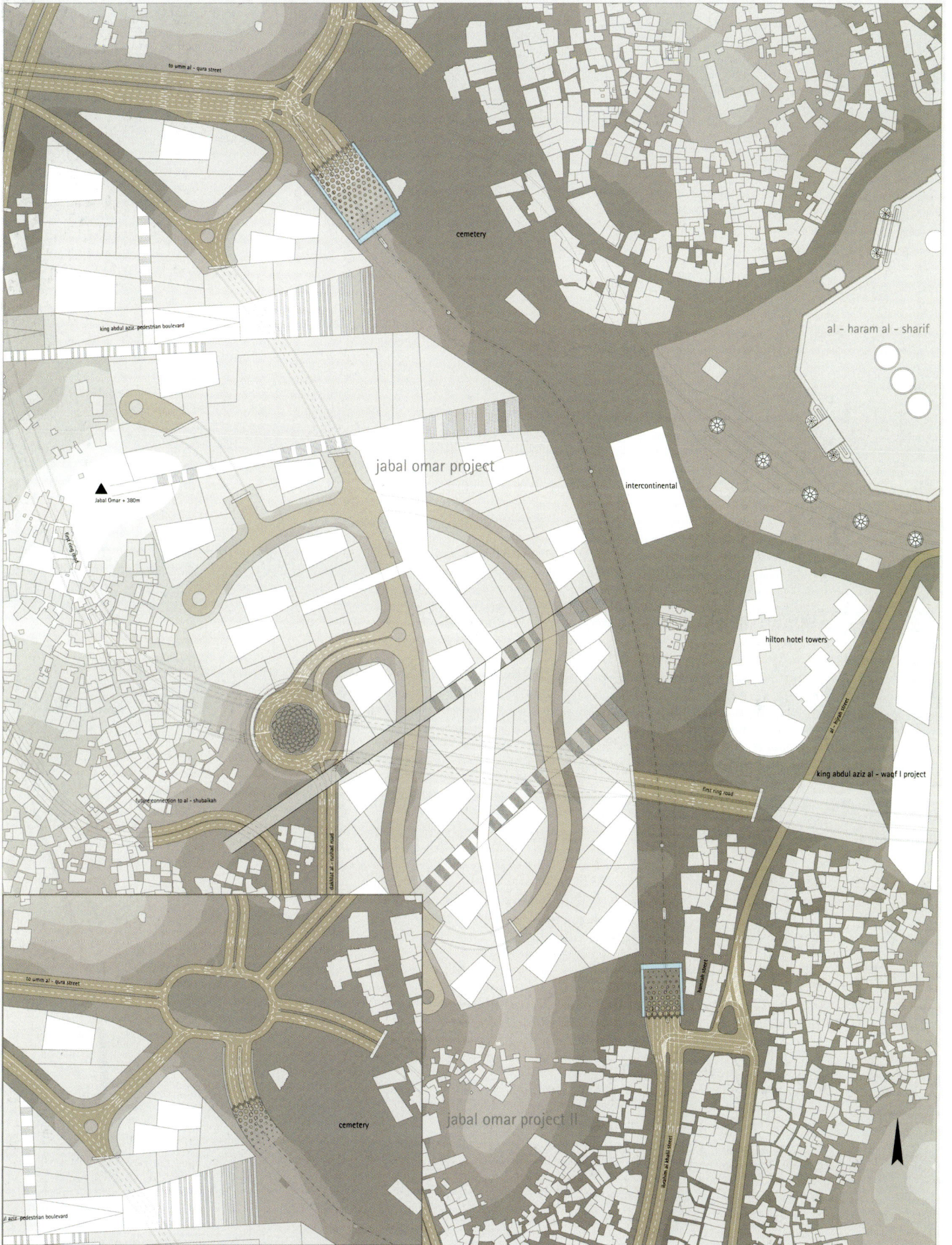

cemetery

to umm al - qura street

king abdul aziz - pedestrian boulevard

Jabal Omar + 380m

first ring boulevard

jabal omar project

al - haram al - sharif

intercontinental

hilton hotel towers

al - hijrat street

future connection to al - shubaikah

shakkat al - rushat road

first ring road

king abdul aziz al - waqf I project

dahhas al khali street

hamzah street

to umm al - qura street

cemetery

jabal omar project II

al aziz - pedestrian boulevard

Buchner Bründler

lee + mundwiler

Swiss Pavilion at the World EXPO 2010 in Shanghai

2010中国上海世博会的瑞士展馆

group8

Lehmann Fidanza

Swiss Pavilion at the World EXPO 2010 Shanghai

2010世博会的瑞士展馆　上海

Anonymous interdisciplinary two-stage project competition for architects, interior designers, engineers, exhibition designers, communication consultants, artists, and stage designers

为建筑师、室内设计师、工程师、展览设计师、沟通顾问、艺术家和布景设计人举办的项目竞赛，匿名、综合，分为两个阶段。

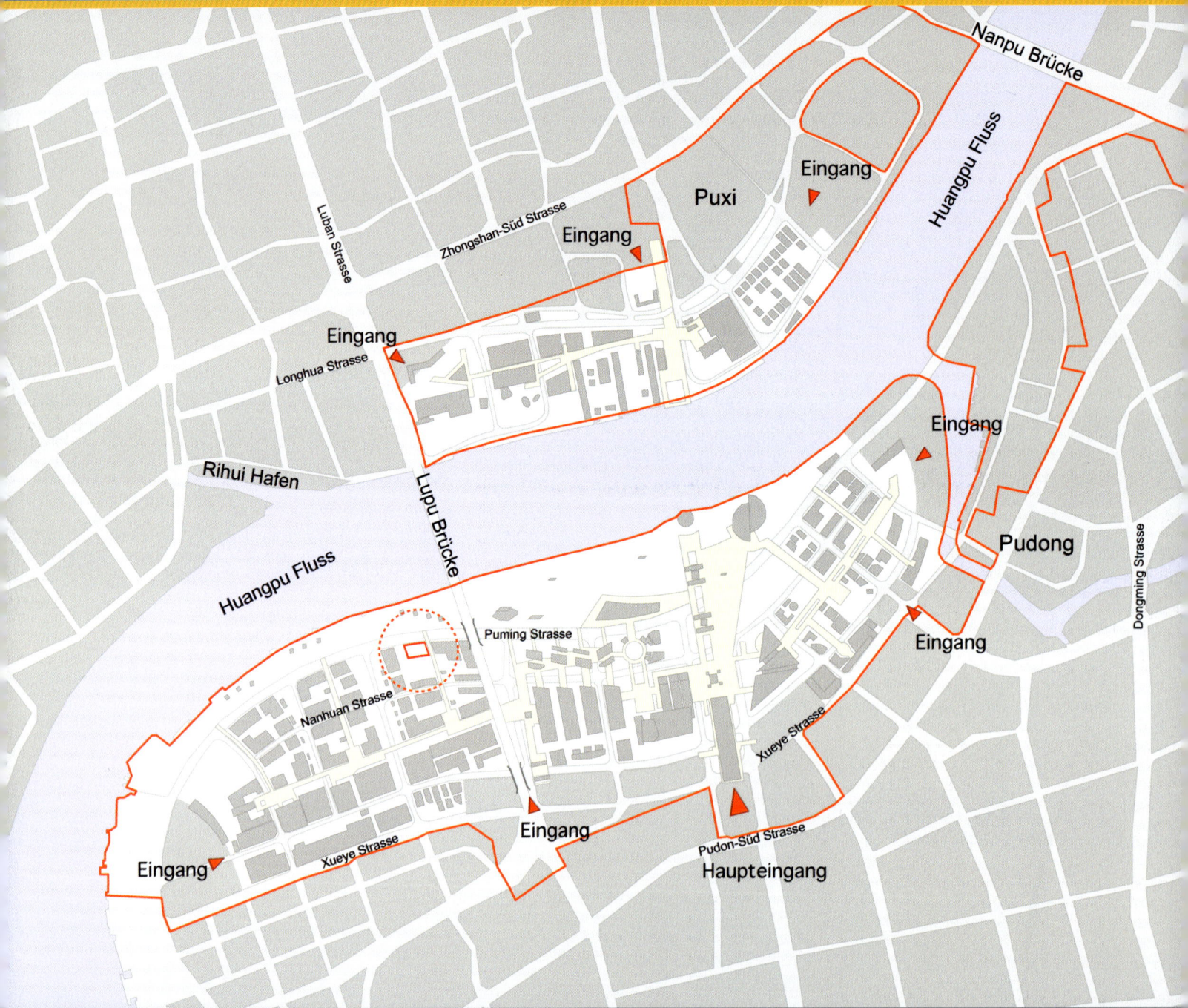

Nanpu Brücke

Puxi

Eingang

Huangpu Fluss

Luban Strasse

Zhongshan-Süd Strasse

Eingang

Eingang

Longhua Strasse

Eingang

Rihui Hafen

Lupu Brücke

Eingang

Pudong

Huangpu Fluss

Puming Strasse

Dongming Strasse

Eingang

Nanhuan Strasse

Xueye Strasse

Eingang

Xueye Strasse

Eingang

Pudon-Süd Strasse

Haupteingang

地点 Shanghai　时间 12/2006–05/2007　主办方 Swiss Confederation represented through presence Switzerland
人数 1st stage: 104 competitors, 2nd stage: 12 competitors　面积 approx. 4,000 sq m　竞赛费用 CHF 200,000
专业评奖委员会 Prof. Marc Angélil, Zurich/Los Angeles; Prof. Dr. Yongjie Cai, Shanghai; Dr. Stefan Nowak, NET Nowak Energy + Technology, St. Ursen; Andreas Reuter, Basel; Dr. Uli Sigg, president of the jury, former Swiss ambassador to Peking, Mauensee; Juri Steiner, director Paul Klee Zentrum, Bern　其他评奖委员会 Eva Brechtbühl, Switzerland Tourismus, Zurich; Ruth Grossenbacher, president presence Switzerlar Bern; general-commissioner for Switzerland, EXPO 2000 Hannover; Susan Horváth, director Swiss-Chinese Chamber of Commerce, Zurich; Pius Knüsel, director Pro Helvetia, Bern; Dr. Thomas Wagner, president Swiss-Chinese Association, Zurich

EXPO 2010, the world fair, will be held in Shanghai from May to October that year. Its theme is "A Better City, A Better Life". The EXPO 2010 fair grounds will be on both banks of the Huangpu River, to the south of the inner city of Shanghai. At present, docks and industry are the defining features of the area. The Swiss Pavilion will be at a 4,000 square metres site on the river's south bank in "Zone C" where there will also be pavilions from other European countries, from America, and Africa. The competition task was to come up with an overall concept that would integrate the pavilion itself, seen as building or object, with the exhibition on display.

The Pavilion, no more than 4,000 square metres, sould provide space for receptions, events and include places to pass the time, a restaurant, retail and office areas, as well as the exhibition area. The competition was held as an open two-stage procedure, open to architects, exposition designers, interior architects, communication experts, visual artists and stage designers. At least one member of the design team had to be a Swiss citizen or resident. Only teams consisting of at least one architect, a mechanical and a structural engineer, plus a participant qualified in one of the five disciplines mentioned above, were admitted to the second phase. Like other invitees in such situations, Switzerland faced the challenge of presenting its own cultural identity and societal perspectives to a wider, mainly Asian public. On the other hand, it was important that Switzerland, Swiss citizens and Swiss companies be able to relate to the exhibition within the context of their own culture.

2010年世博会这一世界盛会将于2010年5月至10月在上海举行，主题是"城市，让生活更美好"。场地将坐落在上海内城南面的黄浦江岸上，该区的定义特征是码头和工业。瑞士展馆占地4000平方米，位于黄浦江南岸的C区，该区还有欧洲其他国家、美洲和亚洲国家的展馆。竞赛任务提出要将展馆自身建筑物与展览和谐统一起来。

这个仅4000平方米大的瑞士展馆将提供接待处、活动区、休闲区、餐厅和办公区域及展区。竞赛分为两个阶段，对建筑师、展会设计师、室内建筑师、沟通专家、视觉艺术家和布景设计师完全开放。设计团队中至少要有一名成员是瑞士人。只有包括至少一名建筑师、技术工程师和土建工程师加上在上述五种专业领域中一方面合格的参赛者的设计团队才被允许进入第二阶段。和其他受邀参展单位一样，瑞士也面临着向更多的人、主要是亚洲人来展示他们自己文化特点和社会视角的挑战。另一方面，瑞士、瑞士人和瑞士建筑事务所能在自己文化中找到与展会相关联的内容也是非常重要的。

Aerial view　鸟瞰图

Competition site (rendering)　竞赛地点

View of the city　城市景色

View from "Three on the Bund", Shanghai　"外滩三号"景色

swiss pavillon expo 2010 shanghai — 01010 — **1**

080324 — **2**

3

4

aussenraum — **5**

103414 — **8**

Schweizer Pavillon an der EXPO 2010 in Schanghai — 201001 / 1 — **9**

10

ÉCHANGE — 58512 — Best of Swiss — Neue Switzerland — **11**

CITYSCAPE — 44798 — **12**

13

Blatt 1 — 734198 — **14**

Schweizer Pavillon in Shanghai. Sinneswandeln — 2542 — **15**

006906 — **16**

PAVILLON SUISSE a Shanghai 2010 — **17**

Strasse der wahr gewordenen Wünsche — Shop im EG — **20**

Plan 2 — VIP Top Lounge — Huang-Pu View — 188 123 — Wassergarten — Night-View — **21**

081548 — **22**

23

Pavillon EXPO 2010 Shanghai — **24**

27

117107117 — **28**

264653 — **29**

121372 — **30**

IN THE MIDDLE OF THE CENTER — EXPO Shanghai 2010 — 1 — **31**

Schweizer Pavillon an der EXPO 2010 in Shanghai — Blatt 1 — 642822 — **34**

01 07 01 — **35**

36

360360 — **37**

306035 — **38**

Shanghai 2010 Kristallsystem — 140723 — **41**

20027 — **42**

211272 — **43**

386108 — **44**

45

SWISS READY MADE 1 — 628889 — **48**

2010 瑞士世展馆 suisse expo — 030203 — **49**

50

647464 — **51**

698869 — **52**

6

7

18

19

25

26

32

33

39

40

46

47

53

54

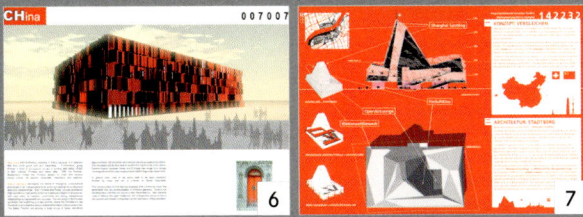

Qualified participants 1st stage 第一阶段合格参赛者

1 re-urbanism, Philippe Müller, Basel, Switzerland **2** group8, Daniel Zamarbide, Geneva, Switzerland **3** Agirbas/Wienstroer, Architektur & Stadtplanung, Neuss, Germany **4** Hauswirth Keller Branzanti, Zurich **5** Daniel Crone, Lucerne, Switzerland **6** lee + mundwiler architects, Santa Monica, CA/USA **7** oos ag, Christoph Kellenberger, Zurich **8** ARGE Lussi+Halter/Velvet Creative Office, Thomas Lussi, Lucerne, Switzerland **9** Buchner Bründler, Basel **10** Christian Bandi, Zurich **11** Lehmann Fidanza, Zurich/Fribourg **12** mischa badertscher architekten, Zurich

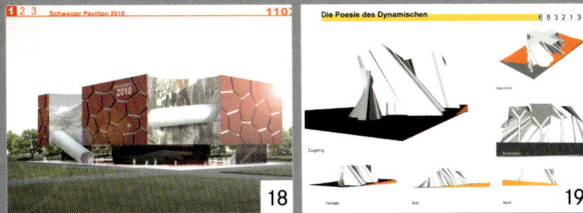

Not qualified participants 1st stage 第一阶段不合格参赛者

13 Franz Wimmer, Munich **14** Prof. Jürg Steiner, Wuppertal, Germany **15** büro blickpunkt, Berlin **16** Stauffenegger + Stutz, Basel, Switzerland **17** Jafar Tukan, Amman, Jordan **18** Jacopo Venerosi Pesciolini, Florence **19** tectur- planung & projektsteuerung, Berlin **20** Martin Rohr, Bellprat Xavier, Siegfried Meier, Michael Gruber, Winterthur, Switzerland **21** Marco Canonica, Liestal, Switzerland; Beat Pfenniger, Lucerne, Switzerland **22** ARCADD, Hisham N. Ashkouri, Massachusetts; CCHE Architecture, Erich Hauenstein, Lausanne, France **23** Wesam Nassar, Vienna; Eric Martinet, Collonge-Bellerive, Switzerland **24** Heinle Wischer und Partner, Till Behnke, Berlin; Mettler Landschaftsarchitektur, Gossau, Switzerland **25** Totems, Gerard de Gorter, Theresia Leuenberger, Simone Königshausen, Hoofddorp, Netherlands **26** Riccardo Pedrazzoli, Bologna, Ares Bolognesi, Concorezzo, Italy **27** Raum B Architektur und Gestaltungskonzepte, Daniela Saxer, Zurich **28** eob, Eduard Baumann, Zurich **29** Markus Ducommun, Solothurn, René Zäch, Biel; Heinz Isler, Burgdorf, Switzerland **30** Bernhard Tatter, Leipzig, Germany; Pascal Storz, Basel, Switzerland **31** MAFEU, Marc Feustel, Berlin; Simon Beer, Zurich **32** avcommunication, Günter Rein, Norbert W. Daldrop, Ludwigsburg, Germany; Hans Peter Weiss, Maussane, France **33** dai AG, Florin Baeriswyl, Zurich **34** Frank Boehm, Milan; Christiane Rekade, Berlin; Marco Bay, Milan **35** Prof. Johann Eisele, Bettina Staniek, Claus Staniek, Prof. Markus Gasser, Darmstadt/Zurich **36** Carmine Abate, Bassano del Grappa, Italy **37** Benjamin Bastianello, Zurich **38** Fürst Architects, Düsseldorf, Germany **39** Ausland Ventures, Emmanuel Ventura, Sophie Bouvier Ausländer, Lausanne, Switzerland **40** rothenhöfer_schlumberger, Stuttgart, Germany **41** Architekturbüro Rusch, Lucerne, Switzerland **42** atelierbrückner, Stuttgart, Germany; integral ruedi baur, Zurich; Axel Steinberger; Zurich **43** nijo, Nina Lippuner, Johannes Wick, Zurich **44** Atelier 10:8, Georg Rinderknecht, Jürg Senn, Perimeter-Stadt; Prof. Dr. Angelus Eisinger, Meylan Lang, Stephan Meylan, Zurich **45** EM2N, Mathias Müller, Daniel Niggli, Schmid Staffelbach, Nüssli International, Rutz Architekten, Zurich **46** Vomsattel Wagner Architekten, Visp, Switzerland **47** Markus Alder Architekten, St. Gallen, Switzerland **48** team stratenwerth, Basel; Andreas Fuhrimann, Gabrielle Hächler, Architekten, Zurich **49** Herren + Damschen Architekten + Planer, Bern **50** Alex Herter, Meilen, Switzerland **51** ab-architekten, Adrian Bühler, Thun, Switzerland **52** Boris Graf, Bern, Thomas Schmid, Köniz, Switzerland; palma 3, Dr. Andreas Schwab, Bern **53** schindlerarchitekten, Natalie Jürgens, Claudia Furrer, Stuttgart, Germany **54** GKS Architekten und Partner, Lucerne, Switzerland

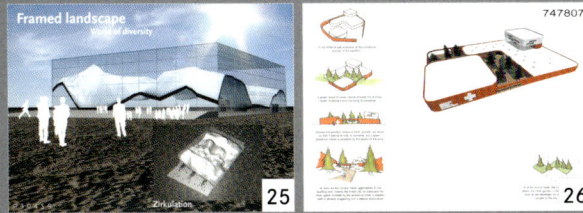

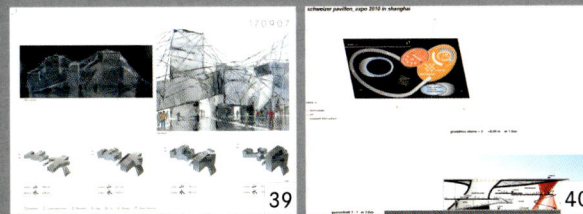

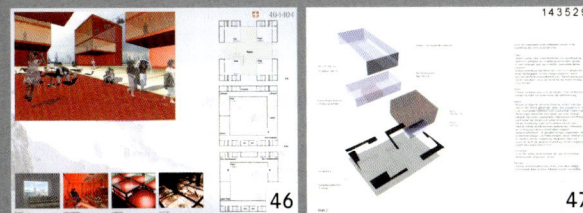

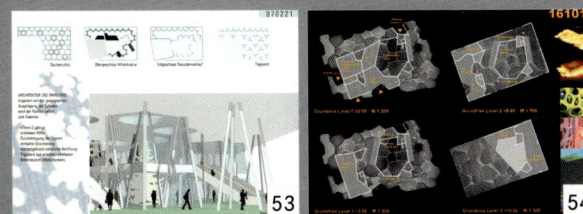

55 56 57 58 59

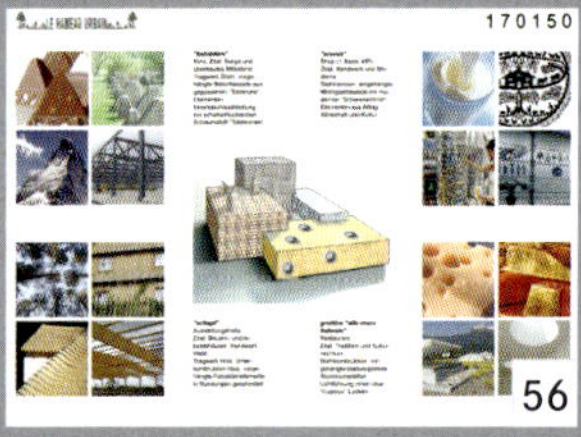

62 63 64 65 66

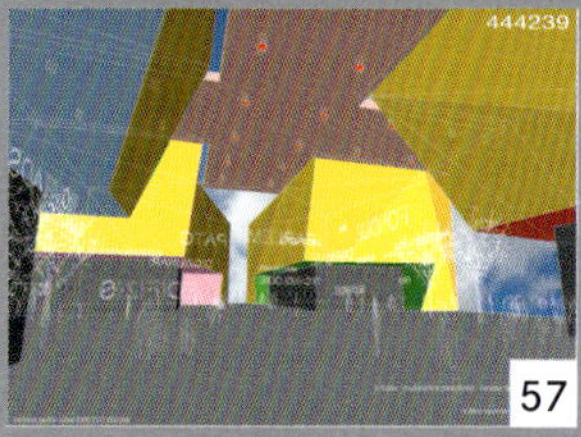

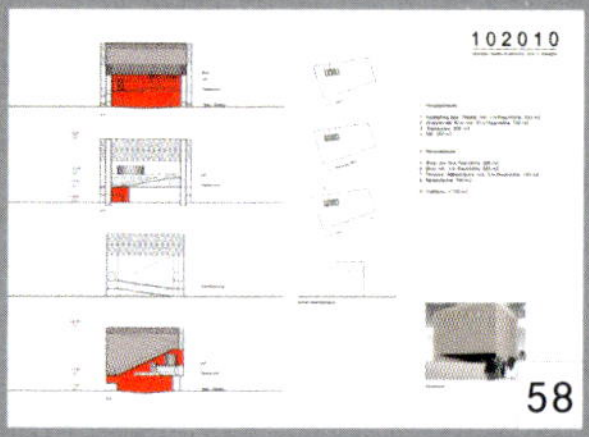

69 70 71 72 73

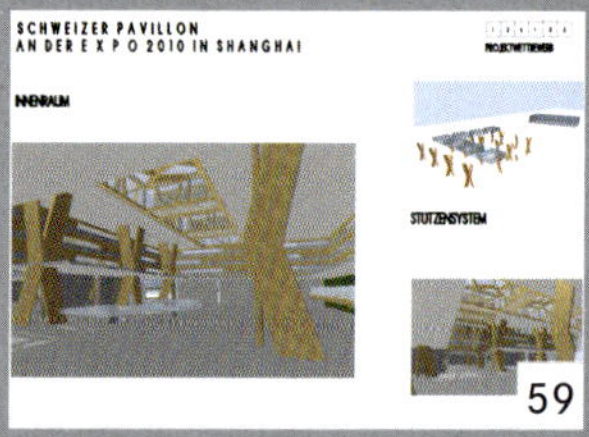

75 76 77 78 79

81 82 83 84 85

87 88 89 90 91

93 94 95 96 97

99 100 101 102 103

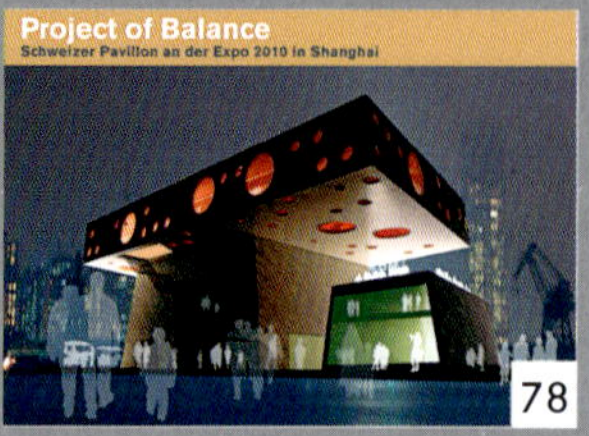

Project of Balance
Schweizer Pavillon an der Expo 2010 in Shanghai

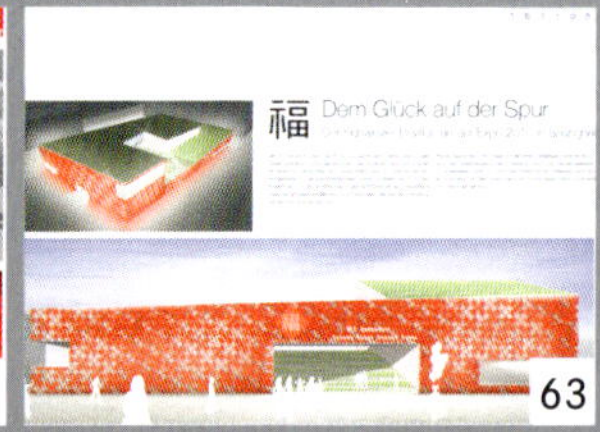

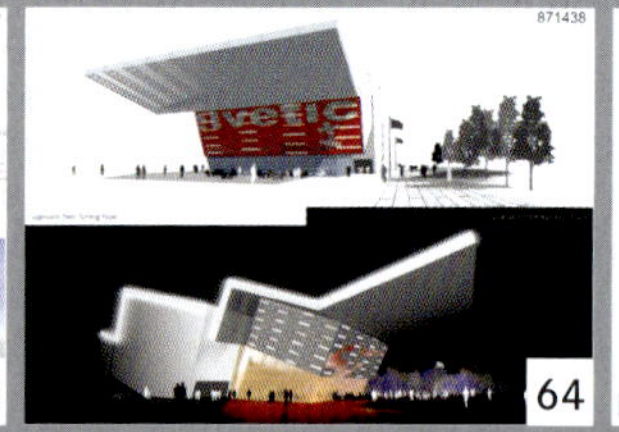

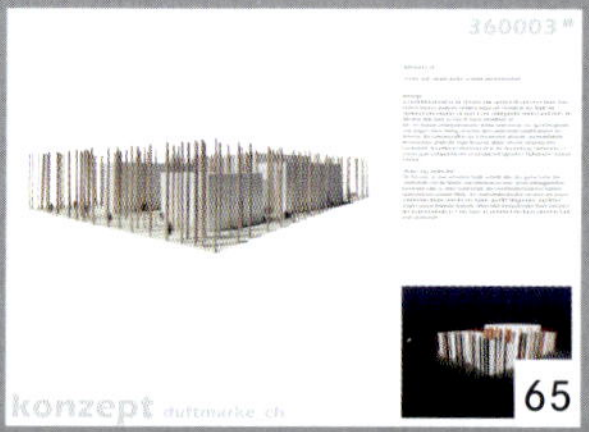

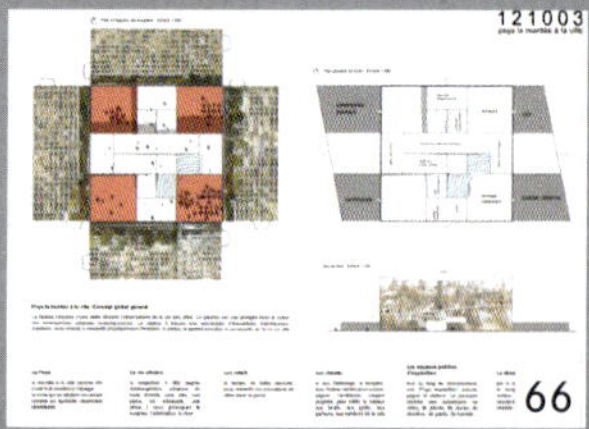

SCHWEIZER PAVILLON AN DER EXPO 2010 IN SHANGHAI

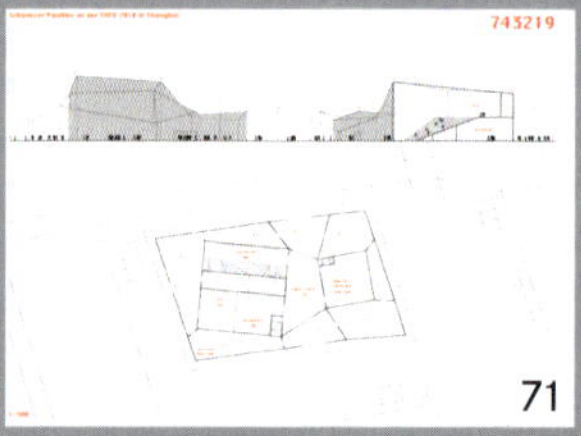

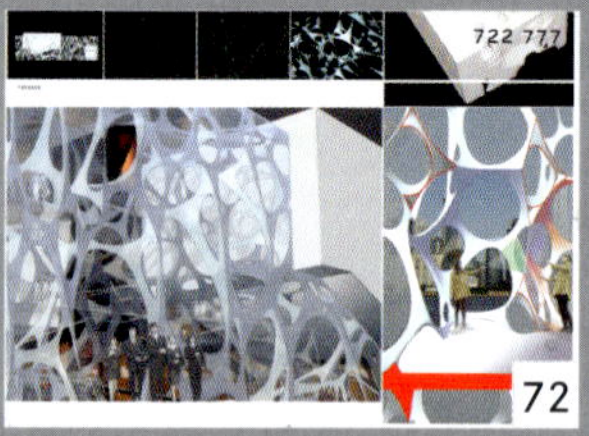

Expo Shanghai 2010 "swiss made" Projekt 200710

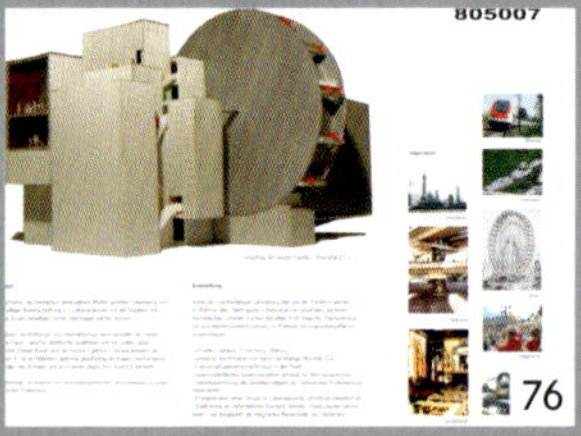

Dem Glück auf der Spur

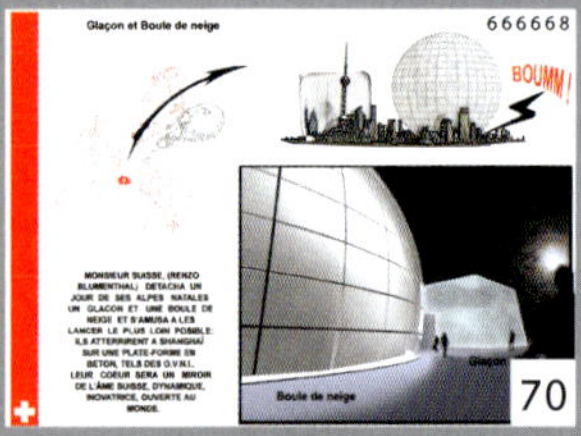

Glaçon et Boule de neige

supernatur

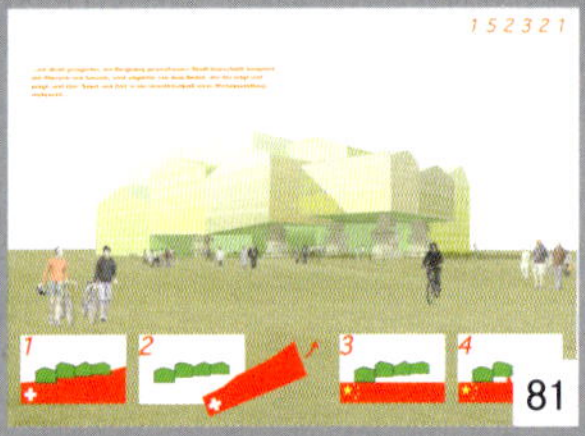

Der Weg durchs Glück

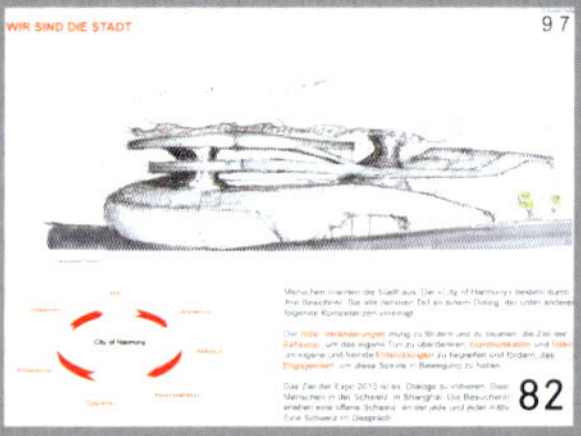

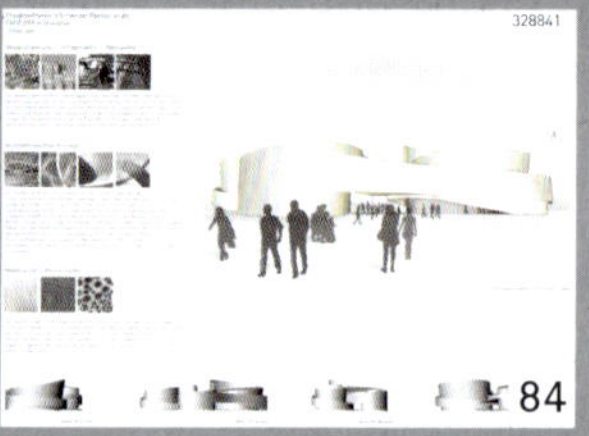

Switzerland 瑞士

WIR SIND DIE STADT

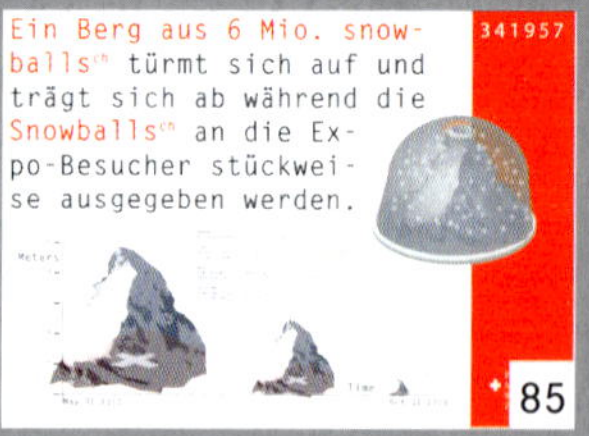

Ein Berg aus 6 Mio. snowballs türmt sich auf und trägt sich ab während die Snowballs an die Expo-Besucher stückweise ausgegeben werden.

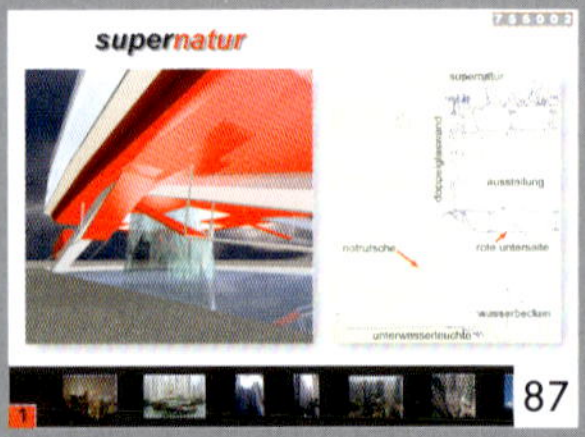

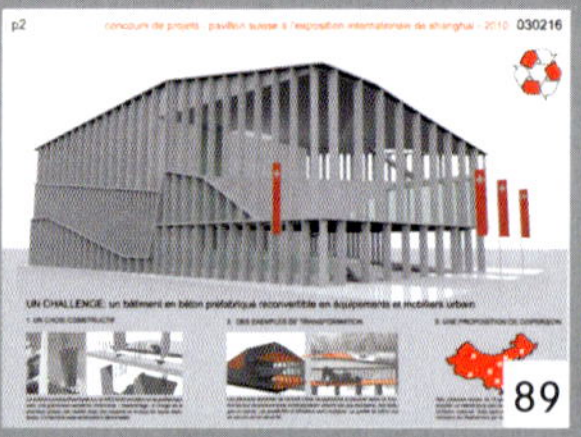

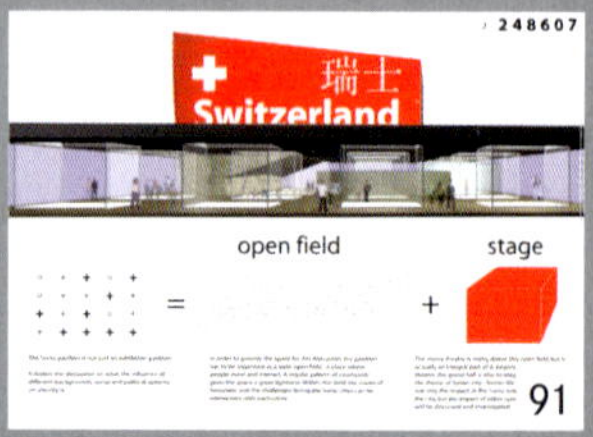

Schweizer Pavillon an der EXPO 2010 in Shanghai

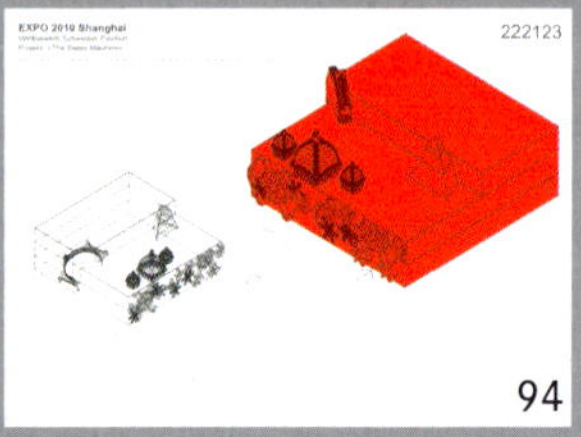

EXPO 2010 Shanghai

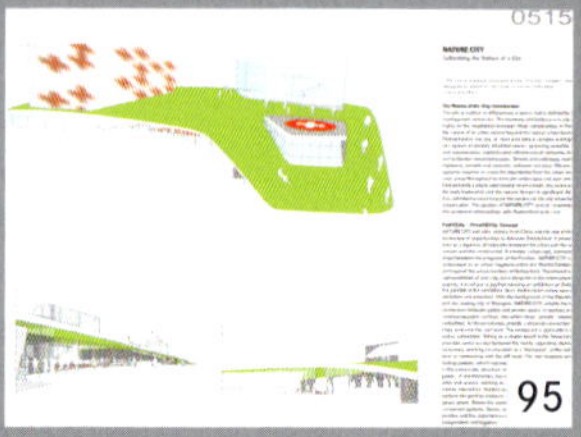

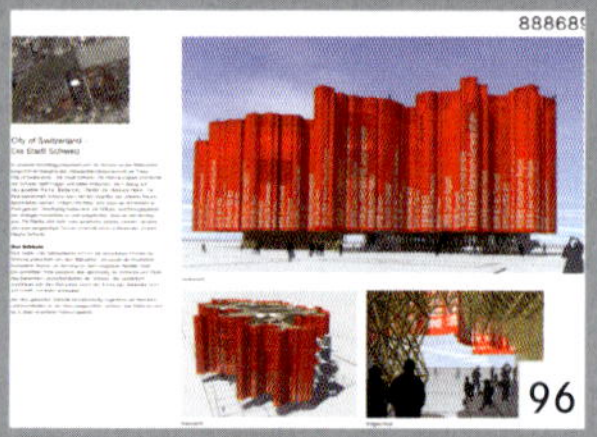

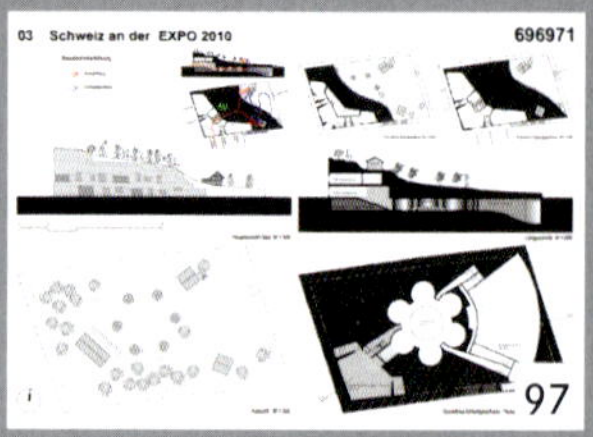

Schweiz an der EXPO 2010

«swiss livingroom»

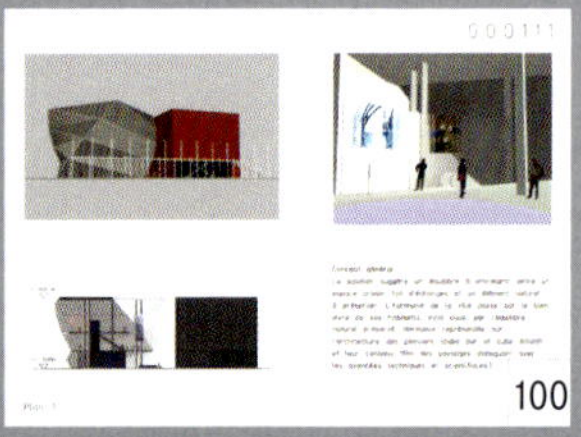

open field + stage

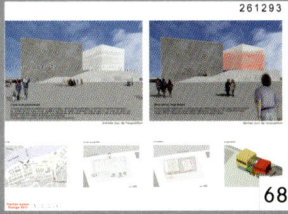

55 Gautschi Storrer Architekten, Zurich 56 IAAG Architekten, Andreas Stebler, Tröhler und Partner, Bern; Esthermaria Jungo, Fribourg, Switzerland 57 A Dettling JM Pélèraux Architectes, Lausanne, Switzerland 58 Mario Lins, Sennwald, Switzerland 59 Michael Kober, Horn, Switzerland; Kaufmann Holzbau AG 60 Walter Hunziker Architekten AG, Bern 61 Budik.Liechti, automatic/Anouk Kuitenbrouwer, Zurich; GVA Studio, Alban Thomas, Hervé Rigal, Geneva 62 Konzeptgruppe Paraform, Architekturbüro Buser, Basel; Boris Brüderlin, Lausanne, Switzerland; Reto Keller, Basel 63 Bluetrac Eventtechnik AG, Andreas Schmucki, Wetzikon, Switzerland; Bednar Albisetti Architekten, Peter Wehrli, Bednar Albisetti Architekten, Ina Walden, Dominic Schmid, Winterthur, Switzerland 64 n-body Architekten AG, Delphine Ammann, Ramón Gómez Larios, Zurich 65 Markus von Grünigen, Marcel Hari, Thun, Switzerland; Bruno Marti, Uetendorf, Switzerland 66 Paysagestion, Laurent Salin, Olivier Lasserre, Lausanne, Switzerland 67 Keller Landolt Partner AG, Zurich; Heinrich Ehrensperger, Hausen, Switzerland 68 Atelier d'architecture Nicoucar & Steininger, Acacias/ Geneva, Jean-Claude Deschamps, Lausanne, Switzerland 69 Devanthéry & Lamunière Architectes, Carouge, Switzerland; Gérant Confino SARL, Francois Confino, Lussan, France 70 CBC.Swiss China Business Center, Charles Merkle, Shanghai/Bern 71 Fawad Kazi, Zurich 72 Team sacs/ Bergit Hillner/Philipp Husistein, Aarau, Zurich; Artificial Intelligence Laboratory University of Zurich, Jonas Ruesch, Weiersmüller, Bosshard Grüninger WBG AG; Markus Bosshard, Zurich 73 ajk architekten, Andreas J. Kellert, Frankfurt, Germany 74 Sonja Berthold, Stephan Brunner, Florian Eyerer, Zurich 75 Prof. Miroslav Sik, Zurich 76 Fehr & Scherrer Architekten, Zurich; Fabienne Hoelzel, Basel 77 Mario Geisser + Andreas Lüdi, Markus Heiniger, Silvia Luckner, Zurich 78 LOST Architekten, dz - productions, Daniel Zimmermann, Basel 79 Raymond Theler, Brig-Glis, Switzerland 80 Tran Hin Van, Designer, Claudia Meier, Giulio Kofi, Armin Wagner, Zurich 81 ramser schmid architekten, Philipp Esch, Zurich 82 firstplayers Architekten, Frank Baeumner, mjm.cc, Martin Matt, Claudio Miozzari, Renato Soldenhoff, Basel 83 huggenberger_GmbH & Erika Fries Arch GmbH, Zurich 84 Holzer Kobler Architekturen, Zurich 85 pha architektur, Arndt Hermann, Ribbeck, Germany; Zetra International, Dr. Ralf Hermann, Zurich 86 Ivica Brnic, Zurich 87 ajö tristess AB, Ingrid Reppen, Evata Finland Oy, Kai Wartiainen, Stockholm 88 Jacques Rordorf, Zurich 89 Steeve Ray & Associés, Geneva 90 phalt *the spatialists, Mike Mattiello, Designrichtung, Christof Hindermann, Zurich 91 Güller Güller architecture urbanism, Zurich 92 Boltshauser Architekten, Zurich 93 Urs Esposito, Zurich 94 Patrick Urben, Rosa Cassano, Basel 95 Philipp von Dalwig, Kit Cheung, New York; Reto Geiser, Basel 96 o-liv brunner volk architekten, Zurich 97 Andres Sabbadini Architekten, Zurich 98 Froelich & Hsu Architekten, Zurich 99 Fiechter & Salzmann, Zurich 100 Henchoz Atelier D'Architectes, Nyon 101 CLR Chevalley Longchamp Russbach Architectes, Geneva; eggs Architects, Tokyo; denninger scholz architekten, Cologne, Germany 102 Dolenc Scheiwiller Architekten AG, leman architects, Zurich; Vincent Zhang, Shanghai; Daniel Pauli, Villnachern, Switzerland 103 Raoul Sigl, Maria Conen, Armon Semadeni, Zurich 104 Park Architekten AG, Peter Althaus, Markus Lüscher, Zurich; AS-IF Architekten, Stephanie Kaindl, Paul Grundei, Berlin

1

2

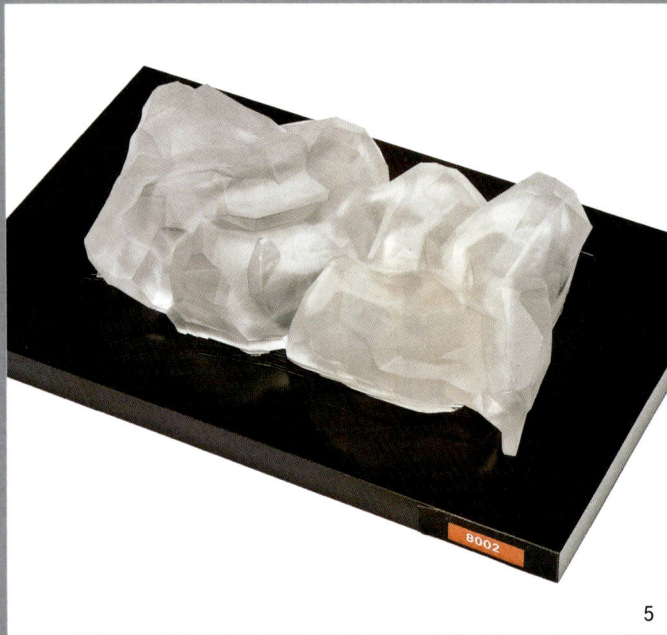

5

6

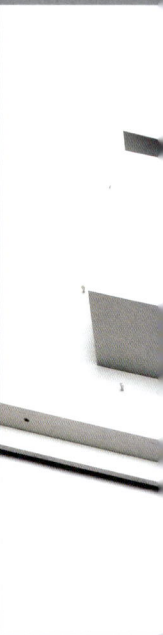

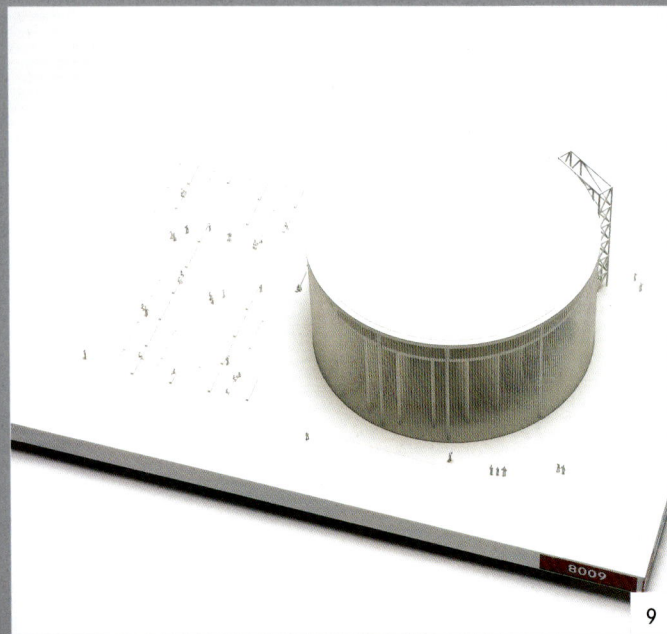

9

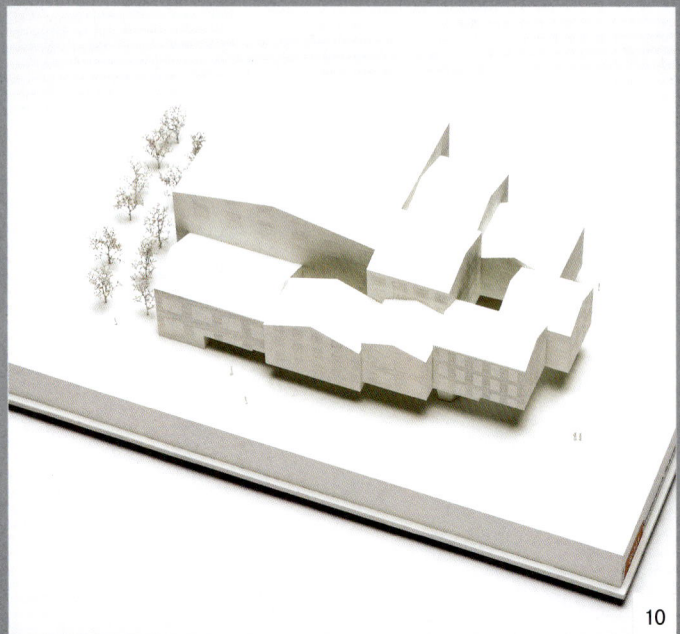

10

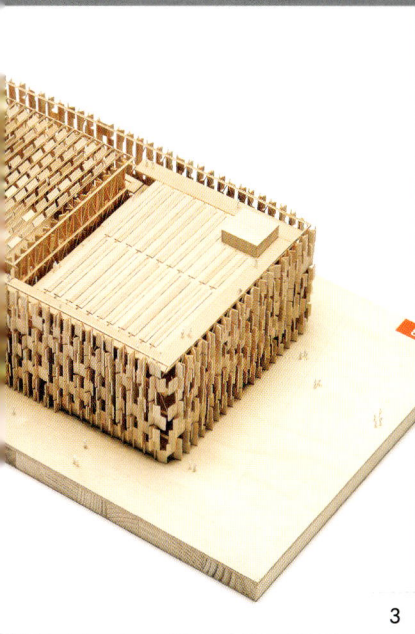

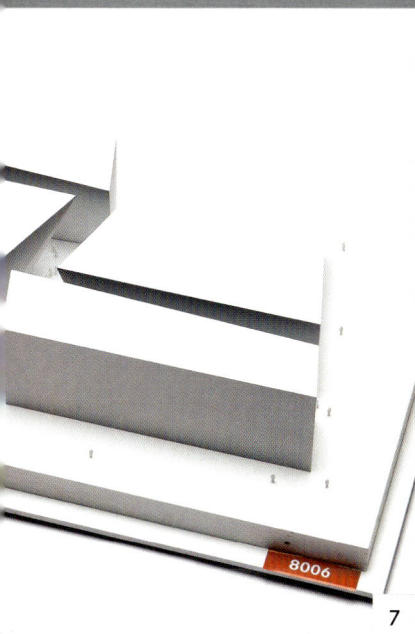

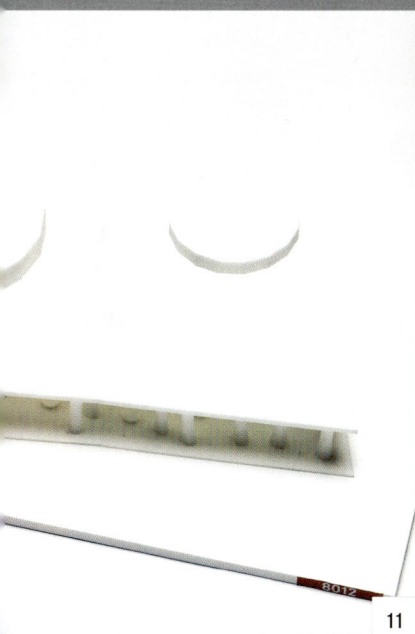

Buchner Bründler AG Architekten BSA Basel
Buchner Brundler AG建筑事务所 巴塞尔

"Climate change – Sustainability – Knowledge transfer"
"气候变化——可持续性——知识转移。"

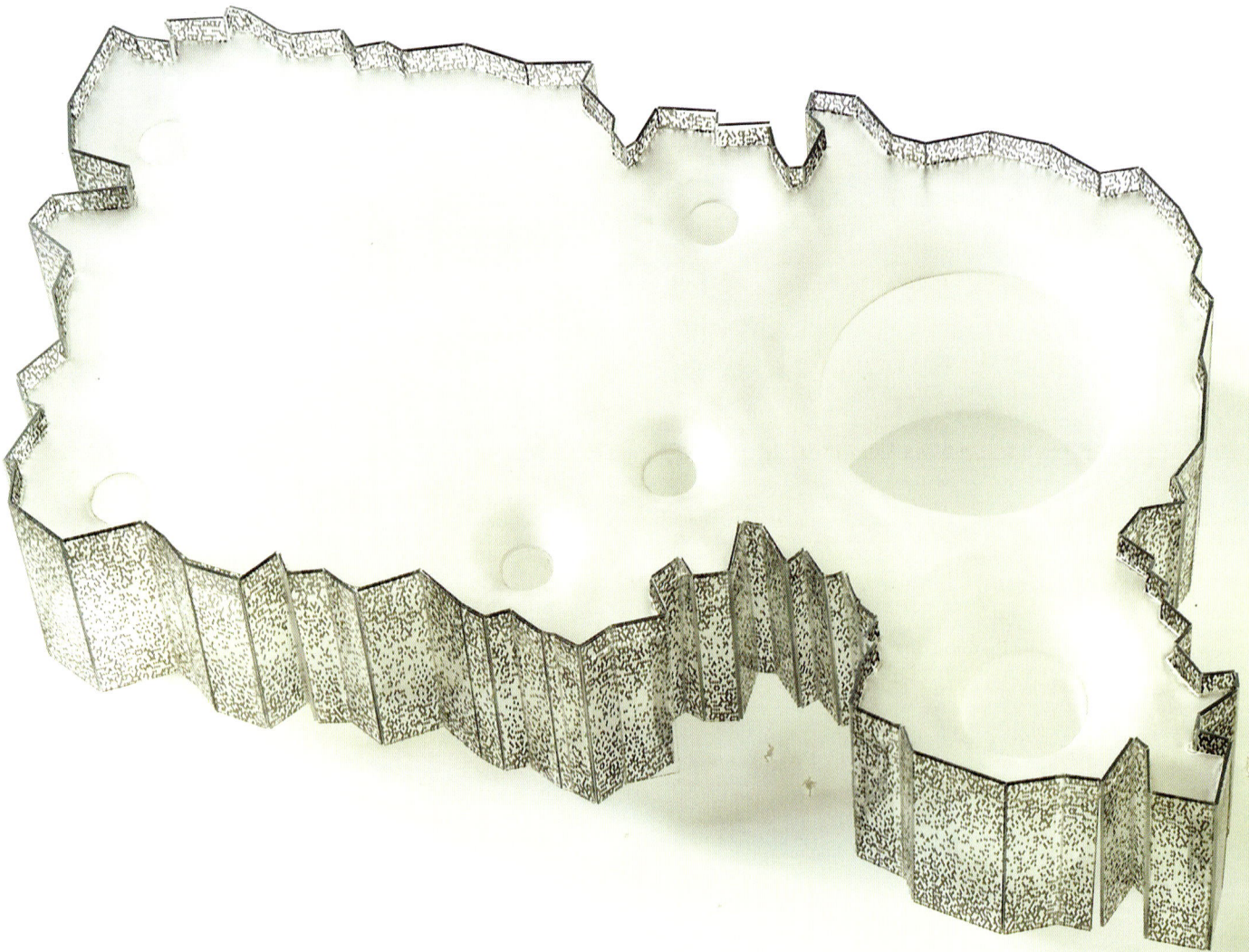

作者 Andreas Bründler   合作伙伴 Buchner Bründler AG Architekten: Nick Waldmeier, Bülend Yigin, Hellade Miozzari, Ewa Misiewicz, Sandra Gonon, Caesar Zumthor, Nicole Johann   专家 interior design: Element GmbH, Basel; Andreas Hunkeler; building services: Waldhauser Haustechnik AG, Basel; Werner und Marco Waldhauser; structural engineers: Huerzeler Holzbau, Magden; Roland Huerzeler, Brigitte Münch, Nicolas Zeuggin, Thomas Brosemer; sinology: Barbara Jenni, Zurich

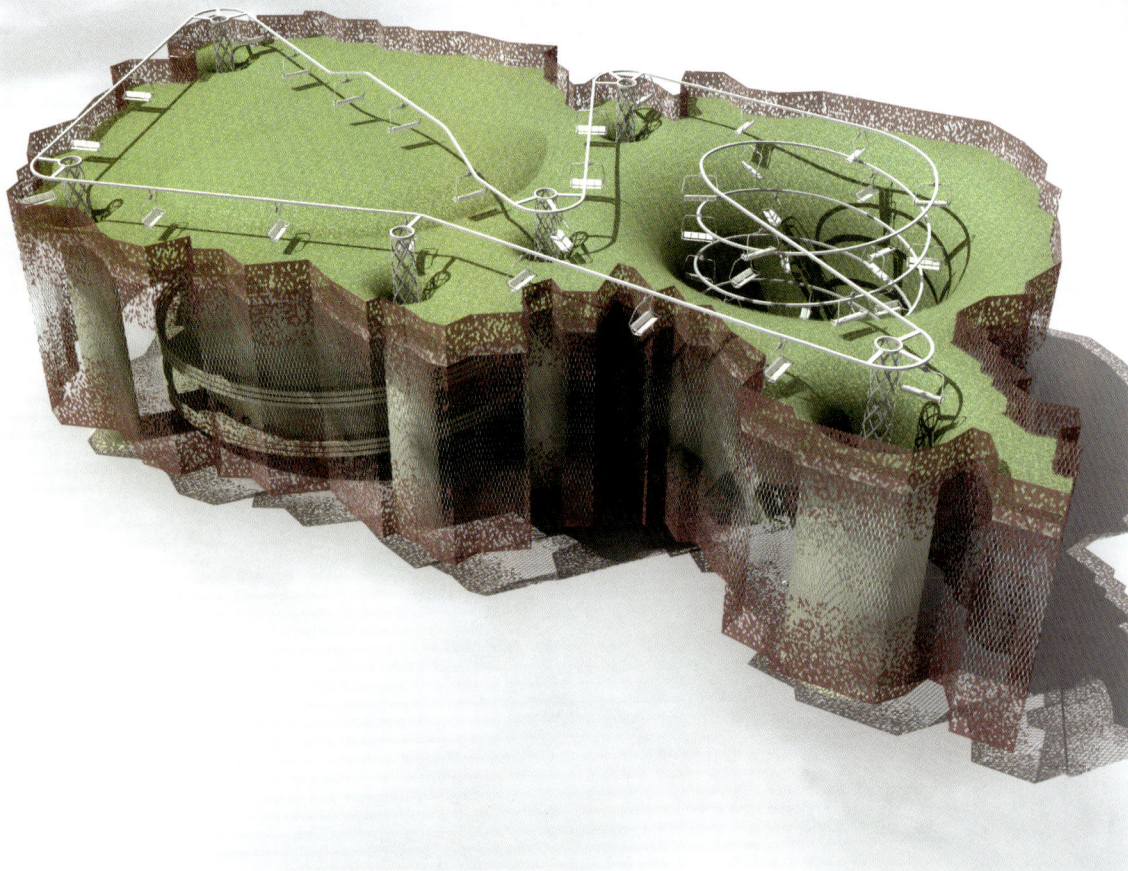

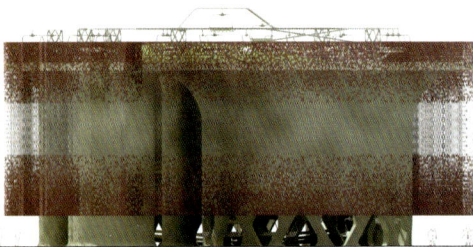

ostfassade

südfassade

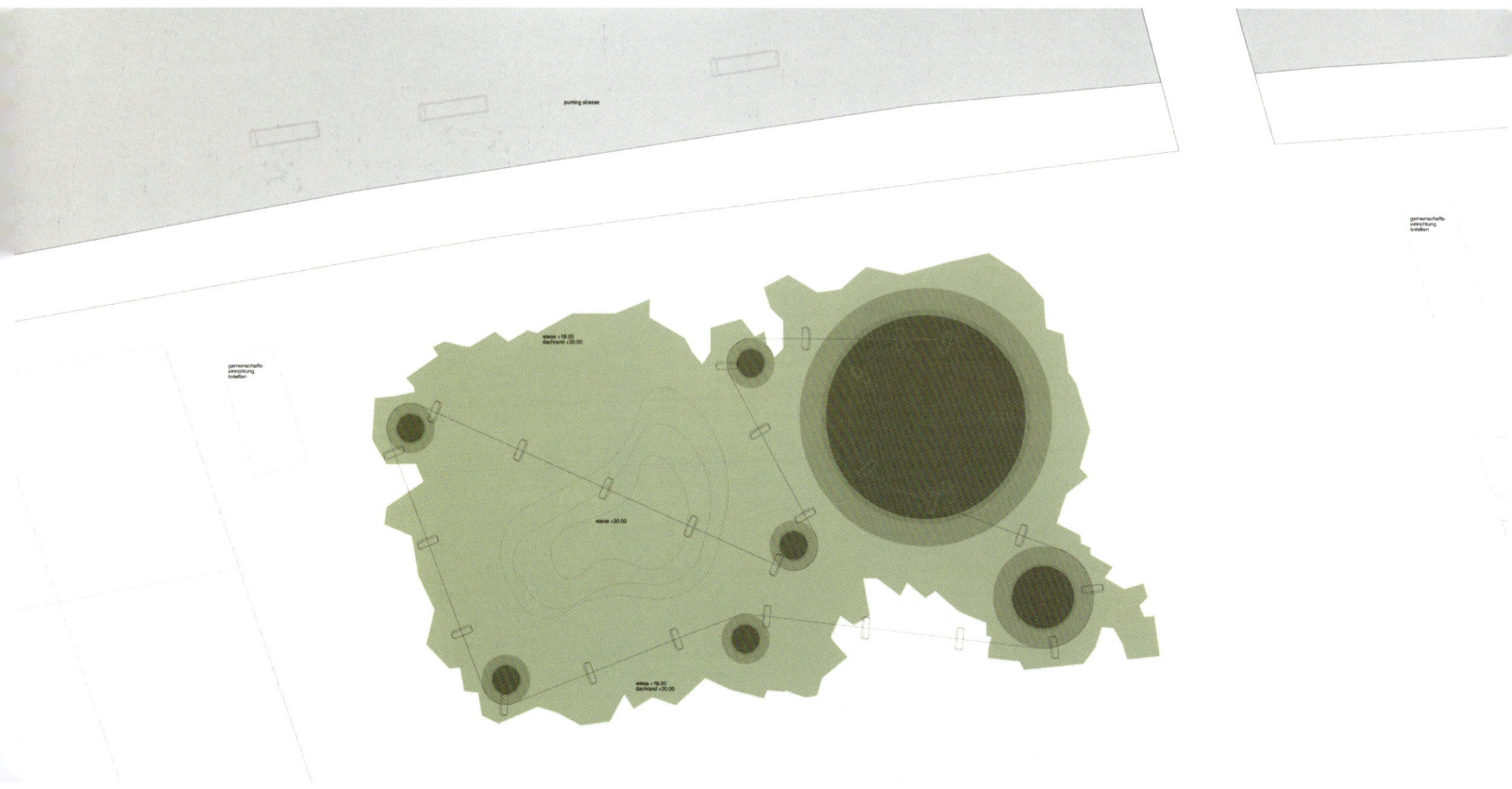

querschnitt

längsschnitt

zugangsniveau +0.00

zwischenniveau 1 +4.00

NACHHALTIGE GEBÄUDETECHNIK

Das während der Ausstellung sehr warme und – vor allem – sehr feuchte Aussenklima hat für die Raumklimatisierung nach europäischem Standard einen enormen Energieverbrauch zur Folge. Deshalb ist für die publikumsintensiven Bereiche (Rampe und Kino) vorgesehen, mit dem Einsatz vorhandener Energiequellen (Flusswasser, ev. Grundwasser?) ein Raumklima zu generieren, welches gegenüber dem Aussenklima einen spürbar besseren Komfort bietet, jedoch nicht den europäischen Vorstellungen von 26.5°C/60%rF entspricht.

Die Wärmeabgabe des Menschen erfolgt im Idealfall (22°C/50%rF) beinahe gleichmässig über Strahlung, Konvektion und Verdunstung. Bei Aussenklimakonditionen von 32°C/70%rF kann bei ruhender Luft über Konvektion und Verdunstung nicht mehr viel Wärme abgegeben werden, sodass die Bedeutung des Strahlungsanteils zunimmt.

Bewachsener Turm TOWER OF WIND
Mit dem Fahrwind (1m/s) der Seilbahn erhöht sich der Konvektionsanteil der Wärmeabgabe, sodass trotz hoher Lufttemperatur und -feuchte eine leichte Komfortsteigerung während der 10 Minuten dauernden Fahrt spürbar sein wird. Zusätzlich könnte, um den Strahlungsanteil erhöhen zu können, die Rampe mittels dem "Fluss"wasserkreislauf gekühlt werden / Abkühlungen betreffend Aufwand / Erfolg in der nächsten Planungsphase. Der Energieaufwand für die Seilbahn ist übrigens halt so gross wie er für einen "klimatisierten" Rampenturm wäre.

Ohne grossen Energieeinsatz ist eine Komfortsteigerung gegenüber dem Aussenklima spürbar.

Kino WATER STONE
Der für die Luftqualität erforderliche Aussenluftwechsel wird während den "Pausen" durch Öffnungen in Bodennähe und im Dach (Entrauchungsöffnungen) infolge der Raumhöhe effizient natürlich (ev. mit Unterstützung von Abluftventilatoren => nächste Planungsphase be- resp. entlüftet. Die "Kontrollkühlung" erfolgt über

Strahlung,
die Umschliessungsflächen werden mittels "Fluss"wasser gekühlt => erhöhte Wärmeabgabe durch Strahlung.

Konvektion
durch die kühlere Oberfläche der Umschliessungswände resultiert ein "Fall-wind", welcher den konvektiven Anteil der Wärmeabgabe erhöht.

Verdunstung
die Oberflächentemperatur der Umschliessungsflächen liegt bei ca. 20°C, die Taupunkttemperatur der Aussenluft ca. 25°C => Kondensation des in der Raumluft enthaltenen Wasserdampfes => Wärmeentzug aus der Umschliessungsfläche und der Raumluft => Entfeuchtung und Abkühlung der Raumluft.

Leitung
die Sitzstufen werden mit Wasser umspült und gekühlt => Wärmeabgabe über Wärmeleitung.

Ohne grossen Energieeinsatz ist eine Komfortsteigerung gegenüber dem Aussenklima spürbar.

Restaurant, Küche, Büros etc.
Diese Bereiche werden nach noch zu definierenden Raumkonditionen "konventionell" be- und entlüftet resp. klimatisiert.

Essbare Hülle aus Biorosin
Dieser aus Fäden der fermentierten Sojabohne bestehende Harz ist ungiftig, extrem abreissbar und leicht zu verarbeiten. Bakterien zersetzen Biorosin in Wasser und Kohlendioxid in der natürlichen Umwelt. Die Rückstände können als Dünger verwendet werden. Zur Herstellung der Hülle und als Innenwand im Kino werden Platten gegossen und thermisch verformt.

Farbstoff Solarzelle
Grätzel-Zelle) dient der Umwandlung von Lichtenergie in elektrische Energie. Michael Grätzel (EPFL, Lausanne) hat sie Anfang der 1990er Jahre entdeckt und 1992 patentieren lassen.
Mit der sonnenlicht-Photosynthese wird in Pariskolonnaturfarben (Beerensaft) Licht absorbiert und in Elektronen umgewandelt, die dann dem Leitungsband eines hohen Flächenhalbleiterlimes beitreten und sich weiter durch einen externen Stromkreis bewegen und somit Licht in Grätzel Energie umwandeln.

Die Vorzüge der Grätzel-Zelle können in den prinzipiell niedrigen Herstellungskosten und in der geringen Umweltbelastung bei der Herstellung liegen. Die Zelle kann diffuses Licht im Vergleich zu den herkömmlichen Solarzellen gut nutzen (Schanghai). Die Entwicklung wird es in zwei Jahren ermöglichen DSC Zellen in kleinen Paneelen einzusetzen.

TOWER OF WIND

WATER STONE

group8 Geneva
group 8建筑事务所 日内瓦

"Swiss identity is rooted in the Swiss landscape and the country's ability to bring people together and conduct discussions on a global scale."

"在瑞士的景观里深深表现出瑞士的特点。"

作者 Daniel Zamarbide, Adrien Besson, Christophe Pidoux, Francçois de Marignac, Laurent Ammeter, Tarramo Broenimmann, Manuel Der Hagopian, Grégoire Du Pasquier, Oscar Frisk   合作伙伴 group8: Thomas Sponti, Grégoire Thomas; norm: Ludovic Varone Guscetti & Tournier: Jérôme Pochat, Marcio Bichsel, Claudio Pirazzi, Marco Andrade   专家 building services: Weinmann - Energies SA, Echallens, Enrique Zurita; structural engineers: Guscetti & Tournier, Ingenierie Civile, Carouge; Gabriele Guscetti; landscape architecture: Hager Landscape Architecture AG, Zurich; Pascal Posset; building physics: Estia SA Physique du bàtiment et développement durable, Lausanne; communication: Norm, Manuel Krebs & Bruni Dimitri, Zurich; surveying: Christian Haller, Acacias; Olivier Bolay, Veyrier

Image du Pavillon (Sud)

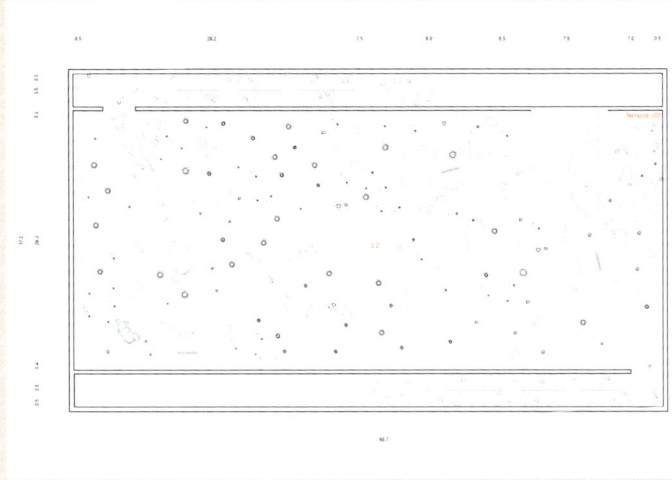
Plan du niveau + 20.00 (1/200)

Image du Pavillon (Est)

Expérience forêt

Scan 3D - forêt de Livenald - Nuage de points

Entretien

Transfert d'un paysage

Expérience
FORÊT

Liste de plantes - forêt de Livenald

Livenald

Répertoire botanique

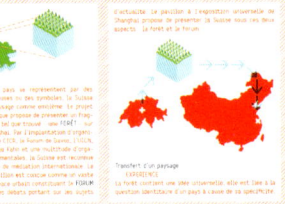

Foyer - Restaurant

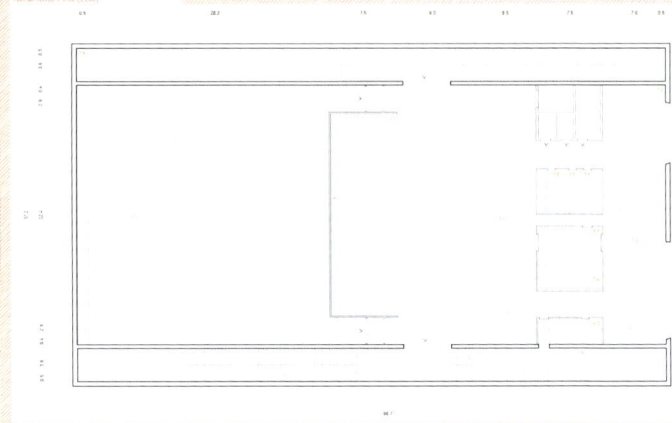
Plan du niveau + 0.00 (1/200)

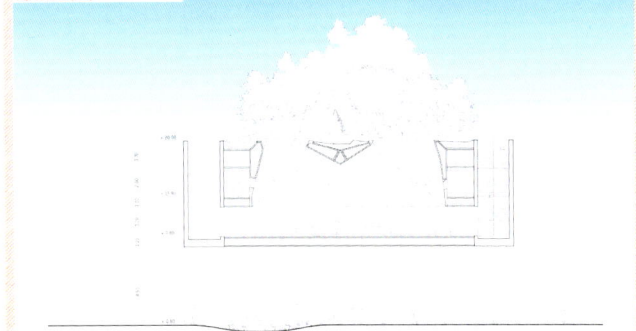
Coupe transversale (1/200)

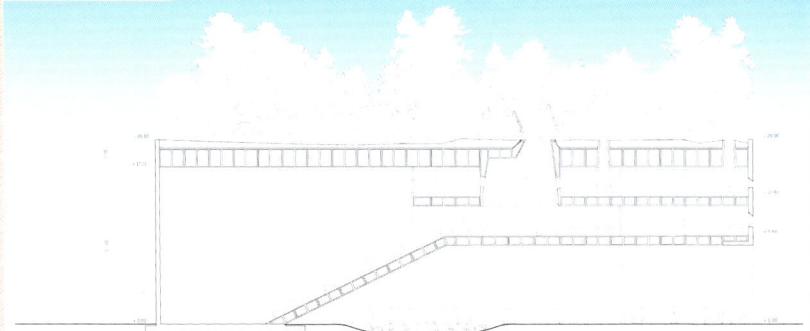
Coupe longitudinale (1/200)

Structure
RATIONALITÉ

Principe structurel

Répartition des efforts dans la structure

Circulation visiteurs

Circulation exposition

Matières
BOIS - BÉTON

Topographie

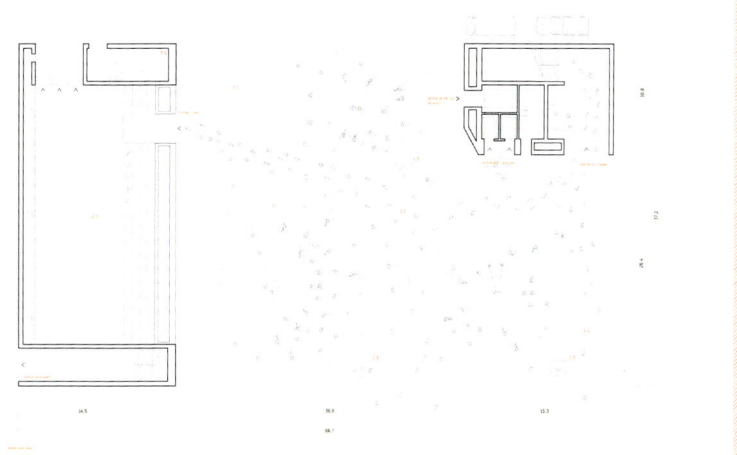

Mediation
FORUM

Helvedata
LECTURE PAYSAGE

EXPERIENCE

HELVEDATA

MEDIATION

Better city better life

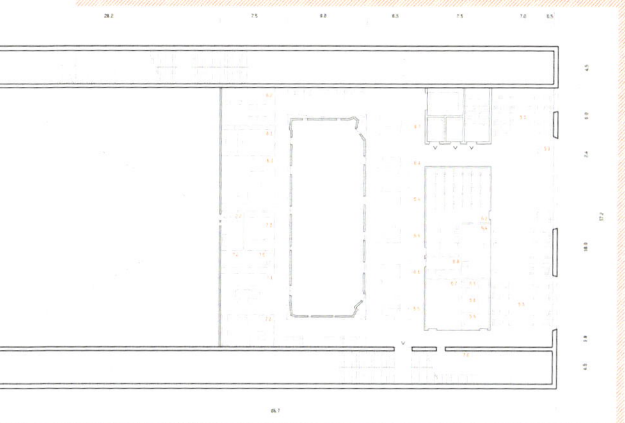

ENERGIE

Principe de développement durable

Recyclage de béton

lee + mundwiler architects Santa Monica
lee+mundwiler建筑事务所 圣莫尼卡

"In China, red doors are typically loaded with meaning. They let fortune, life and warmth enter the house and embody positive energy."

"在中国，朱红大门寓意典型，代表财富、生命、温暖入户和活力。"

作者 Stephan Mundwiler   合作伙伴 lee + mundwiler architects: Cara Lee, Chris Oberli, Nadya Aguilar, Tinka Rogic, Gustavo Santos, Elvis Arevalo, Michael Bennett, Ayako Takaishi   专家 building services: Ingenieurbüro Hausladen GmbH, Kirchheim/Munich; Prof. Gerhard Hausladen; structural engineers: Prof. T. Jeff Guh, Los Angeles, CA/USA; visual layout: Stauffenegger + Stutz GmbH; visual design: HFG, Basel; Christian Stauffenegger; building services: Philipp Dreher; art: Ralf Gschwend, Palm Beach, Florida; structural analysis: Gordon Polon, Santa Monica, CA/USA

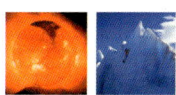

Energie Konzept, von der Sonne zu Eis

Die elektrische Energie für den Betrieb der Türen wird über Photovoltaik, angeordnet über dem Innenhof, erzeugt.

Der gesamte Bedarf zur Kühlung des geschlossenen Gebäudes wird nachhaltig über regenerative Energien gedeckt. Dies geschieht über Solare Kühlung mit einem Feld von Parabolspiegel-Konzentratoren auf dem Dach. Über zwecsstufige Absorptionskältemaschinen wird damit 8°C kaltes Kühlwasser erzeugt für die raumlufttechnischen Anlagen, sowie Kühlwasser mit -3°C für den Betrieb einer spektakulären vereisten Restaurantrückwand, welche die Verbindung zum gezeigten Film über die Alpenwelt herstellt.

Die Eiswand im Restaurantbereich ist nicht nur visuell und haptisch ein Erlebnis, sondern lädt auch zum interaktiven Umgang ein. Zusammen mit den sich bewegenden Türen, werden diese Attraktionen neben dem im Kinosaal gezeigten Film zur Ausstellung. Das Gebäude präsentiert sich selbst als Organismus im Zusammenspiel mit den Elementen.

Solare Kühlung

1 Parabolspiegel - Konzentratoren
2 Zuführung von Leitmedium
3 Absorptionskältemaschinen
4 Eiswand
5 Plenum zur Kaltluftführung
6 Absaugung verbrauchter Luft
7 Türgallerien mit passiver Kühlmöglichkeit

Bauweise

Der CHina Pavillon ist als temporärer Bau konzipiert. Es werden nur "trockene" Baugrösse angewandt. Die gesamte Pavillon agiert als Zwischenlager für die verwendeten Baumaterialien. Sämtliche Teile werden also nach der Ausstellung wieder verwendet. Der Begriff "Baumaterialleasing" kann hierfür kreiert werden.

Die Stahlkonstruktion ist bewusst sehr einfach und repetitiv gestaltet und besteht aus Standardträgern. Verbindungsplatten, Zügelten etc. welche nur mittels speziell entwickelten Friktionsverbindungen zusammengefügt. Damit eliminieren wir Schweissverbindungen und Bohrungen, welche die Wiederverwendung der Baumaterialien erschweren würden.

Da wir alle Bauteile nur für maximal ein Jahr (Aufbau, Betrieb und Abbau) benötigen, schlagen wir vor, sämtliche Materialien nur für diese Zeit zu "leasen". Nach der Expo können alle Teile als vollwertige Baumaterialien in China wieder verwendet werden.

Mit der vorgeschlagenen einfachen Bauweise und "Baumaterialleasing", ergänzt durch attraktive Sponsoring-Möglichkeiten, können die Baukosten niedrig gehalten werden.

PERSPEKTIVE VON SUEDEN

Raumkühlung und Bewegung der Türen

Das Kino, das Restaurant, der Shop und VIP Bereich sowie die BGos werden mit solar entfeuchteter und gekühlter Luft mit einem 2-3 fachen Luftwechsel versorgt. Die Luftführung erfolgt über einen Zwischenraum im Plenum des Kino Auditoriums. Die Luft für das Kino strömt durch die Stufen der aufsteigenden Sitzreihen und steigt von dort über natürliche Konvektion zur Decke. Zuluftleitungen in meckanischen Sinne entziehen unter Decke wird die verbrauchte Luft abgesaugt und der Wärme bzw. Kälterückgewinnung zugeführt.

Die Eiswand

Die Eiswand bildet die Rückwand des Restaurants, ist aktive Raumkühlung wird breitet und den Film ab, welcher im Kinosaal zähriger gezeigt wird. Zudem bietet diese Wand, bestehend aus verfahrenden Betonplatten, ein attraktives Statement für den Hauptbereich.

Die hägenden Gärten

Hängende Gärten aus Hopfenstauden kreieren ein angenehmes Mikroklima im offener, beschatteten Innenhof. Wenn vorhanden, wird dort Restkälte der Absorptionskältemaschinen durchgeleitet. Dadurch entsteht Kondensat aus der feuchten Luft, welches sich an den Pflanzen festsetzt und dadurch zur Verbesserung der Behaglichkeit führen kann.

hängende Hopfenstauden
gefilterte Luft
Kondensat

Prozip Stahlkonstruktion

Speziell entwickelte Friktionsverbindungen erlauben es, alle Stahlteile nach dem Abbau an einem anderen Ort wieder zu verwenden. Es sind keine Schweissungen oder Bohrungen notwendig.

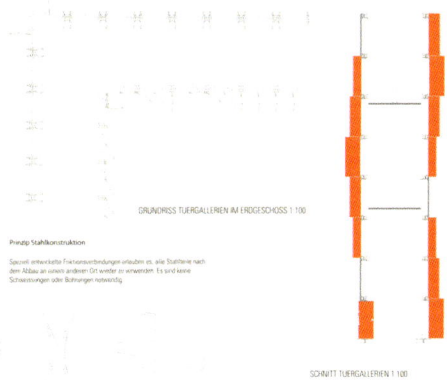

GRUNDRISS TUERGALLERIEN IM ERDGESCHOSS 1:100

SCHNITT TUERGALLERIEN 1:100

Das Gebäude als Ausstellung

Das Konzept basiert auf der Aussage, dass sich der/ die durchschnittlicher Besucher/in nur sehr kurz im Pavillon aufhält, und ungewöhnlich grosse Besucherzahlen erwartet werden.

Wir wollen inmitten der Informationsfülle dieser Expo ein einprägsames Bild des CHina Pavillons kreieren, das sich in der Erinnerung der Besucher/innen festsetzt und positive Assoziationen mit der Schweiz hervor ruft. Wir glauben, dass sich vor allem Bilder und Erlebnisse, als Kommunikationsmittel unabhängig von Sprache, Kultur und Alter, in der Erinnerung der Besucher festsetzen. Wenn wir erreichen, dass sich auch die Kinder der Besucher an diesen CHina Pavillon erinnern, haben wir unser Ziel erreicht.

Die Aussage muss deshalb einfach und einprägsam sein. Eine Ausstellung der traditionellen Art mit Objekten und Lesetafeln etc. macht hier unseres Erachtens keinen Sinn. Der Pavillon, als erfahrbares, spielerisches, sich immer wieder veränderndes Gebäude, wird selbst zur Ausstellung. Dabei soll auf selbstverständliche Art der Umgang mit technologischen Errungenschaften der Energieerzeugung und Effizienz, ausgehend von Schweizer Produktion und Forschung, veranschaulicht werden.

Kreativer Umgang mit Energien

Die Schweiz hat in den letzten Jahren bewiesen, wie sie mit Projekten von "2000 Watt Gesellschaft" oder dem Solar Impulse Flugzeug kompromisslos und kreativ eine Führungsrolle beim Thema Energie übernommen hat. Effizienter Umgang mit Energie und alternativen Energiequellen sind Themen, die der CHina Pavillon auf spielerische Art versucht zu verständlichen. Der Pavillon beinhaltet verschiedene aktive (technologische) und passive (natürliche) Elemente des Umgangs mit Energie

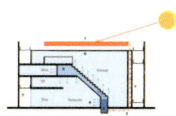

Schützen und Atmen

Die Gebäudehülle erfüllt, ähnlich der Haut in der Natur zwei Hauptaufgaben: Schützen und Atmen. Mit mehreren hundert beweglichen Türen wird der CHina Pavillon zu einem lebendigen Gebäude, von der Natur inspiriert. Ein Gebäude, welches sich an seine Umwelt anpasst, animiert von Sonne und Wind, Wärme, Tageslicht, Besucherfluss etc. Die Bewegungen der Türen sind steuerbar, oder laufen in zufälliger Folge ab. Bei einem speziellen Anlass können die Türen strømen offen stehen, oder die Öffnungen der Türen können eine Art Relief annehmen, bis hin zu einem bildlichen Schweizer Kreuz am 1. August.

Die elektrische Energie für den Betrieb der Türen wird über Photovoltaik, angeordnet über dem Innenhof, erzeugt.

Der gesamte Bedarf zur Kühlung des geschlossenen Gebäudes wird nachhaltig über regenerative Energien gedeckt. Dies geschieht über Solare Kühlung mit einem Feld von Parabolspiegel-Konzentratoren auf dem Dach. Über zwecsstufige Absorptionskältemaschinen wird damit 8°C kaltes Kühlwasser erzeugt für die raumlufttechnischen Anlagen, sowie Kühlwasser mit -3°C für den Betrieb einer spektakulären vereisten Restaurantrückwand, welche die Verbindung zum gezeigten Film über die Alpenwelt herstellt.

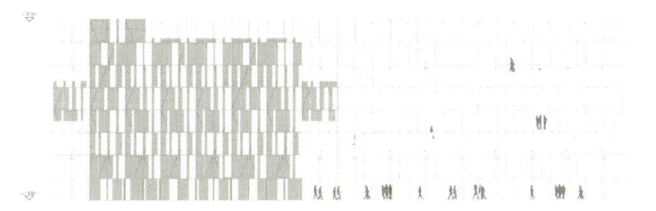

FASSADE NORD 1:200

FASSADE OST 1:200

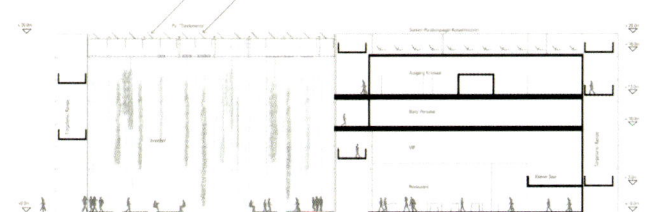

LAENGSSCHNITT 1:200

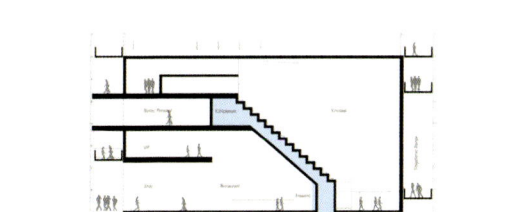

QUERSCHNITT 1:200

EINGANGSSITUATION

DARSTELLUNG VERSCHIEDENER "TUERSTELLUNGEN"

Lehmann Fidanza Zurich/Fribourg
Lehmann Fidanza建筑事务所 慕尼黑/弗里堡

"'Echange' – With its very structure, the Swiss Pavilion communicates Switzerland's openness to ideas, people, and cultures"

"'交换'——瑞士展馆表现出瑞士对思想、人们和文化的开放性。"

作者 Philipp Lehmann, Alain Fidanza　专家 set design: Pius Tschumi, Zurich; artist: Christian Waldvogel, Zurich;
building services: HPS Energieconsulting AG, Küsnacht; Daniel Heule; structural engineers: Dr. Hanspeter Kaiser, Freiburg;
communication consultant: advocacy AG, Mathis Brauchbar, Zurich; real estate management: Roger Gort, Büro für Bauökonomie, Lucerne

GENUSS IM RESTAURANT

Im Restaurant wird Wert auf das Credo des Austausches und den Bezug zur Chinesischen Küche gelegt. Neben den Schweizerischen Klassikern werden auch Köstlichkeiten aus der zeitgenössischen «Cuisine Suisse» angeboten.

Die Vitrinentische enthalten Exponate und sind für Gruppenbesuche und Gesellschaften ausgelegt. Für den Besucher sichtbar wird in der verglasten Küche das Essen auf Schweizer Art zubereitet.

ERINNERUNG IM SHOP

Auch im Shop stehen Vitrinentische. Die käuflichen Exponate aus der Ausstellung sind ergänzt durch Schweizer Lebensmittel, Brands und Designartikel. Zwischen Sackmessern, Farbstiften, Spielen, Schokolade, Uhren, DVDs, T-Shirts und Naturkosmetik finden sich Originalgegenstände mit Bezug zu den präsentierten Schweizer Innovationen.

BAUGESTALTUNG

Der zeitgenössisch-pavillonäre Ausdruck des Gebäudes tritt als glänzender Korpor in Dialog mit der Stadtsilhouette Shanghais.

Die architektonische Ausformulierung des Pavillons und das szenographische Konzept bedingen sich gegenseitig und werden unter einem Dach emblematisch vereint.

Die Dachform lässt sich als Abstraktion einer Dächer- oder Naturlandschaft interpretieren und ist in ihrer Erscheinung «vertikalsymmetrisch» erkennbar umkehrbar.

Die Topographie des Bodens unterstützt die Erfahrbarkeit des Pavillons mit seinen unterschiedlichen Raumhöhen in Unterstreichung der programmatisch-szenographischen Komponenten wie Grossraumkino und Capsules.

Die Abschliessbarkeit des Pavillons wird mittels eines dachrandperipheren Gitternetzes erreicht. Die geforderte Betonwand wird am VIP-Eingangsbereich aussenseitig hin zum Restaurant und an der Mittelachse des Pavillons erfahrbar sein.

KLIMA- UND ENERGIEKONZEPT

Zeitraum: Mai – Oktober
Ort: Shanghai

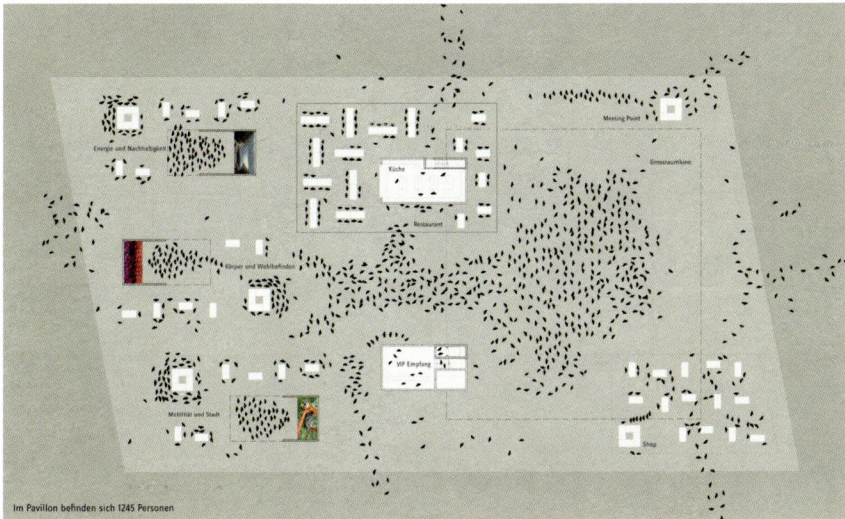

Der Schweizer Pavillon ist in drei «Klimaschalen» aufgebaut. Die Haustechnik nutzt die standortspezifischen Besonderheiten sowie die natürlichen Umstände: die Druckunterschiede zwischen Innen und Aussen aufgrund der Dachform, die leichten Winde, die Tag-Nacht-Schwankungen der Temperatur und die Regenperioden.

WIND

In den grossflächigen Ausstellungsflächen und im Kino wird durch die Automatisation der seitlichen Dachöffnungen, die sich je nach Bedarf öffnen oder schliessen, die steigende Warmluft entweichen und dabei kühle Luft aus dem Erdkanal nachziehen.

Die innenliegenden, zur Ausstellungsfläche abgeschlossenen Räume (Restaurant, Küche, VIP-Bereich, Büro- und Nebenräume)

HAUPTFILM IM GROSSFORMATKINO

Das Kino ist als Steh- oder Anlehnkino ausgelegt, welches schnell und von allen Seiten betreten und verlassen werden kann. Für Ältere und Müde finden sich Sitzgelegenheiten.

Während der Filmvorführung wird der Raum mit einem Vorhang allseitig geschlossen. Danach stehen die Zuschauer in Sekundenschnelle im Zentrum, beim Restaurant oder im Shop.

TRAGSTRUKTUR UND MATERIALITÄT

Der mehrstöckige, zentrale Bereich und die Stützen unter den hohen Kino-Aussenwänden werden auf Mikropfählen fundiert. Der restliche Bereich des Daches ruht auf Einzelfundamenten.

Die Gesamtstabilität (Wind und Erdbeben) wird über drei Betonscheiben des mehrstöckigen Bereichs und über geneigte Stahlstützen sichergestellt. Sowohl die Decken als auch das Dach bestehen aus einer Sandwichstruktur mit Holzpanelen und integrierten Stahlträgern.

Das Dach ist auf seiner Aussenseite mit perlmuttweiss glänzenden Dachziegeln gedeckt. Die Innenseite ist fugenlos seidenweiss glänzend gestrichen. Der topographische Boden wird als dezent farbiger Asphaltbelag in einem hellen Warmton ausgeführt.

Das Gebäude kann nach Abschluss der Ausstellung auf einfache Weise rückgebaut und das Material grösstenteils wiederverwendet und rezykliert werden.

werden zusätzlich mit schlanken Lüftungsanlagen gekühlt und mit der nötigen Hygieneluft versorgt.

SONNE

Mit Fotovoltaik (Strom von der Sonne) wird der elektrische Bedarf für die Antriebe und mit Solarkollektoren (Wärme von der Sonne) das Warmwasser für die Küche bereitgestellt.

REGEN

Das Regenwasser aus den drei Regenperioden wird auf ausgewiesenen Dachflächen aufgefangen und für die Bewässerung der Aussenanlagen und für die WC-Spülung verwendet.

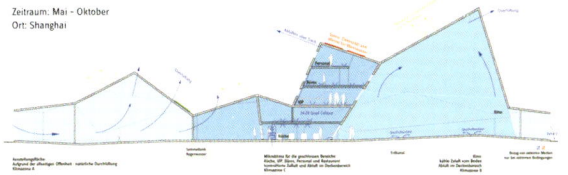

L'OE

Centola & Associati,

Ingenhoven Architekt

UF

Holcim Awards 2006
霍尔森可持续建筑大奖2006

Mariagiovanna Riitano

en

royectos Arqui 5

Holcim Awards 2006

霍尔森可持续建筑大奖 2006

Two-stage award competition (first stage in five regions, second stage global with all regional award winning competitors)
竞赛分为两个阶段，第一阶段在五大地区进行，第二阶段在全球范围内进行，由所有地区获奖竞赛者参加。

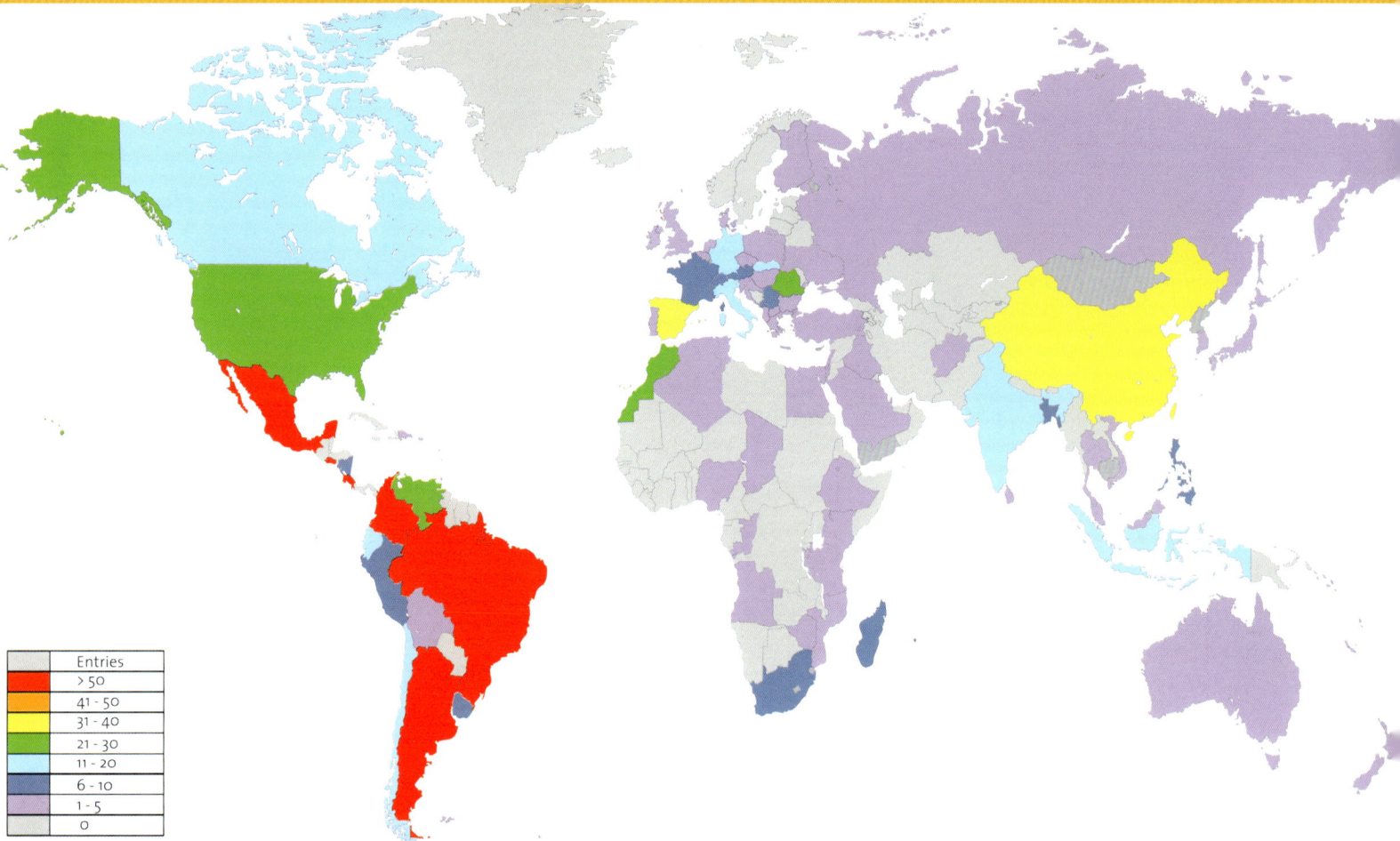

Entries
> 50
41 - 50
31 - 40
21 - 30
11 - 20
6 - 10
1 - 5
0

时间 2006   主办方 Holcim Foundation   人数 3,000 entries from 120 countries   竞赛费用 2 million USD
评奖委员会 see images on the opposite from each regional jury

Sustainable development is crucial to our future Based in Switzerland, the Holcim Foundation for Sustainable Construction encourages innovative approaches to construction and projects contributing to a more sustainable built environment. It pursues this policy in three ways: At regular intervals it hosts the Holcim Forum, a scientific conference attended by experts from all over the world. It also promotes the publication of relevant trade literature. And every three years since 2004 it offers the Holcim Award, the most coveted prize for sustainable construction worldwide. Anyone involved in developing building materials and in building actual projects is eligible, provided that these are future projects.

The procedure is a worldwide, open two-stage competition run in five regions (Europe, Africa & the Middle East, Asia & the Pacific, North America, and Latin America). The winners chosen by each of the five regional juries may then take part in a global competition held the following year with a jury of its own. In the first season, over 1,700 projects were entered. Projects are evaluated on the following internationally applied criteria for sustainable construction:

1) innovation and transferability
2) social equity
3) resource conservation
4) economic viability, and
5) architectural quality and contextual impact.

Holcim Foundation organises the competition jointly with scientists from ETH Zurich, the Technical Competence Center at Boston's MIT, and the architectural firm [phase eins].

可持续发展对我们的未来至关重要。霍尔森可持续建筑基金会，总部设在瑞士，鼓励新建筑方法和可持续的建筑项目。该基金会定期主持霍尔森论坛邀请世界各地专家参加，还推动相关贸易文献的出版。从2004年开始每三年提供一次霍尔森奖金，该奖金是全世界可持续建筑的最丰厚的奖金。只是要未来建筑项目，无论是开发建筑材料还是建设实际项目，任何人都有资格参赛。竞赛是全球性公开进行的，分为两个阶段。第一阶段在五大地区进行，由此选出的获奖者可以参加次年举办的全球性竞赛。第一赛季的参选项目多达1700个。项目评估按照国际标准进行：

1 创新性和可转移性
2 社会公平性
3 资源节约
4 经济生存能力
5 建筑质量和连贯的影响

竞赛由Holcim基金会与苏黎世的ETH、波士顿麻省理工大学和技术能力中心和建筑咨询公司[phase eins]的专家们一起组织的。

Jury Europe　　评奖委员会（欧洲）

Jury North America　　评奖委员会（北美）

Jury Latin America　　评奖委员会（拉丁美洲）

Jury Africa/Middle East　　评奖委员会（非洲/中东）

Jury Asia　评奖委员会（亚洲）

Ingenhoven Architekten Stuttgart
Ingenhoven建筑事务所 斯图加特

"The project places the railway station underground to recover land to create a new urban area, combining structural and landscape aspects"

"该项目把地铁站置于地下来回收土地，从而创造出一个新的市区，把建筑结构和景观结合起来。"

作者 Christoph Ingenhoven   专家 structural engineering: Ingenieurarbeitsgemeinschaft Tragwerksplanung S21, Hauptbahnhof GbR Leonhardt, Andrä und Partner, Stuttgart in corporation with Happold Ingenieurbüro GmbH, Berlin; advice structural form: Frei Otto, Leonberg with SL-Sonderkonstruktion und Leichtbau, Leinfelden; building services: NEK Ingenieure, Frankfurt/Main; building physics: DS-Plan GmbH, Stuttgart; façade planning: DS-Plan GmbH, Stuttgart; ventilation analysis: IFI Institut für Industrieaerodynamik, Aachen; transportation planning: Ingenieurgruppe für Verkehrsplanung und Verfahrenstechnik, Aachen; fire protection: BPK Brandschutz Planung Klingsch GmbH, Düsseldorf; lighting: Tropp Lighting Design, Feldafing; landscape architecture: Ingenhoven Architects, Düsseldorf

Proyectos Arqui 5 Caracas, Venezuela

Proyectos Arqui 5 加拉加斯，委内瑞拉

"Urban improvement project including social aspects of a large shanty town in Caracas, Venezuela."

"城市改善项目包括委内瑞拉加拉加斯大片栅户区的改造。"

ESCALERA 28 DE NOVIEMBRE

作者 Silvia Soonets, Isabel Cecilia Pocaterra, Maria Ines Pocaterra, Victor Gastier　专家 hydraulic facilities: Ahmed Irazabal; road engineering advisor: Freddy Iriza; ground engineering advisor: José Francisco Martínez; electrical facilities coordinator: Pedro Luis Diaz; structural design coordinator: José Luis García Conca

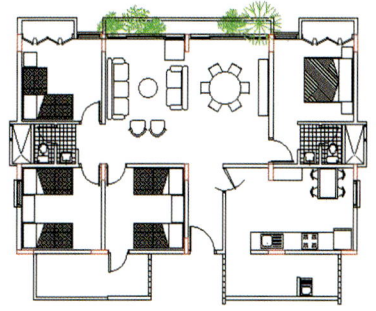

120 m2

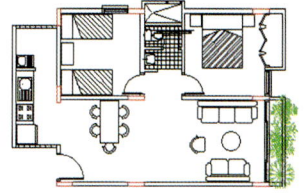

60 m2

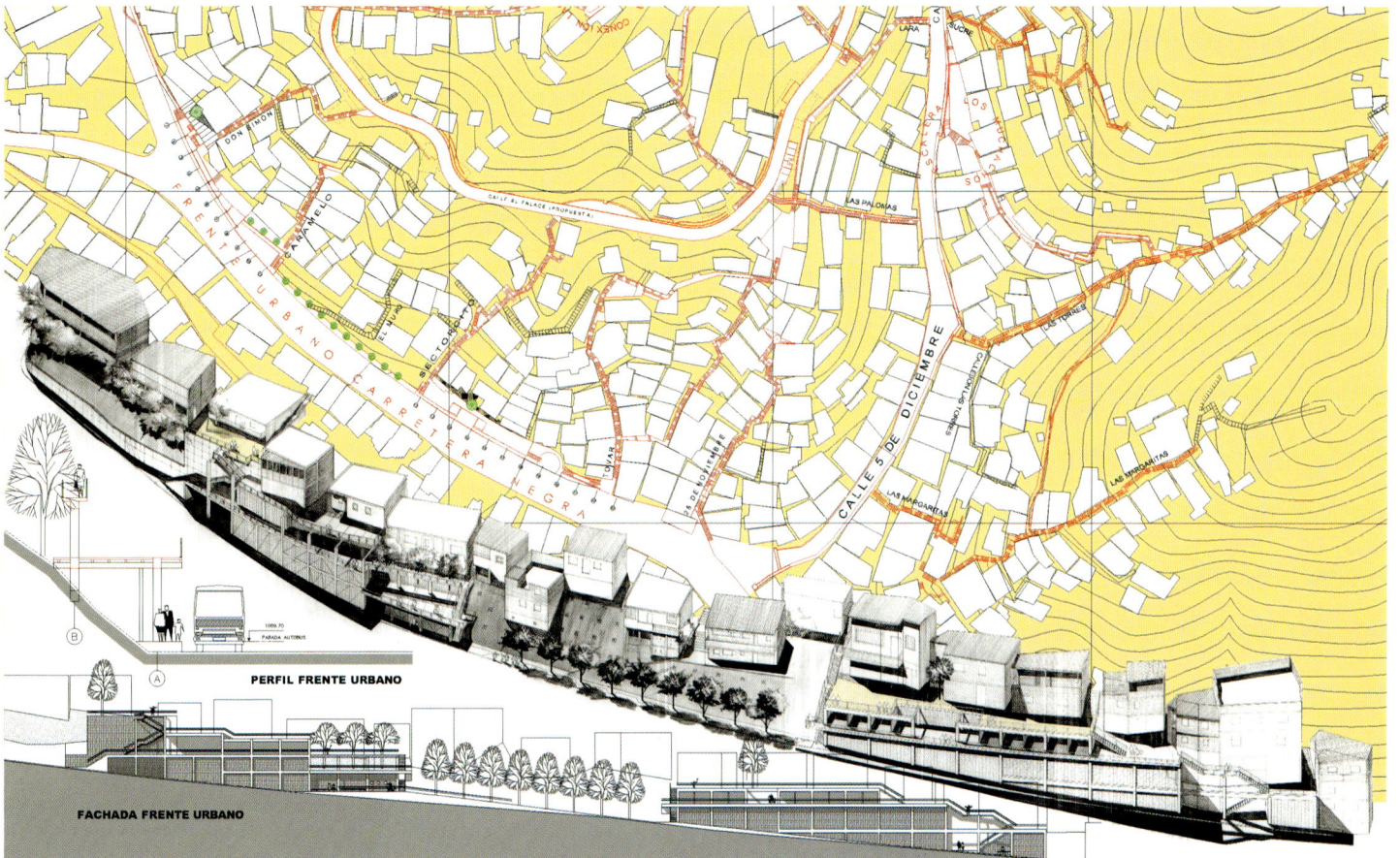

PERFIL FRENTE URBANO

FACHADA FRENTE URBANO

Centola & Associati, Mariagiovanna Riitano Rome/Fisciano, Ita[l]

Centola & Associati, Mariagiovanna Rittano 罗马/Fisciano，意大利

"The project uses water as the central theme with the historic preservation of a number of heritage buildings and maintaining aesthetic balance between existing and new structures."

"该项目用水来展示传统建筑的历史主题，保持现有建筑和新建筑之间的艺术的平衡。"

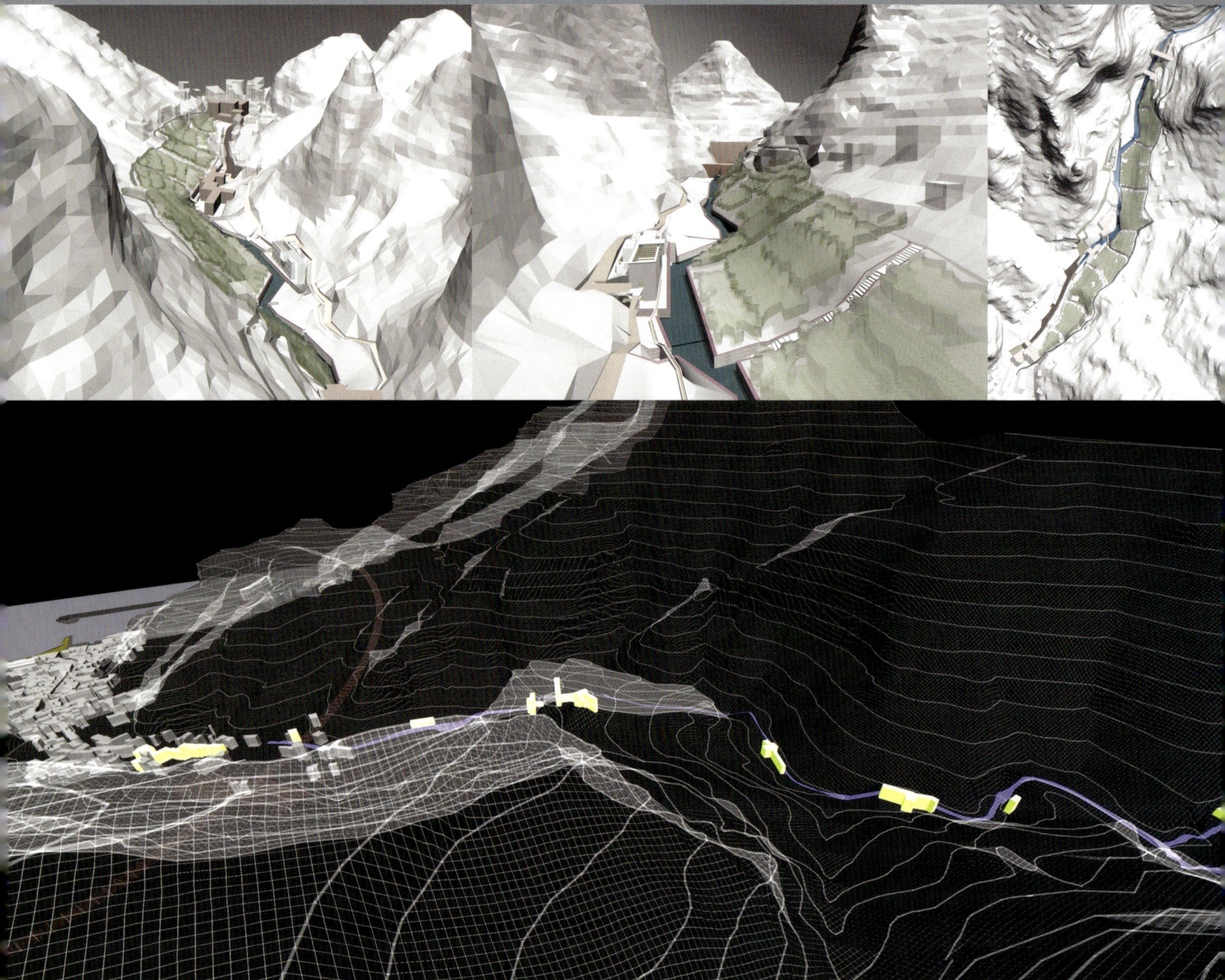

作者 Luigi Centola – Centola & Associati (Rome); Mariagiovanna Riitano – University of Salerno (Fisciano, Italy)   其他作者
J. King, R. Roselli, S. Marra, P. Latella, L. Polimeni, B. Tagliabue, C. Stevens, M. Molè, M. Cicalese, C. de Vargas M., G. Rastrelli, A. Credazzi,
D. de Seta, D. Morentin, A. Simons, I. Pérez Arnal, E. Bueno, K. Jofre, M. Nocedal, G. Pérez, D. Dollens, C. Stolte, S. Ferrini, A. Stella, F. Isidori,
M. C. Clemente, A. Yau, J. Lundberg, A. Scheer, C. Lucchesi, F. Giordano, A. Liuzzo, C. Notaristefano, L. Colosi, G. Naselli, I. Faranda,
M. Cataudella, T. Amodio, R. M. Romaniello

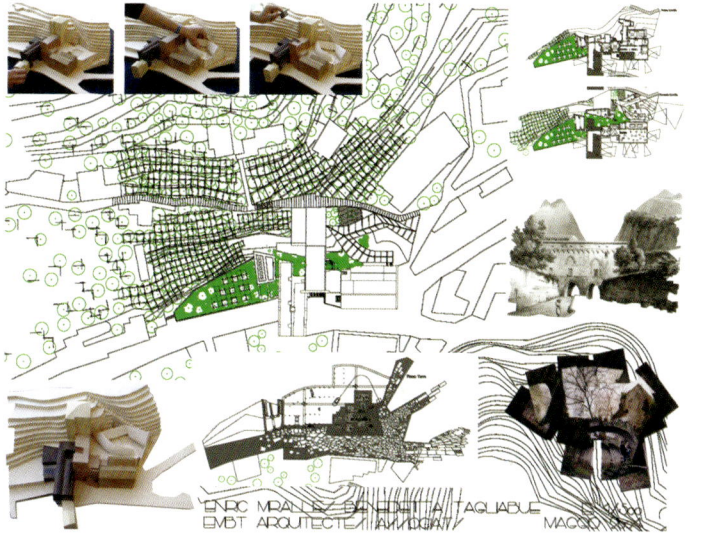

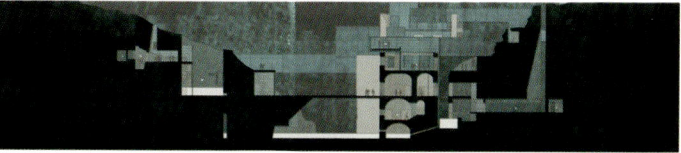

L'OEUF Montreal, Canada
L'OEUF 蒙特利尔，加拿大

"Urban, landscape and architectural project for the sustainable construction and renovation of 187 housing units on four adjacent properties in Montreal."

"该项目对在蒙特利尔四个相邻地产上的187年建筑物进行可持续建筑和整修，是一个城市、景观和建筑的综合项目。"

Chez Soi

Zoo

HCNDG

Building Facilities

Water Services

Energy Services

community infrastructur

作者 Daniel S. Pearl with Mark Poddubiuk and Bernard Olivier, L'OEUF, Montreal   合作伙伴 Sudhir Suri, Corinne Farazli, Luc Doucet, Marie-France Bourassa, Martin Beauséjour, Guillemette de Monteil, Serge Gascon, Tony Round, Lucie Babin, Aradhana Gupta   专家 bio-climatic engineering: Martin Roy et associés Groupe Conseil inc. – collaborators: Gaetan Bélanger, Francois Pelletier, Claude Yelle, Gordon Buckingham; consultants: Michaël Kummert, Prof. Michel Bernier; external review of bio-climatic engineering design: Stefan Holst; electrical engineering consultant: Marc Vacquerie; civil engineers: Mario Gendron, Gabriel Pilon; water suply: Pierre Bertrand, Priscilla Fortier

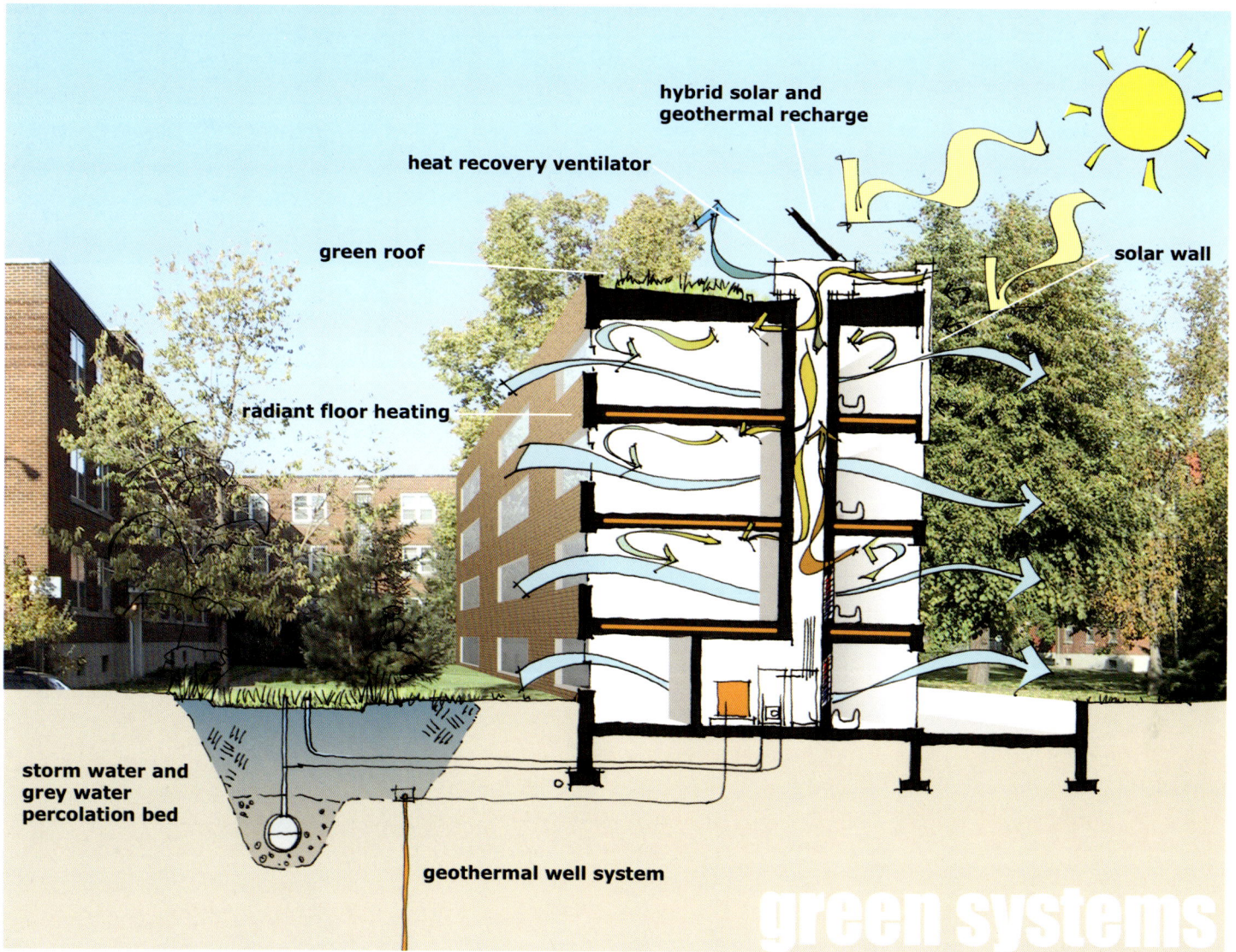

hybrid solar and
geothermal recharge

heat recovery ventilator

green roof

solar wall

radiant floor heating

storm water and
grey water
percolation bed

geothermal well system

green systems

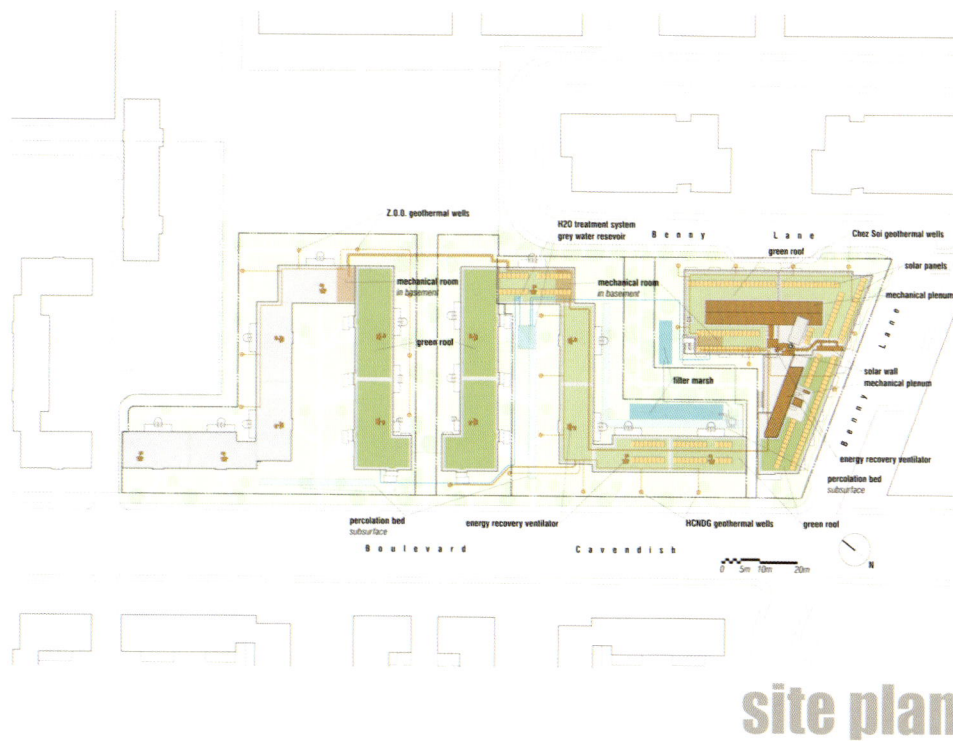

Z.O.O. geothermal wells

H2O treatment system
grey water resevoir

B e n n y   L a n e

Chez Soi geothermal wells

green roof

solar panels

mechanical room
in basement

mechanical room
in basement

mechanical plenum

green roof

solar wall
mechanical plenum

filter marsh

B
e
n
n
y

L
a
n
e

energy recovery ventilator

percolation bed
subsurface

percolation bed
subsurface

energy recovery ventilator

HCNDG geothermal wells

green roof

B o u l e v a r d

C a v e n d i s h

0 5m 10m   20m   N

site plan

Manfred Nagel mit

Ch

Brüning Klapp Rein

JSWD Architekten + Pla

HPP Hentrich – Petsch

h4a Gessert +

KSP Engel un

HBT
ix & Morel et Associés

ThyssenKrupp Quarter in Essen
蒂森克虏伯办公建筑，德国埃森

ner

igg

Randecker

Zaha Hadid Architects

Zimmermann

ThyssenKrupp Quarter Essen

蒂森克虏伯办公建筑 德国埃森

Open two-stage project competition with cooperative second stage
公开的竞赛，分为两个阶段，第二阶段为协商阶段。

Wettbewerbsgebiet

Kreuzgebäude

地点 Essen
时间 06/2006–11/2006
Auslober ThyssenKrupp AG, represented through ThyssenKrupp Real Estate GmbH
参赛者 1st stage: 106 competitors; 2nd stage: 11 competitors
面积 40,000 sq m
竞赛费用 260,000 Euro
专业评奖委员会 Hans-Jürgen Best, head of city planning department, City of Essen;
Prof. Rebecca Chestnutt, Berlin;
Prof. Jochem Jourdan, Frankfurt/Main;
Kaspar Kraemer, Cologne;
Prof. Ulrike Lauber, Munich;
Hartmut Miksch, Düsseldorf;
Simone Raskob, head of construction department, City of Essen;
Prof. Peter Zlonicky, Munich
专家评奖委员会 Dr. Ekkehard D. Schulz, chairman of the management board ThyssenKrupp AG, Düsseldorf;
Ralph Labonte, member of the management of ThyssenKrupp AG, Düsseldorf;
Dr. Martin Grimm, management board, ThyssenKrupp Real Estate GmbH, Essen;
Oliver Wittke, minister for building and traffic, North Rhine-Westphalia;
Dr. Wolfgang Reiniger, mayor of the City of Essen;
Dr. Oliver Scheytt, head of culture department of the City of Essen;
Prof. Hans-Jörg Bullinger, president Fraunhofer Gesellschaft, Munich

The "ThyssenKrupp Quarter" in Essen will house Thys-senKrupp's headquarters for worldwide activities, reflect the corporation's self-image and establish its public brand. This administrative center must reflect Thys-senKrupp's international and historic significance and besides improving staff flexibility, performance and motivation, it must also constitute a paradigm of corpo-rate architecture. The site for Krupp's headquarters is approx. 16 hectares in the west of Essen The result was an inner-city wasteland which is now to be revived and enhanced in value So as well as signifying the corpora-tion's roots and now its international reach, the Quarter is to contribute to a sense of spatial and cultural conver-gence on a site rich with symbolic and historical value. The expansive area in Essen's inner city will accommo-date administration buildings with a gross floor area of 175,000 square metres. Apart from the headquarters for the corporation's top management and other office blocks, it will also house its conference center (to be used for holding events of various kinds), a hotel and ancillary service facilities. Another challenge posed to the planners was the urbanistic and architectural inte-gration into the new Quarter of an extant structure, the so-called "Cruciform Building". With its procedural choice of an open competition, ThyssenKrupp lives up to its ambition of taking on societal responsibility for sustainability and openness, for which it was "rewarded" with an outstanding design.

埃森的蒂森克虏伯办公建筑将成为公司在全球开展活动的总部，反映公司形象，创立自己的公众品牌，因此建筑必须体现蒂森克虏伯的国际和历史重要性，提高公司员工的灵活性、工作表现和积极性，还要成为公司建筑物的典范之作。该项目位于埃森西部一块约16公顷的地点上。结果是城内大片荒地的价值得以复活和增强。该项目还有助于在一个富于象征意义和历史价值的地点上产生空间和文化的集中感。行政办公大楼总建筑面积将为17.5万平方米。除了公司高层管理总部和其他办公大楼以外，还有会议中心、一家酒店和辅助服务设施。竞赛还为规划者提出另一个挑战，现有建筑与新办公大楼在市政和建筑方面必须结合，即所谓的"十字形建筑物"。由于选择公开竞赛程序，蒂森克虏伯承担起可持续性和公开的社会责任，也因而获得出色设计的回报。

Aerial view　鸟瞰图

Competition site　竞赛地点

Competition site　竞赛地点

Colosseum Theatre Essen, former Krupp factory hall
埃森圆形大剧场，原蒂森克虏伯工厂礼堂

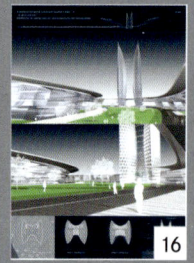

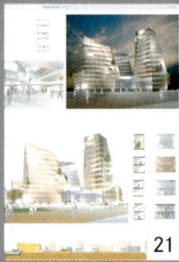
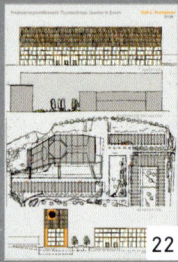

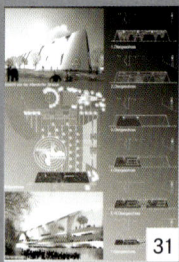
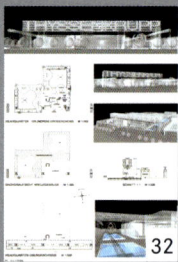
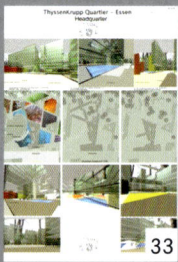

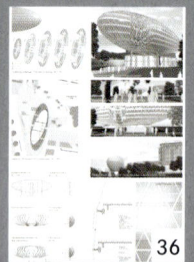

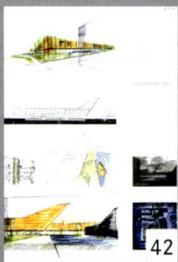
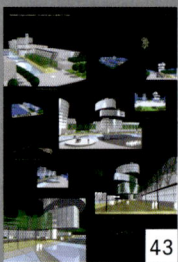

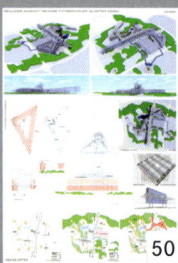

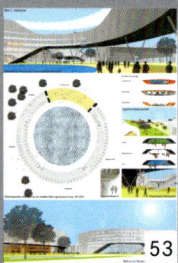
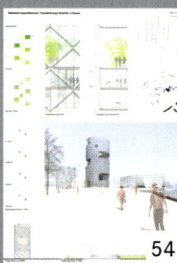

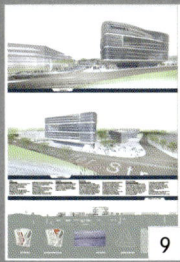

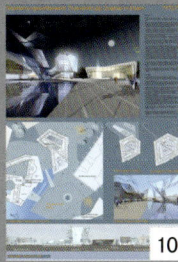

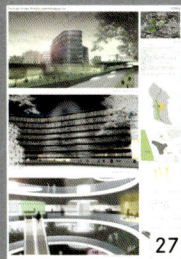

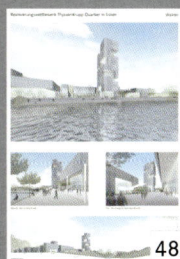

Qualified participants 1st stage 第一阶段合格参赛者

1 J.S.K.-SIAT International, Michael Winkelmann, Frankfurt/Main **2** JSWD Architekten + Planer, Konstantin Jaspert, Cologne **3** Prof. Helge Bofinger & Partner, Wiesbaden **4** HPP Hentrich-Petschnigg & Partner, Düsseldorf **5** Fürst Architects, Düsseldorf **6** Architekten Brüning Klapp Rein, Essen **7** Zaha Hadid Architects, London **8** Manfred Nagel, Kiel **9** KSP Engel und Zimmermann, Frankfurt/Main **10** h4a Gessert+Randecker in AG mit Prof. Legner, Stuttgart **11** schmiedeknecht architekten planschmiede berlin, Berlin

Not qualified participants 1st stage 第一阶段不合格参赛者

12 Oliver Feiling, architect and town planner, Cologne **13** Schuster Architekten, Düsseldorf **14** AP Plan Mory Osterwalder Vielmo, Stuttgart **15** Schormann Architekten, Düsseldorf **16** Gewers Kühn & Kühn Gesellschaft von Architekten, Berlin **17** Kleihues + Kleihues, Berlin **18** thoma architekten, Zeulenroda-Triebes, Germany **19** Architekturbüro Sprenger, Hannover **20** Werkstatt für Architektur, Design + Kommunikation, Prof. Gerhard Diel, Berlin **21** Nickl & Partner, Munich **22** Krämer und Susok, Lingen, Germany **23** Böhm-Architekten, Cologne **24** Architectenbureau Paul de Ruiter, Amsterdam **25** Schleich & Schleich Architekten, Viersen, Germany **26** Staubach+Partner Architekten, Ingenieure, Fulda, Germany **27** Sadar Vuga Arhitekti d.o.o., Ljubljana **28** Heinle, Wischer und Partner, Freie Architekten, Berlin **29** raupach architekten, Christian Raupach, Otto Bertram, Munich **30** Kahlfeldt Architekten, Berlin **31** planungsgruppe16-chapman taylor architekten, Mülheim/Ruhr, Germany **32** Büro Architekt Prof. DI Helmut Hodny, Mödling, Germany **33** Grüttner Architekten, Soest **34** ASP Schweger, Hamburg **35** Kauffmann Theilig & Partner, Freie Architekten, Ostfildern **36** Meilenstein Architektur und Design, Berlin **37** Neugebauer + Rösch Architekten, Stuttgart **38** KCAP /ASTOC Architects & Planners, Peter Berner, Prof. Kees K.W. Christiaanse, Prof. Oliver Hall, Prof. Markus Neppl, Cologne **39** schneider + schumacher, Frankfurt/Main **40** Koschany + Zimmer Architekten und Generalplaner KZA, Essen **41** driendl*architects, Vienna **42** Harter + Kanzler, Freie Architekten, Freiburg **43** Goldstein Architekten, Munich **44** alB agiplan Integrale Bauplanung GmbH, Kai-Uwe Lompa, Moers **45** Findeisen & Partner Architekten–Stadtplaner, Cologne **46** LJA Lothar Jeromin Architekt, Essen **47** Ortner & Ortner Baukunst, Berlin **48** Gerber Architekten, Dortmund **49** Architekt Stefan Schausten, Essen **50** Atelier Plötzl Arch+Ing, Linz **51** Ruhnau Architekten, Essen **52** Fürst Architects, Düsseldorf **53** Du Bessct et Lyon Architectes, Paris **54** netzwerkarchitekten, Darmstadt **55** Busmann+Haberer, Alfred Bohl, Stefan Tebroke, Cologne **56** Peter W. Schmidt, Pforzheim **57** SSP Architekten Schmidt-Schicketanz und Partner, Munich **58** Vinzelberg Architekten, Bochum **59** kohl & fromme architekten, Essen

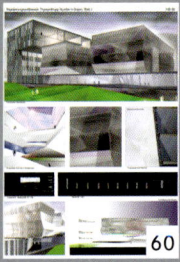

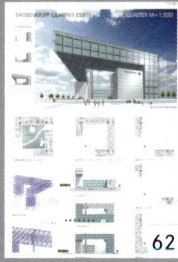

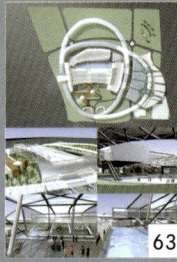

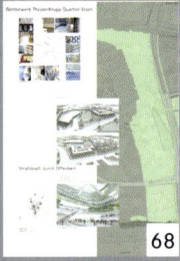

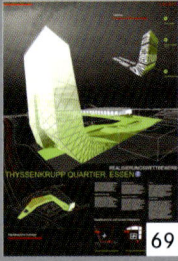

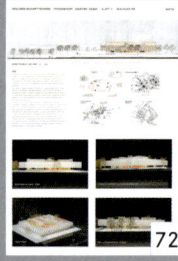

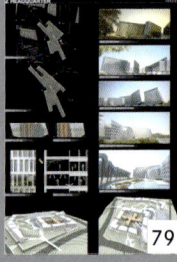

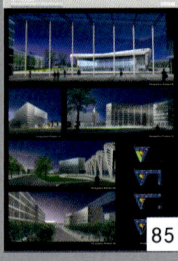

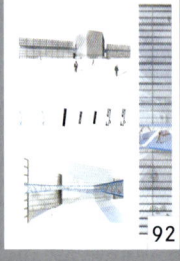

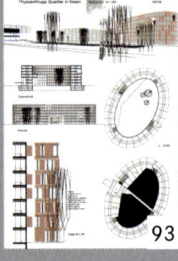

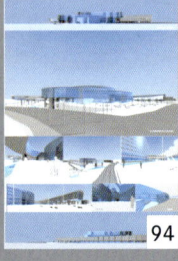

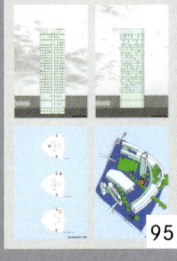

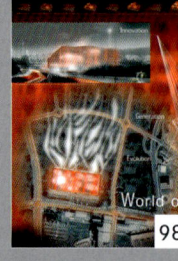

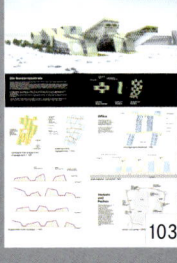

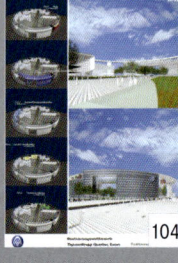

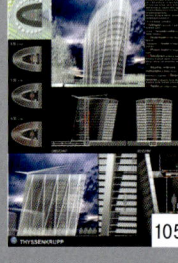

67

75

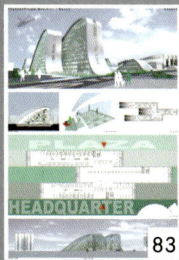

83

91

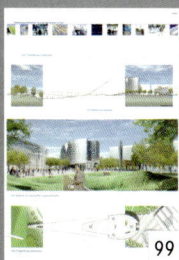

99

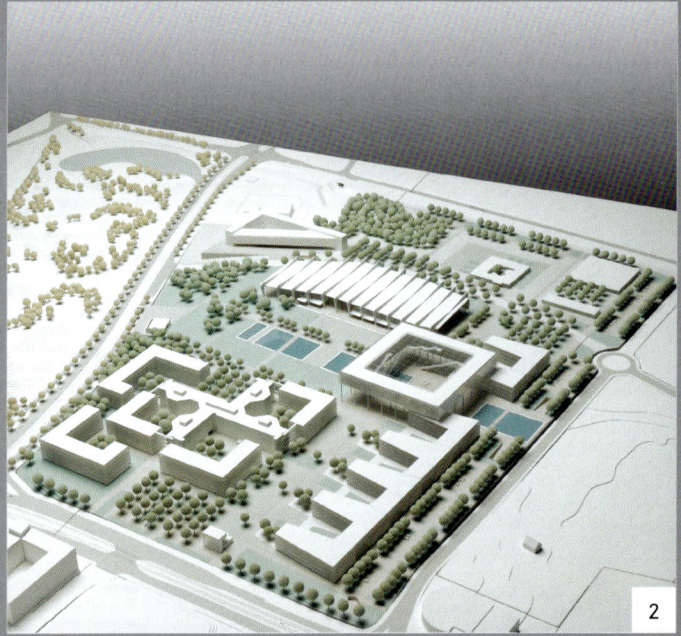

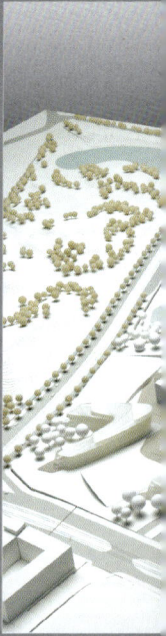

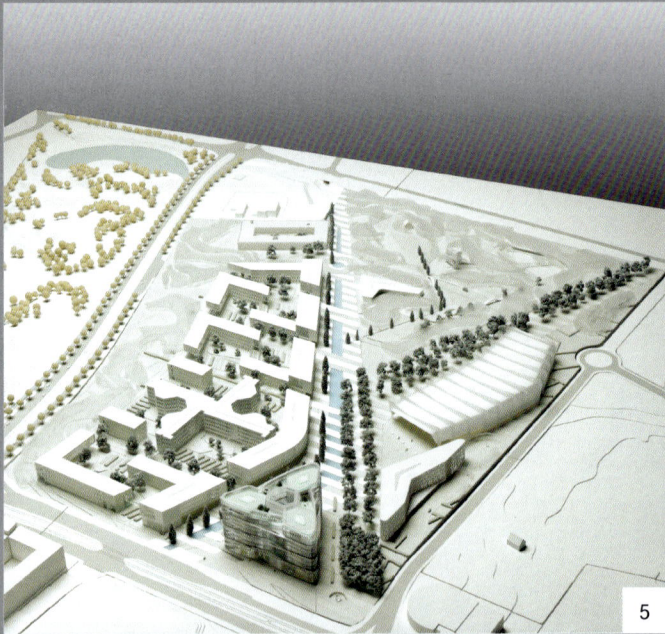

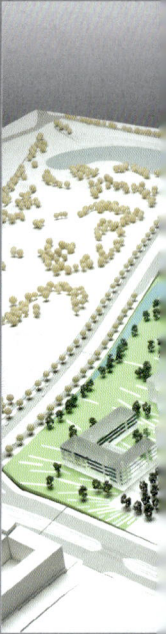

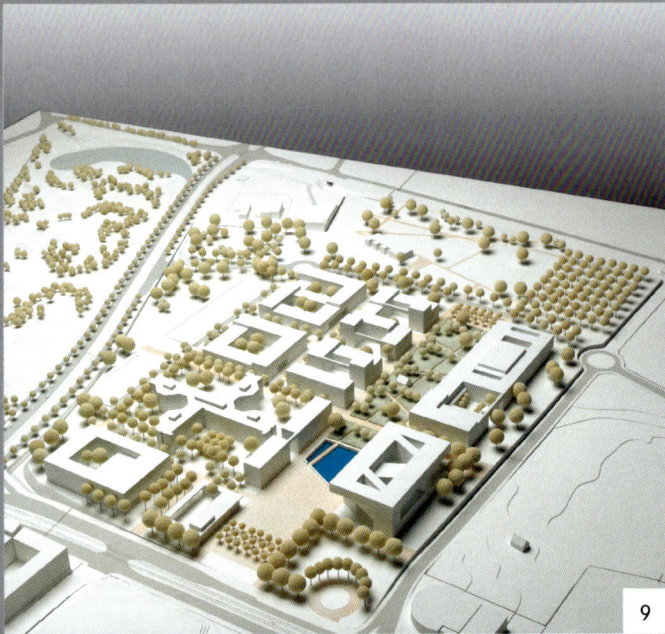

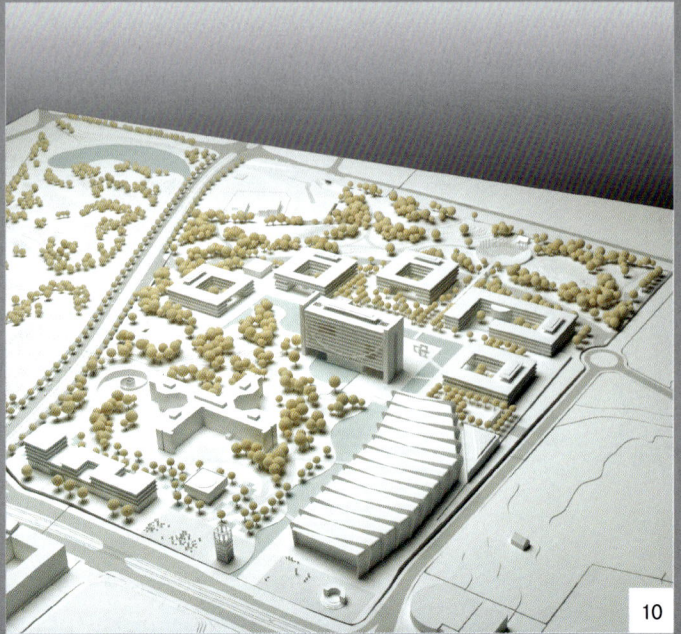

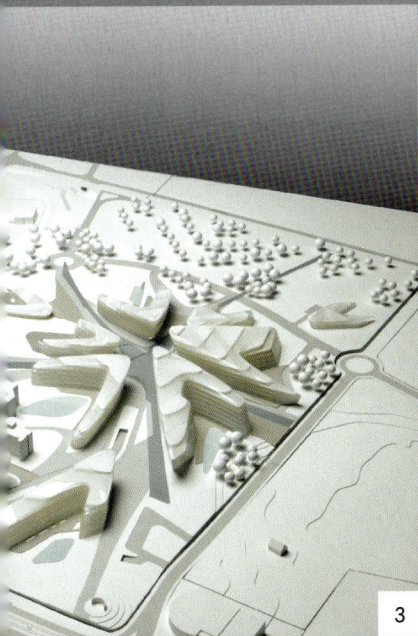

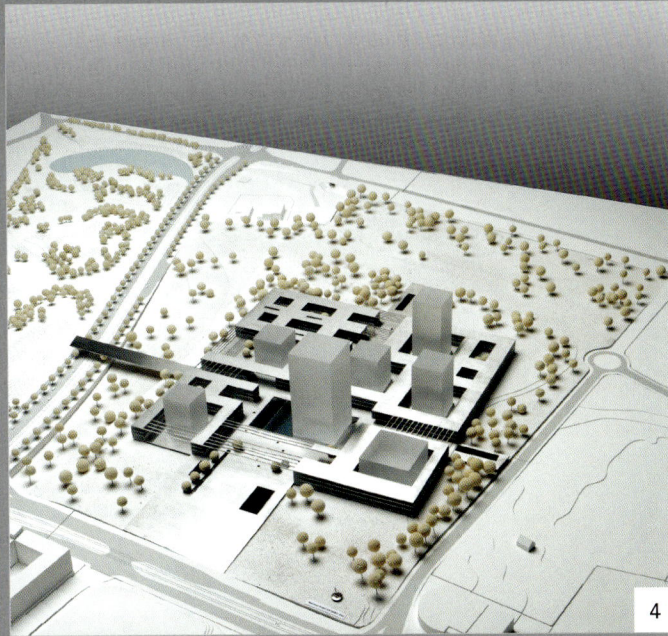

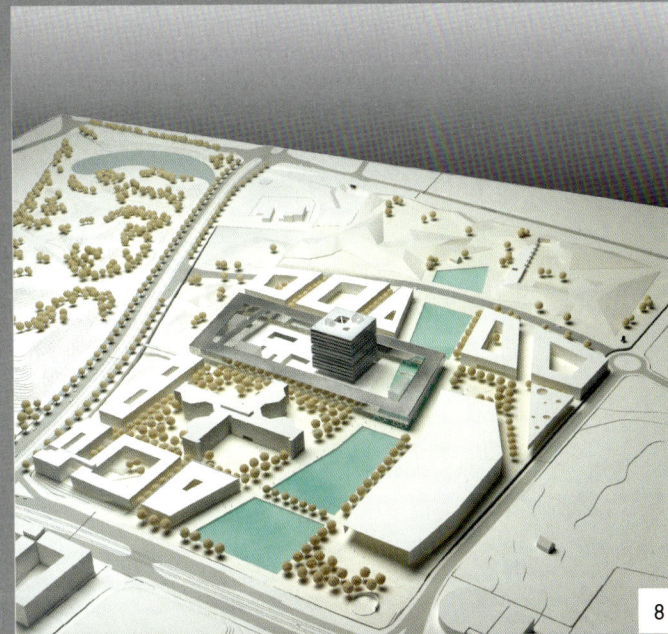

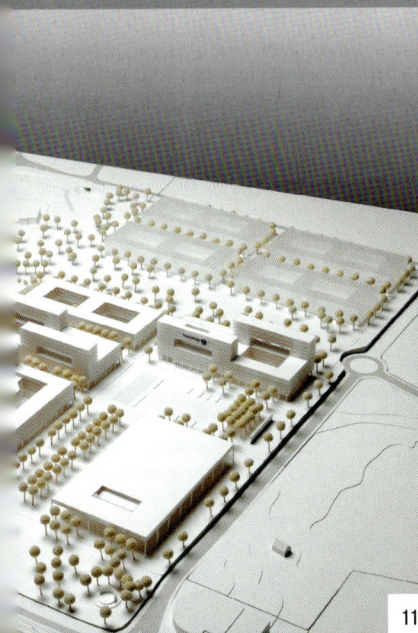

Participants 2nd stage
第二阶段参赛者

1
1st prize   一等奖
Chaix & Morel et Associés, Paris /
JSWD Architekten und Planer,
Cologne

2
2nd prize   二等奖
Architekten Brüning Klapp Rein,
Essen

3
3rd prize   三等奖
Zaha Hadid Architects, London

4
3rd prize   三等奖
Manfred Nagel with DHBT, Kiel

5
5th prize   五等奖
KSP Engel und Zimmermann
Architekten, Frankfurt /Main

6
Further participant   其他参赛者
h4a Gessert + Randecker
Architekten BDA, Stuttgart

7
Further participant   其他参赛者
HPP Hentrich – Petschnigg &
Partner KG, Düsseldorf

8
Further participant   其他参赛者
J.S.K.-SIAT International,
Michael Winkelmann,
Frankfurt /Main

9
Further participant: 其他参赛者
Prof. Helge Bofinger & Partner,
Wiesbaden

10
Further participant   其他参赛者
Fürst Architects GmbH,
Düsseldorf

11
Further participant   其他参赛者
schmiedeknecht architekten
planschmiede berlin, Berlin

Chaix & Morel et Associés/JSWD Architekten + Planer Paris/Cologn
Chaix & Morel et Associes/JSWD建筑规划事务所 巴黎科隆

"... a campus lay-out with free-standing buildings in open-air spaces grouped around the center ..."
"一种大学校园式的规划，户外建筑物自由地矗立在中心周围……"

作者 Konstantin Jaspert in ARGE mit Remy van Nieuwenhove, Philippe Chaix, Walter Grashug　合作伙伴 (JSWD) Maic Auschrat,
Robert Bönsch, Bartosz Czempiel, Stefan Dahlmanns, Christian Mammel, Elmar Schmidt-Bleker, Helmut Schröder, Agi Sobotta; (Chaix & Morel)
Fabian Barthelemy, Grete Lochmann　专家 landscape planning: Frank Flor, Club L94, Cologne; TWP: Andrea De Cillia, InCA-Ingénieurs Conseils
Associés S.à.r.l., Luxembourg; TGA: Klaus Huke, PGH-Planungsgemeinschaft; building services: Dormagen; traffic engineers: Axel Springsfeld,
BSV, Aachen; visualisation: Gaël Morin, Oliver Plou, MyLuckyPixel, Paris

Blick in die Zentrale Mitte

ThyssenKrupp Quartier
Essen, Realisierungswettbewerb 2.Phase

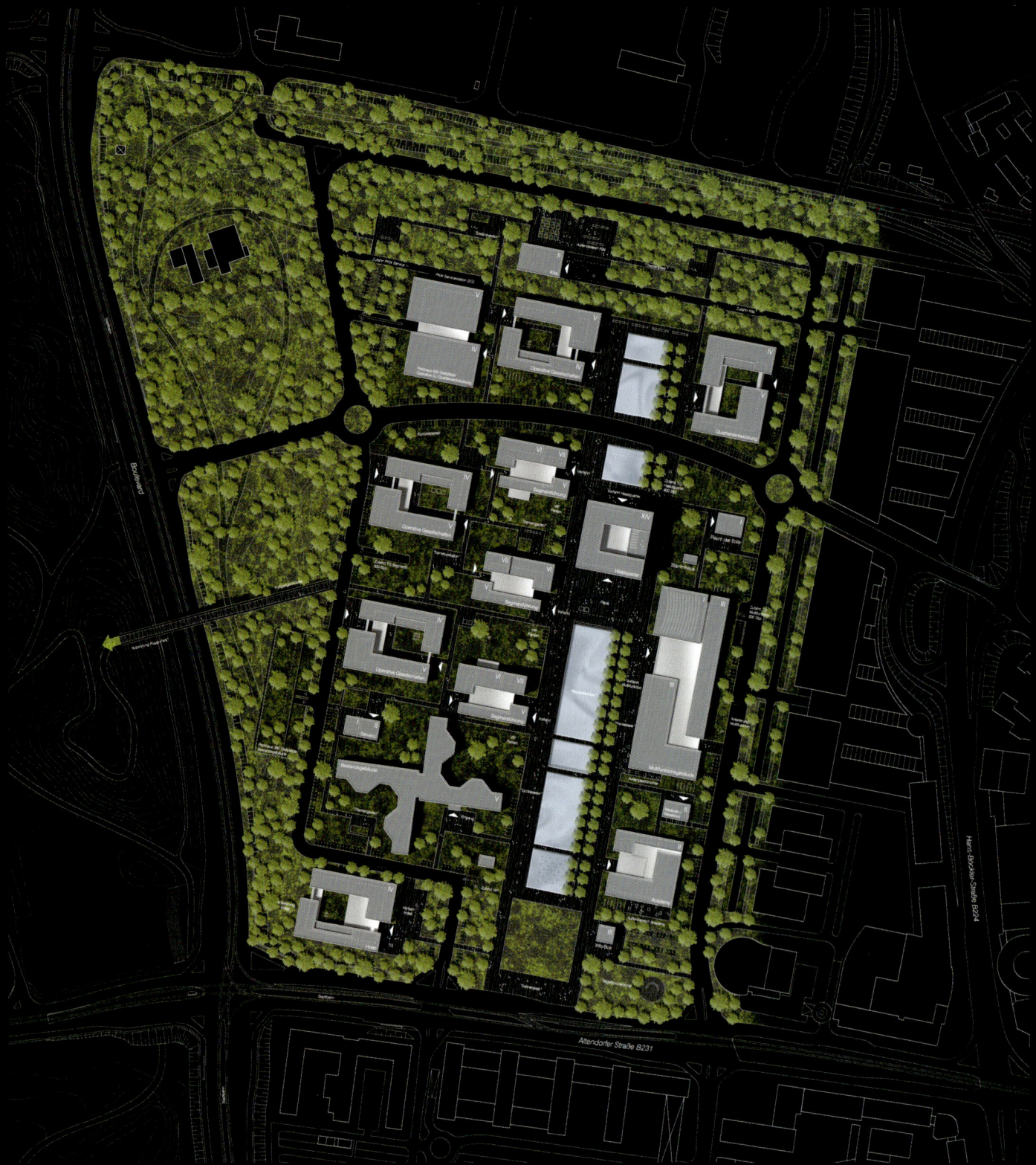

ThyssenKrupp Quartier M1/1000

Strukturplan

Bauphasen

Erschließung

Blick in den Park

Materialität

Fassadenansicht M1/50

Fassadenschnitt M1/50

Ansicht Nord M1/500

Ansicht West M1/500

Schnitt 1-1 M1/200

Ebene 01,02 M1/500
Büro

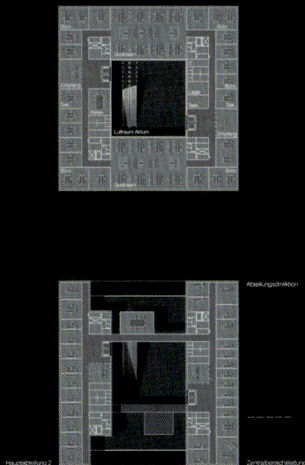

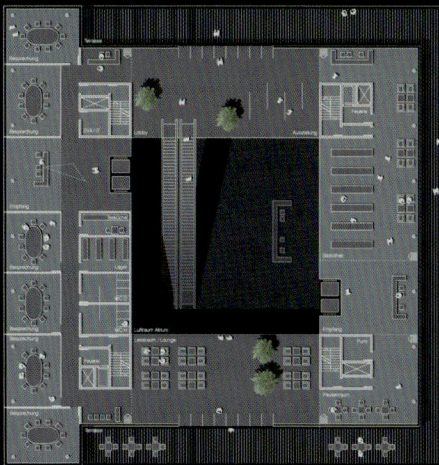

Ebene 04,07,08 M1/500
hier: Zentralbereich 1

Ebene 03 M1/200
Lobby und Bibliotheksebene

Ansicht Segmentführungsgesellschaften/ Operative Gesellschaften M1/500

Obergeschoss Operative Gesellschaften/ Segmentführungsgesellschaften M1/500

Schnitt Segmentführungsgesellschaften/ Ansicht Operative Gesellschaften M1/500

Blick in den Business-Club

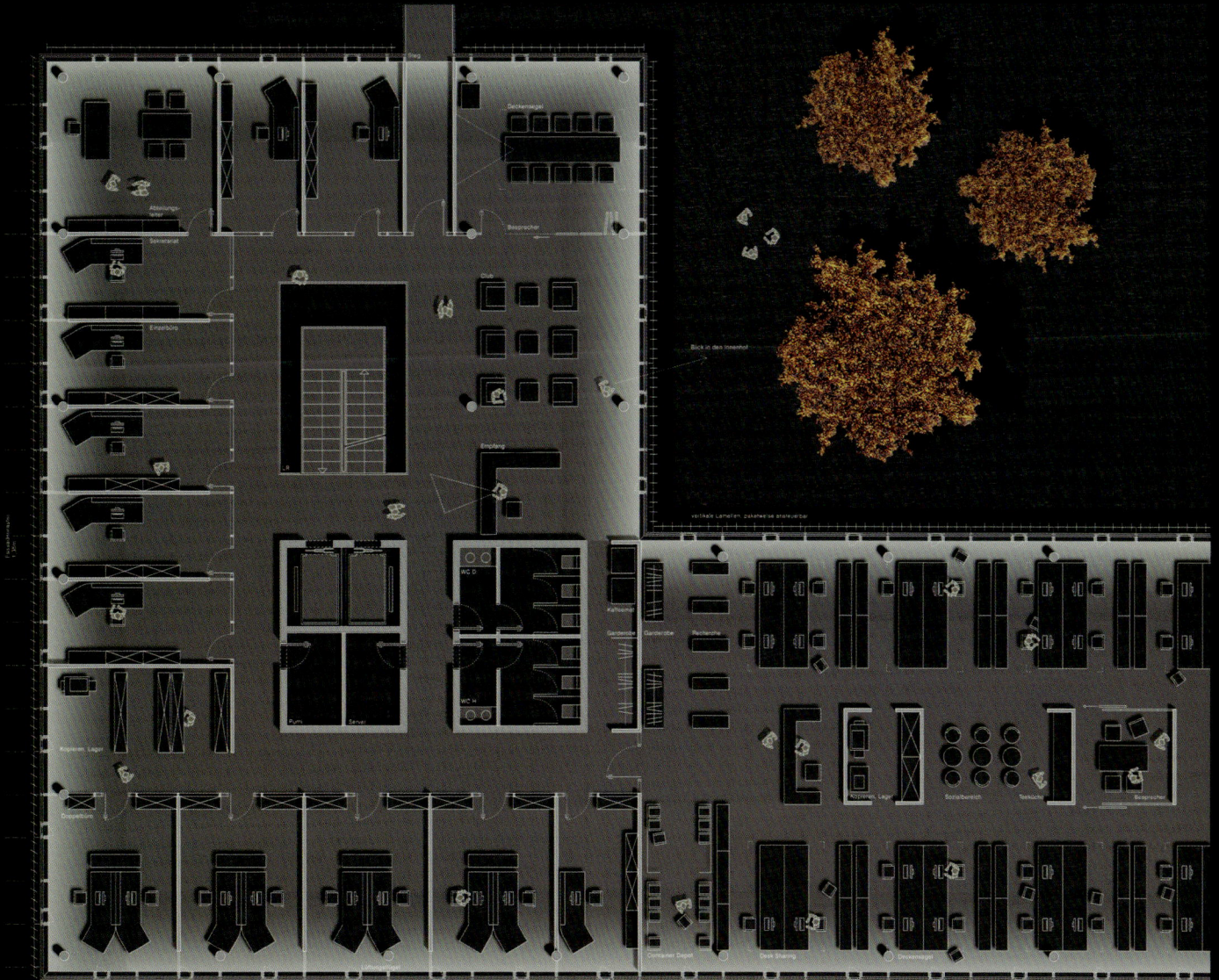

Bürokonzeption M1/50

Chaix & Morel et Associés, Paris/JSWD Architekten + Planer, Paris/Cologne

ThyssenKrupp Quartier, Essen

Architekten Brüning Klapp Rein Essen
Bruning Klapp Rein建筑事务所 埃森

"Within the city, an inviting open area is created reflecting the societal values cherished by the corporation."
"在城内创造出一片诱人的开放区反映公司珍视的社会价值观。"

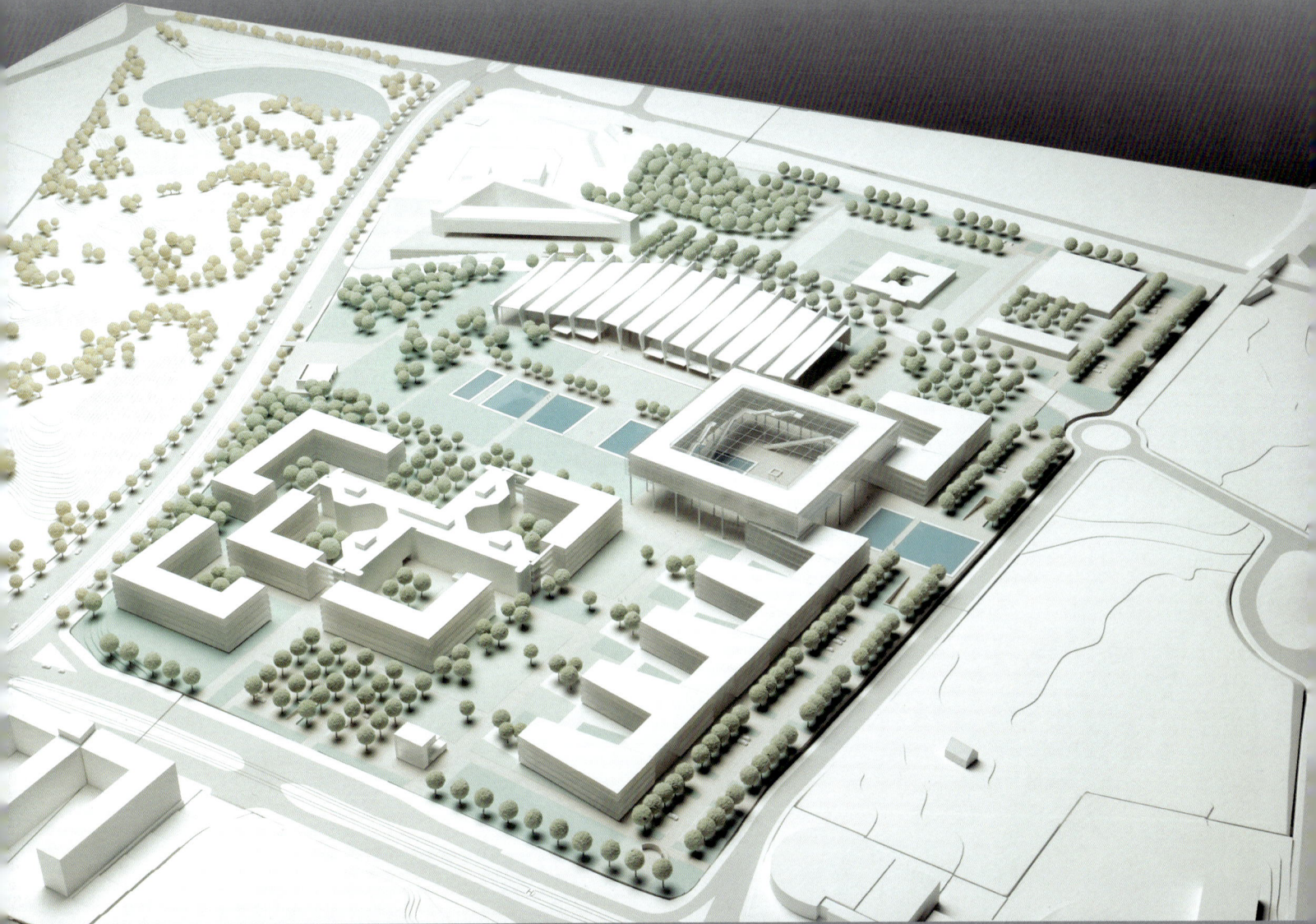

作者 Arndt Brüning, Eberhard Klapp, Volker Rein   合作伙伴 Susanne Knuth, Georg Thomys, Lars-Phillip Rusch, Tanja Brotrück, Alexander Struck, Sven Heinelt   专家 landscape planning: Jan Wehberg, Lützow 7, Berlin; structural analysis: Martin Gersiek, HEG Beratende Ingenieure VBI, Dortmund; TGA: Bernhard Pfeifer, Gerd Behrens, Zibell &Willner Ingenieurgesellschaft für Technische Gebäudeausrüstung, Bochum; modelmaking: Modellbau Römer, Essen

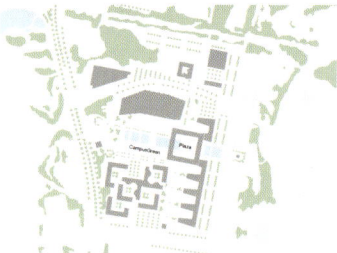

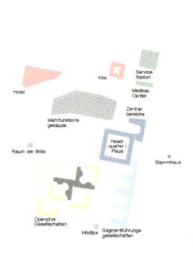

Pferdebahntrasse

RvR Radweg

120 P

PKW Service Station   II
Parken 173 P

Kita
10 P

West-Ost Strasse

Medical Center / Reisebüro   I
17 P

75 P

Autohäuser

Terrasse

TG 10 P
TG 218 P

Kongress
Multifunktions-
gebäude

Akademie
Akademie-
gärten

Gesundheit

Terrasse

Headquarter   IV

22 P

Zentral-
bereiche

Platz

Headquarter   VI bis VIII

Vorfahrt

Becken

Stammhaus

Raum der Stille

Freisitz

CampusGrün

Hängegärten

TG 40 P

Quartiersverdichtung   V

Bestandsgebäude   V

Hofgärten   TKT

Hofgärten   TKT

TG 148 P

Segmentführungs-
gesellschaften   IV   TKX

Operative Gesellschaften   V

Besucherparken 54 P

Infobox

TG 148 P

Autohäuser

Porsche

Audi

Tiefgaragen-
einfahrt

Haltestelle

Altendorfer Strasse B231

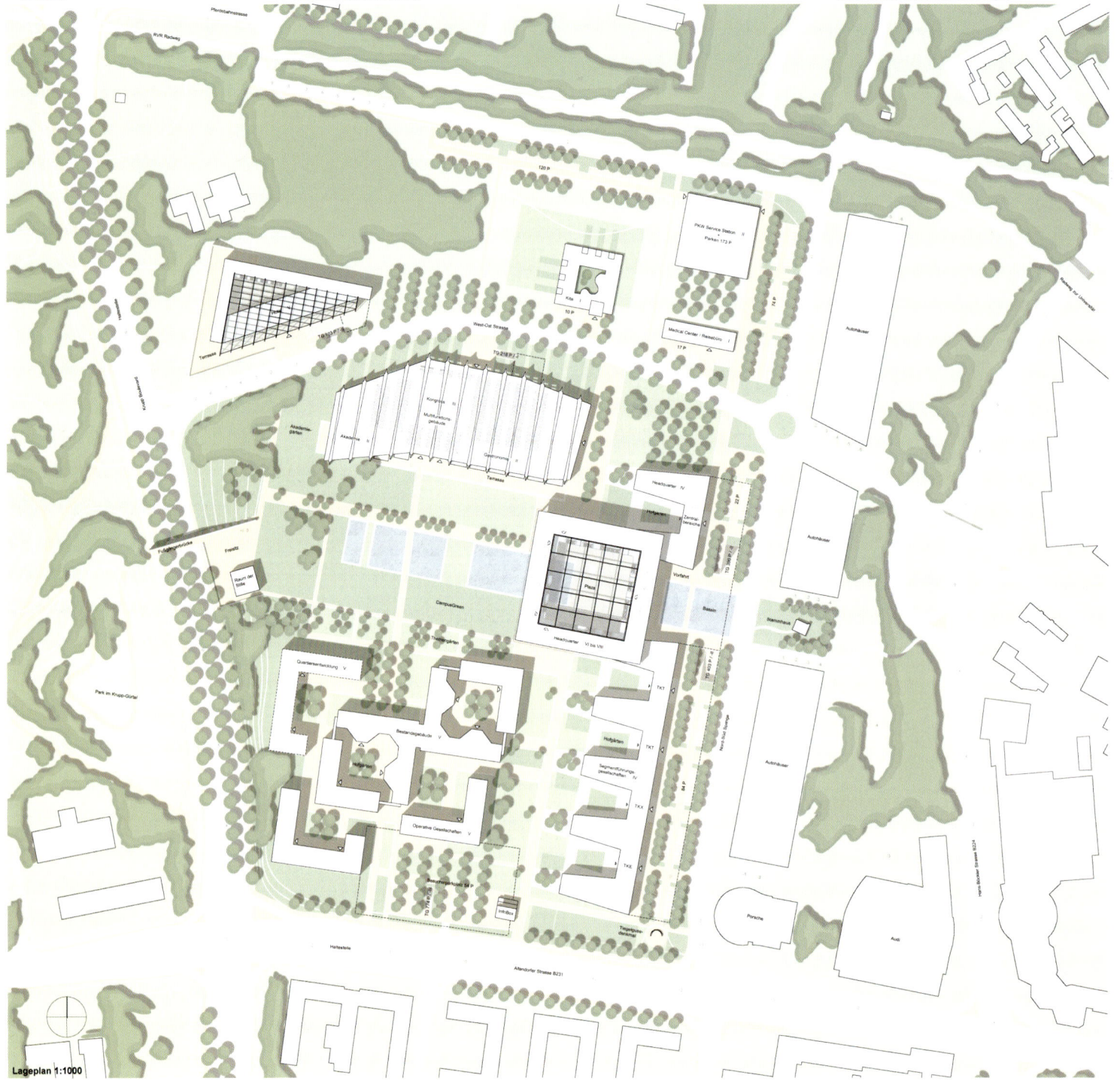

Lageplan 1:1000

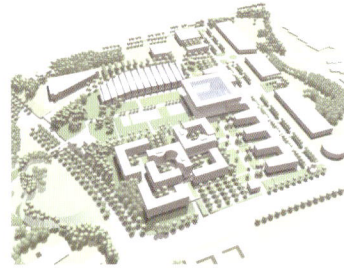

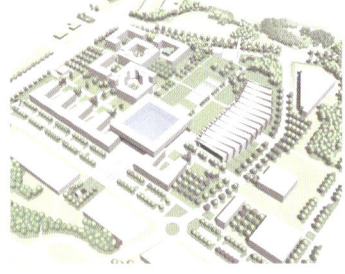

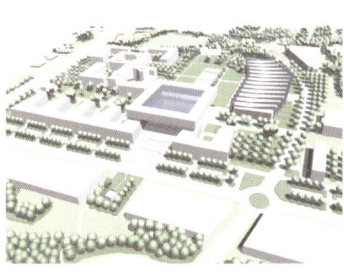

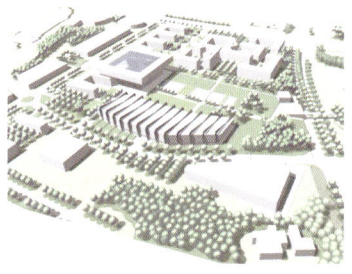

Winterfall   Sommerfall

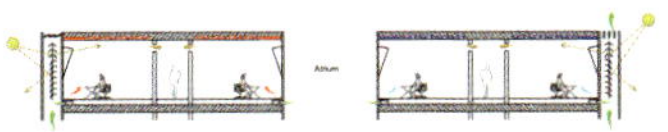

TGA-Konzeption Büro Headquarter

Winterfall   Sommerfall

TGA-Konzeption Büro Segmentführungsgesellschaften

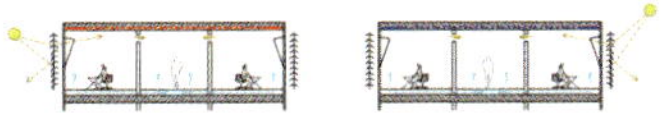

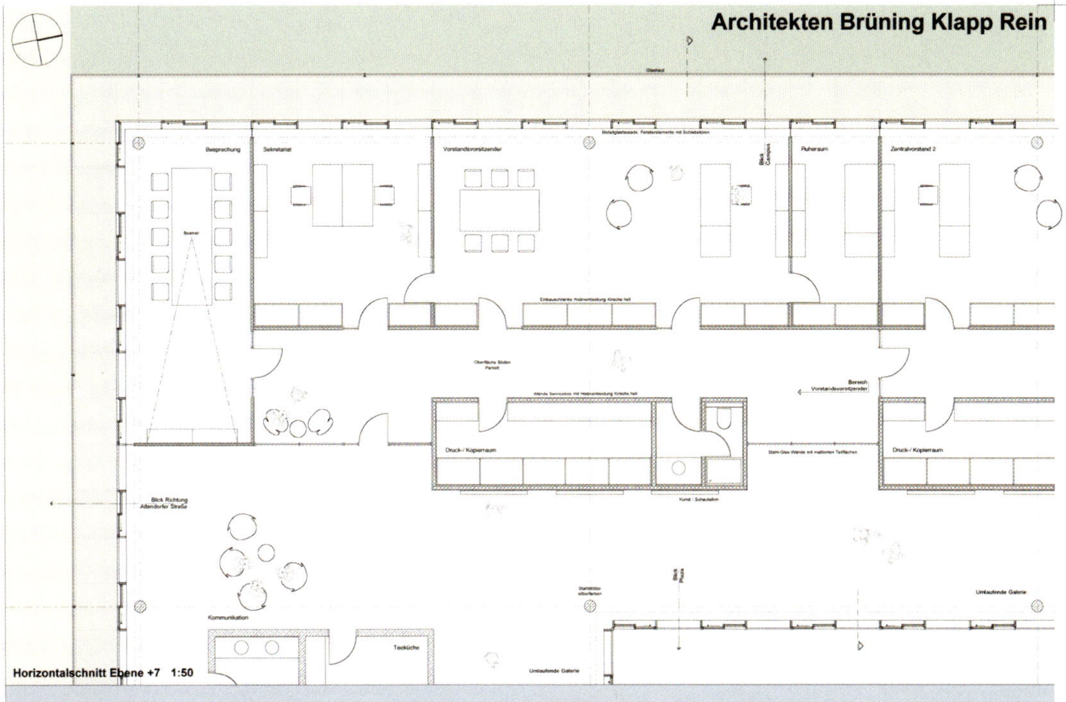

Winterfall   Sommerfall

Wettergeschützte Plaza Headquarter

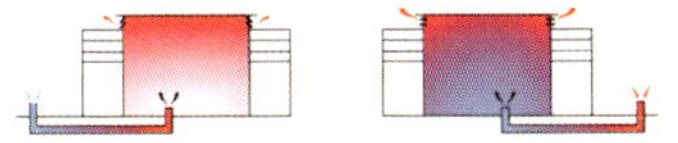

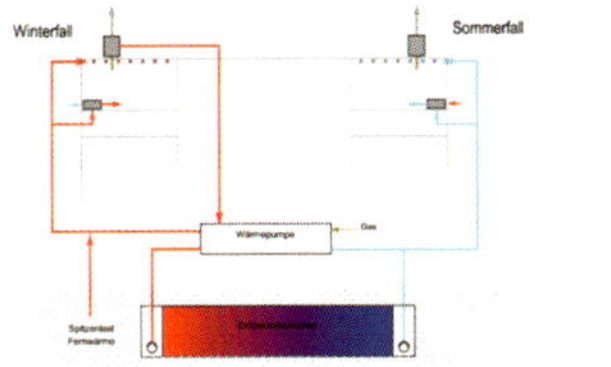

Winterfall   Sommerfall

Energiekonzept

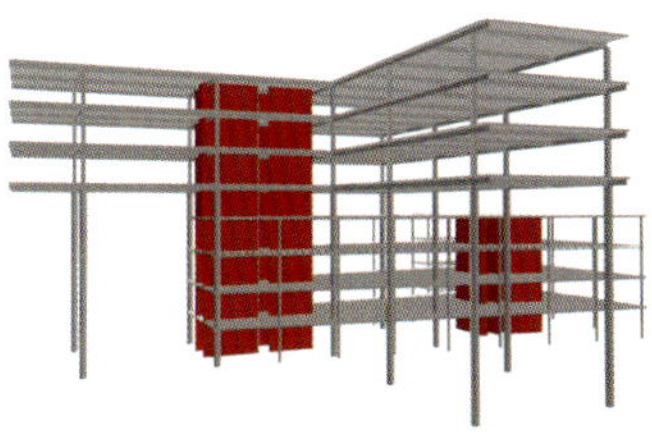

Tragwerk / Aussteifung

Horizontalschnitt Ebene +7 1:50

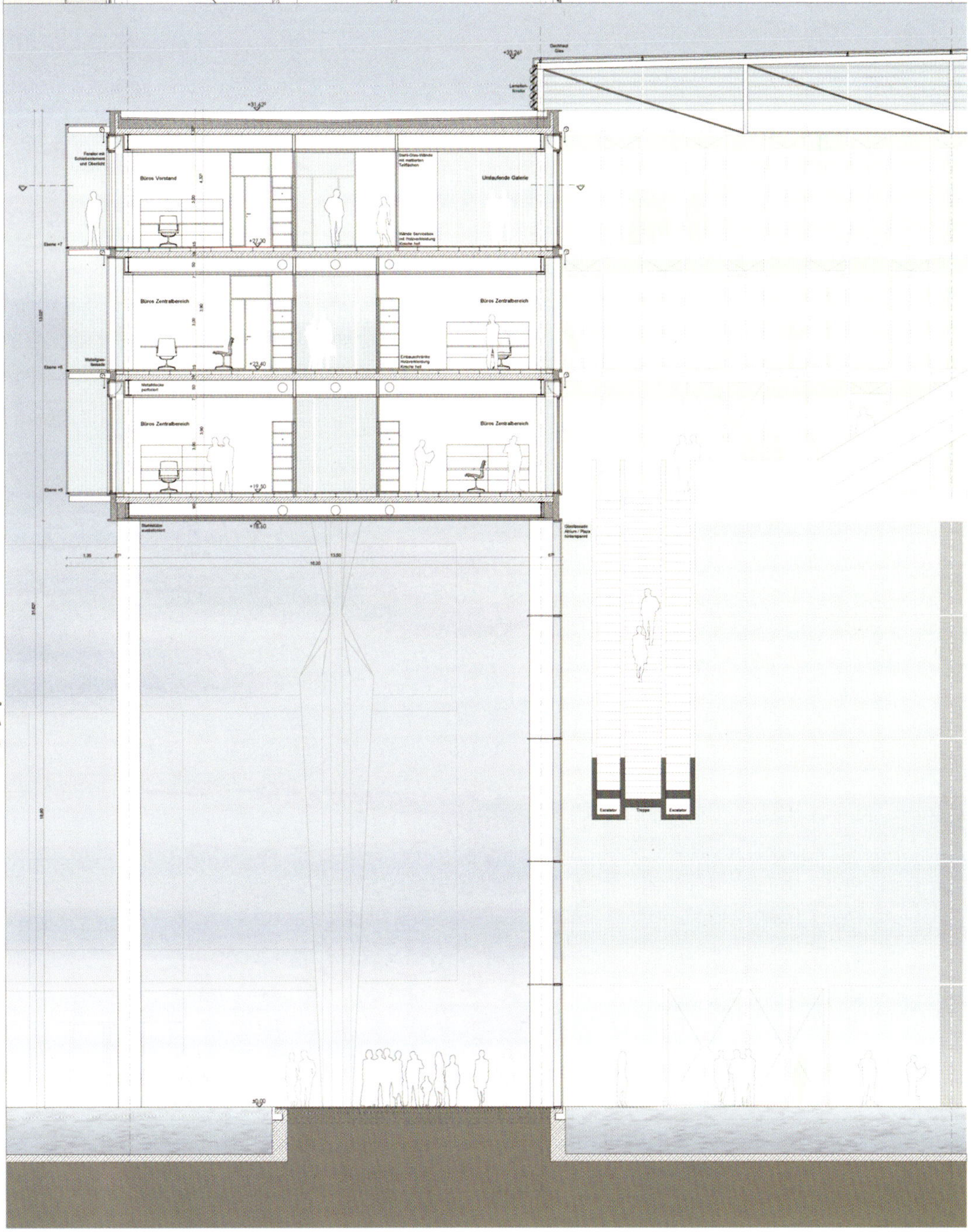

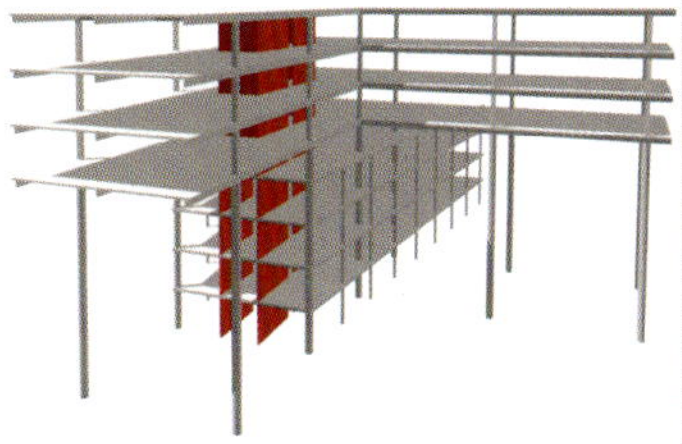

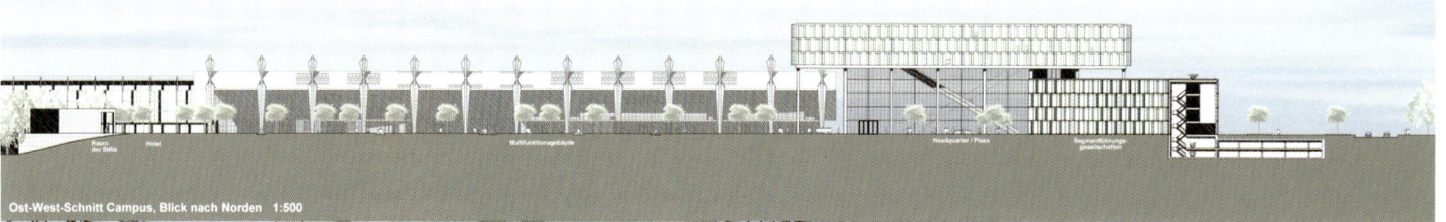

Ost-West-Schnitt Campus, Blick nach Norden 1:500

Grundriss Headquarter und Segmentführungsgesellschaften Eingangsebene 1:500

Grundriss Headquarter und Segmentführungsgesellschaften Ebene +1 bis +3 1:500

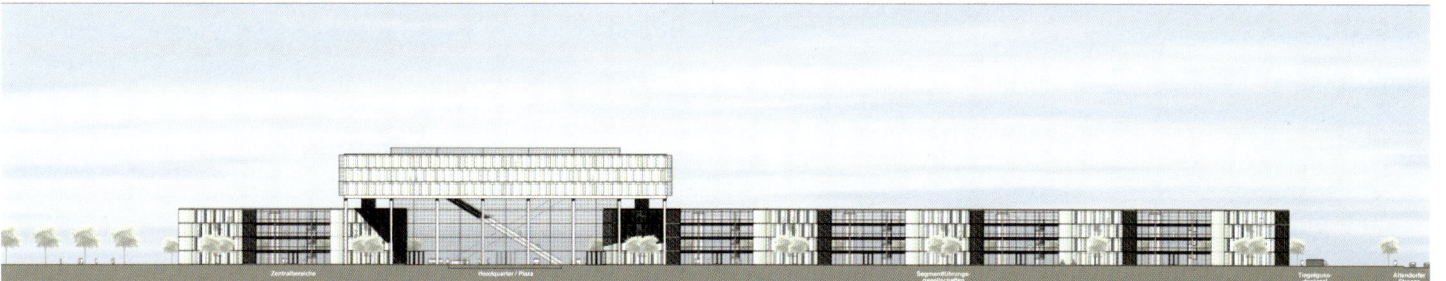

Nord-Süd-Schnitt Campus, Blick nach Osten 1:500

Headquarter und Segmentführungsgesellschaften Blatt 3

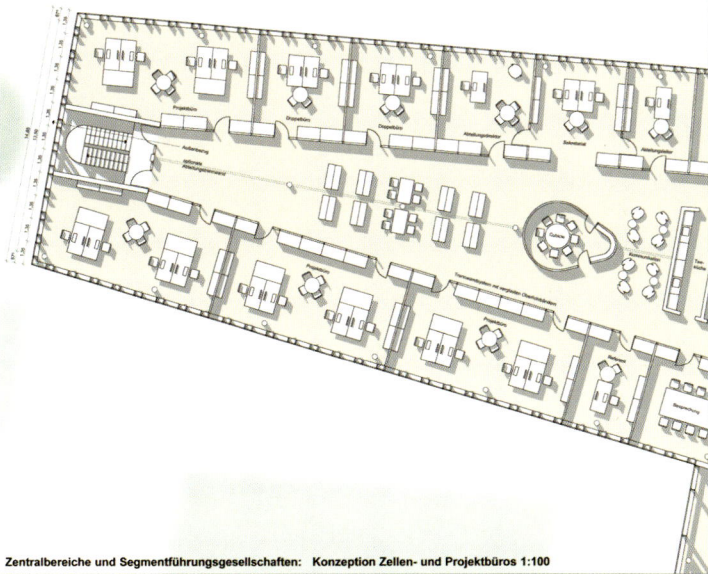

Zentralbereiche und Segmentführungsgesellschaften: Konzeption Zellen- und Projektbüros 1:100

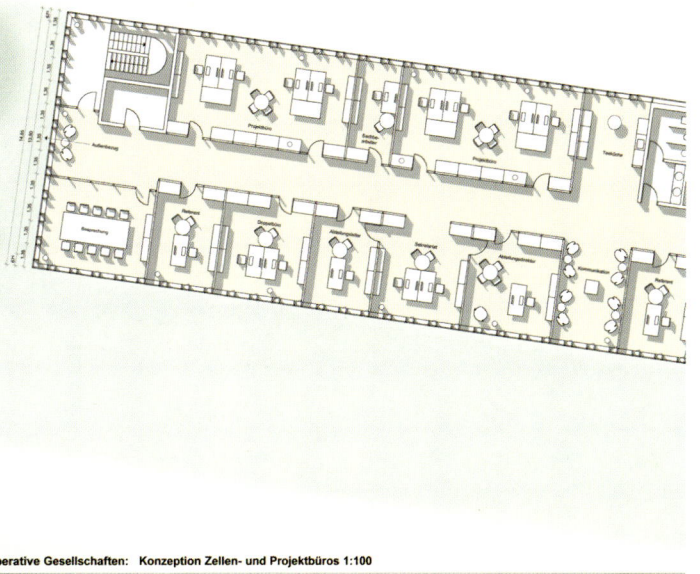

Operative Gesellschaften: Konzeption Zellen- und Projektbüros 1:100

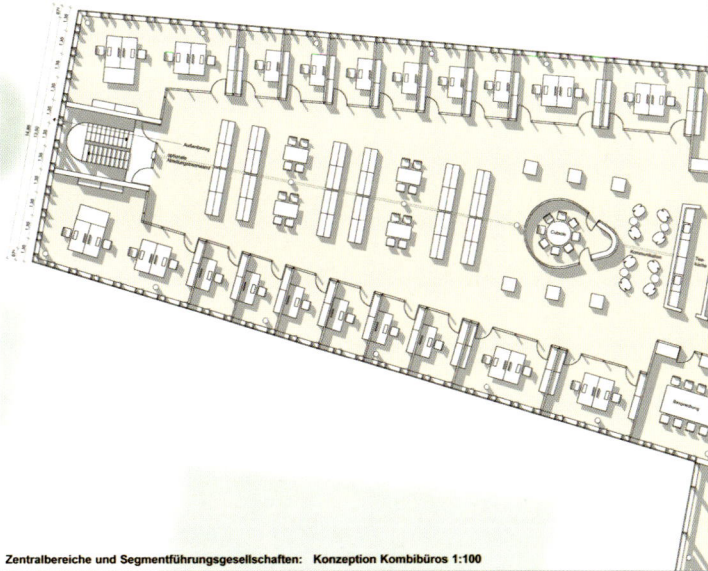

Zentralbereiche und Segmentführungsgesellschaften: Konzeption Kombibüros 1:100

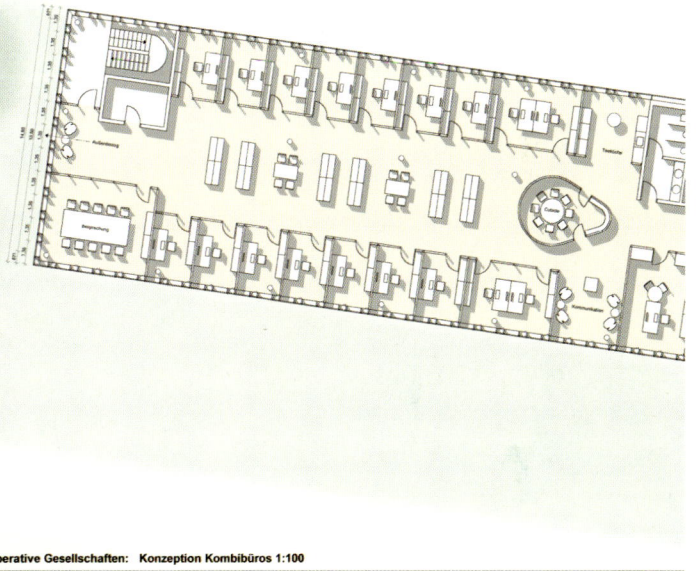

Operative Gesellschaften: Konzeption Kombibüros 1:100

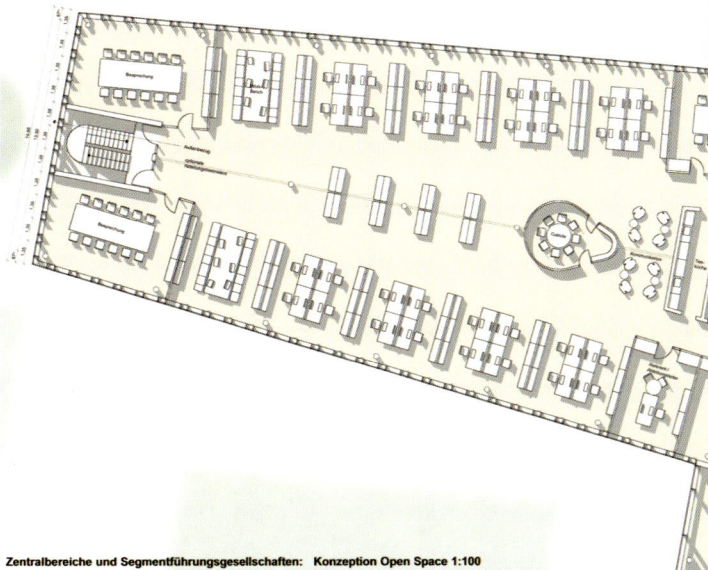

Zentralbereiche und Segmentführungsgesellschaften: Konzeption Open Space 1:100

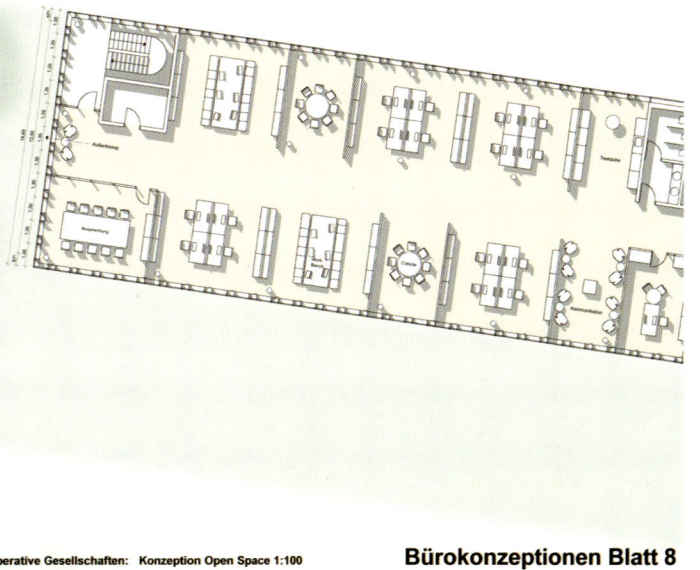

Operative Gesellschaften: Konzeption Open Space 1:100

Bürokonzeptionen Blatt 8

Zaha Hadid Architects London
Zaha Hadid建筑事务所 伦敦

"... to create a showcase architectural structure that stands out like a landmark across the quarter's borders and, with its dynamic shape, presents the corporation's calling card."

"创造一个展示性的建筑结构，像一个地标横穿蒂森克虏伯地产引人注目，同时以其充满动感的外形，打出一张公司的名片。"

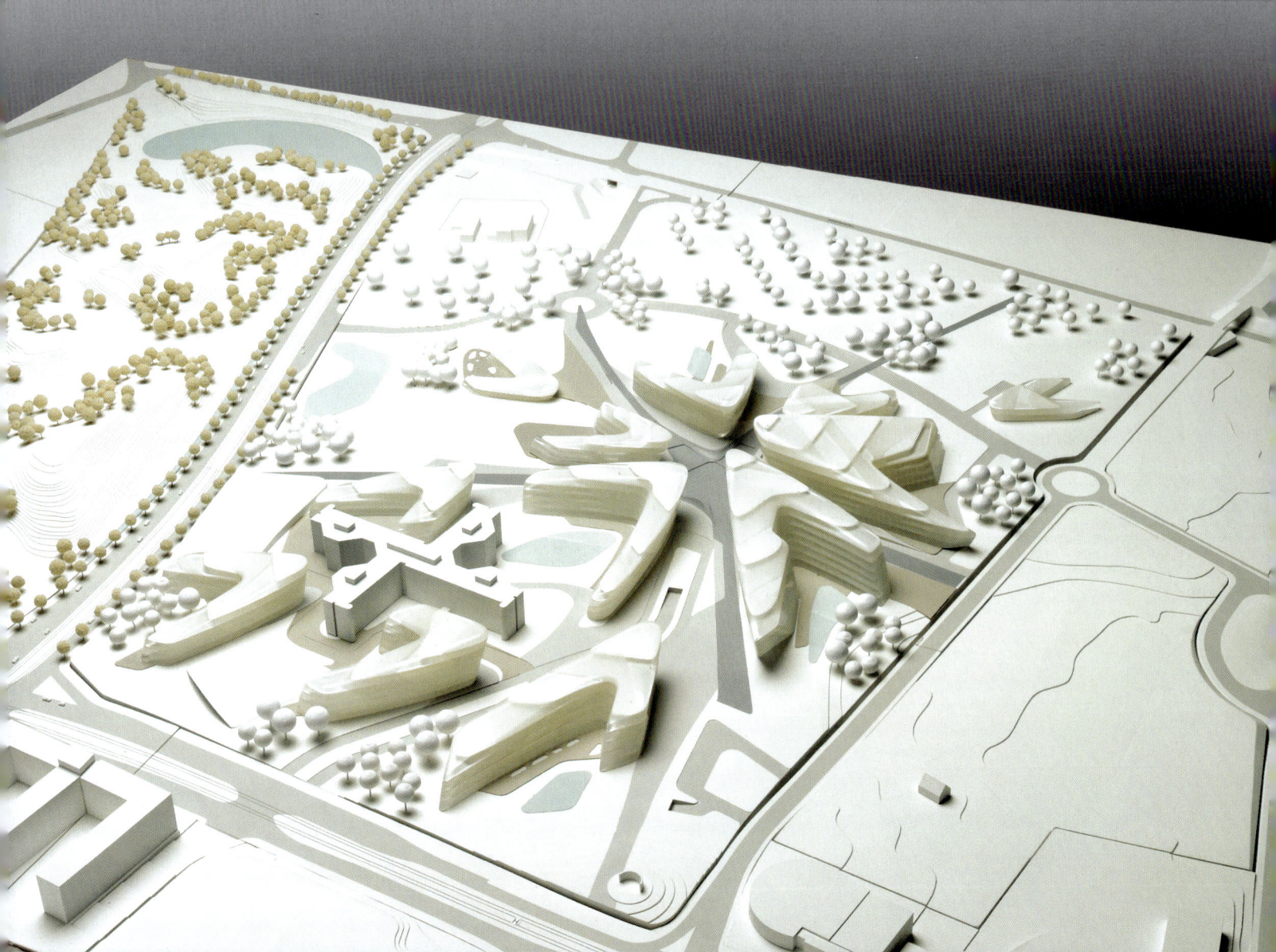

作者 Zaha Hadid, Dr. Patrik Schuhmacher   合作伙伴 Cornelius Schlotthauer, Jan Hübener, Dillon Lin, Federico Dunkelberg, Gonzalo Carbajo, Liat Muller   专家 landscape planning: Tobias Micke, Björn Bodem, ST Raum A, Berlin; structural analysis: Dr. Volker Schmid, Florian Schenk, Rudi Scheuermann, Arup Berlin; energy consulting: Prof. Volkmar Bleicher, Transsolar Stuttgart; traffic engineers: Martin Reed, Arup London; fire protection: Heiko Zies, HHP West, Bielefeld

Zaha Hadid Architects, London

237

ThyssenKrupp Quarter, Essen

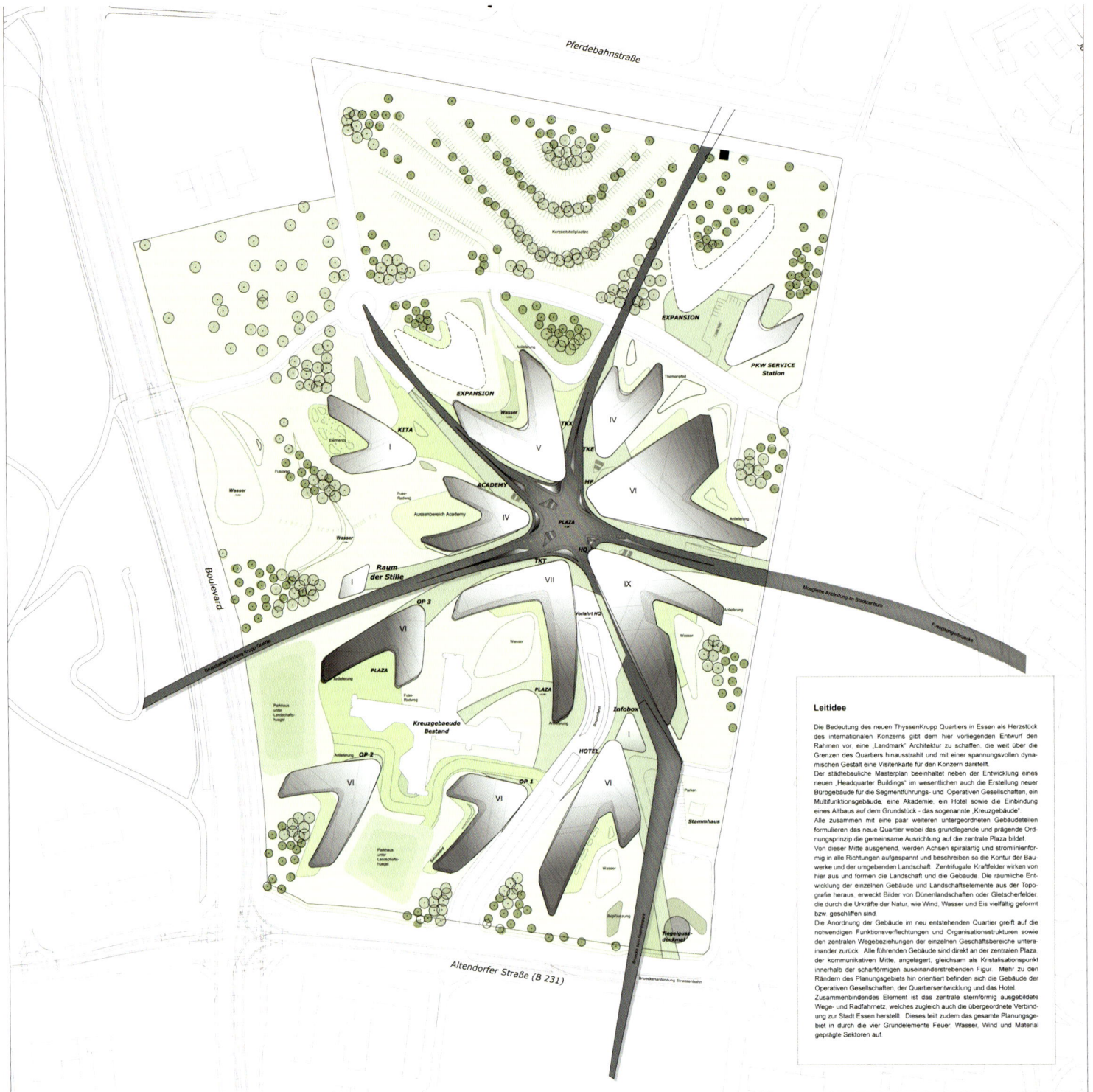

Pferdebahnstraße

EXPANSION

PKW SERVICE Station

EXPANSION

KITA

I

V

TKI

IV

TKE

ACADEMY

MF

VI

Außenbereich Academy

IV

Wasser

PLAZA

HQ

Raum der Stille

I

TKT

VII

IX

Boulevard

OP 3

VI

Wasser

Vorfahrt HQ

Wasser

PLAZA

PLAZA

Infobox

Kreuzgebaeude Bestand

OP 2

HOTEL

I

OP 1

VI

VI

VI

Stammhaus

Parkhaus unter Landschaftshügel

Parkhaus unter Landschaftshügel

Hegelgussdenkmal

Altendorfer Straße (B 231)

Leitidee

Die Bedeutung des neuen ThyssenKrupp Quartiers in Essen als Herzstück des internationalen Konzerns gibt dem hier vorliegenden Entwurf den Rahmen vor, eine „Landmark" Architektur zu schaffen, die weit über die Grenzen des Quartiers hinausstrahlt und mit einer spannungsvollen dynamischen Gestalt eine Visitenkarte für den Konzern darstellt.

Der städtebauliche Masterplan beinhaltet neben der Entwicklung eines neuen „Headquarter Buildings" im wesentlichen auch die Erstellung neuer Bürogebäude für die Segmentführungs- und Operativen Gesellschaften, ein Multifunktionsgebäude, eine Akademie, ein Hotel sowie die Einbindung eines Altbaus auf dem Grundstück - das sogenannte „Kreuzgebäude".

Alle zusammen mit eine paar weiteren untergeordneten Gebäudeteilen formulieren das neue Quartier wobei das grundlegende und prägende Ordnungsprinzip die gemeinsame Ausrichtung auf die zentrale Plaza bildet.

Von dieser Mitte ausgehend, werden Achsen spiralartig und stromlinienförmig in alle Richtungen aufgespannt und beschreiben so die Kontur der Bauwerke und der umgebenden Landschaft. Zentrifugale Kraftfelder wirken von hier aus und formen die Landschaft und die Gebäude. Die räumliche Entwicklung der einzelnen Gebäude und Landschaftselemente aus der Topografie heraus, erweckt Bilder von Dünenlandschaften oder Gletscherfelder, die durch die Urkräfte der Natur, wie Wind, Wasser und Eis vielfältig geformt bzw. geschliffen sind.

Die Anordnung der Gebäude im neu entstehenden Quartier greift auf die notwendigen Funktionsverflechtungen und Organisationsstrukturen sowie den zentralen Wegebeziehungen der einzelnen Geschäftsbereiche untereinander zurück. Alle führenden Gebäude sind direkt an der zentralen Plaza, der kommunikativen Mitte, angelagert, gleichsam als Kristalisationspunkt innerhalb der scharfkantigen auseinanderstrebenden Figur. Mehr zu den Rändern des Planungsgebiets hin orientiert befinden sich die Gebäude der Operativen Gesellschaften, der Quartiersentwicklung und das Hotel.

Zusammenbindendes Element ist das zentrale sternförmig ausgebildete Wege- und Radfahrnetz, welches zugleich auch die übergeordnete Verbindung zur Stadt Essen herstellt. Dieses teilt zudem das gesamte Planungsgebiet in durch die vier Grundelemente Feuer, Wasser, Wind und Material geprägte Sektoren auf.

MASTER PLAN 1:1000

GEBAEUDENUTZUNG   FUSS-/ UND RADWEGE   VERKEHRSERSCHLIESSUNG   GEBAEUDEVORFAHRTEN

LANDSCHAFTSPERSPEKTIVE

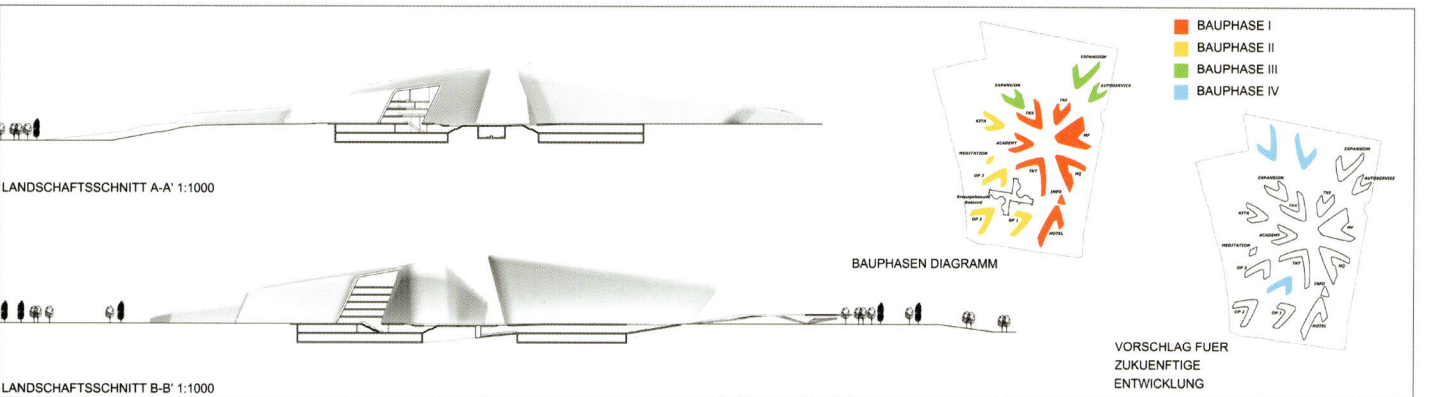

LANDSCHAFTSSCHNITT A-A' 1:1000

LANDSCHAFTSSCHNITT B-B' 1:1000

BAUPHASE I
BAUPHASE II
BAUPHASE III
BAUPHASE IV

BAUPHASEN DIAGRAMM

VORSCHLAG FUER
ZUKUENFTIGE
ENTWICKLUNG

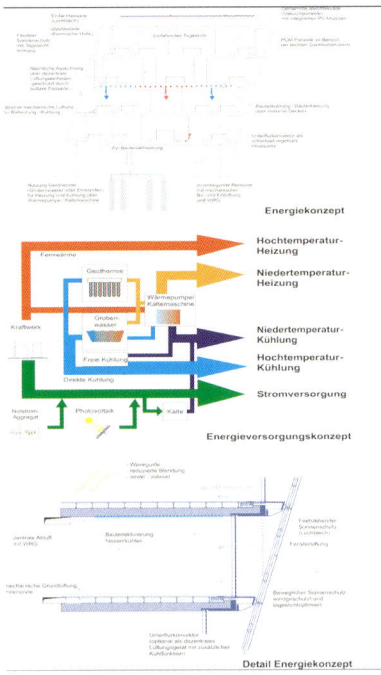

Energiekonzept

Energieversorgungskonzept

Detail Energiekonzept

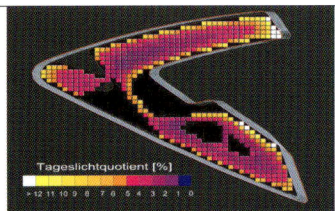

Tageslichtquotient [%]
> 12 11 10 9 8 7 6 5 4 3 2 1 0

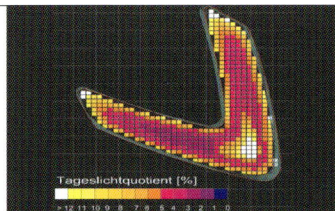

Tageslichtquotient [%]
> 12 11 10 9 8 7 6 5 4 3 2 1 0

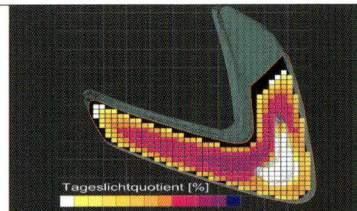

Tageslichtquotient [%]
> 12 11 10 9 8 7 6 5 4 3 2 1 0

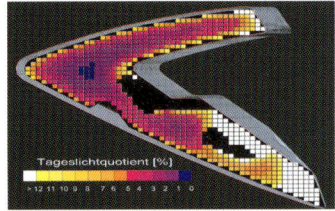

Tageslichtquotient [%]
> 12 11 10 9 8 7 6 5 4 3 2 1 0

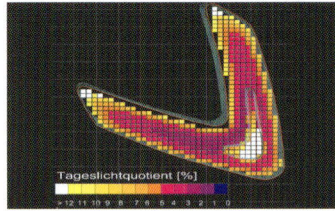

Tageslichtquotient [%]
> 12 11 10 9 8 7 6 5 4 3 2 1 0

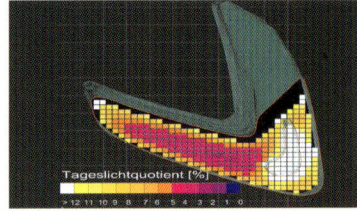

Tageslichtquotient [%]
> 12 11 10 9 8 7 6 5 4 3 2 1 0

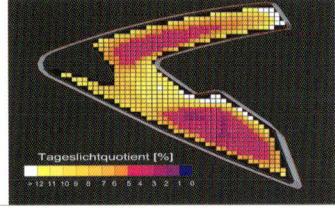

Tageslichtquotient [%]
> 12 11 10 9 8 7 6 5 4 3 2 1 0

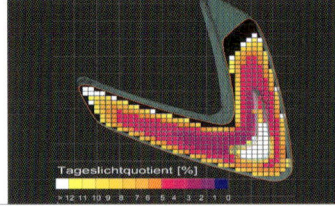

Tageslichtquotient [%]
> 12 11 10 9 8 7 6 5 4 3 2 1 0

TRANSSOLAR

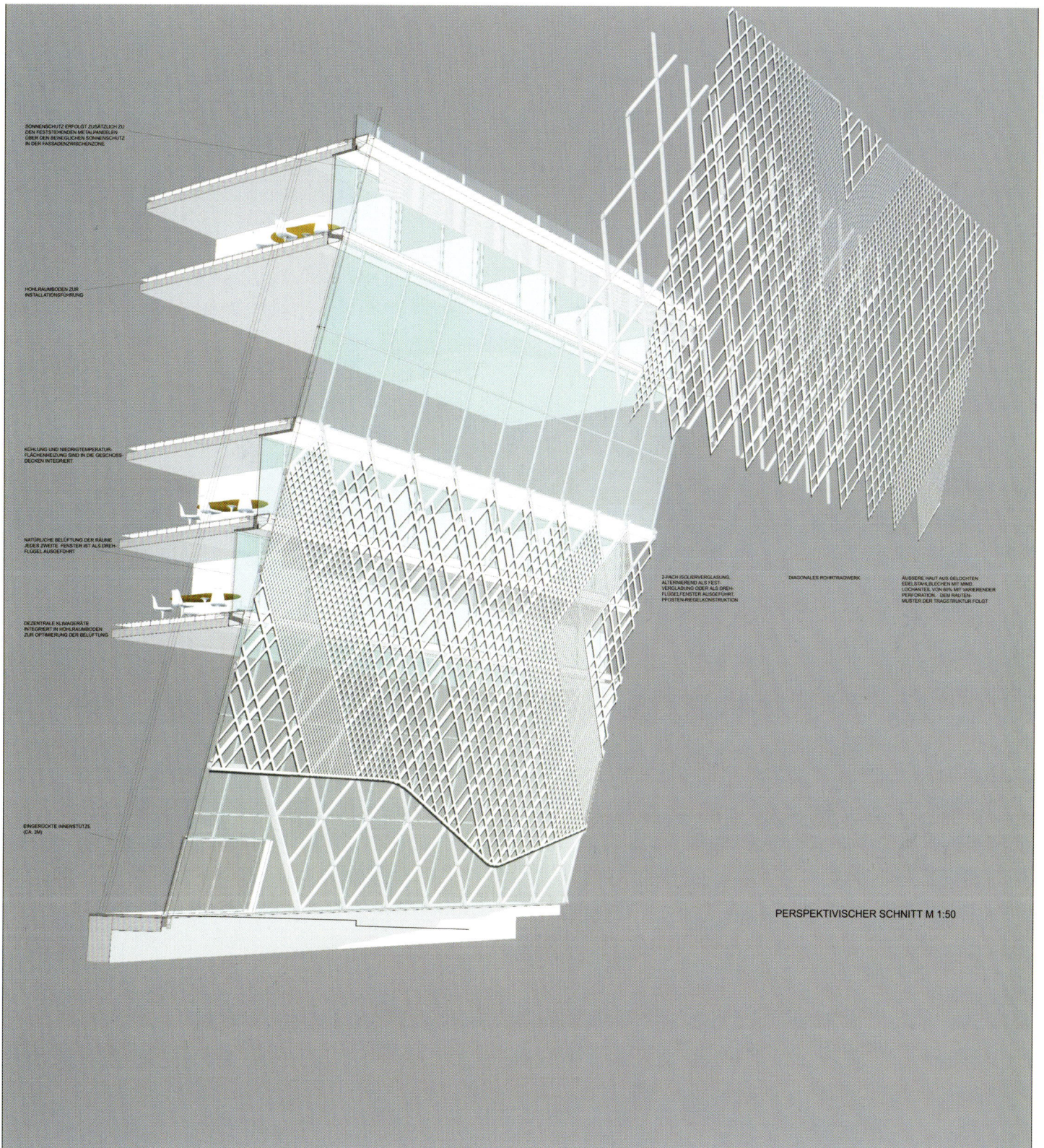

SONNENSCHUTZ ERFOLGT ZUSÄTZLICH ZU
DEN FESTSTEHENDEN METALLPANEELEN
ÜBER DEN BEWEGLICHEN SONNENSCHUTZ
IN DER FASSADENZWISCHENZONE

HOHLRAUMBODEN ZUR
INSTALLATIONSFÜHRUNG

KÜHLUNG UND NIEDRIGTEMPERATUR-
FLÄCHENHEIZUNG SIND IN DIE GESCHOSS-
DECKEN INTEGRIERT

NATÜRLICHE BELÜFTUNG DER RÄUME
JEDES ZWEITE FENSTER IST ALS DREH-
FLÜGEL AUSGEFÜHRT

DEZENTRALE KLIMAGERÄTE
INTEGRIERT IN HOHLRAUMBODEN
ZUR OPTIMIERUNG DER BELÜFTUNG

EINGERÜCKTE INNENSTÜTZE
(CA. 3M)

3-FACH ISOLIERVERGLASUNG,
ALTERNIEREND ALS FEST-
VERGLASUNG ODER ALS DREH-
FLÜGELFENSTER AUSGEFÜHRT,
PFOSTEN-RIEGELKONSTRUKTION

DIAGONALES ROHRTRAGWERK

ÄUSSERE HAUT AUS GELOCHTEN
EDELSTAHLBLECHEN MIT MIND.
LOCHANTEIL VON 80% MIT VARIIERENDER
PERFORATION. DEM RAUTEN-
MUSTER DER TRAGSTRUKTUR FOLGT

PERSPEKTIVISCHER SCHNITT M 1:50

FASSADEN DETAIL 1:50

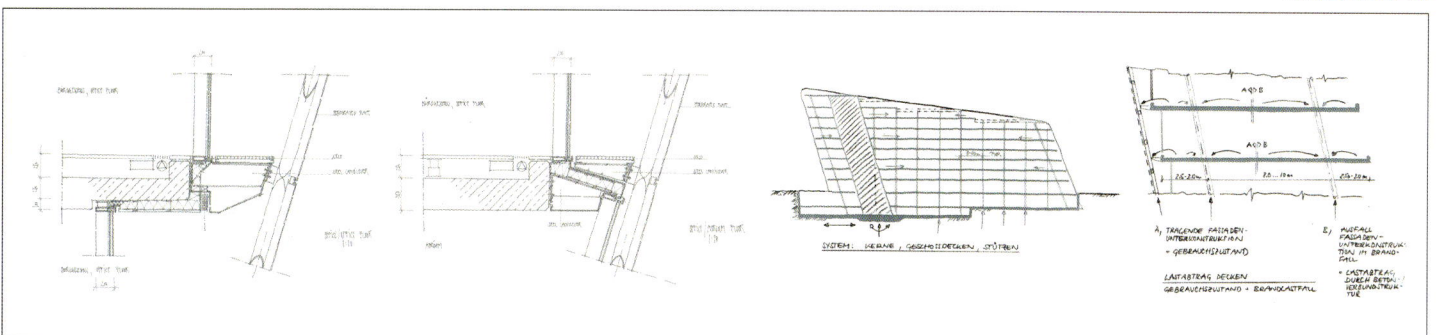

SYSTEM: HÄUME, GESCHOSSDECKEN, STÜTZEN

A) TRAGENDE FASSADEN-
UNTERKONSTRUKTION
+ GEBRAUCHSZUSTAND

LASTABTRAG DECKEN
GEBRAUCHSZUSTAND + BRANDLASTFALL

B) AUSFALL
FASSADEN-
UNTERKONSTRUK-
TION IM BRAND-
FALL
+ LASTABTRAG
DURCH BETON-
KERNDÄMMUNGSTRUK-
TUR

ARUP FASSADEN KONZEPT

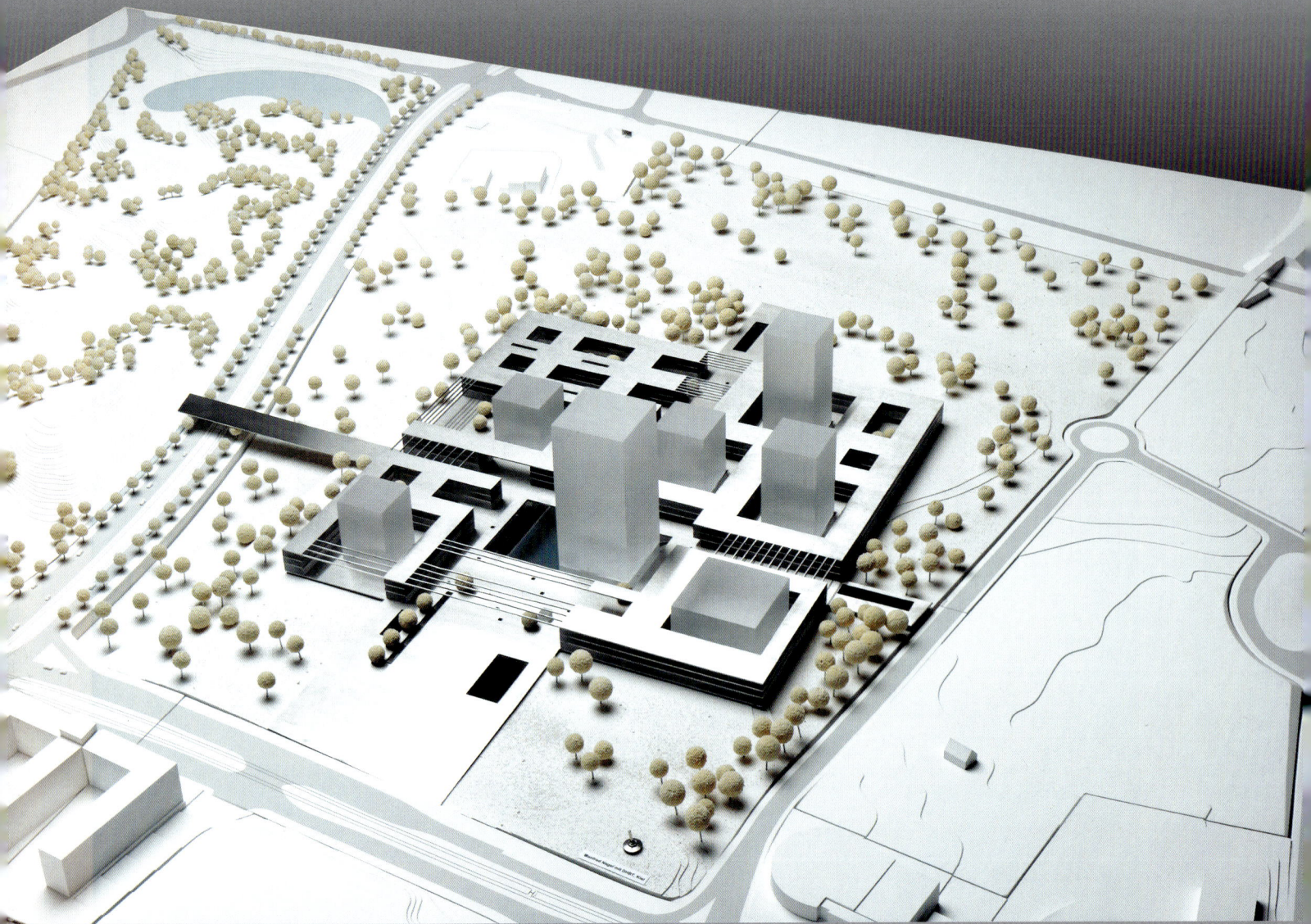

Manfred Nagel with DHBT Kiel
Manfred Nagel with DHBT建筑事务所 基尔

"The design for the ThyssenKrupp Quarter creates a landmark of the polarity between the historical way of thinking and the new, that its corporate concept represents."

"蒂森克虏伯办公建筑的设计在传统的思维方式和新的思维方式之间创造出一个顶极的地标，反映公司的观念。"

作者 Manfred Nagel, Roloff Werner   合作伙伴 Carola Hechtnagel, Kay-Peter Kolbe, Marion Beherens, Kerstin Pulss   自由建筑师 Maik Loss, Bianca Maier   专家 landscape planning: Jens Bendfeldt, Bendfeldt - Schröder - Franke, Kiel; structural analysis: Stephan Schmidt, Christian Drescher, Ing.-Büro Dr. Binnewies, Hamburg; building services: Dr. Torsten Warner, Ebert Ingenieure, Hamburg; façade analysis: Christian Freimann, Prof. Michael Lange Ing.-GmbH, Hamburg; fire protection: Dr. Michael Kiel, HHP-nord-Ost, Braunschweig; lighting: Johann von Bothmer, Ulrike Brandi Licht, Hamburg; consultant: Prof. Martin Jürgen Liemk (Muthesius art college Kiel)

Manfred Nagel with DHBT, Kiel

Image Landmark

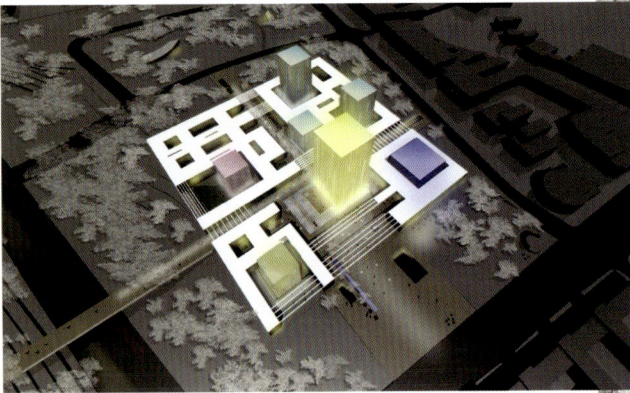

Erweiterung 1:2000

luk digitale Informations und Kommunikationstechnologie

Individualbüros

CHECK_IN

INSTALLATION

Interaktionszonen

Business Club

KOMMUNIKATION

BA

HOF campus

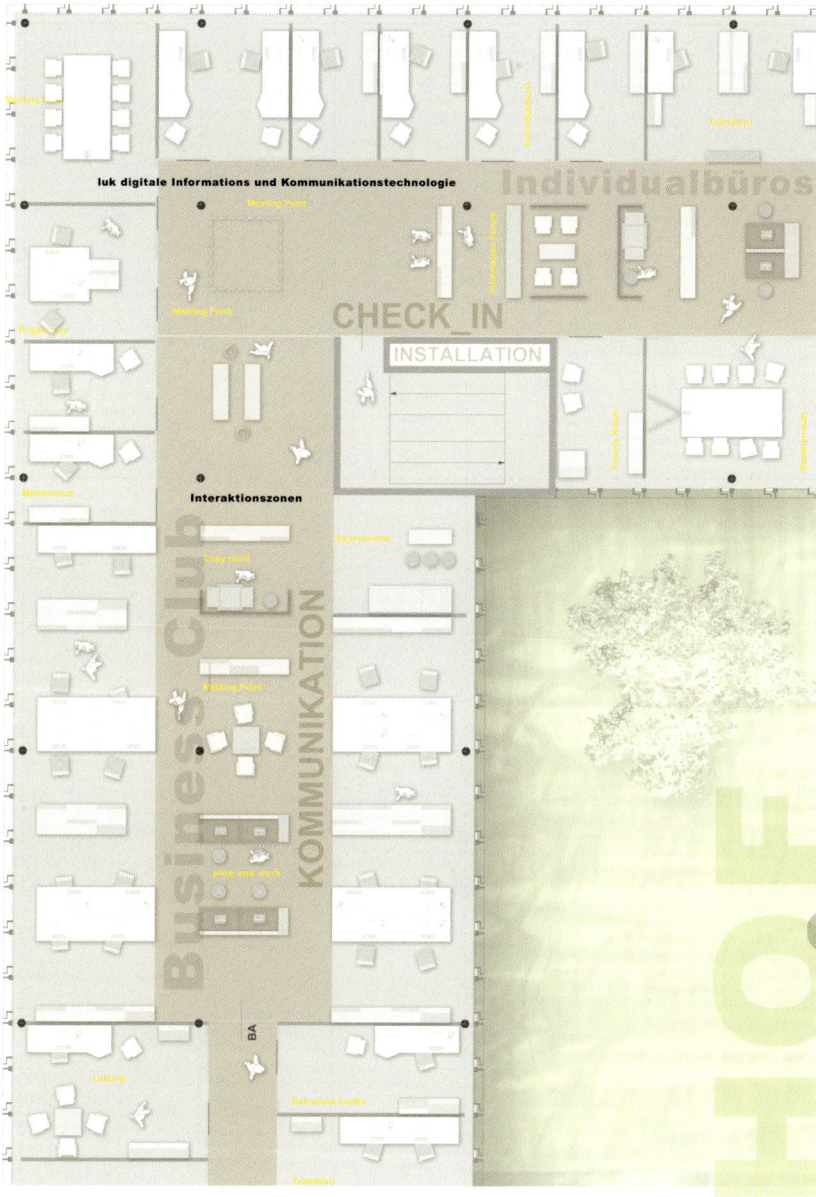

Beispiel Bürostruktur 1:50

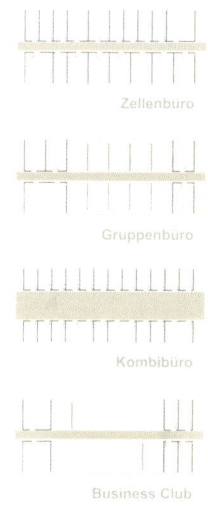

Zellenbüro

Gruppenbüro

Kombibüro

Business Club

Varianten

Kurvenverlauf gemessen in der Mittelachse eines Raumes
Raum ohne gegenüberliegende Bebauung

Mittlerer Tageslichtquotient (TQ) bezogen auf die Gesamtraumfläche:
2,8 % (tageslichtorientiert)

Kurvenverlauf gemessen in der Mittelachse eines Raumes
Raum mit 3 geschossiger, gegenüberliegender Bebauung

Mittlerer Tageslichtquotient (TQ) bezogen auf die Gesamtraumfläche:
1,55 % (ausreichend hell)

Schnitt 1:50

Bereich Tiefgarage

KSP Engel und Zimmermann Architekten Frankfurt/Main
KSP Engel and Zimmermann建筑事务所 德国美因河畔法兰克福

"Two boomerang-shaped landscape elements [...] form the backbone of the ThyssenKrupp Quarter, and embody the idea of the corporation's past, present and future."

"飞镖形状的景观要素构成蒂森克虏伯总部的支柱，体现公司过去、现在和将来的观点。"

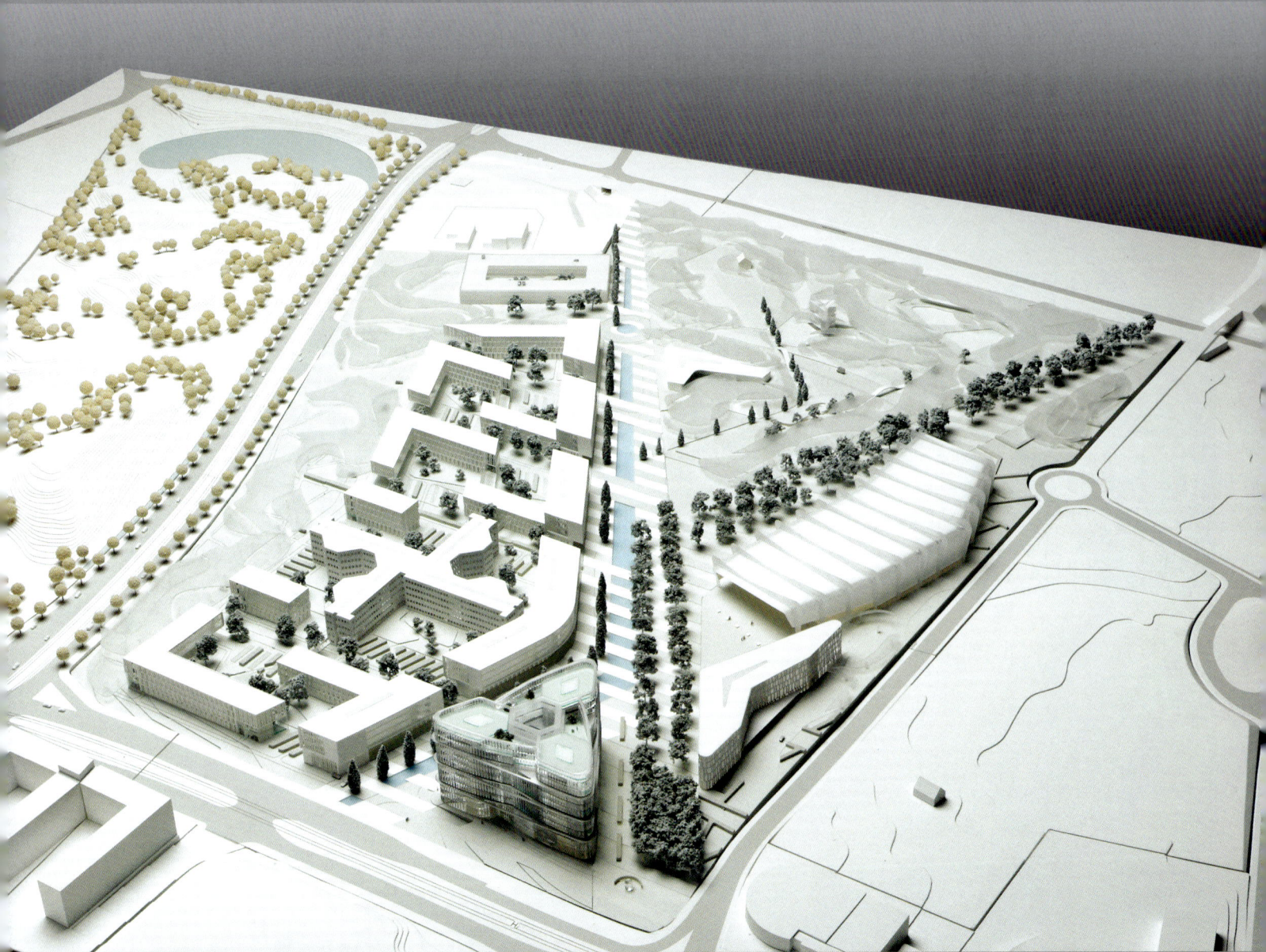

作者 Jürgen Engel   合作伙伴 G. Gutscher, A. Vultaggio, A. Stoyanova, D. Jerke, R. Holubek, S. Güning, N. Bergemann, J. Eichelberger, G. Hohenbleicher, A. Jäger, B. Kroh, R. Lauterbach, L. Lenhard, D. Österbauer, R. Reininger, F. Rudolph, Chr. Schütz, S. Lippert, D. Meier, T. Reinhardt, A. Seibert, M. Khavari   专家 landscape planning: Mr. Fenner, FSW Landschaftsarchitekten, Düsseldorf; support structure: Mr. Plieninger, Schlaich Bergermann und Partner, Stuttgart; TGA: Adrian Altenburger, Amstein+Wahlert AG, Zurich; fire protection: Mr. Dahlitz, hpp Berlin

[blick vom thyssenkrupp forum aus]

ATMOSPHÄRE UND
AUSBLICKE

DEZENTRALE
ERSCHLIESSUNG

FLEXIBILITÄT IM
GRUNDRISS

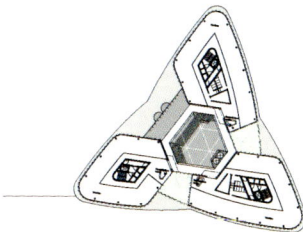

grundriss 1.og

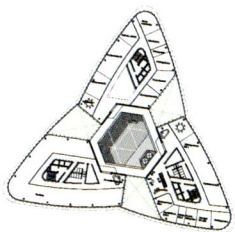

grundriss 4.og (zentralbereich)

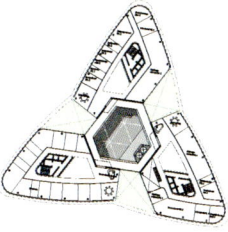

grundriss 5.og (zentralbereich)

grundriss 9.og (vorstand)

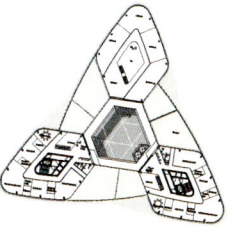

grundriss 10.og (vorstand)

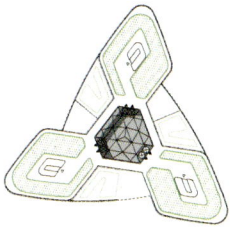

dachgarten

[grundrisse_headquarter] m 1:500

08_gebäudestruktur

Engel und Zimmermann
Architekten

headquater

esplanade

hans-böckler-straße
B224

[büropark]

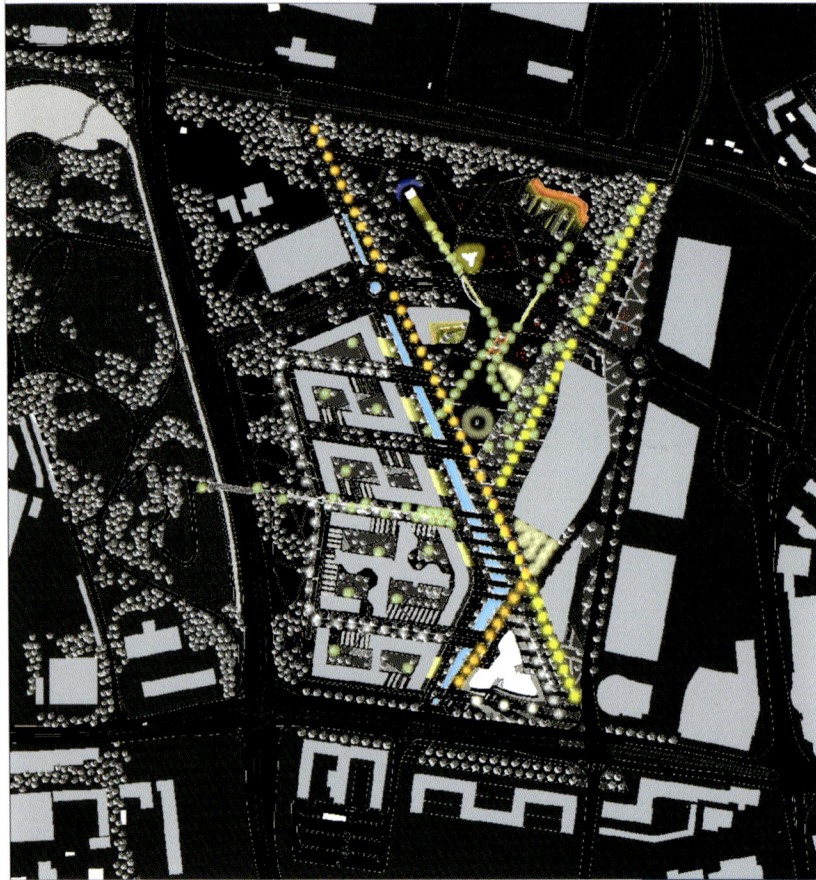
[beleuchtungsplan] m 1:500

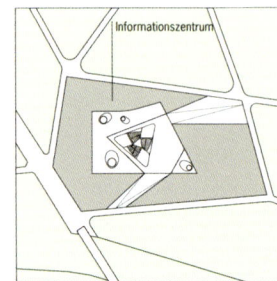
Informationszentrum

11 freiraumplanung

Engel und Zimmermann
Architekten

[schnitt_01] m 1:500

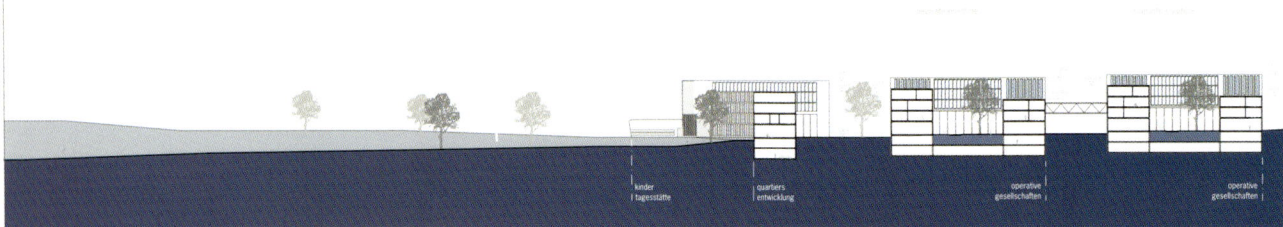
kinder
| tagesstätte

quartiers
entwicklung

operative
gesellschaften

operative
gesellschaften

[atrium headquarter]

[grundriss_vorstand_hq] m 1:50

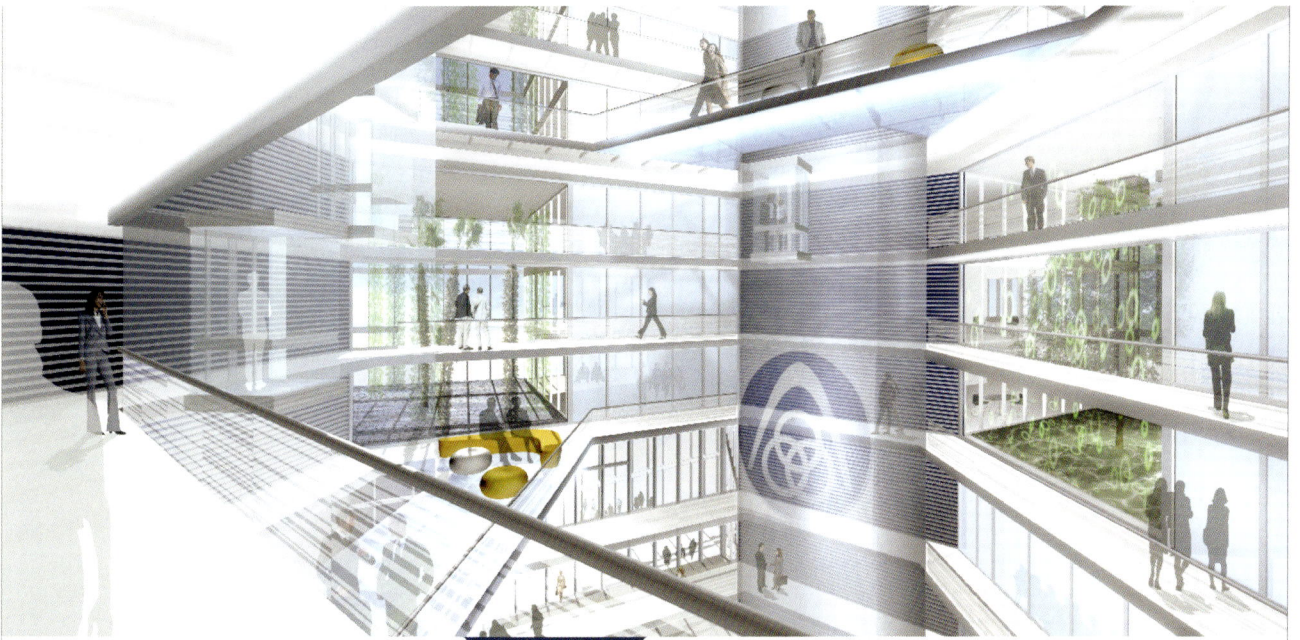

09_atmosphäre

 Engel und Zimmermann
Architekten

[schnitt_02] m 1:500

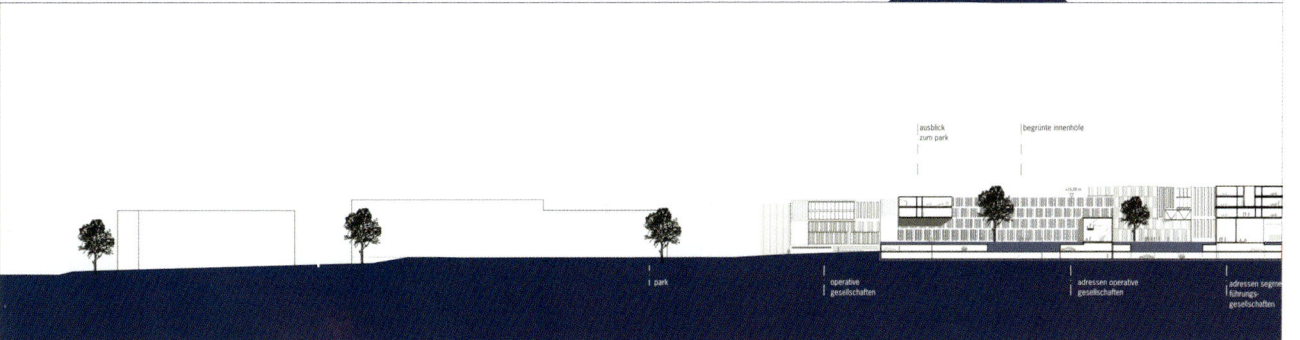

h4a Gessert + Randecker Architekten BDA Stuttgart
h4a Gessert + Randecker建筑事务所 BDA 斯图加特

„... vitality and vision"
"……活力和想象力。"

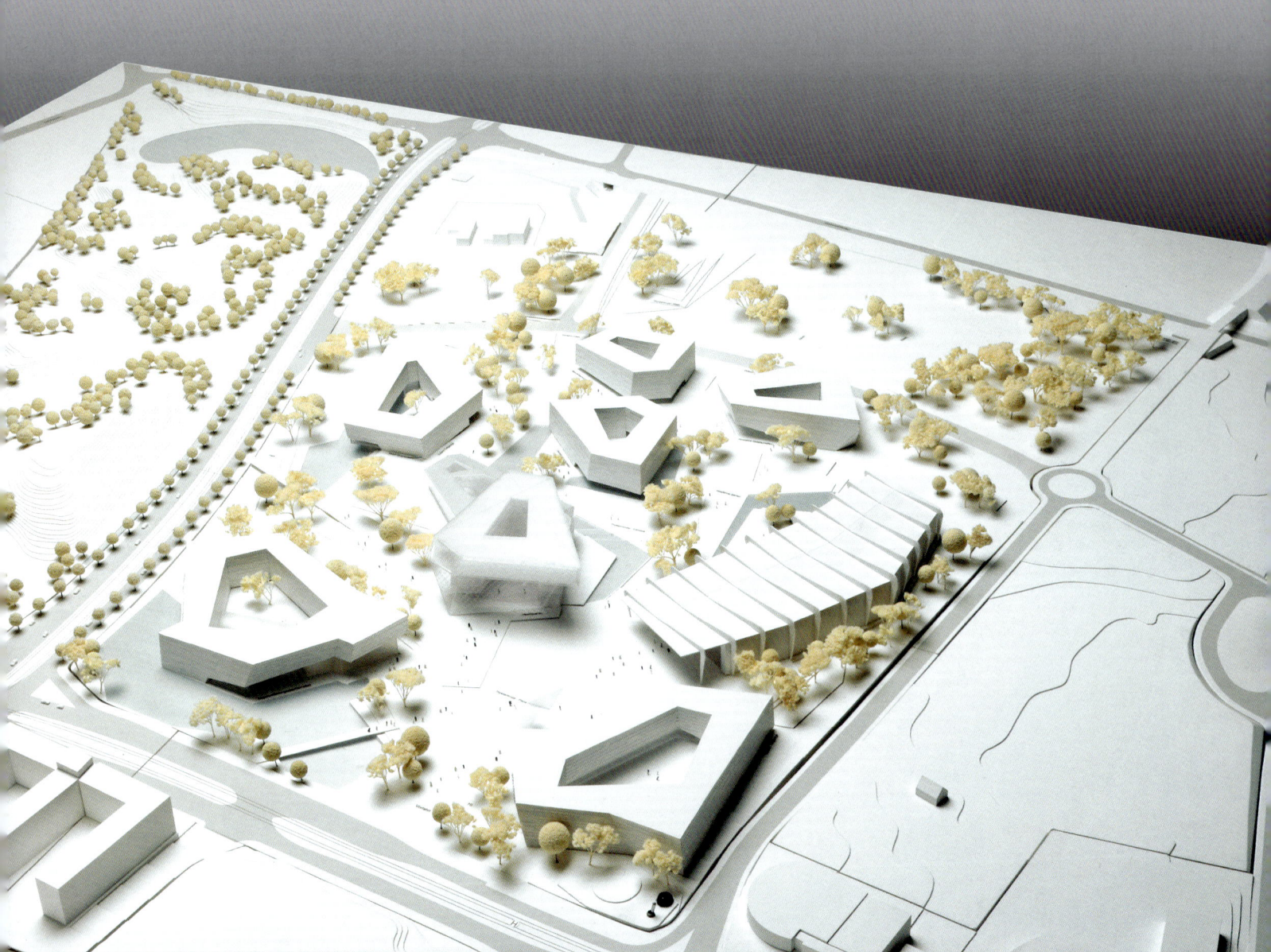

作者 Martin Gessert, Albrecht Randecker, Prof. Klaus Legner   合作伙伴 Dimitri Boikov, Katrin Hauth, Frank Trefzer, Maike Haase, Diana Binder, Bernhard Eppstein, Rabea Stader, Muhammad Fakhri Aulia   专家 landscape planning: Michael Glück, Stuttgart; structural analysis: Knippers Helbig Beratende Ingenieure, Stuttgart; building services: Ebert & Baumann Consulting Engineers Inc., Washington D.C. USA; traffic engineers: Hans Peter Henes, Stuttgart

Perspektive ThyssenKrupp – Plaza – Headquarter

Halle/ Atrium

ThyssenKrupp - Welt

Headquarter

Entsprechend seiner Stellung setzt sich das neue Headquarter aus dem heterogen Umfeld mit seiner Architektursprache deutlich ab. Das Headquarter verkörpert dabei genau jene Position, Innovationskraft zu repräsentieren, ohne abzugrenzen und Tradition widerzuspiegeln, ohne auf den lebendigen Austausch zu verzichten, die ThyssenKrupp so unverwechselbar machen.

Das Headquarter ist als dynamisch geformte Raumskulptur konzipiert. Sie birgt die zentralen öffentlichen Nutzungen und gibt der „Erfahrbarkeit der Dynamik ThyssenKrupp" Raum. Die Basis beinhaltet die öffentlichen Nutzungen und fungiert im Inneren nicht nur als Foyer und Verteiler, sondern eröffnet gleichzeitig auch Einblicke über das zentrale über alle Geschosse reichende Atrium in die ThyssenKrupp Welt.
Über eine großzügige Freitreppe wird man von der Plaza in die Eingangshalle führt. Die Fortführung der Treppenanlage im Inneren erschließt das höher gelegene Galeriegeschoß, das über einen großzügigen Luftraum mit dem Erdgeschoß verbunden ist. Durch die angehobene Eingangssituation wird der Blick frei über die Plaza und den vorgelagerten See.
Der alle Geschosse verbindende Luftraum öffnet zum Galeriebereich der Obergeschosse, und schafft einen spannenden Kommunikations- und Verweilbereich mit großzügigen Freiraumbeziehungen und hoher Aufenthaltsqualität. Hier sind die zentralen Bereiche mit Besprechungs- und Konferenzräumen untergebracht und können so von allen Abteilungen gemeinsam genutzt werden.
Die zentralen Nutzungen, wie Kantine, Bibliothek und Schulungsräume sind im Eingangsbereich untergebracht und stellen die Verbindung zum Erdgeschoß dar.
Die interne Erschließung erfolgt über Aufzüge, Treppen und Stege in der Halle, an die sich die geschlossenen Abteilungen angliedern.

h4a Gessert + Randecker Architekten BDA, Stuttgart

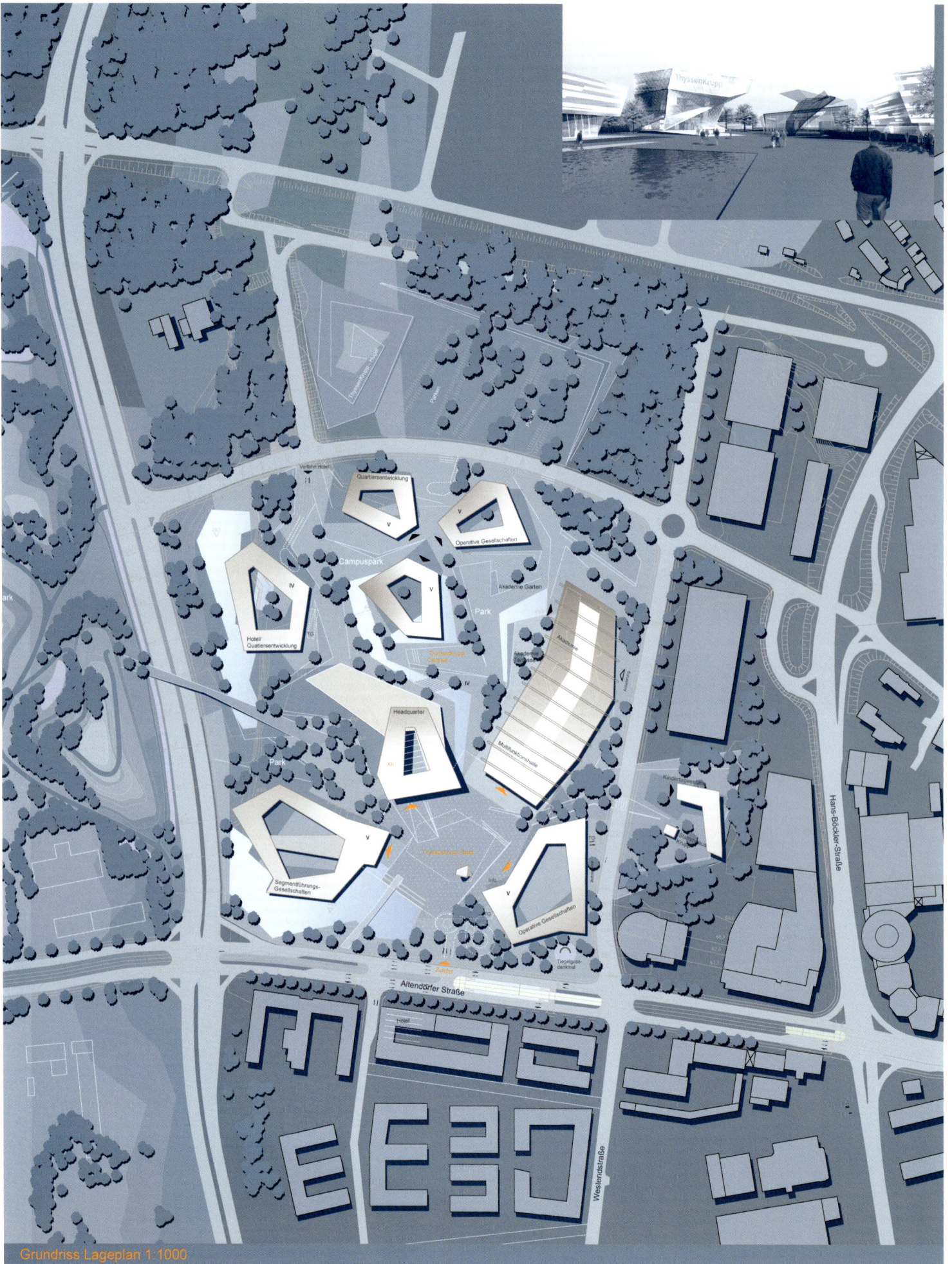

Quartiersentwicklung

Operative Gesellschaften

Campuspark

Akademie Garten

Park

Hotel/
Quartiersentwicklung

Akademie

Akademie

Multifunktionshalle

Headquarter

Park

Kindertagesstätte

Segmentführungs-
Gesellschaften

Operative Gesellschaften

Tiegelguss-
denkmal

Altendörfer Straße

Hotel

Hans-Böckler-Straße

Westendstraße

Schnitt 2 ThyssenKrupp - Campus, Schnitt Headquarter 1:500

Querhauptverwaltung Operative Gesellschaften ThyssenKrupp World Ausstellung Halle / Empfang Headquarter Multi-Funktionshalle ThyssenKrupp Place Operative Gesellschaften Altendorfer Straße

Fassadenschnitt Headquarter 1:50

Sonnenschutz
Sonnenschutz
Dachfläche mit Photovoltaikelementen
Abluft
Abluft
Abluft
Glasdach
Glasdach
Oberlicht
Stahl - Fachwerkrost

Besprechung
Vorzone Vorstand
Büro Vorstand
+32,00

Büro
Oberlicht
Flur / Kombizone
Betonkerntemperierung heizen / kühlen
Büro
+28,50

Büro
Büro
Oberlicht
Flur / Kombizone
Zuluft
Büro
Tageslichtlenkung
+25,00

Meeting Point
Metallbekleidung
+21,50

Halle
+16,50

Erläuterung:

Dachkonstruktion:
Metallpaneele /Photovoltaikelemente
Abdichtungsebene
Dachträger
Dämmung
Thermoaktive Decke mit integrierten
Luftkanälen zur Bauteilaktivierung
und Belüftung

Fassadenkonstruktion:
Pfosten-Riegel Konstruktion
Raumhoch Verglast mit teilweise
bedruckten Paneelen im
Brüstungsbereich
Oberes Lichtband mit Lichtlenklamellen
im Scheibenzwischenraum
Aussenliegender Sonnenschutz mit
wechselnd farblich bedruckten Lamellen

Abgehängte Deckenelemente zur
Verbesserung der Raumakustik
mit integrierten Leuchten und Abluftelement

Treppenkern / Aufzug
+13,00

+9,50

Tragwerk

Tragwerksbeschreibung

Der skulptural geformte Bau wird entsprechend den statisch-konstruktiven Anforderungen in zwei Konstruktionstypen unterschieden.

Die Erdgeschossbebauung wird in konventioneller Stahlbetonmassivbauweise, zum überwiegenden Teil in Ortbeton ausgeführt.
Der aufgeständerte Kopfbau ist als massenminimierter Stahl-Fachwerkrost mit ausbetonierten Trapezblechdecken konzipiert.
Das Primärtragwerk des aufgeständerten Kopfbaus ruht ausschließlich auf den drei durchlaufenden Massivbaukernen auf. Die gesamte Vertikallast kann also zur Unterstützung der Horizontalaussteifung genutzt werden.

Die Trägerebenen des Primärtragwerks werden weitgehend in das Gebäude-raster integriert. Die horizontalen Gurte liegen in den Deckenebenen, die Ver-tikalstreben sind in den Wänden integriert. Die Diagonalen liegen in den Schwer- bzw. Trennwänden oder sind räumlich integriert.
Der räumliche Trägerrost lastet auf den innenliegenden Kernen auf und weist eine Auskragungen von maximal _ der Feldspannweite auf. Damit sind wirt-schaftlich optimierte Bauteilabmessungen realisierbar.

Umlaufend halten die quer abgehängte Zugstreben die Fassadenebene frei von massiven Stahlbauelementen. Dadurch wird eine filigrane Fassadenaus-führung möglich.

Die Gründung erfolgt entsprechend der Baugrundverhältnisse und der einzu-leitenden Lasten. Die Erdgeschossbebauung wird auf konventionelle Streifen-fundamenten aufgesetzt. Die punktuell hohen Lasten der Kerne erfordern eine Einleitung der Lasten in die tieferen und tragfähigeren Bodenschichten. Bei entsprechender Bodenkonsistenz ist dies durch eine vertiefte Flachgründung realisierbar.

Zum Bauablauf:

Gründung, Erdgeschossbebauung und massive Kerne werden in konventionel-ler Stahlbetonbauweise, vorwiegend in Ortbeton erstellt.

Der Stahlträgerrost wird in transportablen „Schüssen" vorgefertigt und auf die Kerne aufgelegt. Auf den überhöht montierten Primärstahlbau werden die Trapezblechschalen aufgelegt, verankert und ausbetoniert. Nach Dachdich-tung und Fassadenmontage erfolgt der Innenausbau.

Mit der vorgeschlagenen Bauweise sind die Bauabschnitte entsprechend der jeweiligen Anforderungen optimiert: eine konventionelle Bauausführung im Erdgeschoss, eine massenreduzierte und montagefreundliche Stahlbauweise mit leichten Decken- und Wandelementen für den aufgeständerten Kopfbau.

Grundriss 6. OG
abgehängte Fassade
KERN
KERN
System A
STAHL
KERN
- Länge- und Querschotten
  in Stahlbauweise
- Einleitung der Vertikallasten
  in die Kerne
- druckgehende Schotten

System A
Fassade Kern Kern Fassade

System B
Fassade Kern Kern Fassade

Schnitt
KERN KERN

Energiekonzept

Energetisches Konzept - Solitäre

Zur Energieversorgung des ThyssenKrupp Areals liegt folgendes energetisches Konzept zugrunde:

- Basiswärme und -kältegewinnung durch das Grundwasser
- Fernwärme mit einem KWK-Anteil von 94% zur Abdeckung der Spitzenheizlast
- Gebäudehülle gemäß Passivhaus-Standard
- Aktive Stromgewinnung durch Photovoltaik
- Betonkerntemperierung

Der Zielwert von 80 kWh Primärenergie pro m2 und Jahr wird jeweils von den Gebäudetypen unterschritten.

Die Betonkerntemperierung ermöglicht eine primärenergetisch effiziente Nutzung des Grundwassers. Das Grundwasser wird im Winter über eine Wärmepumpe als Wärmequelle genutzt. Im Sommer sorgt das Grundwasser in der Betonkerntemperierung für eine milde Kühlung. Der Einsatz von Kältemaschinen ist nicht erforderlich. Das Grundwasser wird über Förderbrunnen hoch gepumpt, in der Wärmepumpe abgekühlt und im Schluckbrunnen wieder der gleichen Grundwasserschicht zugeführt.
Statische Heizkörper ermöglichen es dem Nutzer die Temperatur individuell zu regeln. Das Heizkörpernetz ist an die Fernwärme (Primärenergiefaktor 0,46) angeschlossen.

Photovoltaik-Elemente auf den Dächern (ca. 8.000 m2, Neigungswinkel > 30°) erzeugen primärenergetisch wertvollen Strom für die Pumpen.

Die solaren Lasten werden durch einen dem Sonnenstand nachgeführten lamellierten Sonnenschutz minimiert. Der Sonnenschutz sorgt gleichzeitig für hohe Tageslichtautonomie und geringe Beleuchtungsabwärme.

Energetisches Konzept - Headquarter

Das Headquarter erhält in den oberen Geschossen eine Doppelfassade und einen überdachten Innenhof. Die Doppelfassade ermöglicht eine windunabhängige Fensterlüftung. Der Innenhof ist als Nutzungszone konzipiert.

Für das Gebäude wird eine mechanische Lüftung und Betonkerntemperierung vorgesehen. Die Zuluft wird über Quellluftauslässe im Hohlraumboden eingebracht und zentral einer hocheffizienten Wärmerückgewinnung
(Gegenstromschicht-Wärmetauscher mit WRG-Faktor 80%) zugeführt. Als Wärme- und Kältequelle dient das Grundwasser, ggf. das Erdreich. Eine Wärmepumpe hebt das Temperaturniveau im Winter.

HPP Hentrich – Petschnigg & Partner KG Düsseldorf
HPP Hentrich–Petschnigg & Partner KG建筑事务所 杜塞尔多夫

"An artificial landscape park, 'Magnetic Attraction', is a welcoming gesture to the green campus ensemble of buildings in the new ThyssenKrupp Quarter while opening up to the surrounding landscape with an inviting gesture."

"一个人工的风景公园，'磁力般的吸引力'，对建筑物整体作欢迎姿态，同时向周围景观开发呈邀请姿态。"

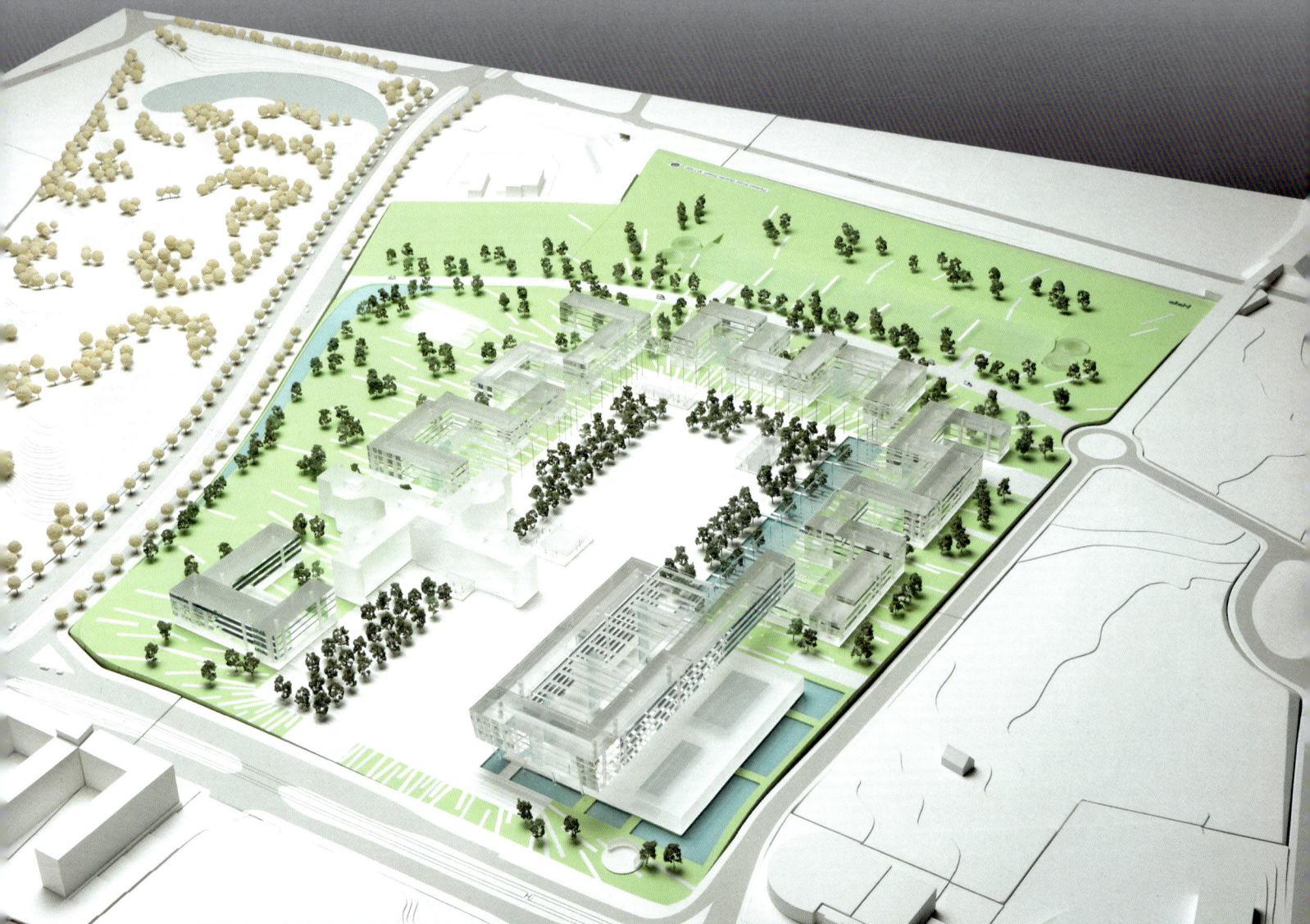

作者 Joachim H. Faust, Volker Weuthen   合作伙伴 Oliver Conzelmann, Sebastian Wiswedel, Stefan Haupt, Götz Schrader, Darius Scheible, Mark Buhrdorf   专家 landscape planning: KuBuS Freiraumplanung, Wetzlar; structural engineers: Werner Sobek Ingenieure, Stuttgart; building services and façade analysis: DS-Plan, Cologne

HPP Hentrich – Petschnigg & Partner KG, Düsseldorf

ThyssenKrupp Quarter, Essen

Perspektive Headquarter - Altendorfer Strasse

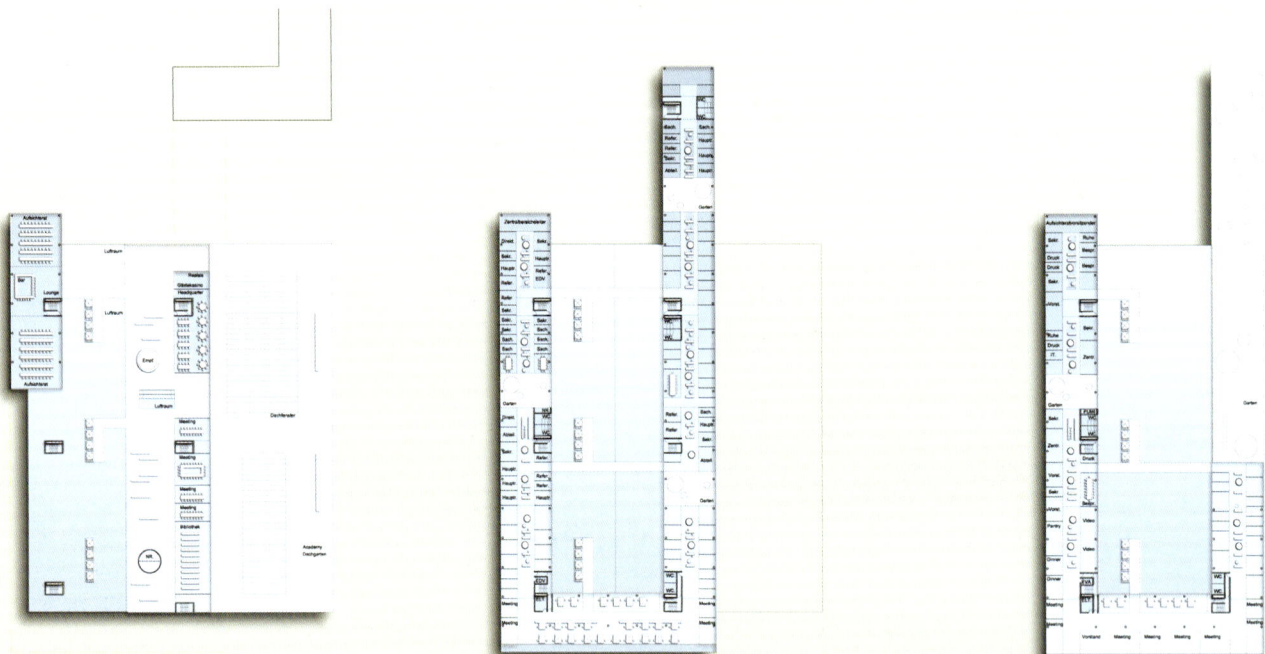

2. OG 1:500

Regelgeschoss 3. OG

Vorstand 7. OG

Ansicht Headquarter - Ost 1:500

Schnitt Campus C - C 1:500

Ansicht Ost / Schnitt Campus D - D 1:500

A

David Chipperfield Ar

Staab Architekten

Königs Archit

SANAA   SCHULTE

aye Associates

hitects

Folkwang Museum
in Essen

埃森Flokwang博物馆

Gigon Guyer

ekten

FRANK ARCHITEKTEN

Zaha Hadid Architects

Folkwang Museum Essen

埃森Flokwang博物馆

Restricted cooperative project competition preceded by an application procedure

限制严格的协商项目竞赛，开始前有申请程序。

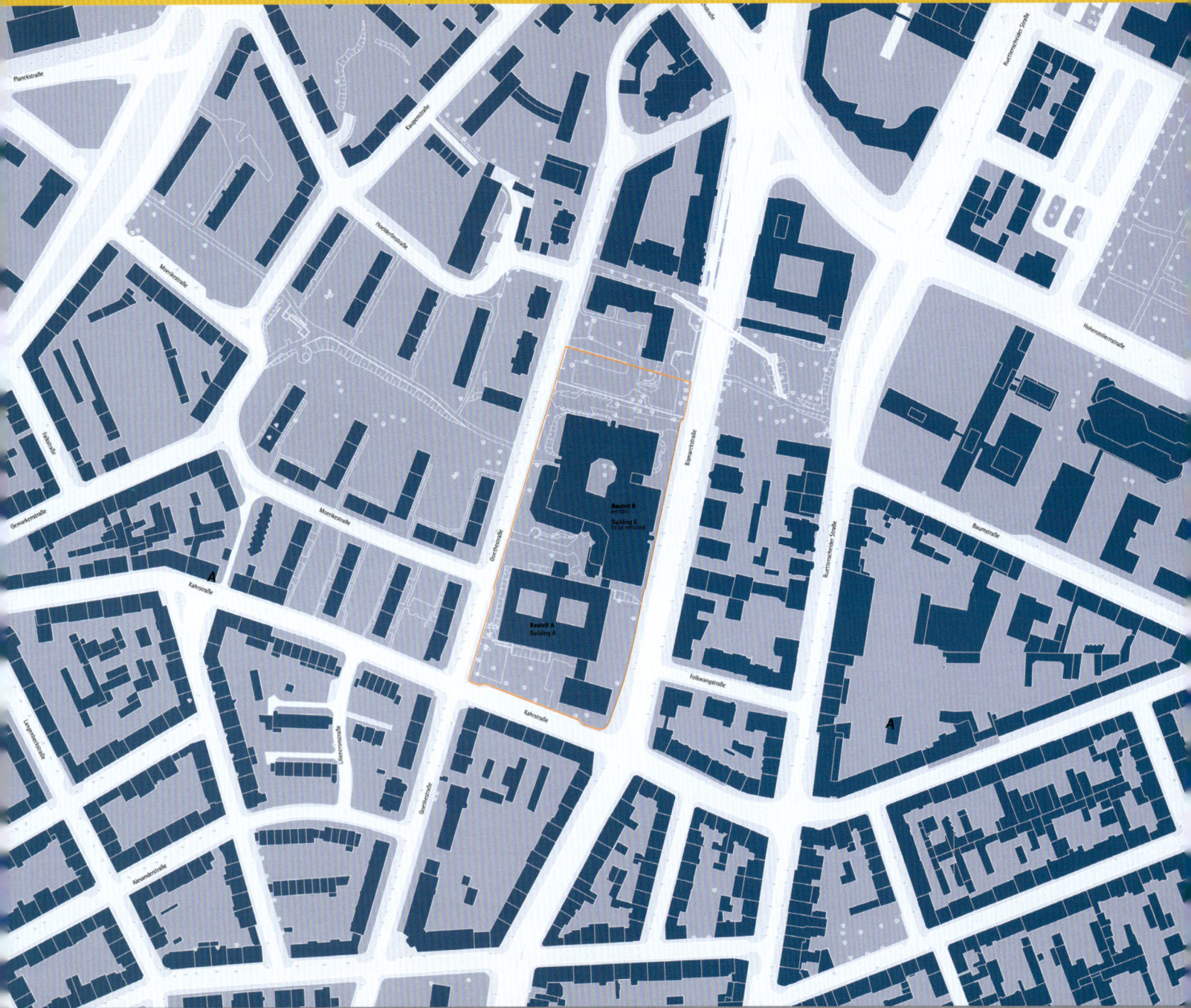

地点 Essen　时间 10/2006-03/2007　主办方 City of Essen, represented by Geschäftsbereich 6A　参赛者 12 competitors
面积 19.000 sq m　竞赛费用 530,000 Euro　专业评奖委员会 Hans-Jürgen Best, head of city planning, City of Essen; Prof. Eckhard Gerber, Dortmund; Fritz Heinrich, Dortmund; Prof. Marcel Meili, Zurich; Manfred Osthaus, Bremen; Simone Raskob, head of environment & construction, City of Essen; Prof. Kirsten Schemel, Berlin/Münster　专家评奖委员会 Prof. Dr. hc. mult. Bertold Beitz, chairman of board of trustees, Alfried Krupp von Bohlen und Halbach-Foundation, Essen; Dr. Hartwig Fischer, director Museum Folkwang, Essen; Dr. Volker Rattemeyer, Director Art Museum Wiesbaden; Dr. Wolfgang Reiniger, mayor, City of Essen; Dr. Oliver Scheytt, head of culture, City of Essen; Dr. Katharina Schmidt, Zurich

In 2010, Essen and the Ruhr area will be the designated European Capital of Culture. The new extension to the Folkwang Museum is one of a small number of newly built projects within the framework of this event, the motto of which, "Change through culture, culture through change", was coined by Karl Ernst Osthaus, the museum's founder. Originally located in Hagen and opened in 1902, the Folkwang Museum was the first in Europe to acquire works by van Gogh and Matisse. After the early death of Osthaus, the collection was transferred to Essen and merged with the Essen Museum of Art to form the new Folkwang Museum. Conceived as a complement to a heritage-protected wing dating from 1960, the new project has a usable area of approx. 12,000 square metres and the new building will house post-World War Two and contemporary art. The project also includes a new entrance zone, spaces for changing exhibitions, educational facilities, offices, workshops, depots and halls for the presentation of those parts of the collection that, due to lack of space, were hitherto inaccessible to the public. The aim was to create an architectural and conceptual urban landmark while integrating the museum and its new wing within the existing urban context.

2010年，德国埃森和鲁尔地区将被指定为欧洲文化之都。Folkwang博物馆的扩建就是该事件框架内的一些新兴项目之一。"通过文化进行改变，通过改变传播文化"是博物馆奠基人Karl Ernst Osthaus创造的理念。该馆最初位于哈根，于1902年对外开放，是欧洲第一家收藏凡高和马蒂斯作品博物馆。在Osthaus英年早逝之后，该馆搬至埃森，与埃森艺术博物馆合并成为现在的Folkwang博物馆。新项目被看成是1960年建成的保护遗产侧楼的补充，使用面积约1.2万平方米，用于收藏二战之后和现代的艺术作品。项目也包括一个新的入口区域、换展区、教育设施、写字间、车间、仓库和临时展厅等。竞赛任务是要在与现有建筑和谐的同时建成一个建筑与概念上的城市地标。

Aerial view　鸟瞰图

Yard of Museum Folkwang　博物馆后院

Entry of Museum Folkwang　博物馆入口

Front view Museum Folkwang　博物馆正面图

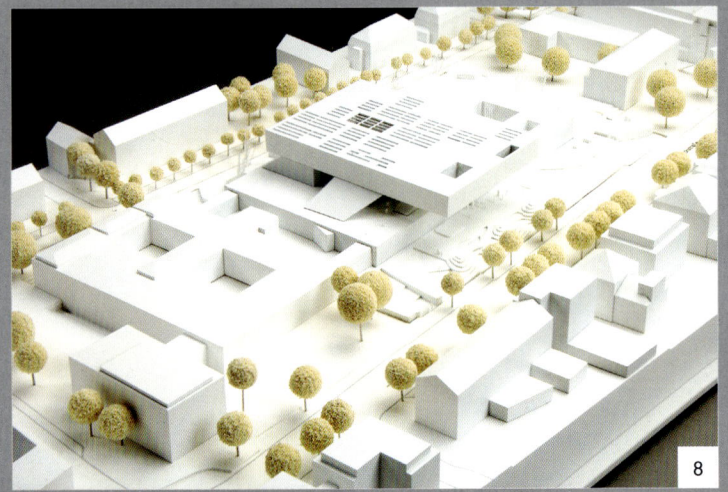

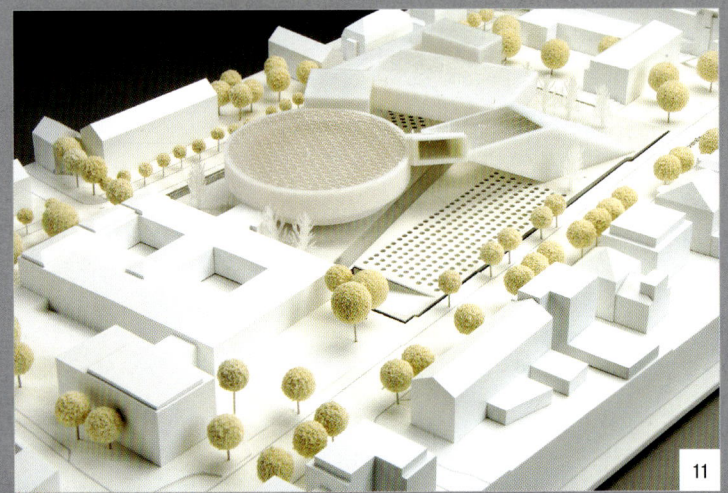

3

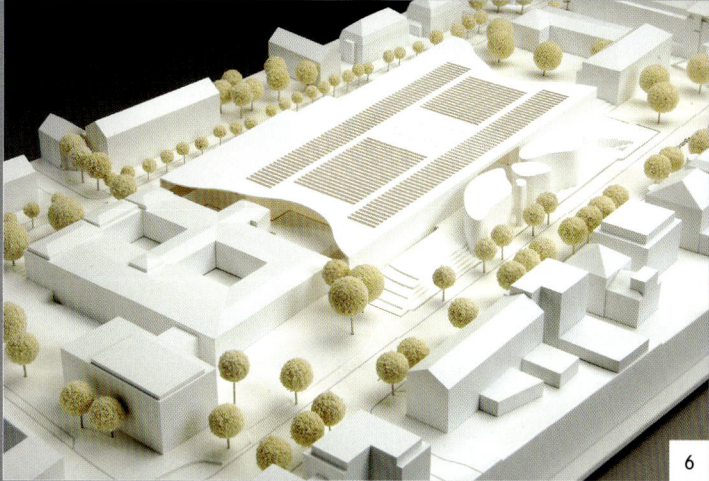
6

9

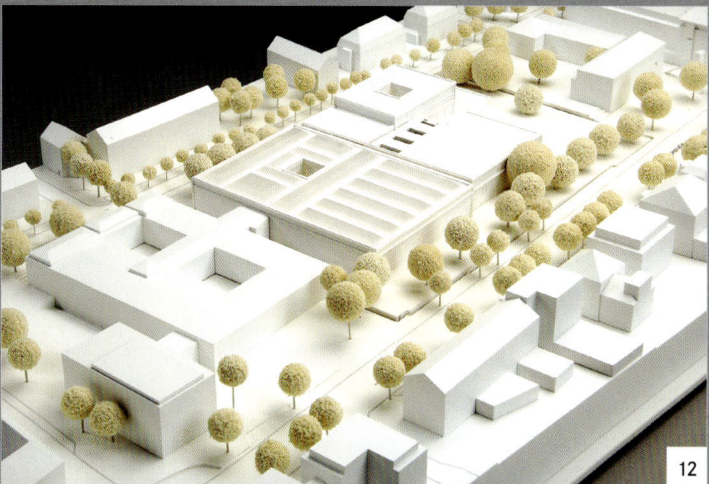
12

Qualified participants
合格的参赛者

1
1st prize　一等奖
David Chipperfield Architects,
London / Berlin

2
2nd prize　二等奖
Adjaye Associates, London

3
3rd prize　三等奖
Gigon Guyer Architekten, Zurich

4
Honorable mention　荣誉奖
Staab Architekten, Berlin

5
Honorable mention　荣誉奖
Zaha Hadid, London

6
Further participant　其他参赛者
SCHULTES FRANK ARCHITEKTEN,
Berlin

7
Further participant　其他参赛者
SANAA, Tokyo

8
Further participant　其他参赛者
Königs Architekten, Cologne

9
Further participant　其他参赛者
Architekten Brüning Klapp Rein,
Essen

10
Further participant　其他参赛者
BOLLES + WILSON, Münster

11
Further participant　其他参赛者
MVRDV, Rotterdam

12
Further participant　其他参赛者
HILMER & SATTLER und ALBRECHT,
Berlin / Munich

1st Prize
一等奖

David Chipperfield Architects London/Berlin
David Chipperfield建筑事务所 伦敦/柏林

"The most important thing by planning the new building of the Folkwang Museum should be the relationship between the museum, its art, and the society."

"规划Folkwang博物馆新楼的重点应该是处理好博物馆、艺术及社会的关系。"

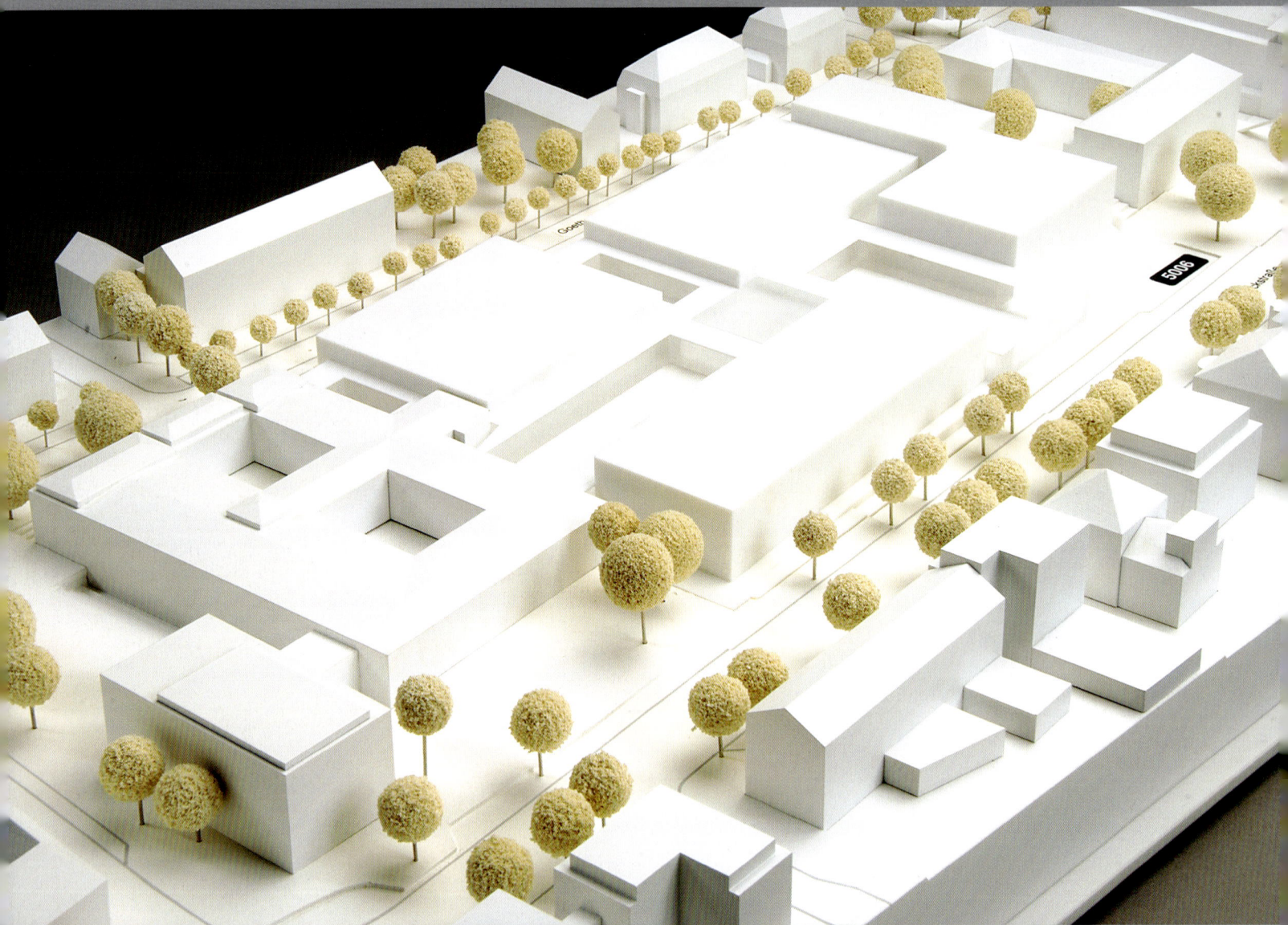

作者 David Chipperfield, Harald Müller, Mark Randel 合作伙伴及建筑系在校生 Alexander Schwarz, Peter von Matuschka, Marika Schmidt, Florian Dirschedl, Barbara Koller, Nicolas Kulemeyer, Ulrike Eberhardt, Antonia Schlegel, Ute Zscharnt, Annette Flohrschütz, Marcus Matthias, Dalia Liksaite 专家 building services: ARUP Berlin; lighting engineers: ARUP Berlin, Andrew Sedgwick; structural analysis: IGB Karlsruhe

David Chipperfield Architects, London/Berlin

Folkwang Museum, Essen

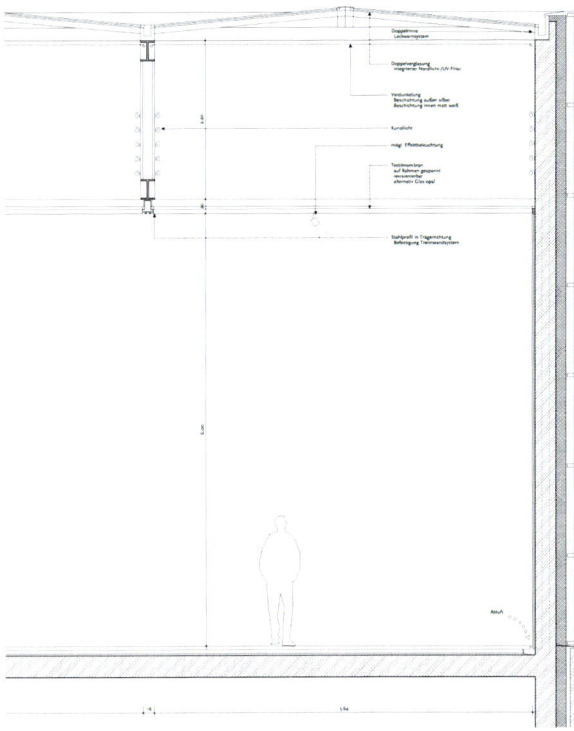

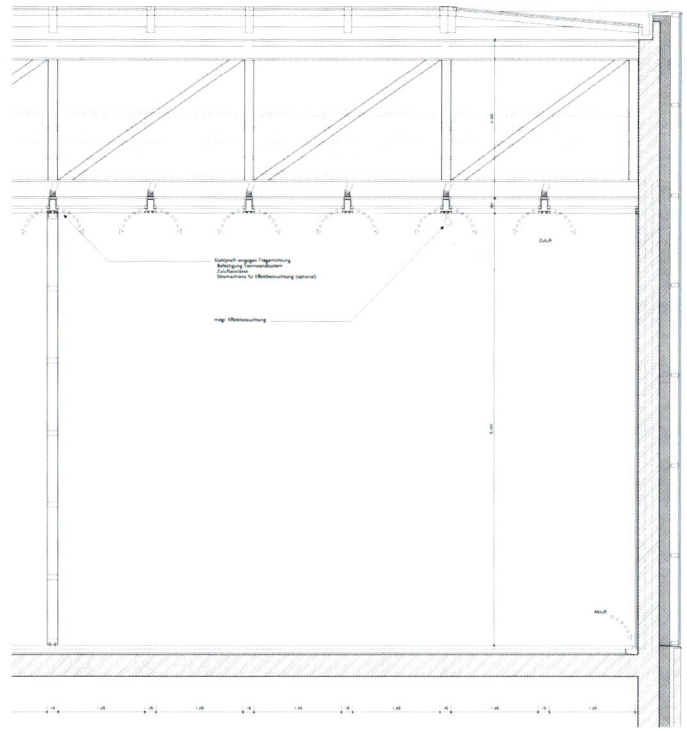

WECHSELAUSSTELLUNG

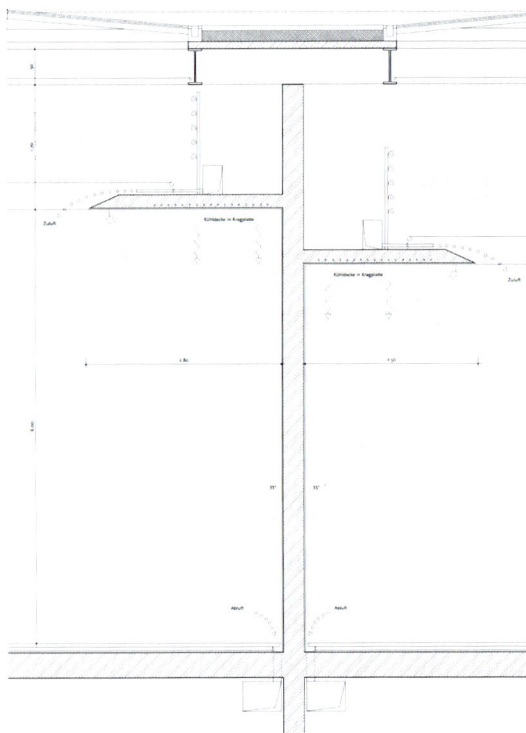

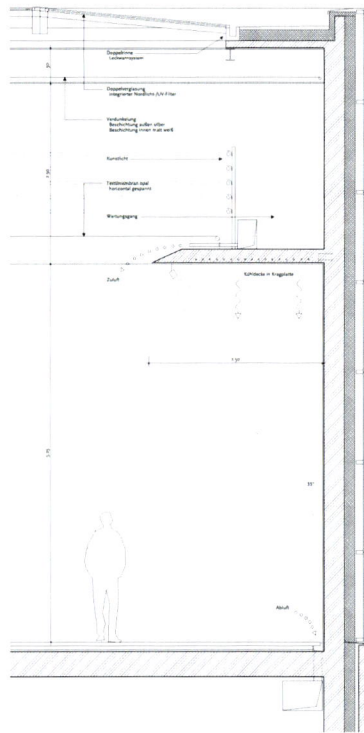

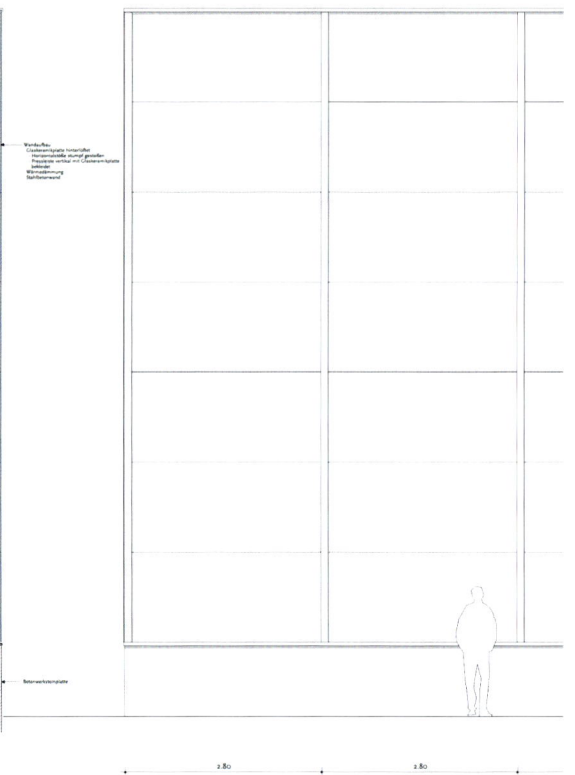

DAUERAUSSTELLUNG

UNTEN: FASSADENMATERIAL GLASKERAMIK

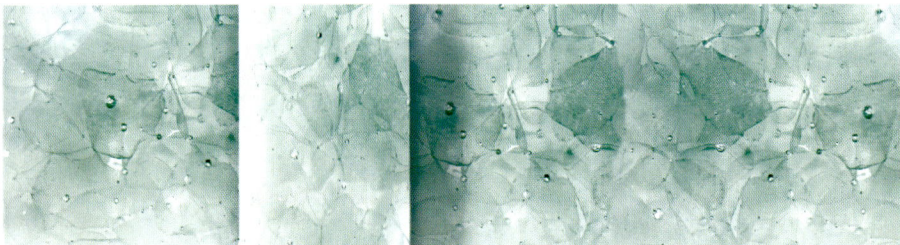

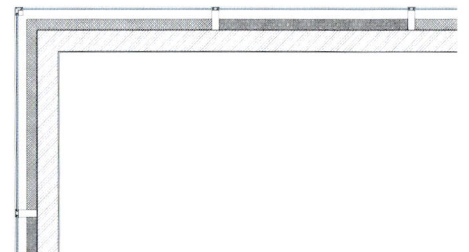

ANSICHT OST

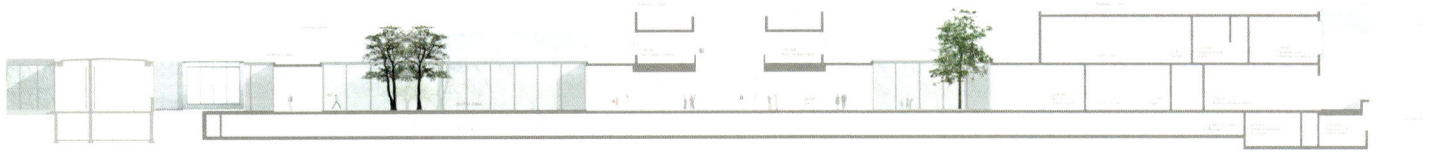

SCHNITT A-A

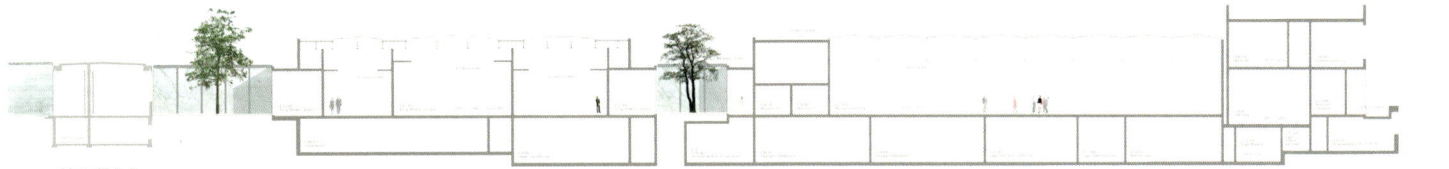

SCHNITT D-D

ANSICHT WEST

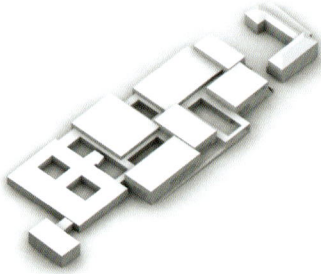

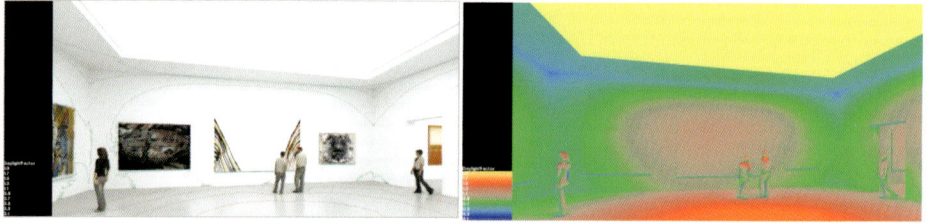

David Chipperfield Architects, London/Berlin

Folkwang Museum, Essen

Adjaye Associates London
Adjaye联合建筑事务所 伦敦

"Our design intends to create a new museum that, in terms of scale, form, materiality and visual relations, respects the existing Folkwang Museum [...] and its built environment."

"我们设计的目的是创造一个在规模上、形式上、重要性上和视觉关系上全新的博物馆，尊重现有的博物馆建筑及其周围环境。"

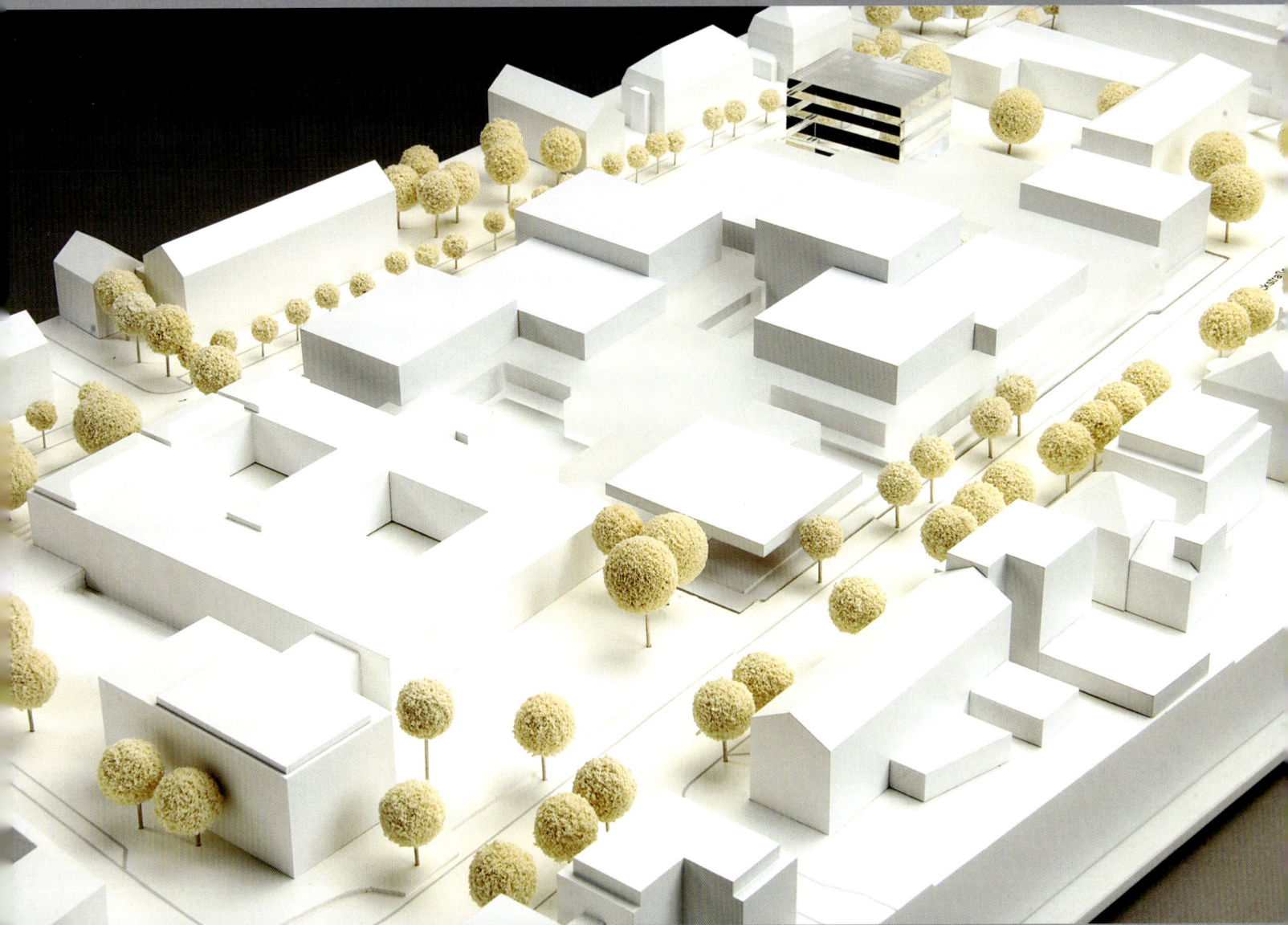

作者 David Adjaye   合作伙伴 Mansour El-Khawad, Jeanne-Francoise Fischer, Jojo Grieger, Teresa Gössl, Nico Leist, Pavandeep Panesar, Mirko Petzold, Roman Piotrowski 专家 lighting engineers: ARUP London; landscape architect: Büro Kiefer, Berlin

HAUPTEINGANG BISMARCKSTRASSE

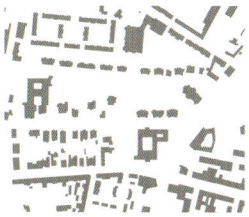

SCHWARZPLAN 1926

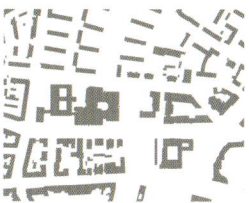

SCHWARZPLAN 2006

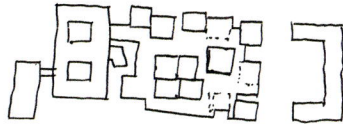

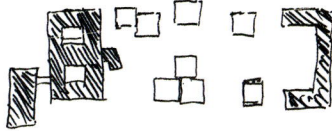

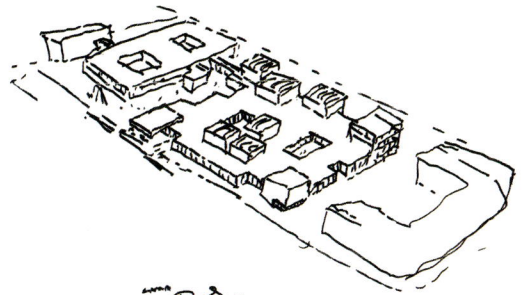

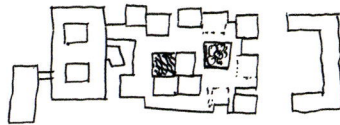

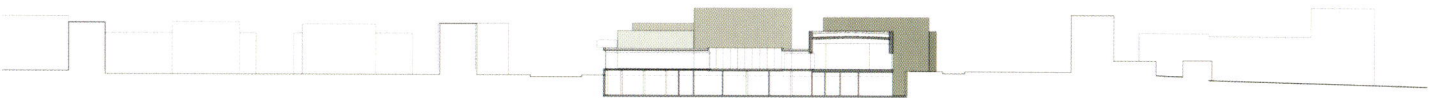

SCHNITT B-B 1:500

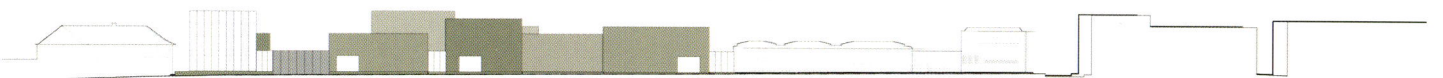

ANSICHT WEST 1:500

1 Haupteingang
2 Eingang
    Museumspädagogik /
    Karl-Osthaus-Saal
3 Nebeneingang
    Multifunktionsraum und
    Bibliothek

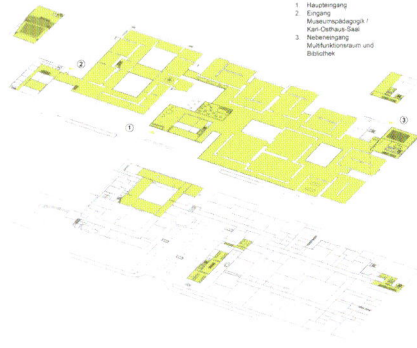

1 Kunstwerke
2 Gastronomie

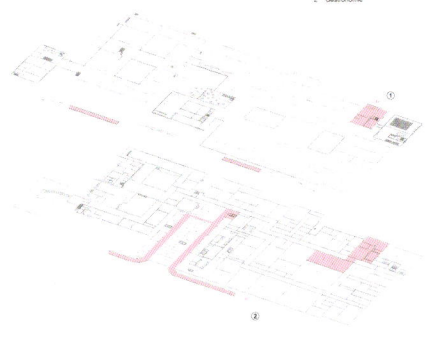

Zirkulation Besucher

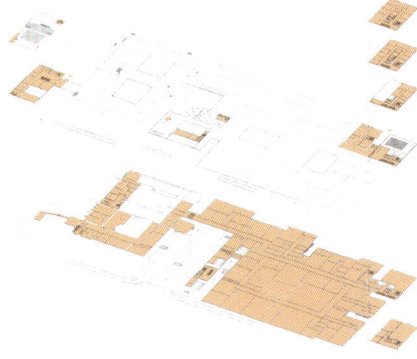

Zirkulation Mitarbeiter

Anlieferung

BLICKBEZIEHUNGEN INNEN / AUSSEN

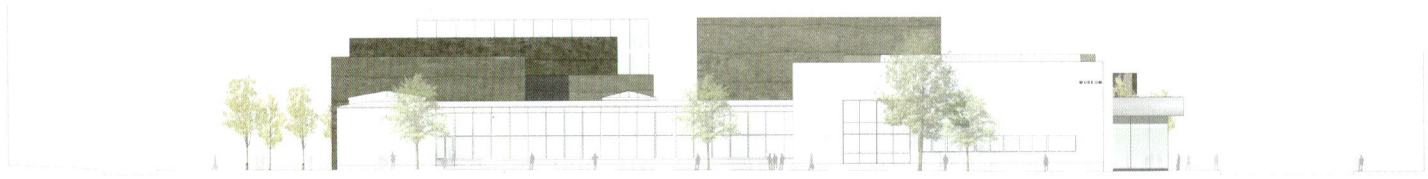

ANSICHT SÜD 1:200

ANSICHT WEST 1:200

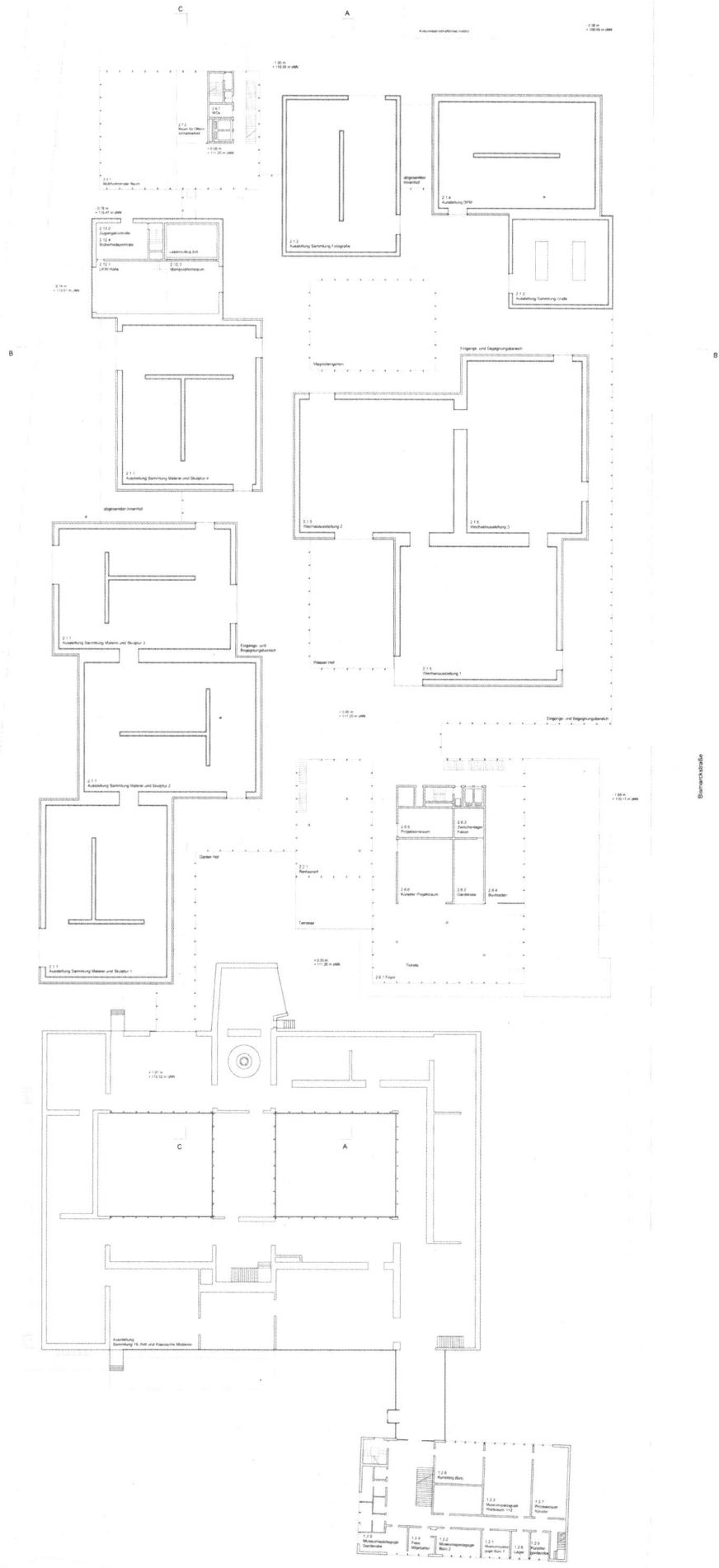

Adjaye Associates, London

273

Folkwang Museum, Essen

GRUNDRISS EG 1:200

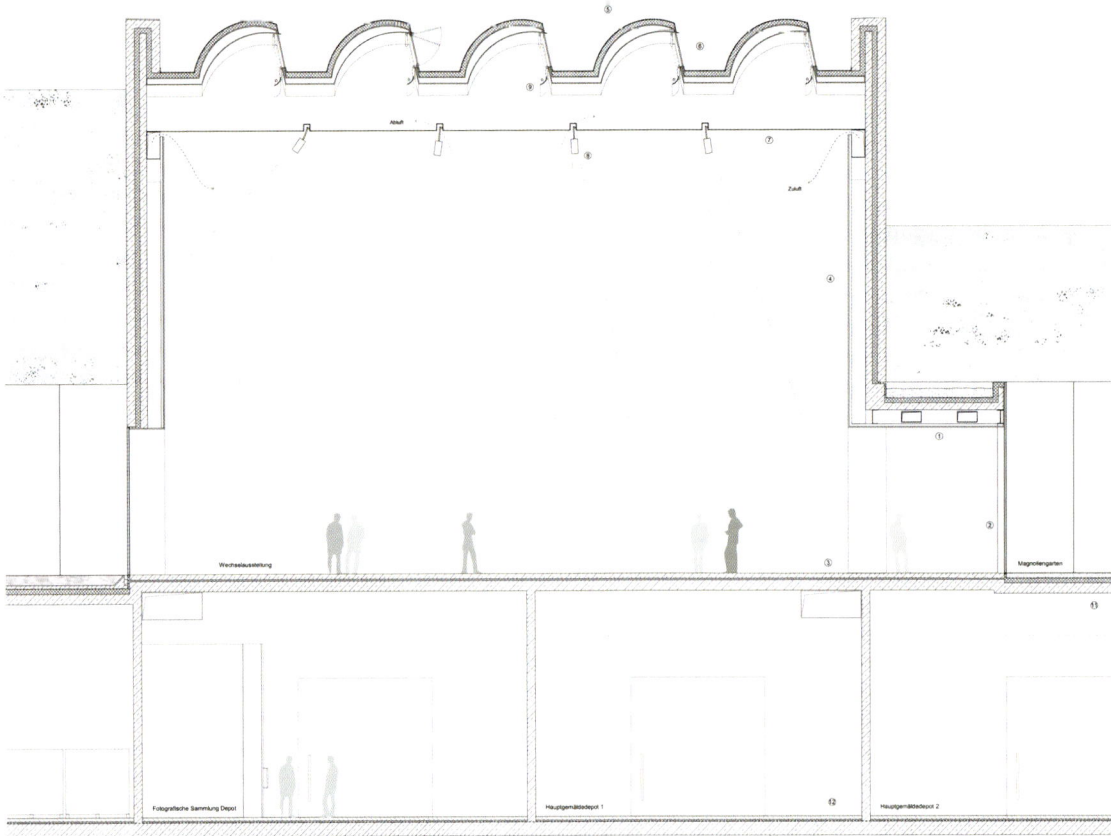

SCHNITT 1:50

Ausstellungsraum: Raumklimakontrolle

Jeder der Ausstellungsräume wird mit einer eigenen Aufbereitungsanlage ausgestattet, die sich im Untergeschoss direkt unter dem jeweiligen Ausstellungsraum befindet. Um einen hohen Konservierungsstandard zu erreichen, wird die einströmende Frischluft behandelt, d.h. Staub und chemische Stoffe herausgefiltert. Anschließend wird die behandelte Luft in die jeweiligen Luftungsanlagen der Ausstellungsräume geleitet und gepresst, um eventuelle Lufteinwirkung von sowohl Außen- als auch Nebenräumen zu reduzieren.

Dieses Verfahren ermöglicht es, die strengen Raumklimaanforderungen auf effiziente wie einfache Art und Weise zu gewährleisten.

Die Primärluft wird von den Primäraufbereitungsanlagen durch eine Vielzahl von Luftungskanälen in alle Räume und Luftungsanlagen verteilt.

Belastbarkeit:

Es wird angenommen, dass ein hoher Umweltschutzstandard aus Konservierungsgründen gefordert ist. Um dem gerecht zu werden, wurde folgendes berücksichtigt:
- jede Frischluftanlage wird mit 2 Anlagen ausgestattet, die je zu 50% Frischluft liefern,
- alle Pumpen operieren mit automatischer Umschaltung unter „Run and Standby".
- Die Luftungsanlagen der Ausstellungsräume werden mit manuellen Steckverbindungen ausgestattet, um im Falle eines Anlagenausfalls 50% Standby zu gewähren.

Energiestrategien:

Ausstellungsgebäude mit hohem Konservierungsstandard haben oftmals einen hohen Energieverbrauch. Um den Jahresenergiebedarf im Verhältnis zu traditionellen Ausstellungsräumen niedrig zu halten, werden die folgenden Entwurfsprinzipien vorgeschlagen:
- Einsatz von Quellluftungssystemen für alle Ausstellungs- und Besucherbereiche, um die Klimaeffizienz der hochfrequentierten Besucherbereiche zu maximieren.
- Die Kühlung der Ausstellungsräume erfolgt mittels Wärmepumpen mit Energierückgewinnungsfunktion.
- Die Erwärmung erfolgt mittels drei unterschiedlicher Methoden durch Abwärme von den Klimaanlagen, einem Grundwasserheizsystem mit integrierten Wärmepumpen, und wenn diese Methoden nicht ausreichen, durch das regionale Fernwärmenetz.
- Alle Pumpen sind mit variablem Antrieb ausgestattet und arbeiten mit 2er-Anschluss Steuerventilen.
- Ausstattung aller Ventilatoren mit variablem Antrieb, um das Luftvolumen bei niedriger Besucherzahl oder bei Nacht zu reduzieren.
- Unabhängig vom Ausstellungszweck werden in allen Räumen, deren räumliche Gegebenheiten es zulassen, Luftungsanlagen mit Freikühlungsfunktion eingesetzt.
- In den Räumen, die nicht Ausstellungszwecken dienen, werden - wenn möglich - Luftungsanlagen mit Freikühlungsfunktion eingesetzt.

Gründach:

Die Ausbildung des Daches oberhalb der primären Erschließungszone und der 'black box' Ausstellungsräume als Gründach hat wichtige Vorteile.
- die thermische Masse des Gründaches erhöht den Dämmfaktor des Gebäudes erheblich und führt zu geringeren Energieverbrauchswerten.
- Die Absorption des Regenwassers in der Dachebene ist ökologisch und reduziert Abwassermengen.
- Das auf dem Dach entstehende Biotop hat mikroklimatische Vorzüge und unterstützt die Ausbreitung der Artenvielfalt.

Raum für permanente Sammlung
Raumklimakontrolle

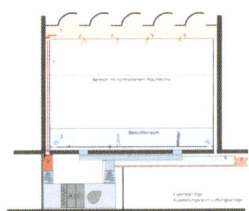

Raum für Wechselausstellungen
Raumklimakontrolle

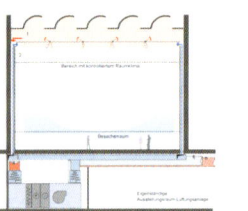

Raum für Wechselausstellungen
Belichtungsprinzipien

BELÜFTUNG UND BELICHTUNG

Materialität

Oberflächen Außen:
a) Dunkeleingefärbter Sichtbeton mit Ruß-, Basalt- und Schwarzglasaggregaten versetzt, sandgestrahlt, unterschiedliche Farbgebung der einzelnen Körper. Horizontale Fugen.
b) Außenbündige, rahmenlose, transparente Isolierverglasung mit dahinter liegenden vertikalen Edelstahlpaneelen.
c) Außenbündige, rahmenlose, transparente Isolierverglasung mit rückseitiger Farbbeschichtung im Bereich des Dachaufbaus und dahinter liegenden bronzefarbenen Profilen / Stützen.
d) Außenbündige, transparente, isolierverglaste Vorhangfassade mit dahinter liegenden bronzefarbenen Profilen / Stützen.
e) Nordgeschnitte, kupferverkleidete Oberlichtsheds.

Oberflächen Innen:
a) Geglätteter Beton in allen Bereichen bis auf Multifunktionsraum (Holzparkett) und Verwaltung (Linoleum).
b) Gipskartonwand in den Ausstellungsräumen und in der Wandelhalle.
c) Gipskartonwand und Sichtbetonstützen / Sichtbetonwand im Kellergeschoss.
d) Glastrennwände mit transparenten und transluzenten Qualitäten in der Verwaltung.
e) Sichtbetondecken in der Verwaltung.
f) Im allgemeinen Sichtbetondecken im Kellergeschoss, teils abgehängte Gipskartondecken.

MATERIALBEISPIELE

WAHRNEHMUNG DER MUSEUMSRÄUME VON DER BISMARCKSTRASS

AUSSTELLUNGSRAUM MIT BLICK AUF DEN WASSERHOF

Gigon Guyer Architekten Zurich
Gigon Guyer建筑事务所 苏黎世

"The new museum building covers the site in the form of a 'single-story' museum facility. Isolated masses are lined along a covered walk [...] together forming a mass-grouped configuration."

"新博物馆大楼以'单层平房'的形式建起，形成由单个建筑串在一起的整体的外貌。"

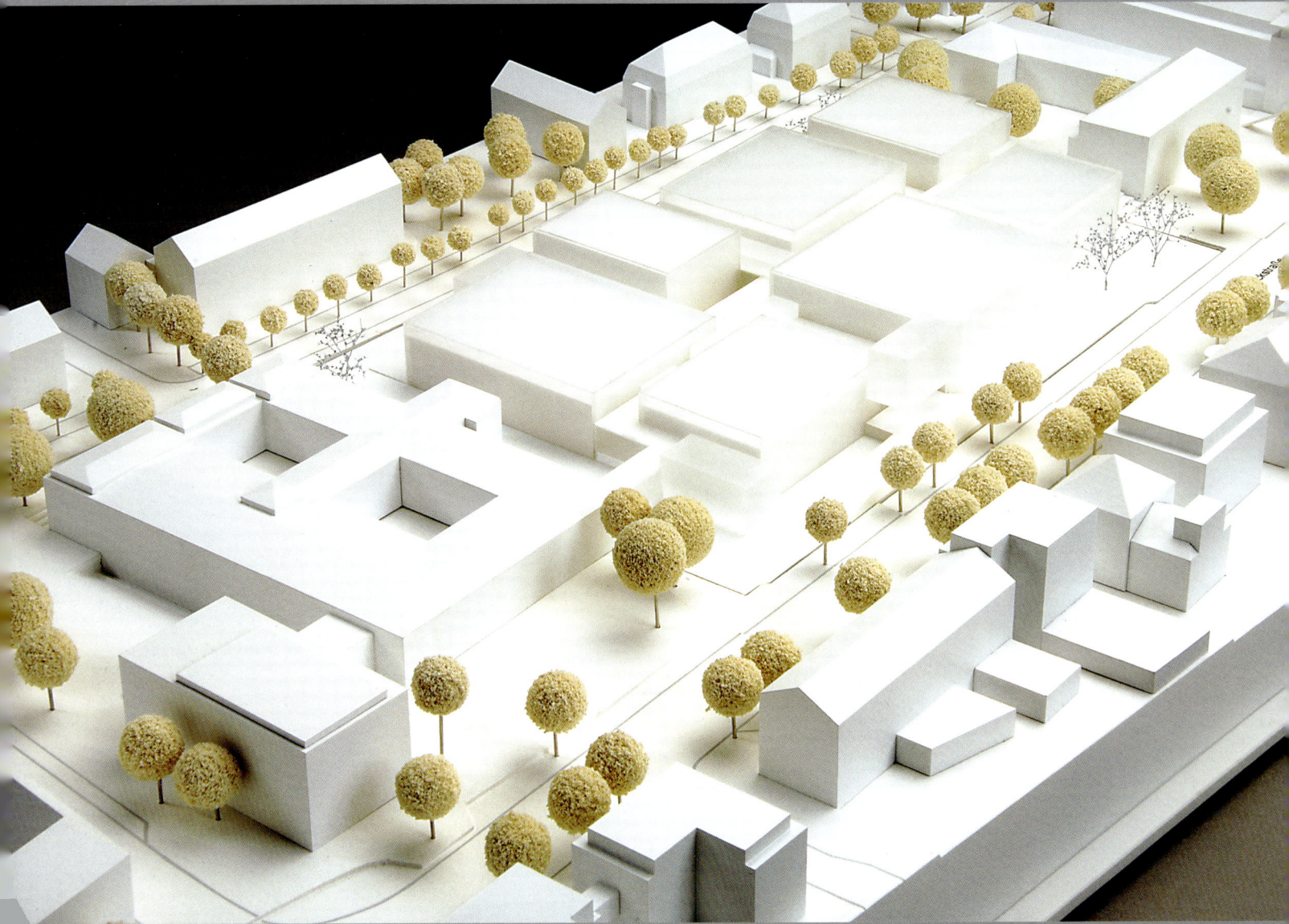

作者 Annette Gigon, Mike Guyer   合作伙伴 Raphaela Schacher, Ivana Vukoja, Brigitte Rüdel, Karsten Buchholz, Andri Gartmann, Raul Mera, Martino Pedroli   专家 HLKKS/MSRL: Waldhauser Building Services AG, Basel; building engineer: Dr. Lüchinger & Meyer, Zurich; landscape architect: Schweingruber Zulauf BSLA, Zurich; natural lighting systems: Institut für Tageslichttechnik, Stuttgart; artificial lighting systems: LICHTDE-SIGN Ingenieurgesellschaft mbh, Cologne

Gigon Guyer Architekten, Zurich

Folkwang Museum, Essen

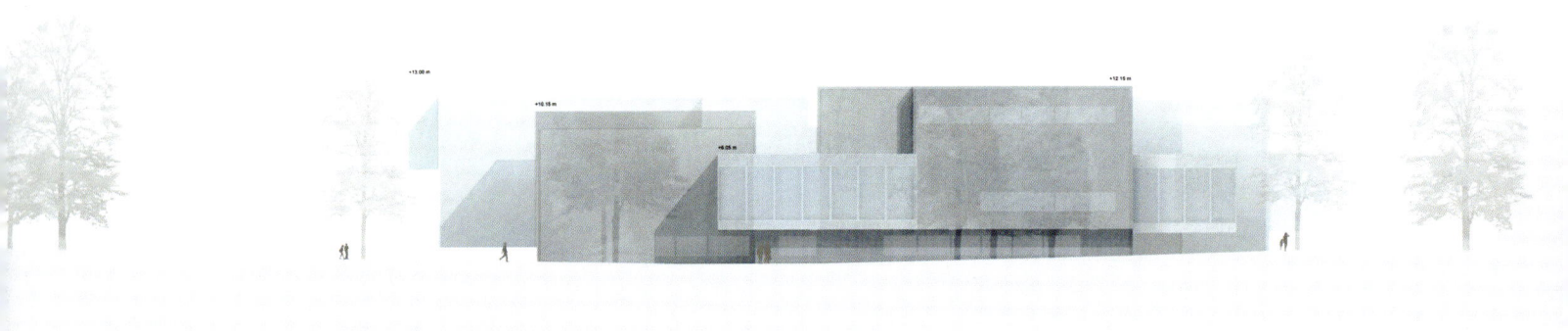

Nordansicht 1:200

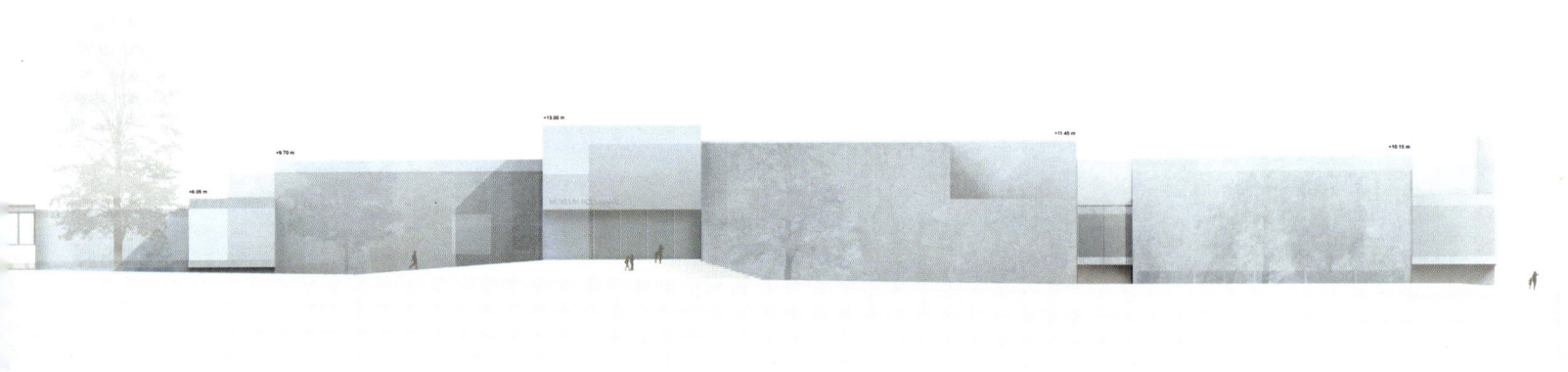

Südansicht 1:200

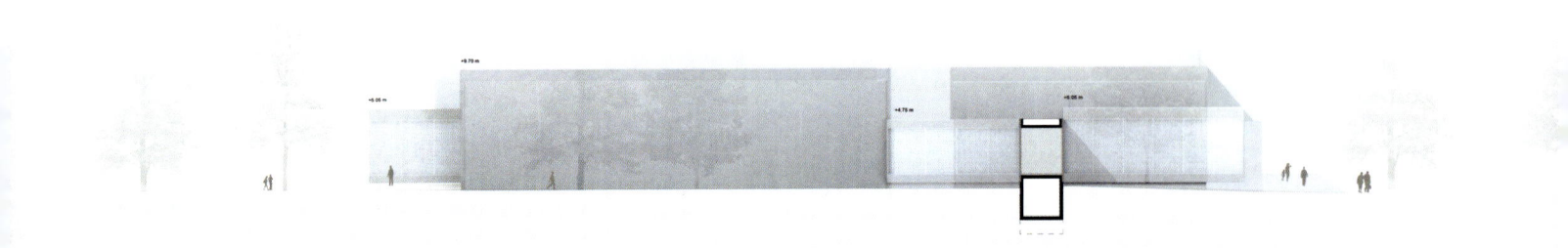

Ostansicht 1:200

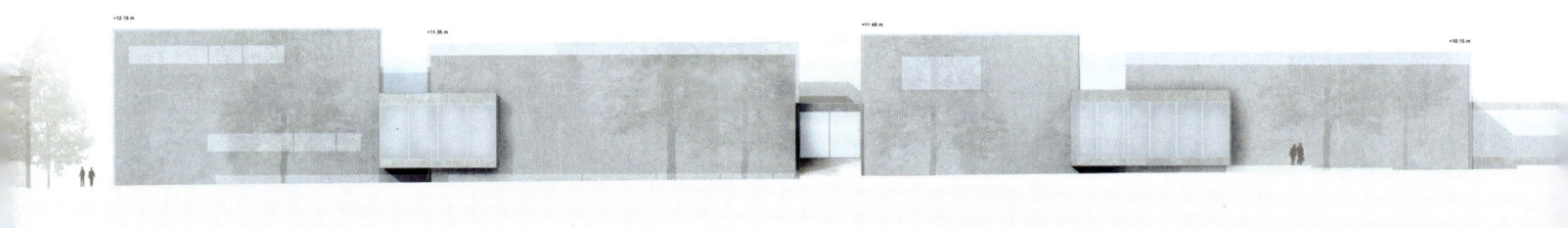

Westansicht 1:200

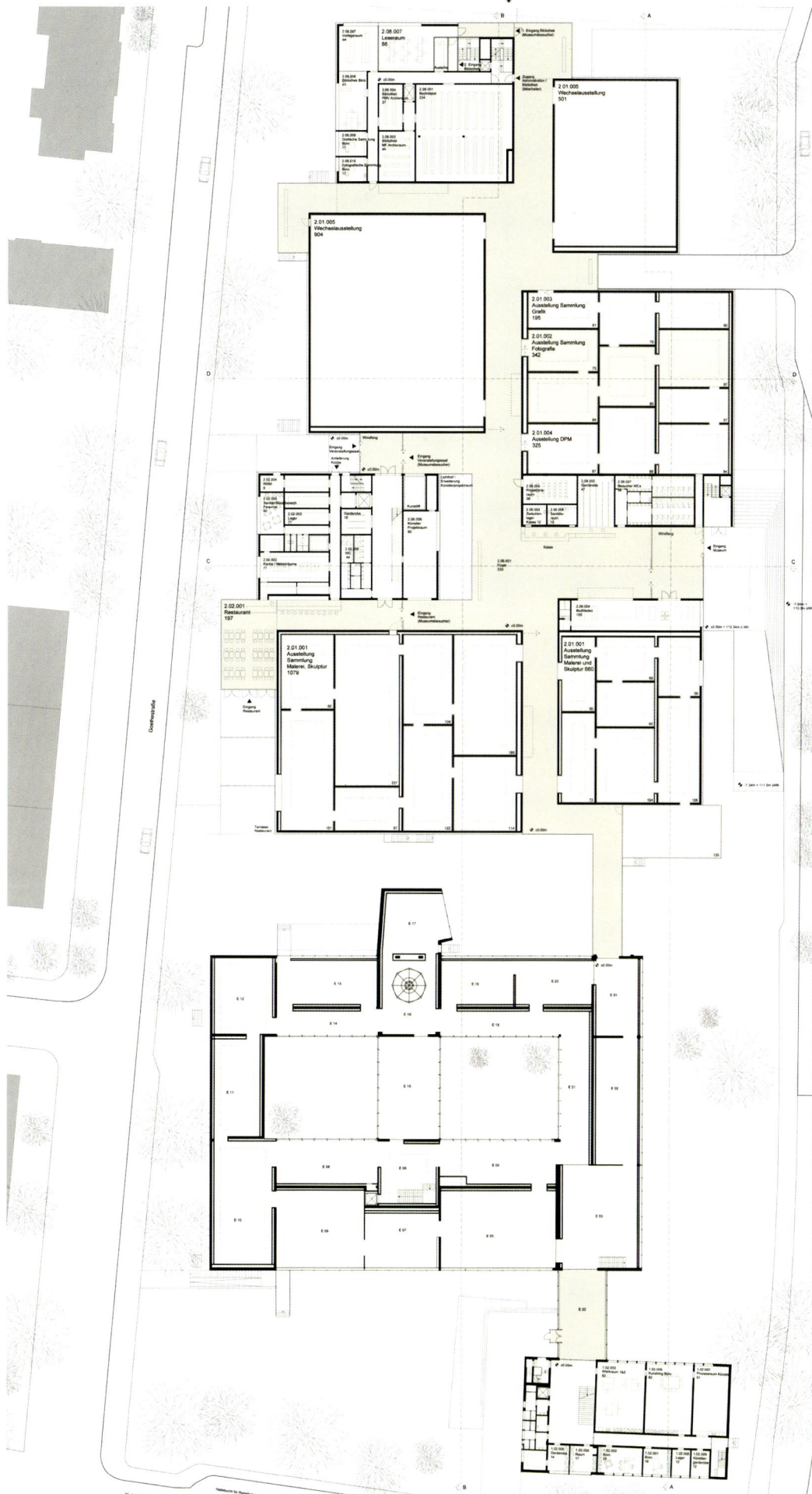

Gigon Guyer Architekten, Zurich

ARCHITEKTENWETTBEWERB MUSEUM FOLKWANG IN ESSEN   23.01.2007

Erdgeschoss und Aussenanlagen 1:200

279

Folkwang Museum, Essen

Staab Architekten Berlin
Staab建筑事务所 柏林

"Both in view of the continuity of the collection halls and in respect of the quality of the extant building, we judged that, instead of an additive extension, continued construction was the adequate strategy."

"在收藏厅的持续性和现有建筑的质量方面，我们认为连续性的建筑，而不是单纯的增加扩展策略可行。"

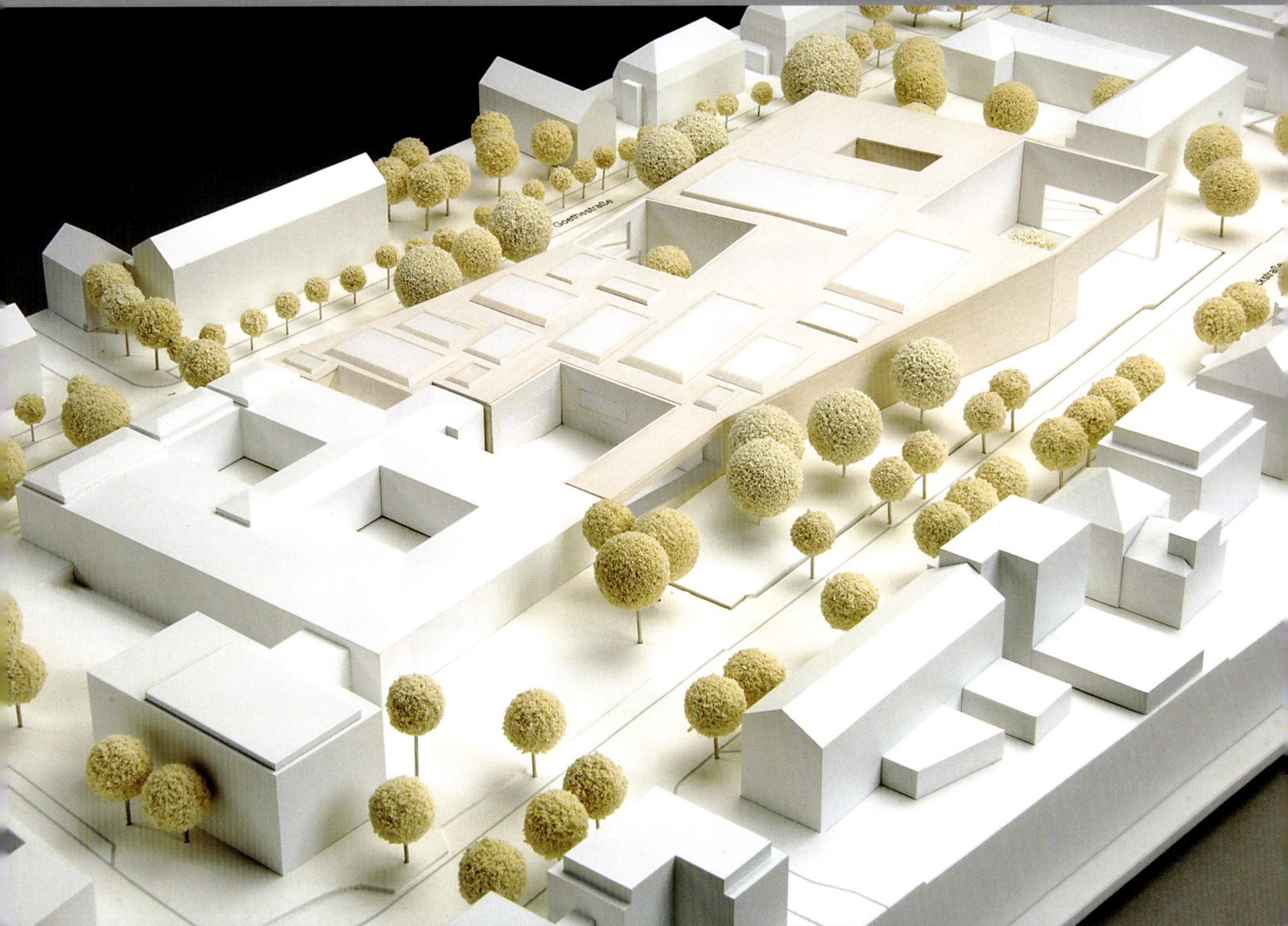

作者 Volker Staab   合作伙伴及建筑系在校生 Antje Bittorf, Per Köngeter, Daniel Verhülsdonk, Petra Wäldle
专家 structural and building services: ARUP Berlin; landscape: Levin Monsigny, Berlin; light: LichtKunstLicht, Berlin/Bonn

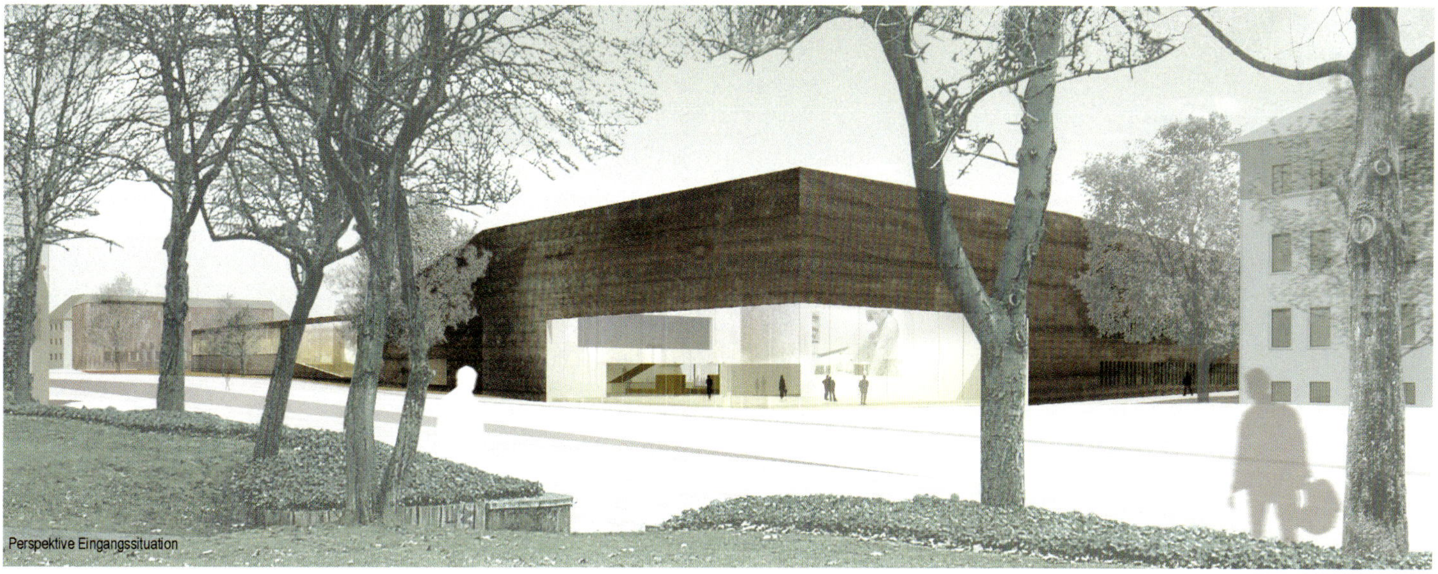
Perspektive Eingangssituation

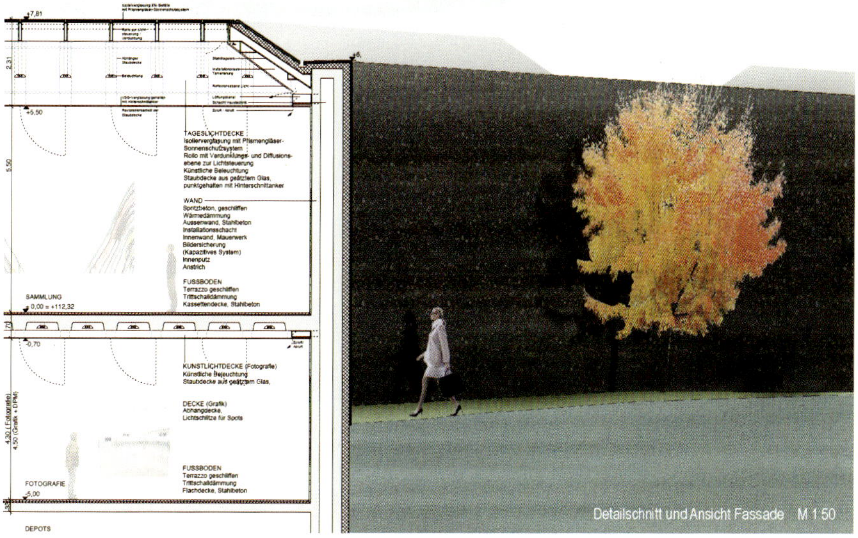
+7,81
2,31
+5,50
5,50
TAGESLICHTDECKE
Isolierverglasung mit Prismengläser-
Sonnenschutzsystem
Rollo mit Verdunklungs- und Diffusions-
ebene zur Lichtsteuerung
Künstliche Beleuchtung
Staubdecke aus geätztem Glas,
punktgehalten mit Hinterschnittanker
WAND
Spritzbeton, geschliffen
Wärmedämmung
Aussenwand, Stahlbeton
Installationsschacht
Innenwand, Mauerwerk
Bildersicherung
(Kapazitives System)
Innenputz
Anstrich
FUSSBODEN
Terrazzo geschliffen
Trittschalldämmung
Kassettendecke, Stahlbeton
SAMMLUNG
± 0,00 = +112,32
70
-0,70
KUNSTLICHTDECKE (Fotografie)
Künstliche Beleuchtung
Staubdecke aus geätztem Glas,
DECKE (Grafik)
Abhangdecke,
Lichtschlitze für Spots
4,30 (Fotografie)
4,50 (Grafik + DPM)
FUSSBODEN
Terrazzo geschliffen
Trittschalldämmung
Flachdecke, Stahlbeton
FOTOGRAFIE
-5,00
DEPOTS
Detailschnitt und Ansicht Fassade M 1:50

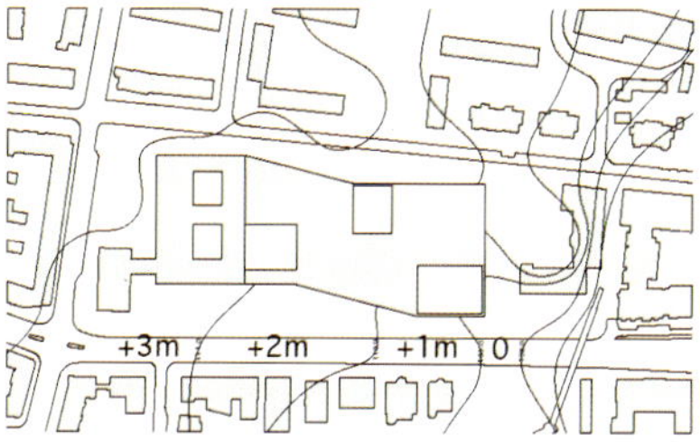
+3m
+2m
+1m
0
Topographie

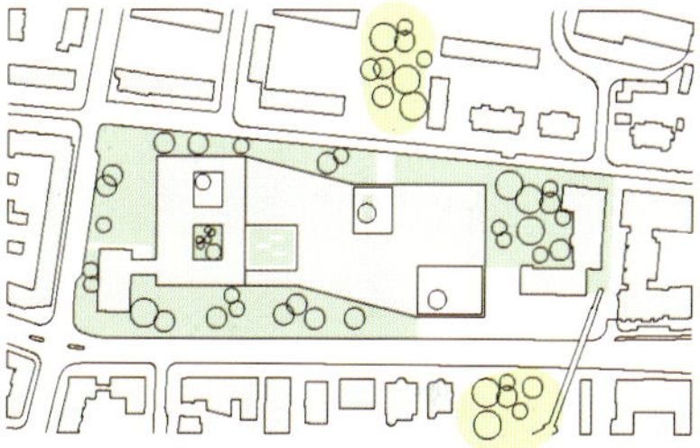
Grünzüge

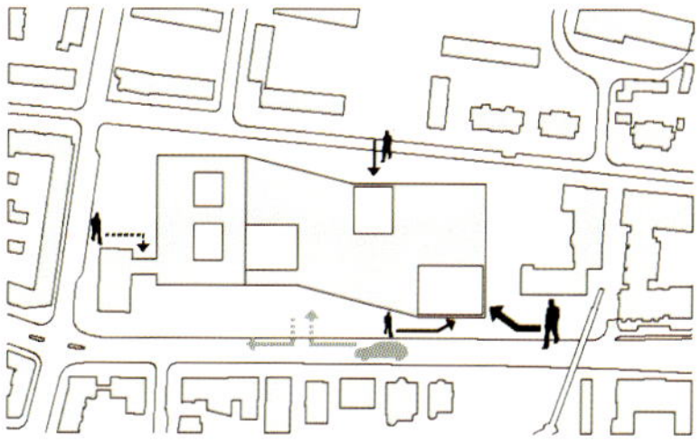
Erschliessung

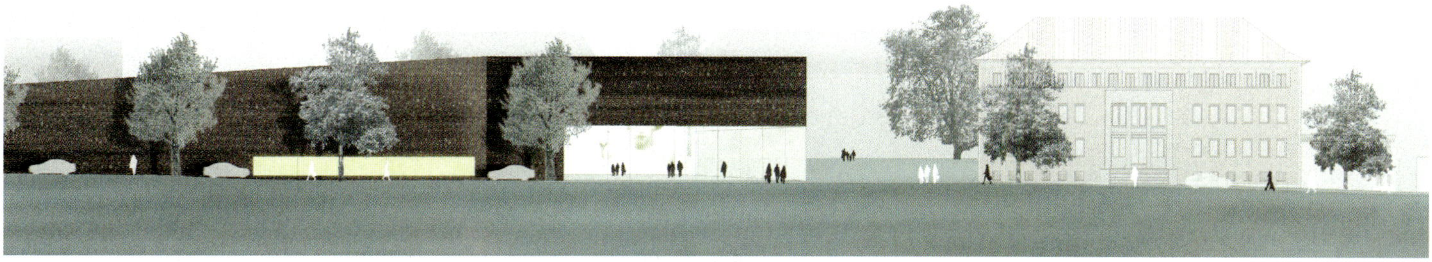

Lageplan M 1:500

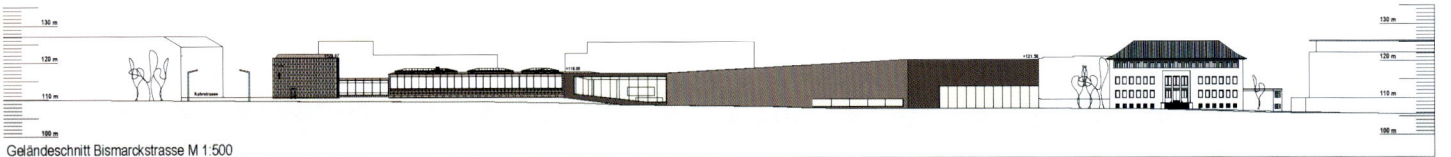

Geländeschnitt Bismarckstrasse M 1:500

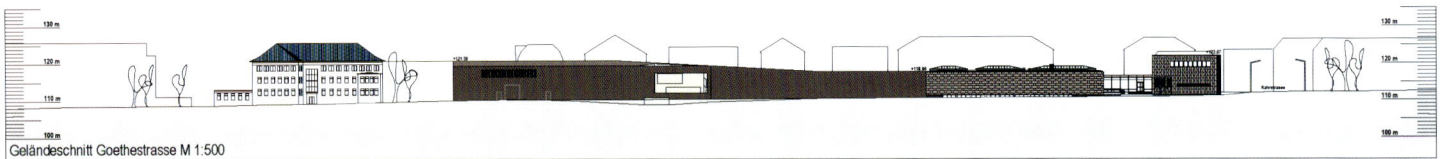

Geländeschnitt Goethestrasse M 1:500

Cafehof

Skulpturenhof

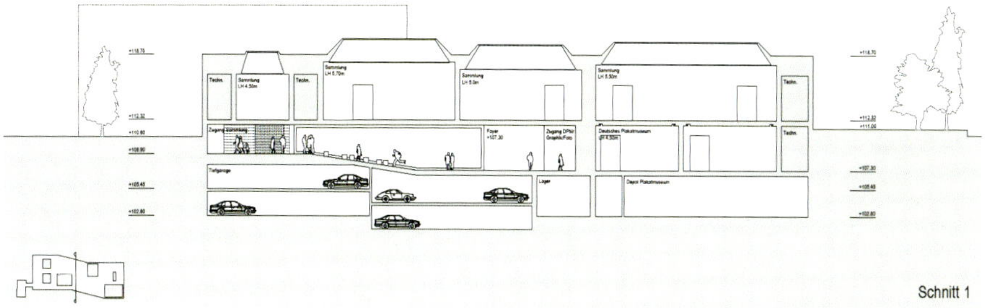

Schnitt 1

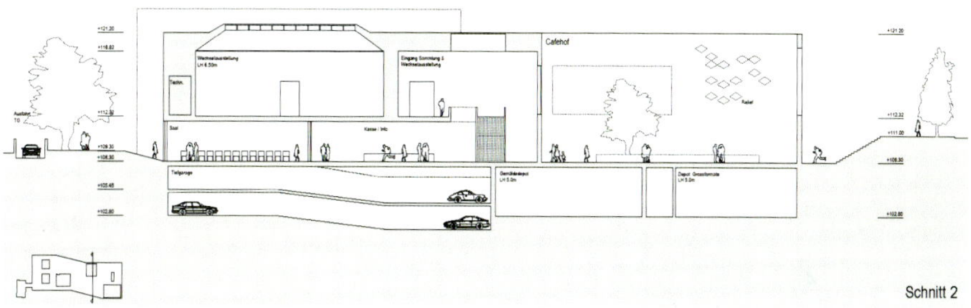

Schnitt 2

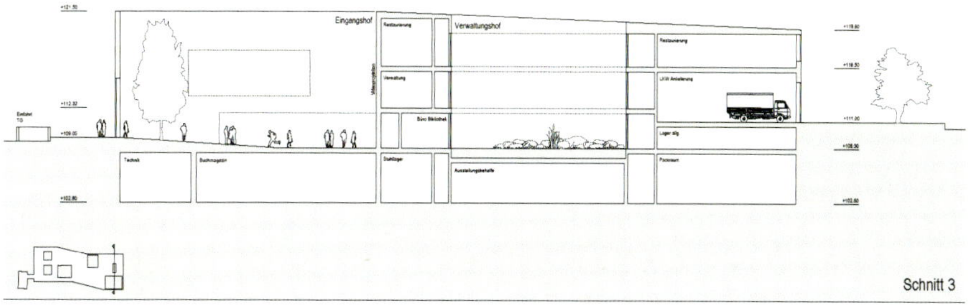

Schnitt 3

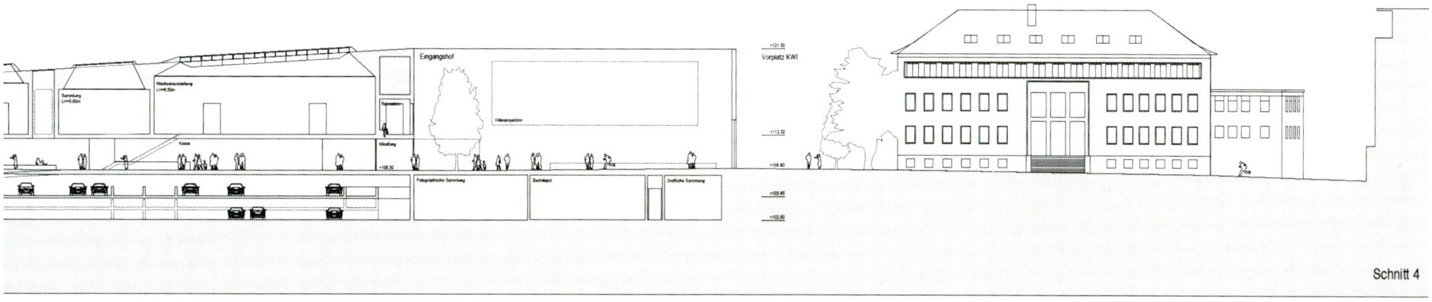

Schnitt 4

Zaha Hadid Architects London
Zaha Hadid建筑事务所 伦敦

"The new museum building [...] pursues two goals: On the one hand to create generously dimensioned exhibition halls directly connected to the extant building, on the other to nest the museum, together with the Institute of Cultural Sciences, within a 'Creative Campus'."

"新博物馆大楼追求两个目的：一是创造出宽阔的展览大厅与现有建筑直接相连，二是把博物馆与文化科学研究院一并收入其中。"

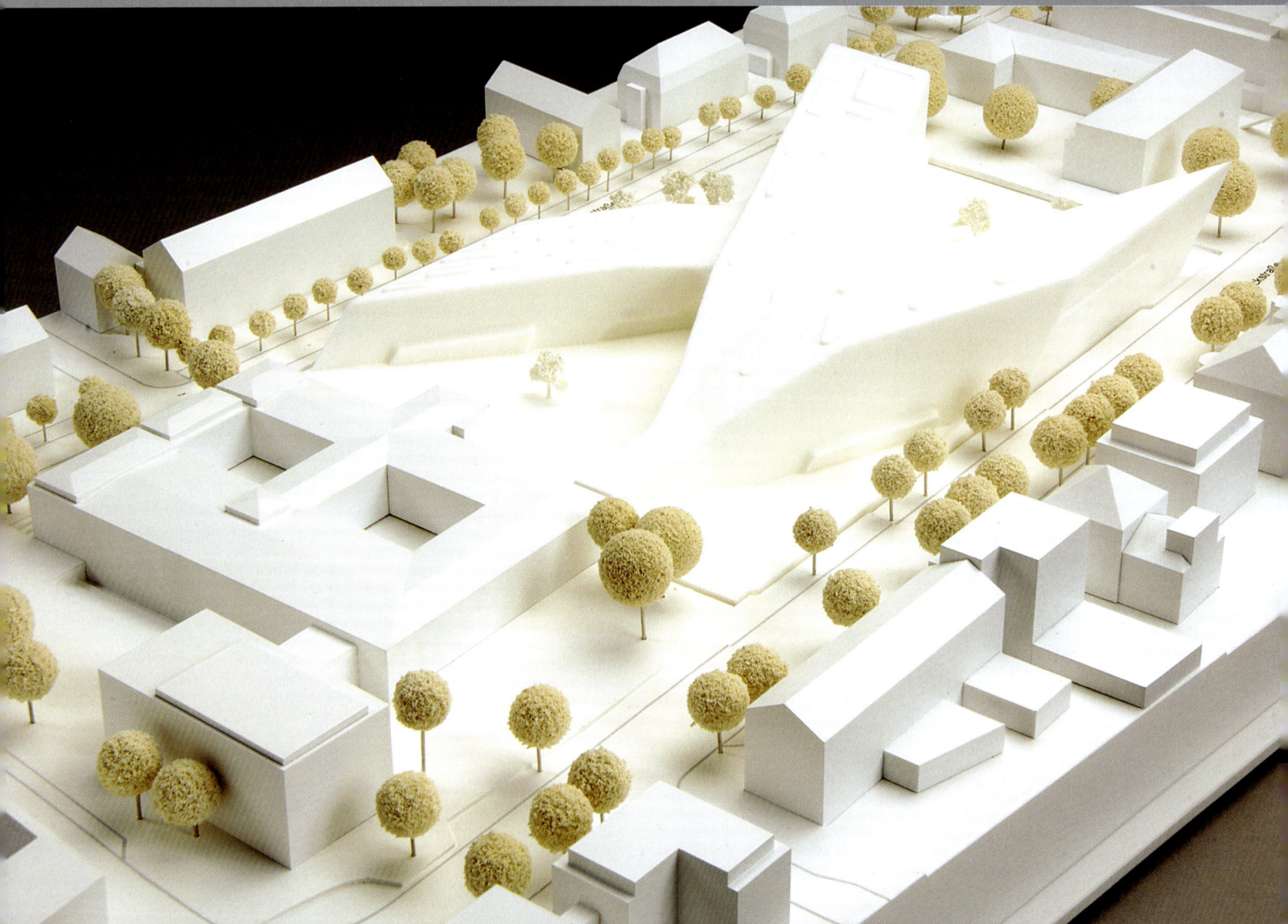

作者 Zaha Hadid, Patrick Schumacher   合作伙伴及建筑系在校生 Jan Hübener, Cornelius Schlotthauer, Patrick Bedarf, Gonzalo Carbajo, Enrico Kleine, Susanne Lettau   专家 structural analysis: ARUP Berlin, Florian Schenk; fire protection: HHP-West, Heiko Zies, Bielefeld; energy: Transsolar Energietechnik, Volkmar Bleicher, Stuttgart; light: Ulrike Brandi Licht, Hamburg; landscape: WES & Partner, M. Kaschke, Hamburg

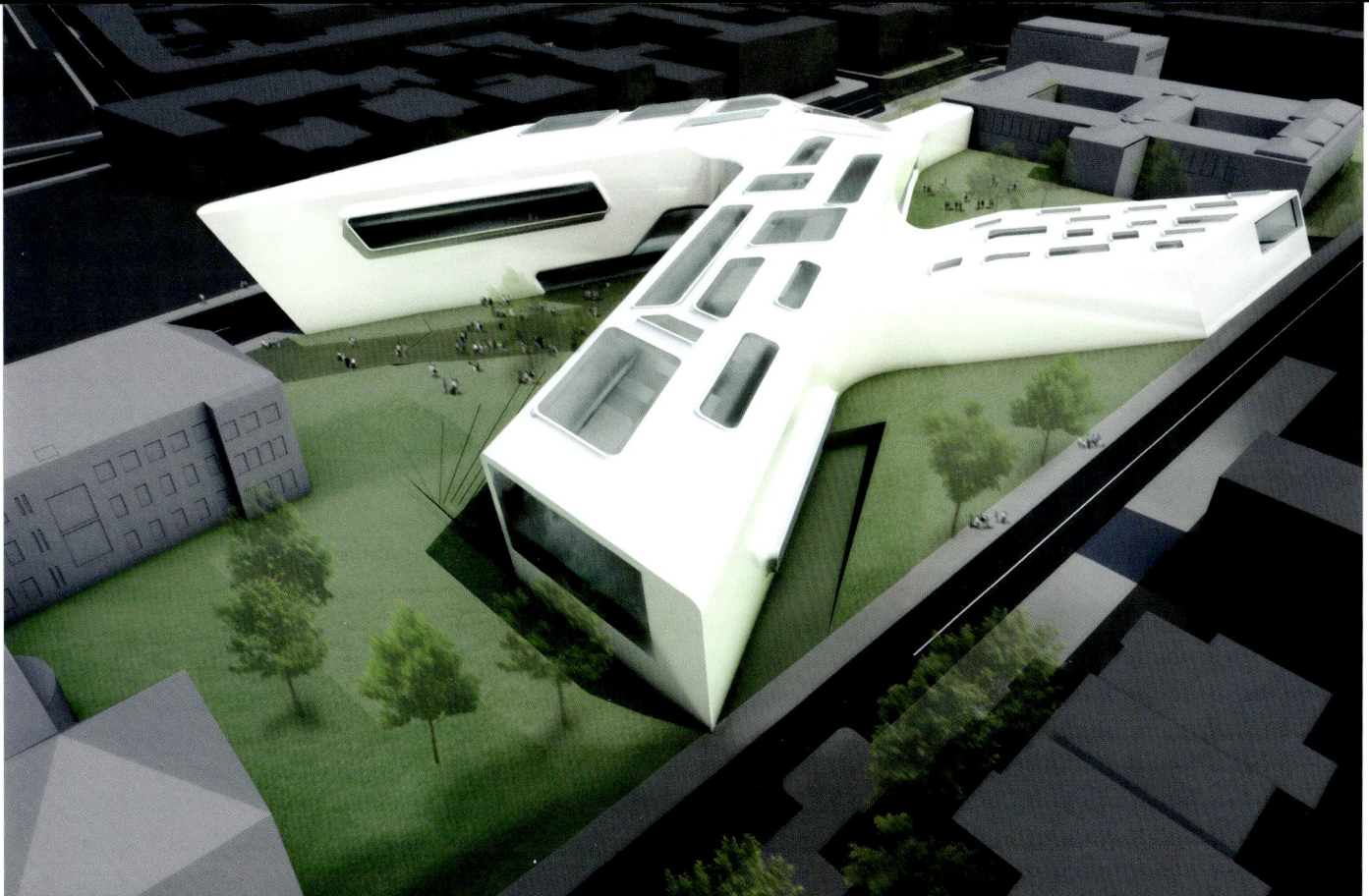

GROBKONZEPTION BRANDSCHUTZ
HHP West Beratende Ingenieure GmbH

Nachstehend sind auf Grundlage des Planstands Wettbewerb wesentliche Punkte des Brandschutzkonzepts textlich zusammengestellt, das System der brandschutztechnischen Unterteilung und der Rettungswegführung ist aus der beiliegenden Planunlage ersichtlich.

1.) Brandabschnitte / Rauchabschnitte
Der geplante Neubau wird als ein zusammenhängender Brandabschnitt mit maximalis Ausdehnungen von ca. 140 m in der Länge und von ca. 80 m in der Breite definiert. Dieser Ansatz wird in Arbeitsschritt der nächstehenden Randbedingungen als vertrebar eingestuft.

Der mehrgeschossige Ausstellungsbereich ad vergleichbare brandlasten und wird zudem mit (natürlichen) Rauchabzugsvorrichtungen ausgestattet. Alle übrigen Gebäudeteile werden – die Tiefgarage ausgenommen – durch raumabschließende Bauteile mit definierten Feuerwiderstand in vergleichsweise kleine Teilbereiche unterteilt.

Die Tiefgarage wird als ein zusammenhängender Rauchabschnitt mit einer Fläche von ca. 3.000 m² definiert und mit einem Schutzventilationssystem ausgestattet, das sowohl für die normale Lüftung als auch für die Entrauchung genutzt werden kann (Sprinklerung wird auch in der Garage nicht vorgesehen).

Der Neubau wird mit einer flächendeckenden automatischen Brandmeldeanlage mit Aufschaltung zur Feuerwehr ausgestattet, so die vorzugweise Ober die Kenngröße Rauch angeschlossen werden.

2.) Rettungswege
Das System der Rettungswegführung ist aus der Anlage zur vorliegenden Grobkonzeption ersichtlich. Zur Rettungswegführung werden folgende Punkte angemerkt:

Alle erforderlichen Rettungswege werden baulich sichergestellt, d. h. eine Rettung von Personen über die Leitern der Feuerwehr ist nicht erforderlich.

Die verklärte Erschließung des Gebäudes erfolgt über insgesamt sechs notwendige Treppenräume, von denen fünf als außenliegend und eine als innenliegend einzustufen sind.

Die baurechtlich zulässigen Rettungsweglängen werden in allen Gebäudeteilen eingehalten bzw. allenfalls in geringem und somit unkritischem Maße überschritten.

3.) Tragwerk
Das Tragwerk des Neubaus wird in der Feuerwiderstandsklasse F 90-A nach DIN 4102-2 ausgeführt, dewwährend ebenso werden die Unterkonstruktion transparente Dachflächen ohne definierten Feuerwiderstand realisiert.

4.) Sicherheitstechnische Anlagen
Folgende sicherheitstechnische Anlagen werden im Neubau vorgesehen:
- flächendeckende Brandmeldeanlage,
- akustische Alarmierungseinrichtungen (flächendeckend),
- Sicherheitsbeleuchtung (in allen den Besuchern zugänglichen Bereichen, in der Tiefgarage, in allen Technikräumen und den Haupterschließungswegen),
- BOS-Funk (soweit im Ergebnis einer Feldstärkemessung nach Schließen der Fassaden erforderlich),
- Blitzschutzanlage (äußerer und innerer Blitzschutz),
- Brandfallsteuerungen für die Aufzüge,
  - natürliche Rauchabzugsvorrichtungen (im mehrgeschossigen Ausstellungsbereich),
- Schutzventilationssystem (in der Tiefgarage),
  - maschinelle Entrauchung für innenliegende Räume mit erhöhter Brandlast, die einzeln eine Fläche von > 100 m² aufweisen,
  - Überdrucklüftungsanlage des innenliegenden Treppenraumes,
- natürliche Rauchabzugsvorrichtungen der außenliegenden Treppenräume und der Aufzugsfahrschächte,
- Handfeuerlöscher,
- Ersatzstromquellen (Batterieanlagen sowie Ersatzstromaggregat)

Licht für das Folkwang Museum Essen   ULRIKE BRANDI LICHT

Tageslicht in den Ausstellungsräumen

In den Ausstellungsräumen dominiert das Tageslicht. Das abwechslungsreiche Spiel des Himmels und des Lichts -4er blauer Himmel, sich verändernde Wolkenbilder, Dämmerung im Verlauf eines jeden einzelnen Tages ist für den Besucher des Folkwang Museums im Innenraum beim Blick hinaus durch die hla vorglasten Deckenöffnungen unmittelbar erlebbar. Großzügige Oberlichter mit einem spezeillen Raster aus stehenden, angeschnittenen weißen Zylindern versorgen die Ausstellungsbereiche mit Tageslicht.

Der analog zur Sonnenbahn in Essen am 21 Juni verlaufende Anschnitt eines jeden Zylinders verbindet eine direkte Sonneneinstrahlung im gesamten Jahresverlauf und ermöglicht der flachen Wintersonne mit weniger Reflexionsverlust in den Raum zu gelangen.

Die Mehrfachreflexionen des direkten Lichts auf den Zylinderinnenwänden erzielen eine gleichmäßig diffuse Tageslichtbeleuchtung der Räume.

Zwei verfahrbare Rollos, eines für vollständige Verdunkelung und eines für zusätzliche Lichtmessung, ermöglichen eine den verschiedenen Ausstellungsanforderung entsprechende optimale Tageslichtregulation in den Sammlungsräumen.

Kunstlicht in den Ausstellungsräumen

In den Sammlungsräumen sind die allgemeine Raumbeleuchtung und die flexible Ausstellungsbeleuchtung für die Anstrahlung der Kunstwerke vollständig in die Architektur integriert.

Ausgewählte Zylinder des Oberlichtsystems ermöglichen jeweils im Zentrum den Betrieb von allen für die modernen Museumsbetrieb notwendigen Strahlern und Flutern mit dem erforderlichen Zubehör. Die Innenräumesser bieten eine gute Erreichbarkeit der Anschlußstellen, sodaß eine einfache Leuchtenmontage, Wartung Austausch und Ausrichtung durch das Museumspersonal gewährleistet sind.

Kunstlicht im Eingangsfoyer

Das Eingangsfoyer verbindet die Außenkante Vorplatz und Innenhof des Museums tiefsend miteinander und soll seinen Außenraumcharakter wahren.

In den Abendstunden und bis Nacht hilf das Licht aus den großflächigen, ausgestülpten Fassadenöffnungen des Gebäudes nach Außen.

Dieses Lichthofra wird in kleinerm Maßstab im Foyer wieder aufgegriffen.

Sich aus der Gebäudegeometrie entwickelnde Ausstülpungen an den Decken erhalten eine Glasabdeckung. Die hinter der Abdeckung positionierten Strahleinrichten und Leuchtstofflampen schaffen eine angenehme helle Atmosphäre im gesamten Foyer.

Eine genau definierte Verbundkonstruktion des Glases steuert den Lichteinprägung und erzeugt einen Fassadenöffnungen ähnlichen Eindruck.

Indirektes Licht aus dem Randbereich der Ausstülpungen erhellt zusätzlich die Decke, mindert den Kontrastunterschied und erweitert dem Raum optisch.

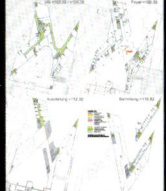

Zaha Hadid Architects, London

285

Folkwang Museum, Essen

LAGEPLAN 1:1500

ANSICHT BISMARCKSTRASSE

ANSICHT GOETHESTRASSE

BLICK VON DER BISMARCKSTRASSE

BLICK VOM KWI

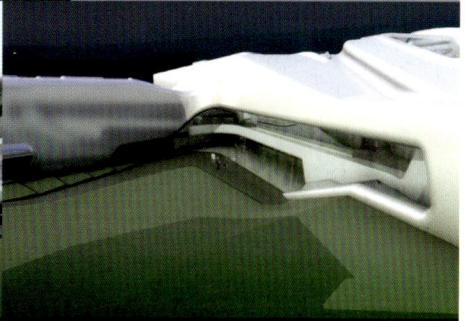

IM GARTENHOF

Energiekonzept Folkwangmuseum · Transsolar Energietechnik GmbH

Ziel
Ziel des Energiekonzepts ist neben der Einhaltung der zukünftigen Klimaschutzvorgaben die Maximierung des Energieverbrauchs sowie der Unterhaltskosten (Energie- + Wartungsaufwand) für den Gebäudebetrieb bei optimiertem thermischen und visuellem Komfort. Gleichzeitig soll die technische Aufwand reduziert werden und die vereinfachte Gebäudetechnik sich in den Gebäudeentwurf als integraler Bestandteil quasi selbstverständlich einfügen.

Beschreibung
Tageslicht
Die Tageslichtversorgung erfolgt bedarfsgerecht. Die Ausstellungsbereiche werden von oben belichtet. Bereiche wie Verwaltung, Veranstaltungsbereiche, etc. werden über Fenster optimal mit Tageslicht versorgt.

Lüftung
Eine mechanische Lüftung mit Wärmerückgewinnung versorgt die Ausstellungsbereiche, Restauration, Depot, Foyer, Veranstaltungsraum und Restaurant. Die Zuluftverteilung kann in 2 unterschiedliche Arten eingeteilt werden.
Quellüftung (Foyer, Mulitstellbereich): Dabei werden Kanäle in die Bodenplatte integriert. Die Zuluft wird impulsarm über die Konvektionsschächte an der Fassade in die Räume eingebracht. Der Zuluftemperator ist stark geringer als Raumtemperatur, so dass eine bodennahe Ausbreitung bis in große Raumtiefen gewährleistet ist. Die Abluft erfolgt punktuell, so dass auf stabile Luftvolgskanäle vollständig verzichtet werden kann.
Mischlüftung punktiche Bereiche): Die Zuluft wird über Deckenauslässe in den Ausstellungsraumweg in den Oberlichten in den Raum eingebracht. Die Abluft erfolgt punktuell an den Luftungsschächten. Hierfür wird die gegenteilige Ebene als Abluftkanal verwendet. Dadurch werden nur geringe Höhen für die Höhungsräche notwendig.
Die unterschiedlichen Bereiche werden mit unabhängigen Anlagen versorgt, so dass eine angepasste Betriebsführung insbesondere stündlich ab- und Entlüftung erfolgt über die zentralen Anlagen.

Heizung
Die Beheizung der Räume erfolgt in Wesentlichen mittels Strahlung. Hierfür wird im Foyer, Restaurant und im Mulitfunktionsraum eine Fußbodenheizung einwendet. Zusätzlich befinden sich an den geschlossenen, Oberflächen Konvektoren so dass Kalhifaktifäll vermieden wird. Der Verwaltungsbereich wird mittels Radiatorflächen beheizt. Nebenräume erhalten, ein erforderlichstatische Heizflächen. Der Ausstellungsbereich wird über die Wände beheizt. Bei räumen mit einem mechanischen Lüftung kann hier zusätzlich bei Bedarf nach geheizt werden.

Kühlung
Die Kühlung erfolgt hauptsächlich die Strahlungskühlung (Fußbodenkühlung bzw. Wandkühlung). Heizfaktifäll, so dass eine angenehme Temperatur bei niedriger Komfort. Zur Minimierung des Kältebedarfs und zur Verbesserung die thermischen Komforbedingungen wird die zur Raum eingesetzte, thermische Gebäudemasse mittels Nacht/Luft gekühlt passivische Grundlinik in den Nachtstunden, außer Ausstellungsbereichen.

Wärme-/Kälteerzeugung
Die Wärme- und Kälteerzeugung erfolgt über eine reversible Wärmepumpe. Diese wird über Erdsonden bzw. über Grubenwasser versorgt. Dadurch wird Kunstanlage als auch Kühlform vermieden, was sich zu eine Minimierung die Wartungsaufwands führt.
Die 2-stufige Wärmepumpe versorgt mittels der ersten Stufe die Fußbodentemperierung während die zweite Stufe die statischen Heizflächen versorgt. Im Kühlbetrieb werden die Zuluftkühlzug mittels die Wärmepumpe mit Kälte versorgt, während die Fußbodenkühlung direkt von den Erdsonden bzw. über das Grubenwasser erfolgt.
Eine grundsätzlich ökologischen Energiebilanz (Wärmeinhalt die aus dem Erdreich im Winter sind "Kälteinhalt im Sommer) gewährleistet, dass sich das Erdreich über die Jahre weder aufheizt noch auskühlt und als Grundlage die Gerechtigkeitsfähigkeit.

Fazit
Alle technischen Anlagen sind bedarf orientiert, so dass wir allem im Außenbereich auf sichtbare Technik vollständig verzichtet werden kann. Die Gebäudebetrieb erfolgt ausschließlich zeitlos energieeffizienten Systemen. Dabei werden natürliche Potentiale (Tageslicht, Kühpotential die Nachtluft und das Erdreichs Wärmepotential die Erdreichs) zur Unterstützung des Gebäudebetriebs genutzt.
Gebäudeheizung und -Kühlung erfolgt auf die selbe Spektro was den Gebäudebetrieb wartungsarm und robust macht.

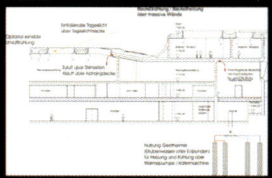

Statisches Konzept · ARUP

Der Neubau des Gebäudes ist eine monolithische Stahlbeton Konstruktion mit zwei Untergeschossen sowie bis zu drei Obergeschossen. Massive Außenwände mit wenigen Öffnungen nehmen die Lasten der Geschossflächen auf und leiten sie in die Gründung weiter. Die weiten ein Lochfassade austeilend zusammen mit den Kernen und Treppenhäusen die vorwiegend im Wurzelbereich sowie den Gebäudeflügeln vorgesehen sind. Innenwände und Stützen werden in Stahlbeton bzw. bei hohen Lasten in Stahlverbundbauweise geplant.

Die Gebäudetiefe beträgt zwischen den parallel verlaufenden Fassaden ca. 20 - 22 m. Die Geschossflächen werden bis zu zwei Spannweiten mit ca. 9m als Flachdecken zwischen als Verbund bzw. Unterzugsdecken vorgespannt. Die Dachdecke mit Oberlichten erhält austeilende Rippen als auch kriegerenförge Unterdrängung um die Öffnungen, wobei die großen Spannweiten mittels eines zwischenliegenden Tragsdecken überbrückt werden können.

Die Oberlichter werden durch eine netzartig hängenden Konstruktion mit aufgestellenden Verglasung geschlossen. Rundstahlbilder verbinden hierbei einzeneinerne die einzelnen Zelostrukturen und bilden durch einen eifenllen krifferenhof zwischen Zyrinderssdecang und Stützparschnitt die Netzzuluftur bis zum starken Öffnungsrand.

Der Wasserbereich / Foyer ist Erschließungs und Luftbau und wird durch ein ca. 45m langes Brückenbauwerk auf Ebene + 112,32m gekrümpft. Dieser Steg, ebenso wie die wird spannende Treppen zwischen den Geschossflächen, werden als Tragganschnitt in Stahlverbundbauweise ausgebildet.

Die neuen Tiefgaragen und Kellerflächen begrenzen im abgeormten mit einem Abstand von ca. 1 m entfernt von den Bestandsbebauung und bilden ca. 3 m in den Boden ein. Dieser Abstand erhöht durch das Volumen die verbleibenden Erdreiches, eine Minimierung der Einflusse aus Verbaubaubuzert und neuen Gebäudelasten. Der Erforderung an das Bestandsbauwerk erfolgt über einen zweigeschossigen Gebäudeflügel der aus dem Neubau austeigt und von dem Altbau endet.

Sämtliche außenliegenden Außenräume werden als wasserundurchlässige (WU-) Betonbauweise mit zusätzlich außen liegender Abdichtung hergestellt. Die Gründung erfolgt die Flächengründung über eine massive Bodenplatte infolge die umfassenden Randbodenfelden die Gebäudebetrieb ermöglicht nach Bodenvolichten mit selbst verbindendlichen Stellgütern Zur Setzungsminimierung zwischen den unterschiedlichen Gebäudeflügeln werden dort Boden schwenwende Maßbahnen erwarkt.

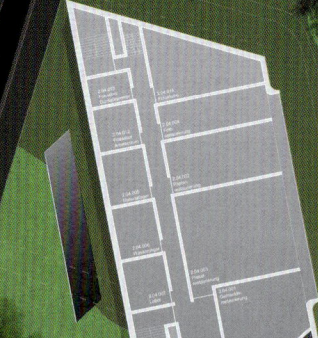

Deckenuntersicht

Perspektive Verschattungs-Zylinder

Ansicht Verschattungs-Zylinder

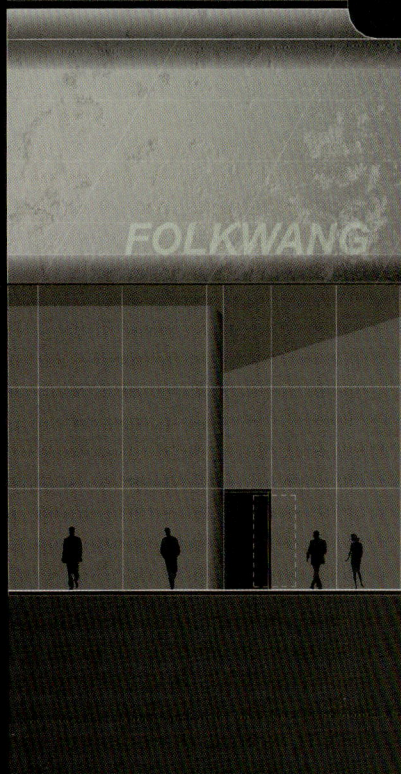
Alugussplatten -
Oberflächenstruktur

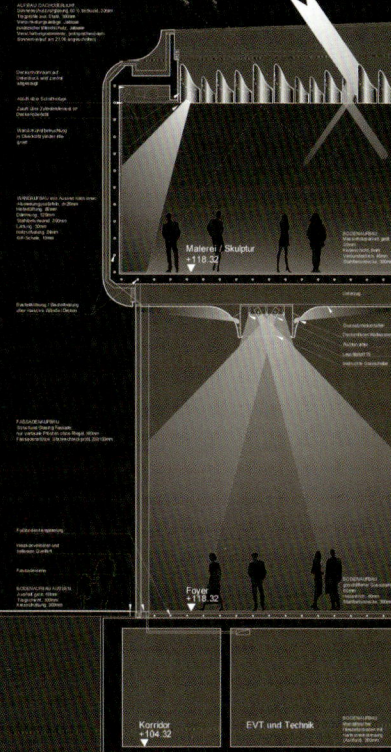
Alugussplatten -
Materialbeschaffenheit / Dicke

Alugussplatten -
Prototypenherstellung

Teilansicht Foyer
M 1:50

FOLKWANG

Fassadenschnitt Foyer
M 1:50

Malerei / Skulptur
+118.32

Foyer
+118.32

Korridor
+104.32

EVT und Technik

Teilansicht Restaurant
M 1:50

Fassadenschnitt Restaurant
M 1:50

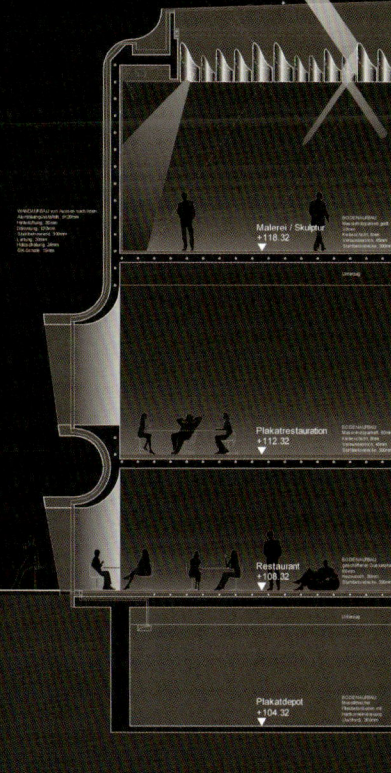

Malerei / Skulptur
+118.32

Plakatrestauration
+112.32

Restaurant
+108.32

Plakatdepot
+104.32

SCHULTES FRANK ARCHITEKTEN Berlin
SCHULTES FRANK建筑事务所 柏林

"With their flexible consistency, the exhibition halls strike up the rhythm that governs the development of 'what remains'."
"展览厅灵活地运用一致性原则，打破了控制原有建筑发展的节奏。"

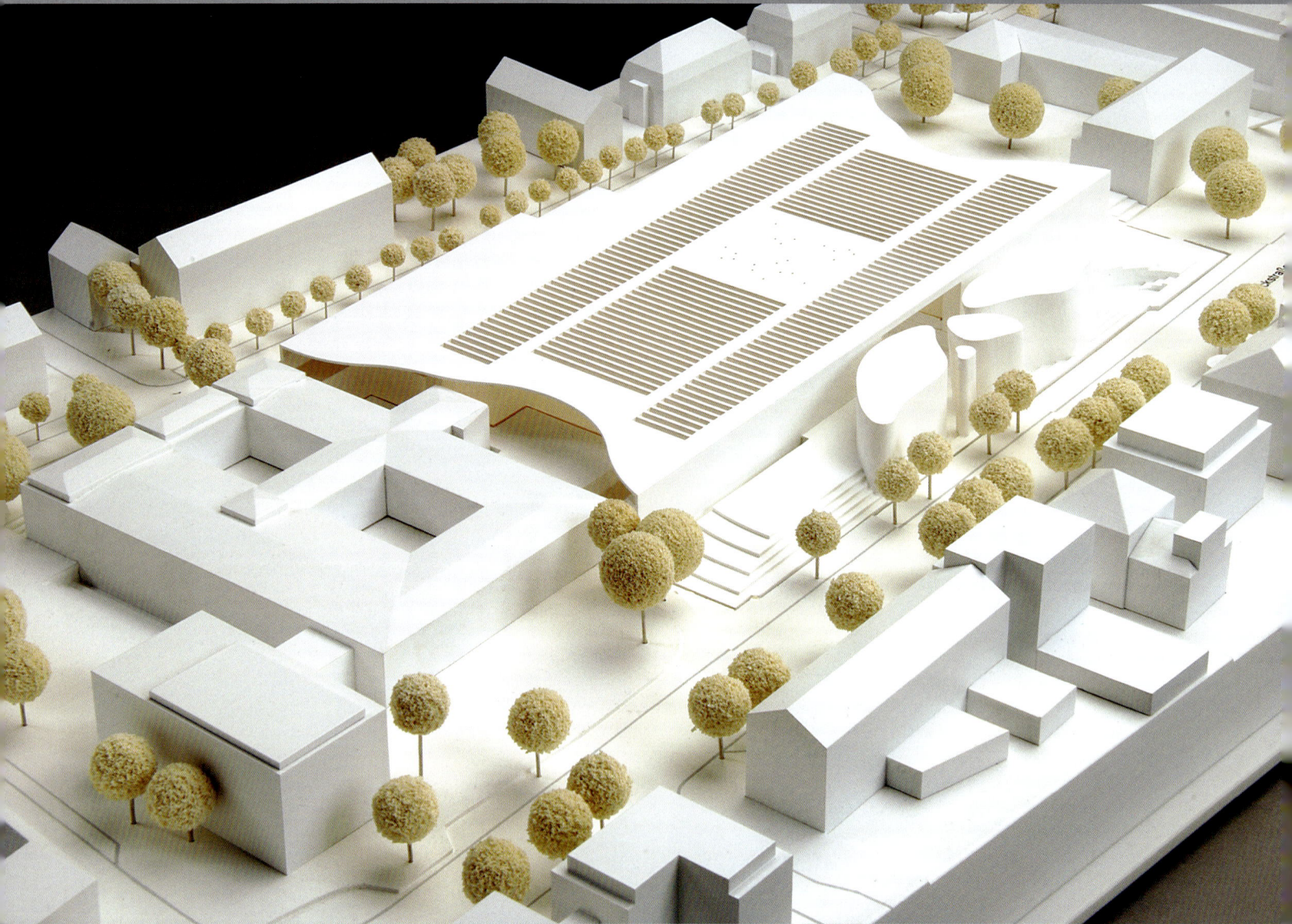

作者 Axel Schultes, Charlotte Frank   合作伙伴及建筑系在校生 Monika Bauer, Fritz Lobeck, Sören Timm, Christian Laabs, Andreas Ulrich, Robert Freudenberg   专者 light: Ing. Martin Klingler MAS, Moosbach; building services: HL-Technik, Prof. em. Dr.-Ing. e.h. Klaus Daniels, Munich; structural analysis: Pichler Ingenieure, Peter Saradshow, Berlin; landscape: Thomas Garten- und Landschaftsarchitekten, Kirsten Thomas, Berlin

Realisierungswettbewerb Museum Folkwang – Essen

Schultes Frank Architekten

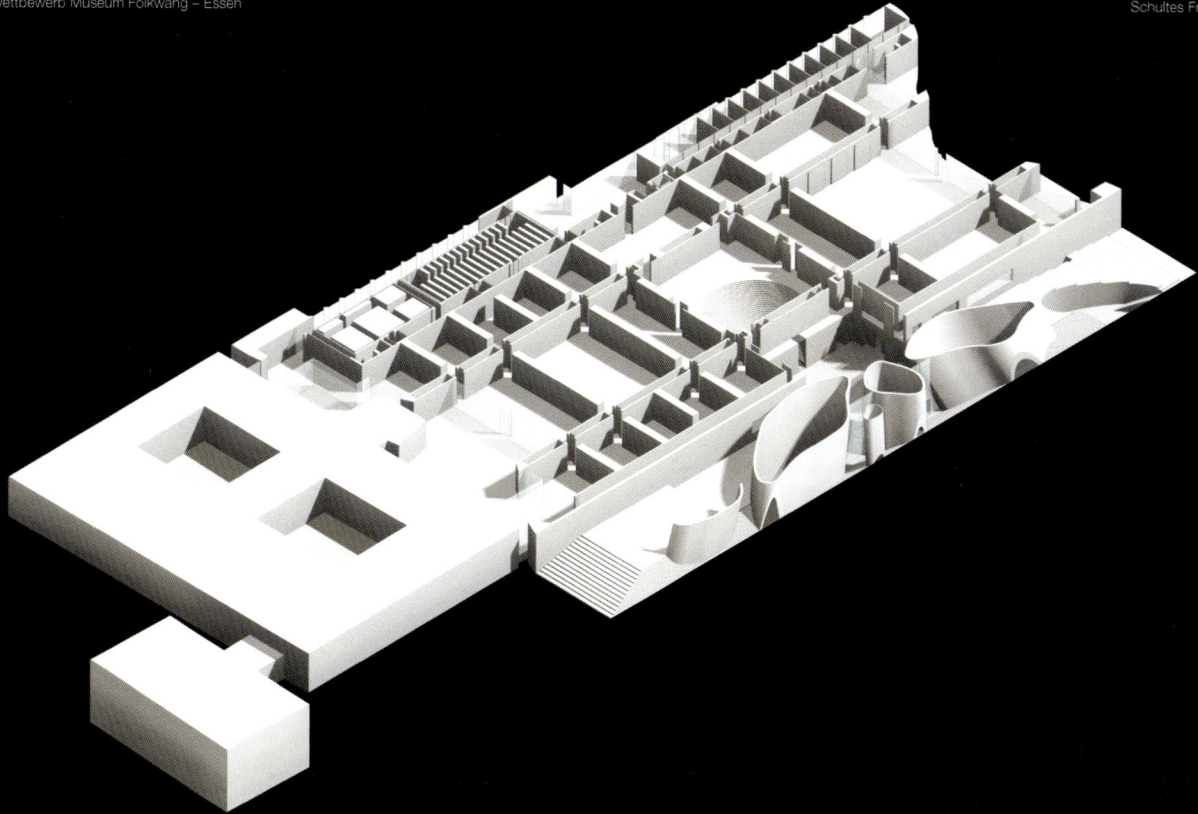

Realisierungswettbewerb Museum Folkwang – Essen

Schultes Frank Architekten

Perspektiven, Tageslichtkonzept

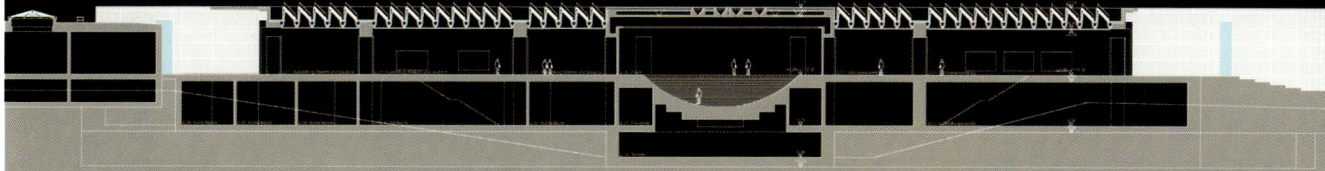

Süd-Nord Schnitt 1:200

West-Ost Schnitt 1:200

West-Ost Schnitt 1:200

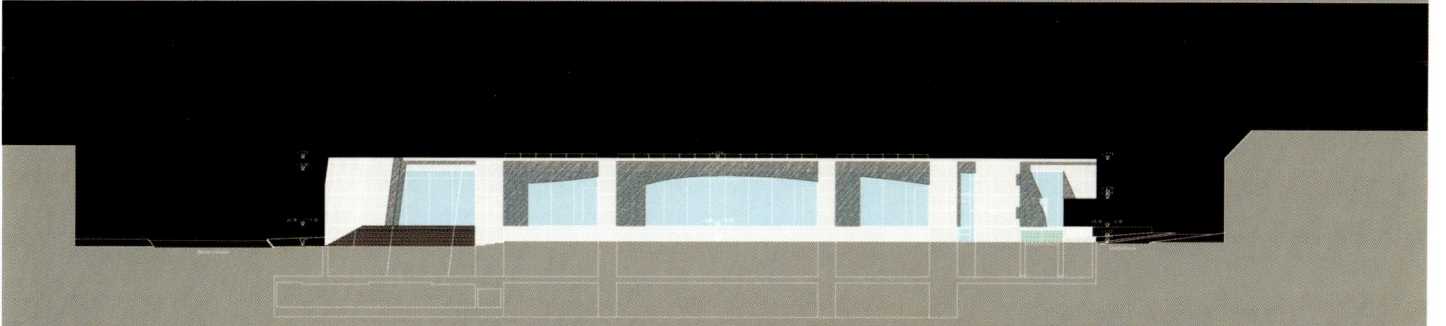

Ansicht von Norden 1:200

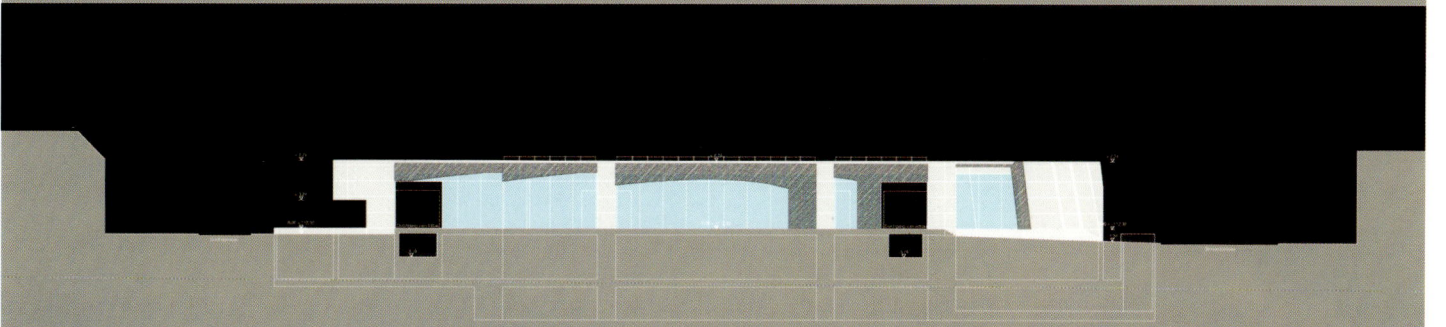

Ansicht von Süden 1:200

SCHULTES FRANK ARCHITEKTEN, Berlin

293

Folkwang Museum, Essen

SANAA Tokyo
SANAA建筑事务所 东京

"Our intension is to open up the site and the Museum towards the city, connecting Bismarckstraße and Goethestraße with green space."

"我们的目的是朝向城市开放的项目地点和建筑物，用绿色空间联结BismarckstraBe和GoethestraBe两地。"

作者 Kazuyo Sejima, Ryue Nishizawa   合作伙伴 Jonas Elding, Johanna Meyer-Grohbrügge, Matthias Härtel, Riken Yamamoto

Folkwang Museum, Essen

SECTION A-A

2.03.005
SEATING
STORAGE
2.03.004
PREPARATION
2.03.002
CATERING
2.03.003
DIRECTOR'S
2.03.001
MULTI-PURPOSE

2.01.001
COLLECTION
2.01.001
COLLECTION
2.01.001
COLLECTION

FOYER

2.01.001
COLLECTION

2.01.001
COLLECTION

CAR
RAMP

EV

PATIO

FOYER

SECTION B-B

PATIO

2.01.006
CHANGING
EXHIBITION

2.01.003
DRAWING
& PRINT

2.01.002
PHOTOGRAPH

EV

2.01.001
COLLECTION

FOYER

SECTION B-B

Gustav-Heinemann-...

VOID
(PATIO
BELOW)

2.01.002
PHOTOGRAPHY

2.01.006
CHANGING
EXHIBITION

2.06.004
BOOKSHOP

Bananen-... straße

2.01.001
COLLECTION

2.01.004
POSTER

2.01.004
POSTER

2.01.001
COLLECTION
2.01.001
COLLECTION

VOID
(PATIO
BELOW)

2.06.001
FOYER

2.07.004
MEETING
(TOUR BRIEFING)

2.07.002
PRESS

LOBBY

TRUCK
RAMP

2.01.001
COLLECTION

2.06.002
PROJECTION

EV
FFV

RESTAURANT
PATIO

2.01.001
COLLECTION

2.06.003
ARTIST'S
PROJECT

2.02.001
RESTAURANT

SERVING
AREA

EV

SECTION A-A

Ground Floor Plan, scale 1:200

Elevation South

Elevation West

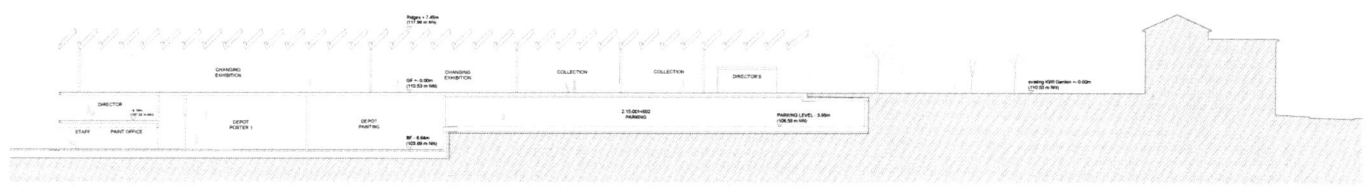

Section A-A

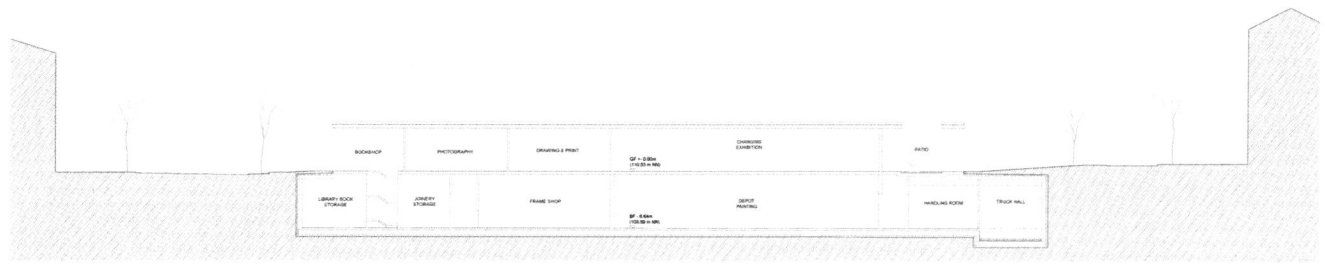

Section B-B

Königs Architekten Cologne
Konigs建筑事务所 科隆

"The Folkwang idea links an assemblage principle, a societal task, and its history to a particular place [...] a holistic view to be developed further [...] integrated with the concept for the new building."

"Folkwang观点把组合原则、社会任务和它的历史与某一地点联系到一起……一个整体深入开发的观点……与新建筑物概念统一整合。"

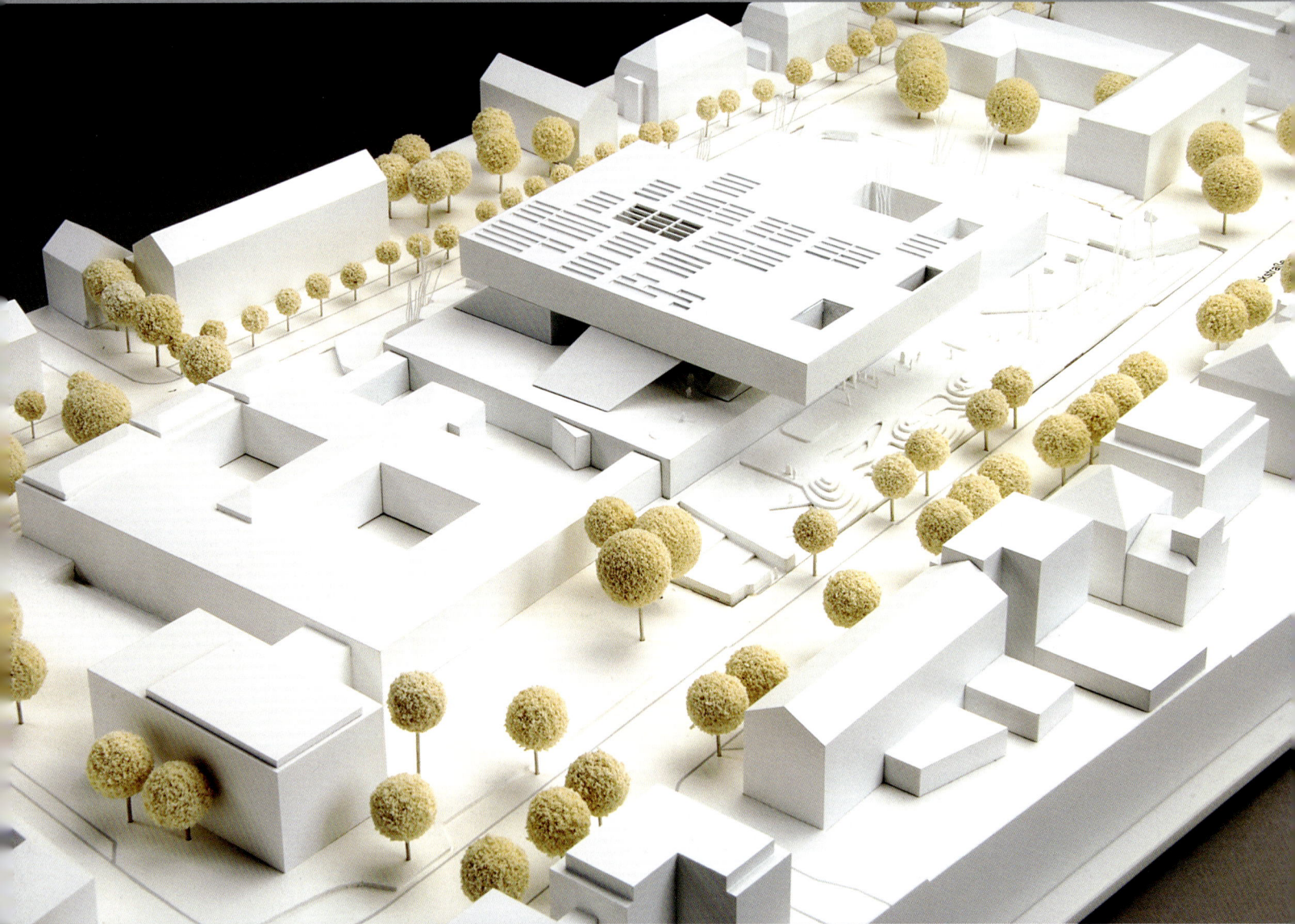

作者 Ulrich Königs, Ilse Königs   合作伙伴 Mirjam Patz, Geraldine Bach, Karla Spennrath, André Rethmeier, Thomas Roskothen
专家 structural analysis: Torsten Schröder-Wilde, ARUP Düsseldorf; TGA: Arwid Theuer-Kock, Brandi IHG, Cologne; lighting: Anette Hartung, Lightning Cologne

Perspektive Vorplatz

Perspektive Großer Saal Amerikanische Nachkriegskunst

Kulturwissenschaftliches Institut

Park

TK + 121.50

TG Einfahrt

Anlieferung

+ 118.90

+ 132.00

Eingang
Museum Folkwang
+ 112.30

+ 118.30

+ 119.00

+ 112.30

+ 123.07

Goethestraße

Bismarckstraße

Kahrstraße

Lageplan 1:500

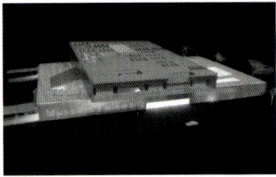

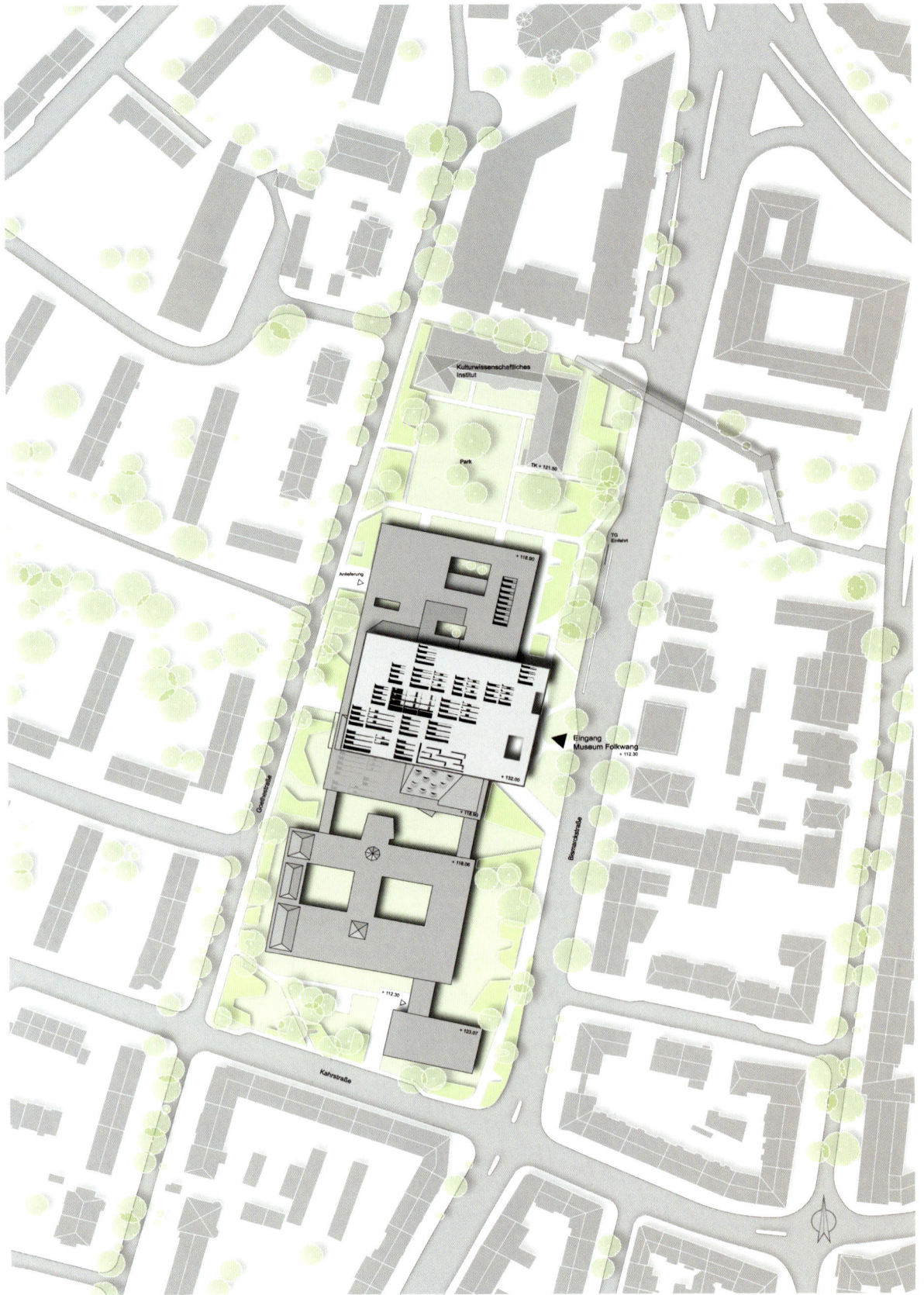

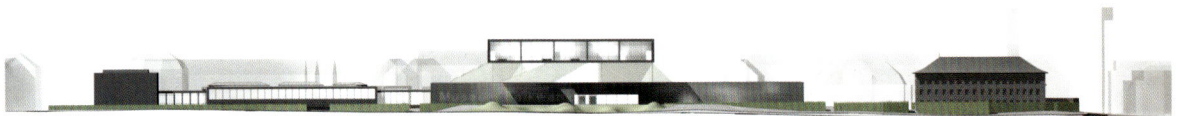

Profilschnitt Bismarckstraße 1:500

Profilschnitt Goethestraße 1:500

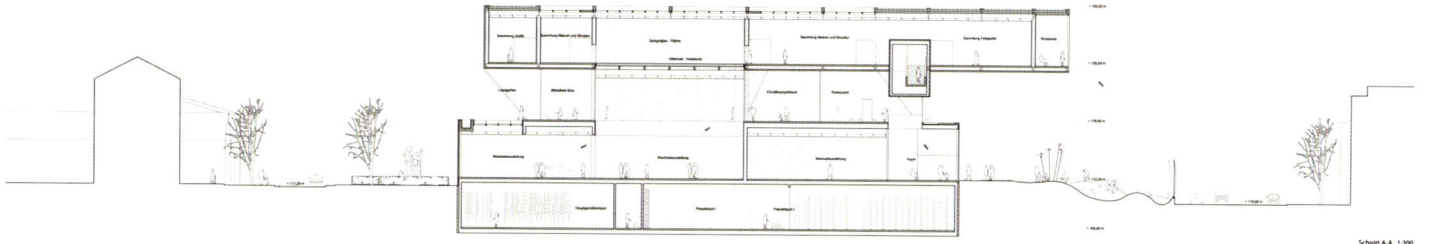

Schnitt A-A 1:200

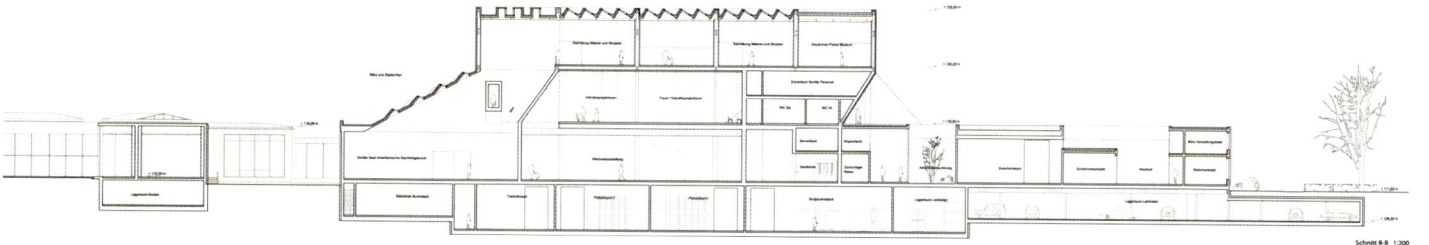

Schnitt B-B 1:200

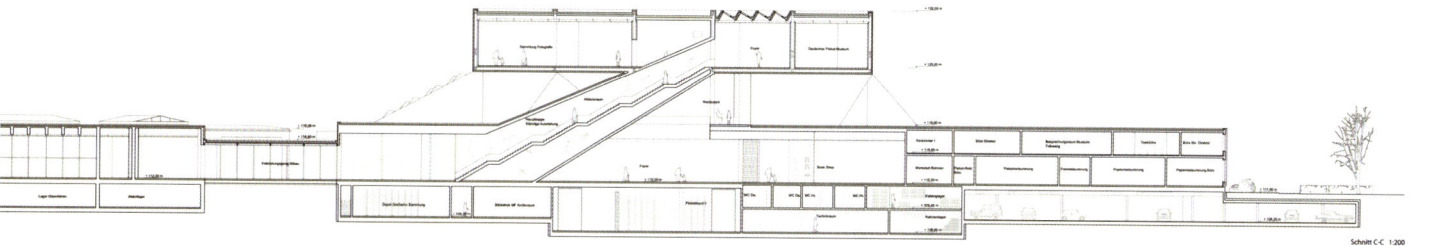

Schnitt C-C 1:200

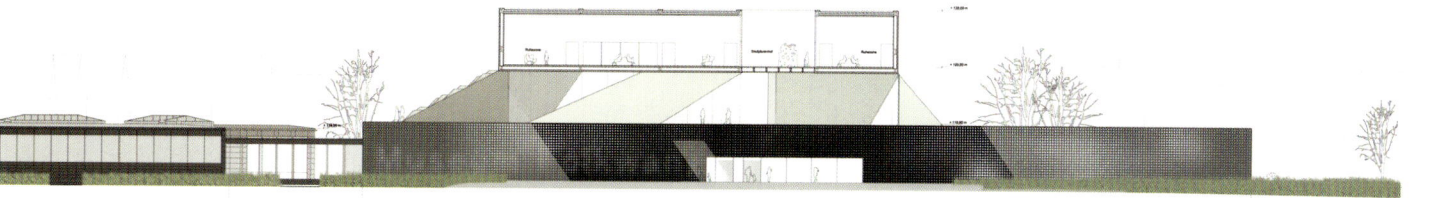

Schnitt D-D / Ansicht Ost 1:200

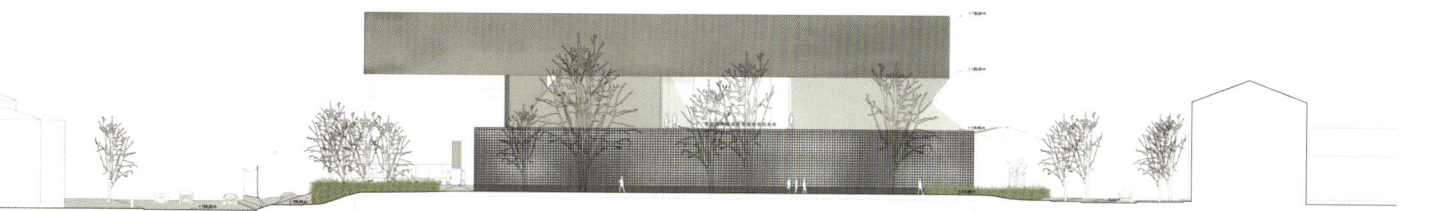

Ansicht Nord 1:200

[phase eins].

The Architecture of Competitions 2

建筑竞赛 2

2006 – 2008

Benjamin Hossbach   **Christian Lehmhaus**

(德) 本杰明·胡斯巴赫   克里斯蒂安·雷姆豪斯   王晨晖 译

辽宁科学技术出版社

keller mayer wittig

Grüntuch E

Modersohn & Freiesle

Lehrecke Architekten

Glass Dairy in Münchehofe

玻璃奶品厂

ist Architekten

en

Glass Dairy Münchehofe
玻璃奶品厂 Munchehofe

Invited competition preceded by an application procedure
邀请赛，开始前有申请程序。

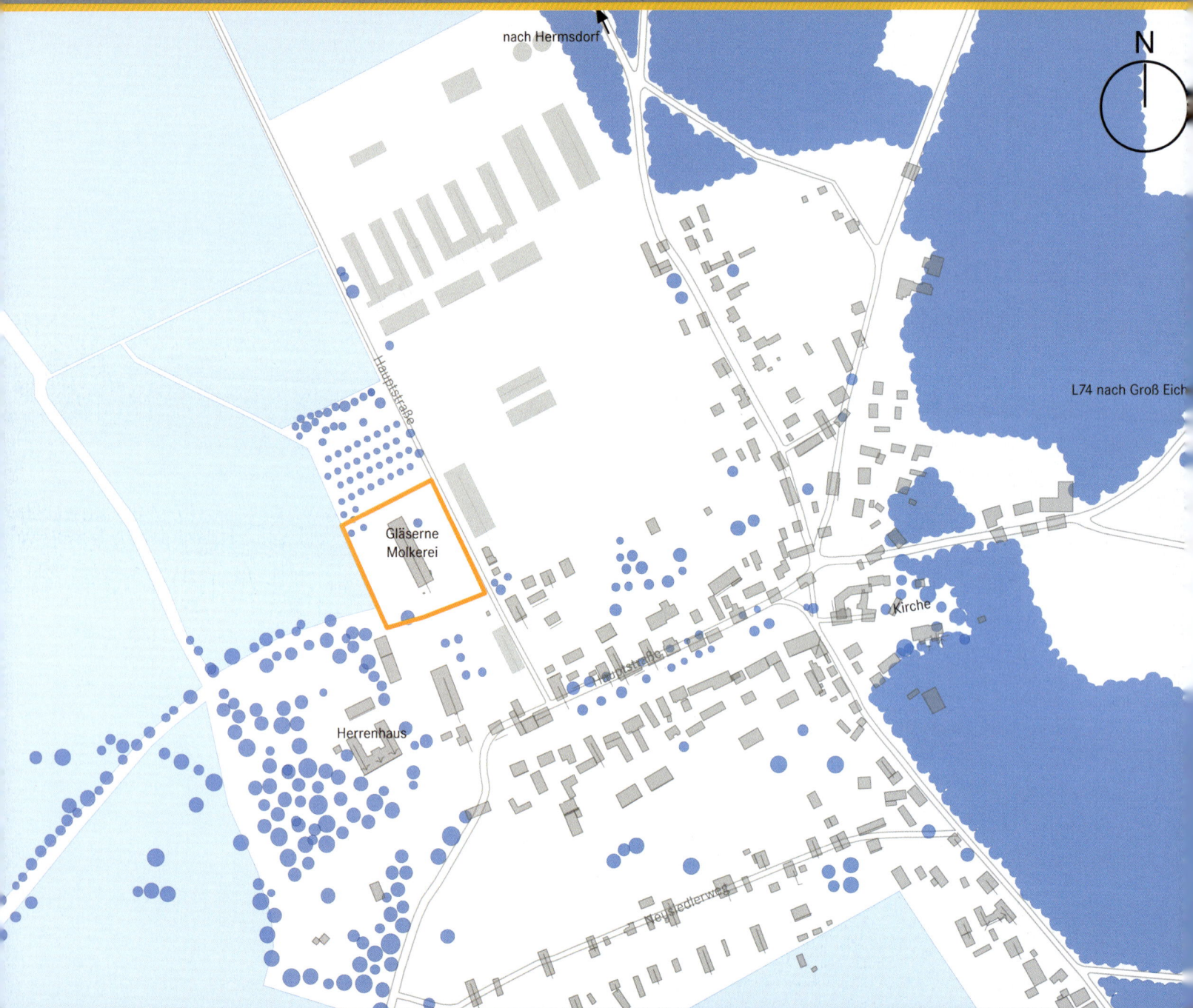

nach Hermsdorf

N

Hauptstraße

L74 nach Groß Eich

Gläserne
Molkerei

Kirche

Herrenhaus

Hauptstraße

Neusiedlerweg

地点 Münchehofe 时间 11/2006-02/2007 主办方 Gläserne Molkerei GmbH 参赛者 23 applicants; 7 participants
面积 about 1,700 sq m 竞赛费用 36,200 Euro 专业评奖委员会 Prof. Gisela Glass, Berlin/Munich; Doris Gruber, Berlin;
Tim Heide, Berlin; Prof. Claudia Lüling, Berlin/Frankfurt/Main 专家评奖委员会 Hubert Böhmann, head Gläserne Meierei GmbH, Upahl;
Dr. Peter Danckert, MdB, Berlin; Meinrad Schmitt; head and holder TERRA Naturkost Handelsgesellschaft, Berlin

In Münchehofe, in the Dahme-Heideseen nature reserve southeast of Berlin, the Gläserne Molkerei GmbH runs an ecofriendly dairy and cheese factory. Here, milk produced by environmental sound methods is packaged or processed into yoghurt, curd, cream and cheese. The name "Gläserne Molkerei" ("The Glass Dairy") reflects standards of transparency and authenticity in its dairy production. The basis of the competition was to extend the dairy and construct a new cheese factory. In keeping with the company's name, the plant conversion was indeed envisaged in terms of a "glass dairy" where visitors can actually witness the manufacturing process. The aim is to promote consumer interest in ecological sensitive approaches to food production and thereby promote customer loyalty. Extant buildings included, the spatial program comprises about 1,700 square metres of useable area, 900 square metres of which are to be new. It also includes building a small conference hall The aim of the competition was to come up with an approach that is both functional and aesthetically striking, one that does not imitate standard patterns for plants of this kind. Accordingly, the new building was to be harmoniously aligned with its environs, the village locale of Münchehofe, as well as add qualitative uplift.

Munchehofe在柏林东南的Dahme-Heideseen自然保护区里，Glaserne Molkerei GmbH开办一家生态环保的乳品和乳酪工厂，把用环保安全方法生产出来的牛奶包装或加工成酸奶、豆腐、冰淇淋和乳酪。"玻璃奶品厂"这个名字反映乳品生产过程的透明性和可靠性的标准。竞赛的目的是扩建奶品厂，建成一家新乳酪工厂。该项目被命名为参观者可以目睹生产过程的"玻璃奶品厂"，目的是提高消费者对食品生产的生态环保方法的兴趣和信任度。包括现有建筑，空间规划包括约1700平方米的使用面积，其中900平方米是新建的，也包括一个小会议厅。竞赛的目的是提出一个与这类工厂标准模式不同的在功能性和美学方面引人注意的方法。因此，新建筑不但会增强质量方面的提升，而且将与周围环境和谐统一。

Aerial view 鸟瞰图

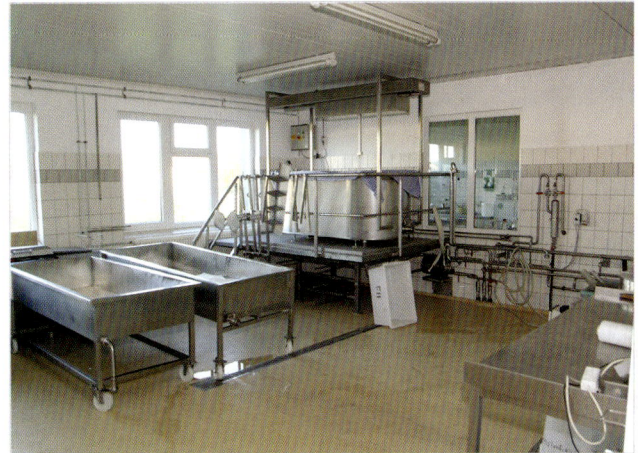

Cheese dairy 乳酪制品

Competition site 竞赛地点

View of the village 村庄景色

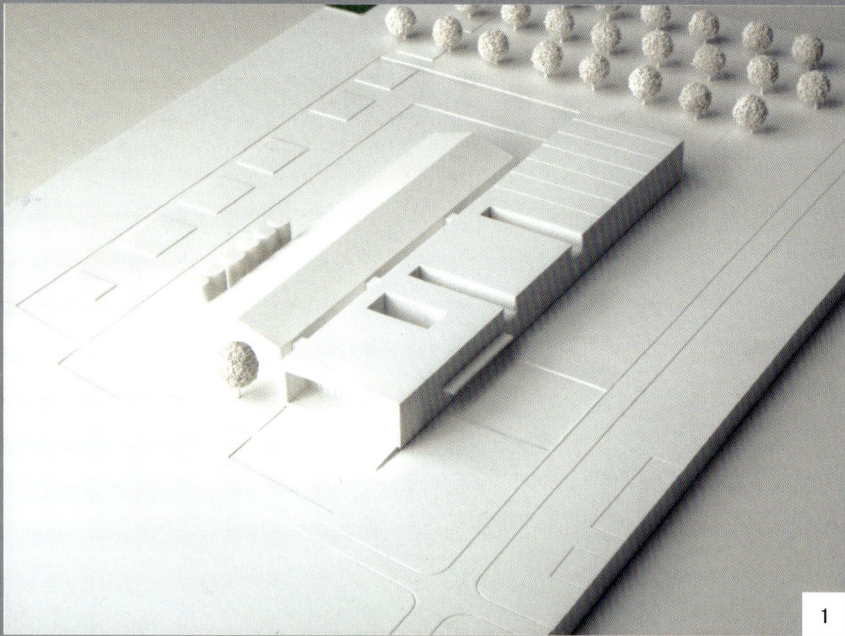

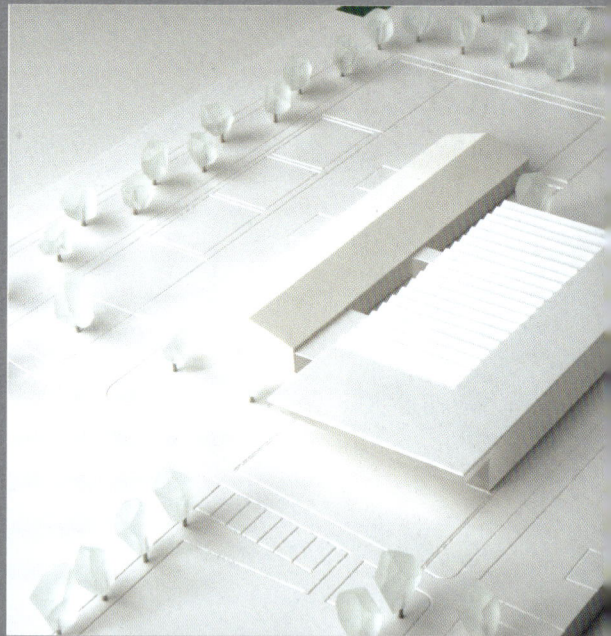

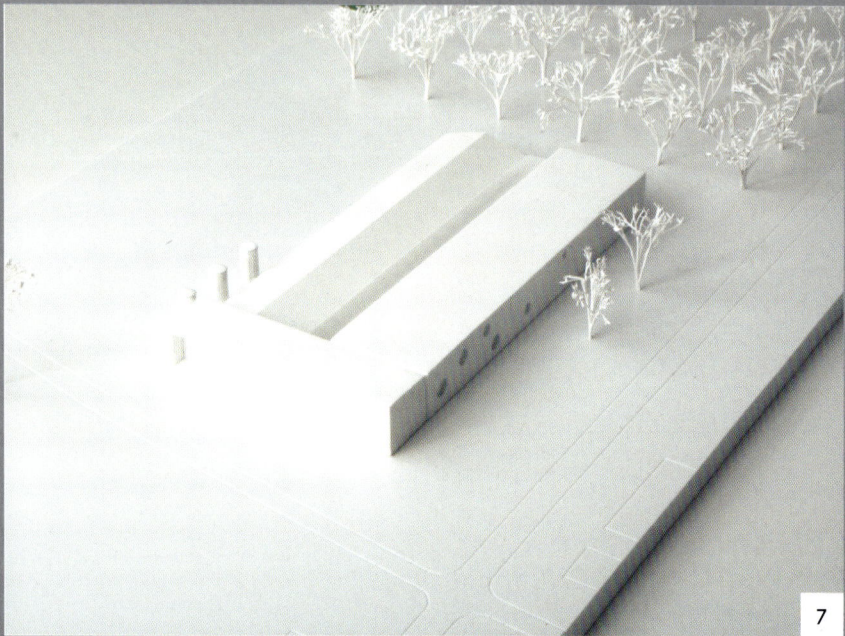

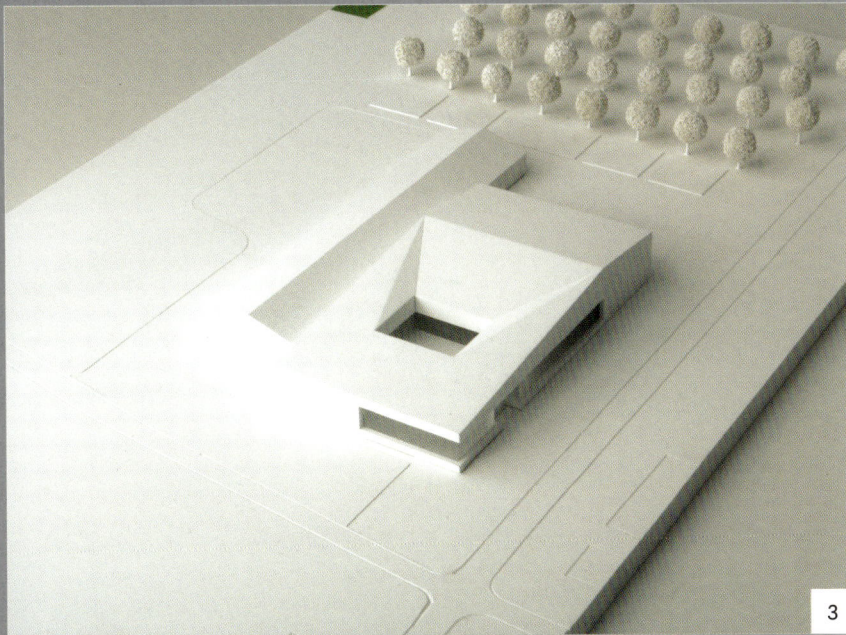

2

3

5

6

Qualified participants
合格的参赛者

1
1st prize 一等奖
Lehrecke Architekten, Berlin

2
2nd prize 二等奖
keller mayer wittig, Cottbus

3
3rd prize 三等奖
Modersohn & Freiesleben
Architekten, Berlin

4
Further participant 其他参赛者
Grüntuch Ernst Architekten, Berlin

5
Further participant 其他参赛者
Anderhalten Architekten, Berlin

6
Further participant 其他参赛者
Kretzschmar & Weber Architekten,
Oranienburg

7
Further participant 其他参赛者
Architekturbüro Kühn-
von Kaehne + Lange, Potsdam

Lehrecke Architekten Berlin
Lehrecke建筑事务所 柏林

"The structure is endowed with [...] an elegant air of sobriety that confidently inserts itself into the context of Münchehofe's agricultural scenery."
"给建筑结构赋予一种严肃的优雅气氛，自信地融入到Munchehofe的农业景色之中。"

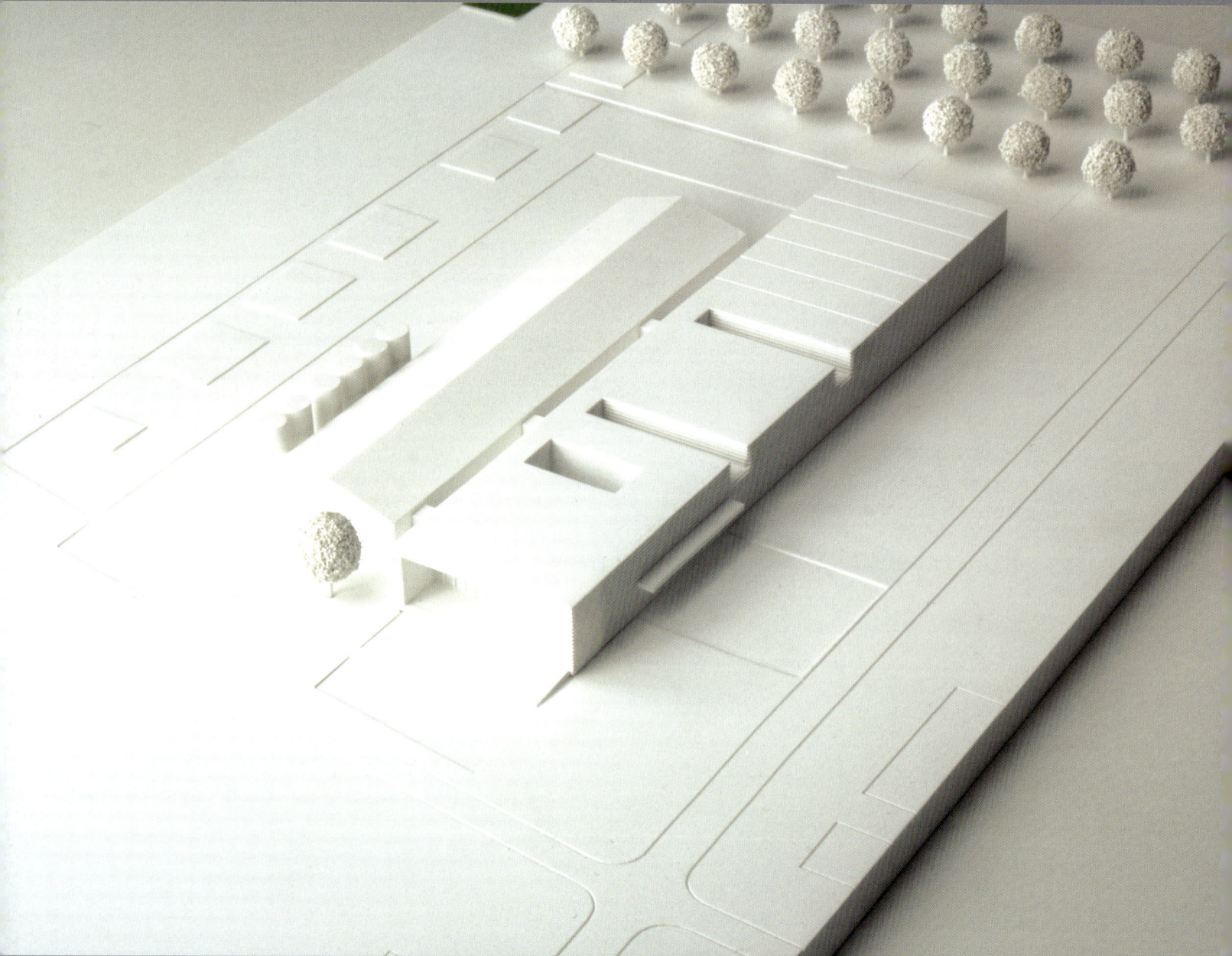

作者 Jakob Lehrecke   合作伙伴 Florian Kammerer, Ralf Tschöpe, Robert Witschurke
专家 structural analysis: Gerd-Walter Huske, Berlin; energy: Transsolar Energietechnik, Stuttgart

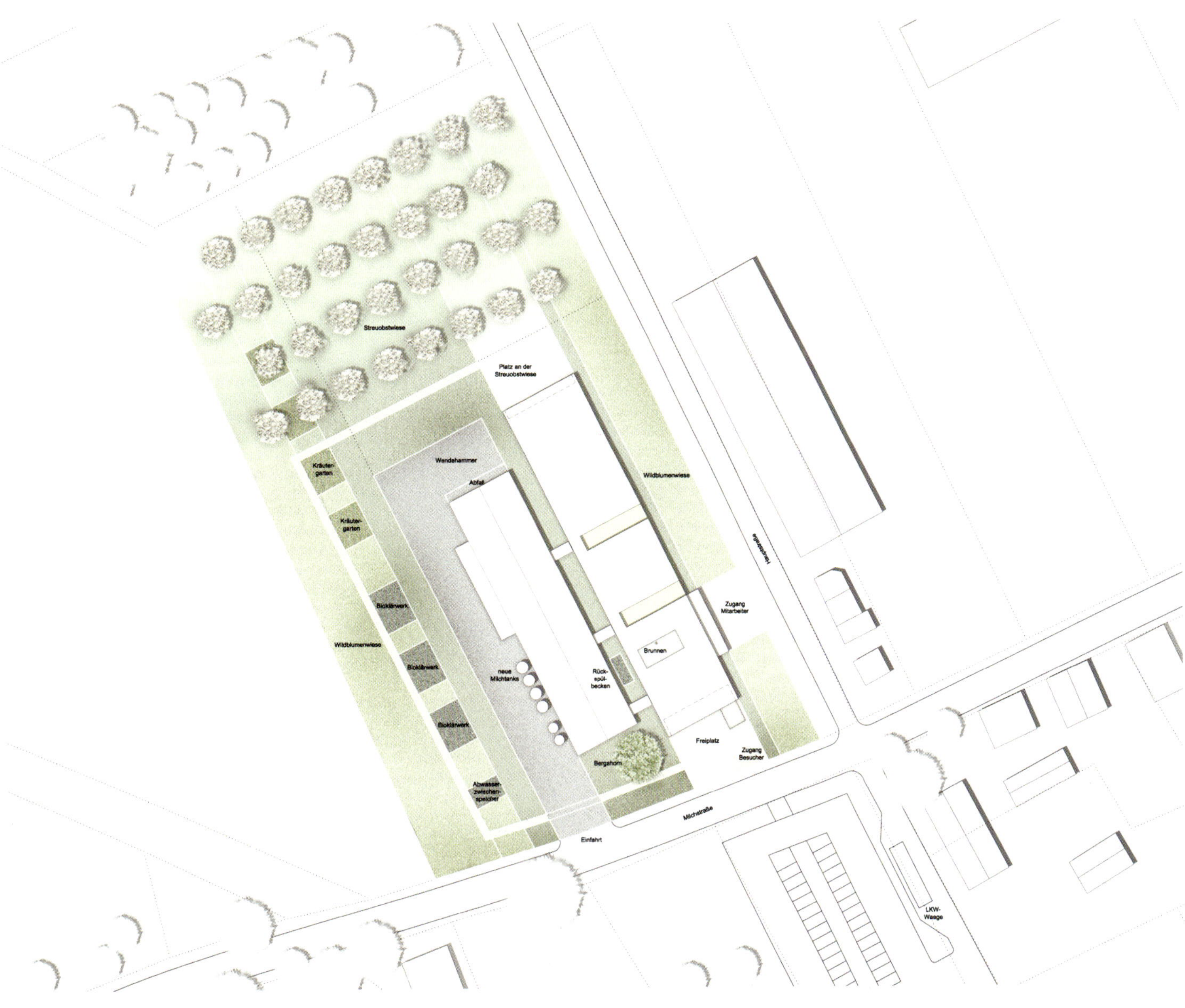
Streuobstwiese
Platz an der Streuobstwiese
Wendehammer
Abfall
Wildblumenwiese
Kräuter-garten
Kräuter-garten
Bioklärwerk
Bioklärwerk
Bioklärwerk
Wildblumenwiese
neue Milchtanks
Rück-spül-becken
Brunnen
Zugang Mitarbeiter
Freiplatz
Zugang Besucher
Bergahorn
Abwasser-zwischen-speicher
Milchstraße
Einfahrt
LKW-Waage

keller mayer wittig Cottbus
Keller mayer wittig建筑事务所 科特布斯

"Its location [...] makes it certain to be the first port of call for every visitor with an urban background, the show operation of the Glass Dairy acting as an attraction in its own right."

"它的地理位置使它成为邀请城里人参观的第一站，玻璃奶品厂的生产过程的展示操作成为一大亮

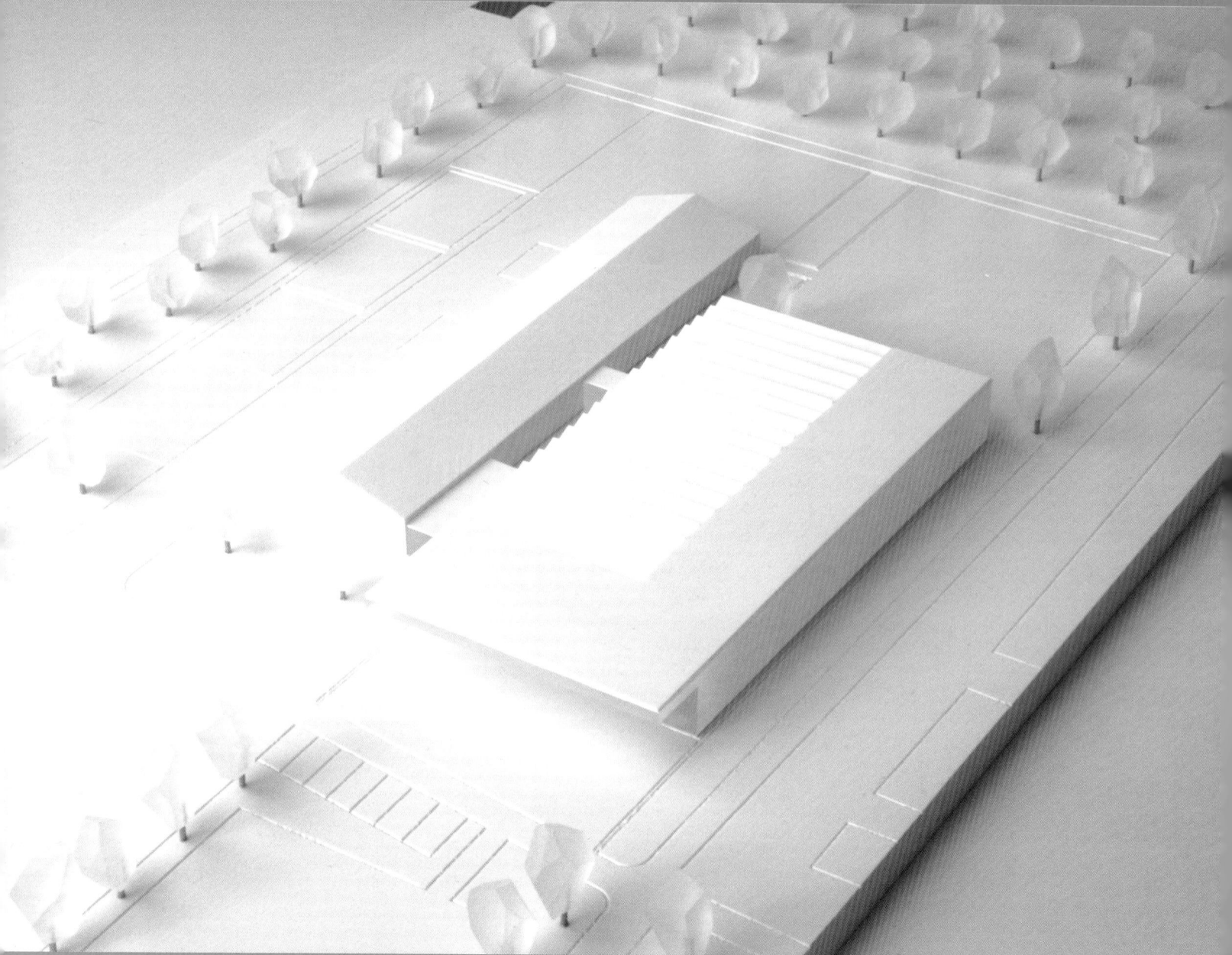

作者 Christian Keller   合作伙伴 Christoph Schulze, Martina Lehmann, André Krämer, Marco Laske   专家 structural analysis and energy: Hendrik Lindner, Prof. Pfeifer und Partner Ingenieure, Cottbus/Darmstadt; lighting: Zumtobel Staff, Berlin

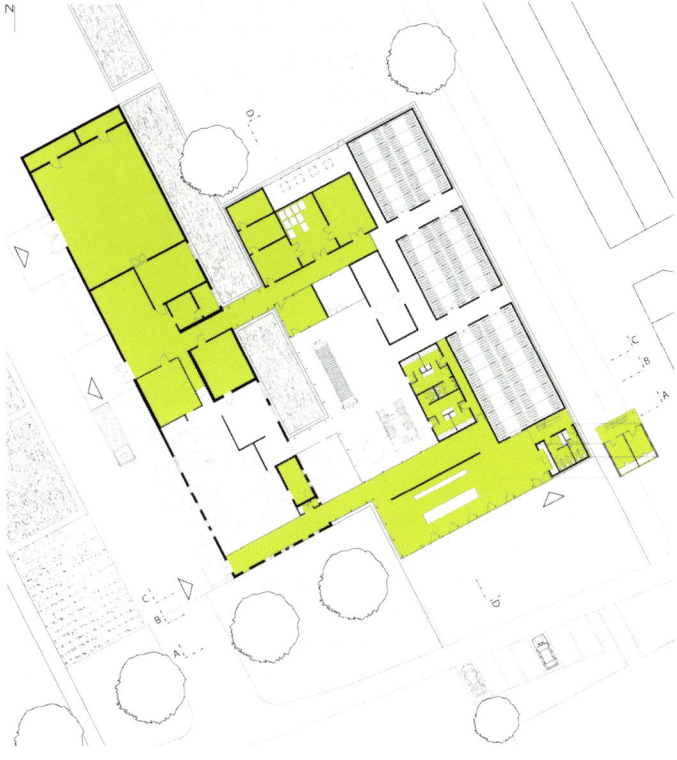
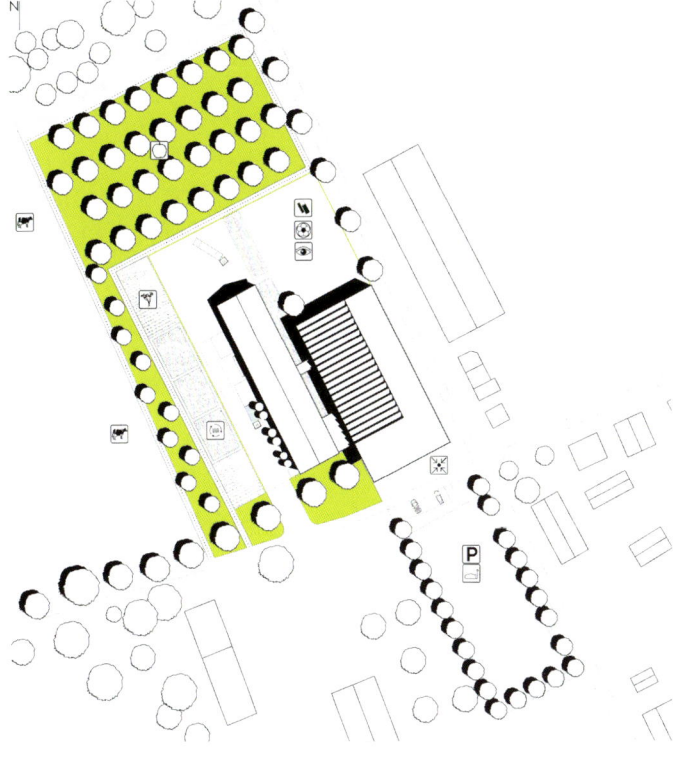

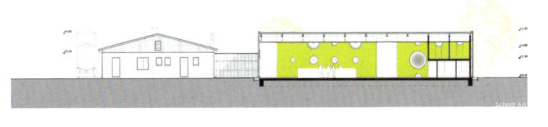

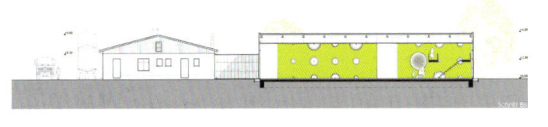

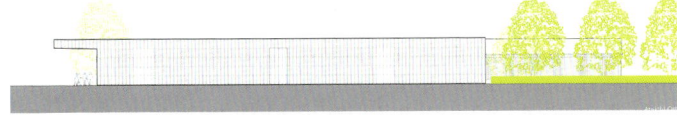
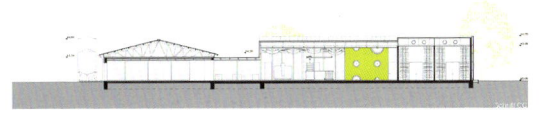

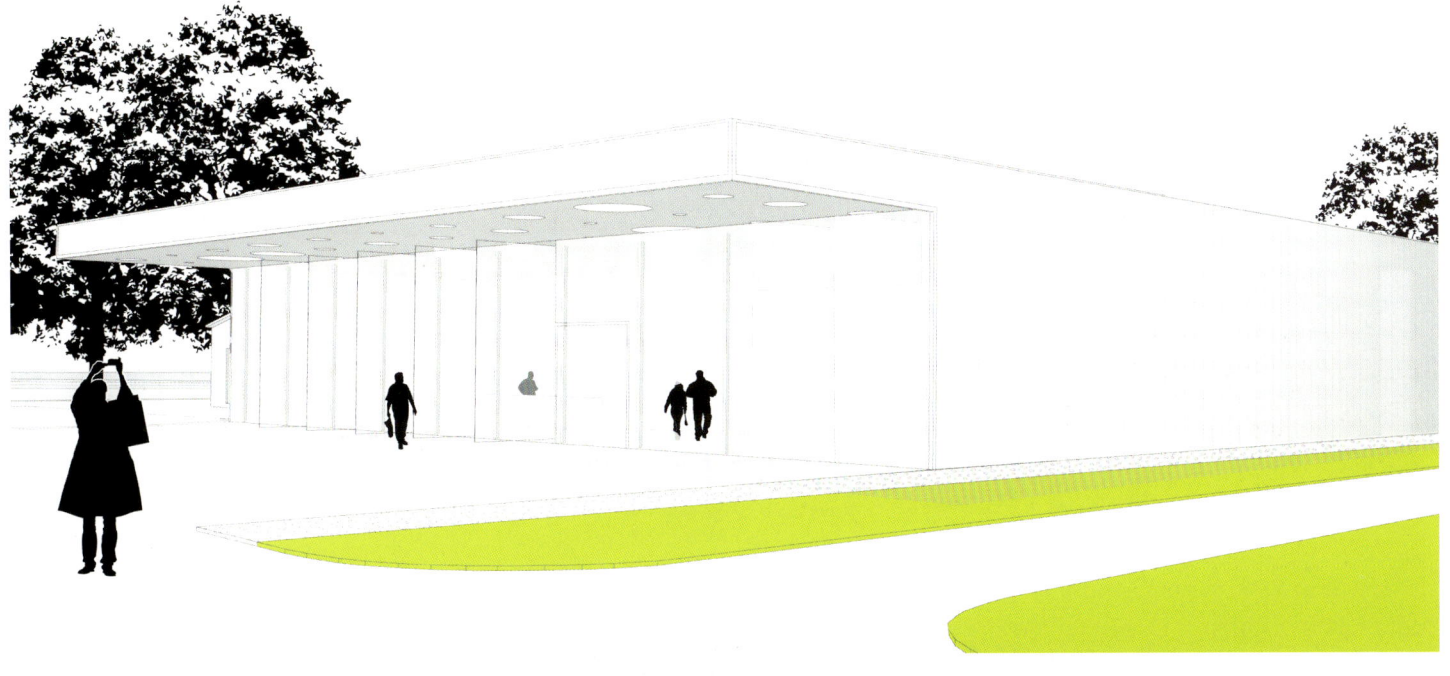

Modersohn & Freiesleben Architekten Berlin
Modersohn & Freiesleben建筑事务所 柏林

"The design presents a roofed building that opens up almost symbolically: A wood structure with wide inviting ribbon glazing, set on an inconspicuous plinth."

"设计一种带屋顶的建筑朝上，具有象征性。带有彩条玻璃窗的木质结构放置在一个不显眼的基座上。"

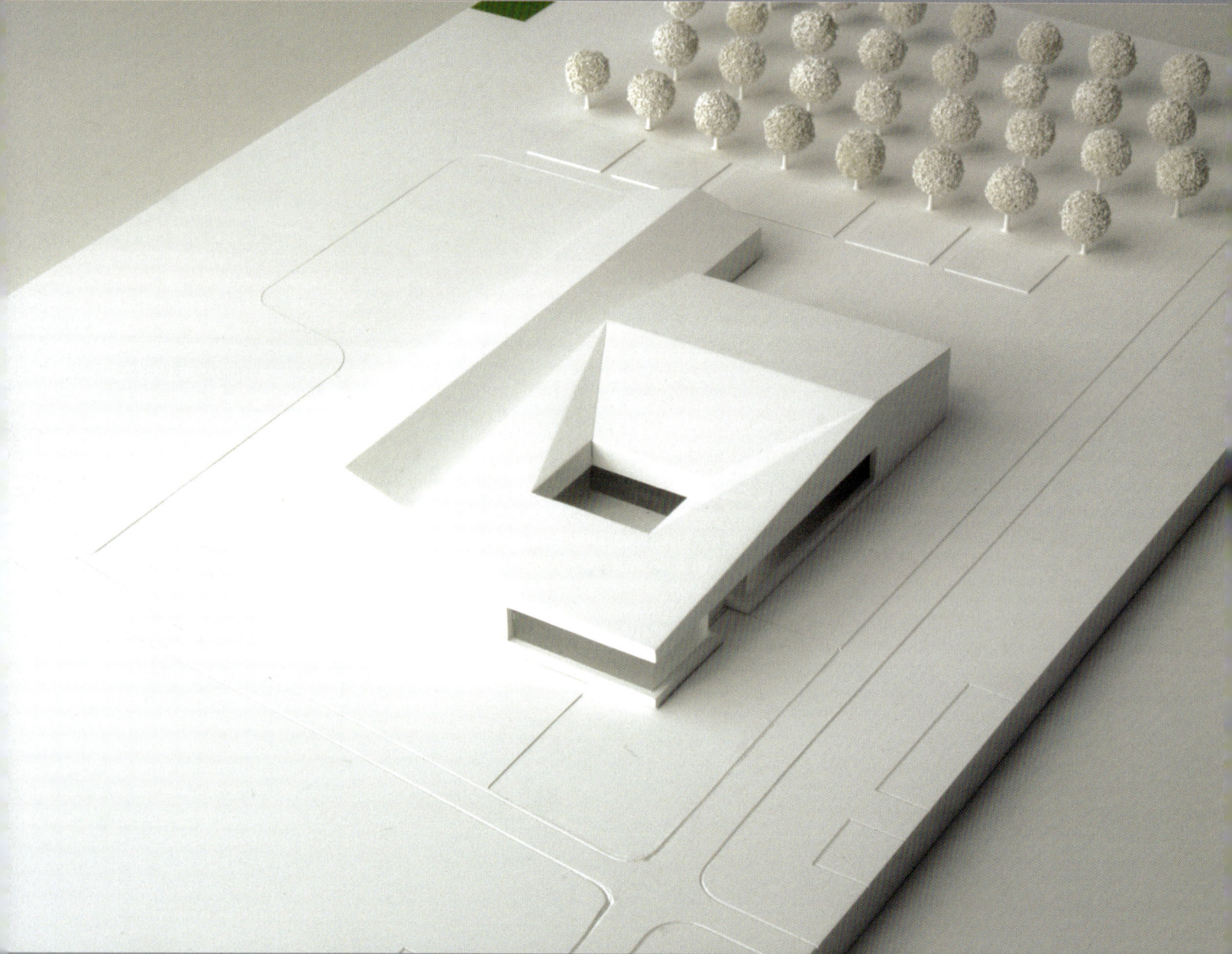

作者 Johannes Modersohn, Antje Freiesleben   合作伙伴 Christian Holthaus, Janine Ritz, Aimée Wolf
专家 structural analysis: Ingenieurbüro für Structural Analysis, Dr. Ing. C. Müller; technical equipment: Gebäudetechnik Dresden

Schwarzplan M 1:10.000

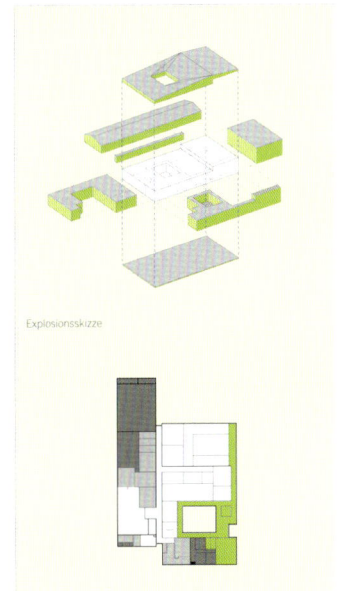

Explosionsskizze

Flächenschema Schwarz-, Grau-, Weißbereich + Besichtigungszone

Lageplan mit Außenanlagen M 1:500

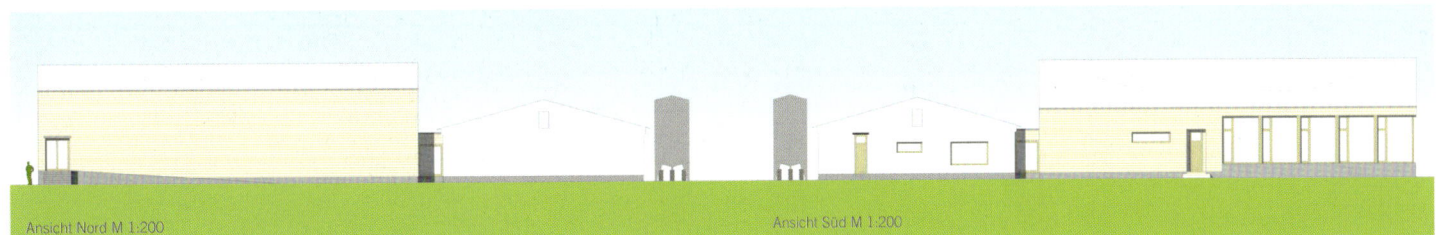

Ansicht Nord M 1:200

Ansicht Süd M 1:200

Grüntuch Ernst Architekten Berlin
Gruntuch Ernst建筑事务所 柏林

"The extension to the Glass Dairy in Münchehofe provides an opportunity to create a distinctive structure, embodying the idea of transparent and credible food processing within a setting of unique and unconventional architecture."

"Munchehofe里玻璃奶品厂的扩建产生一种有特色的建筑结构，用一种独特的非常规的建筑来体现食品加工透明可信的观点。"

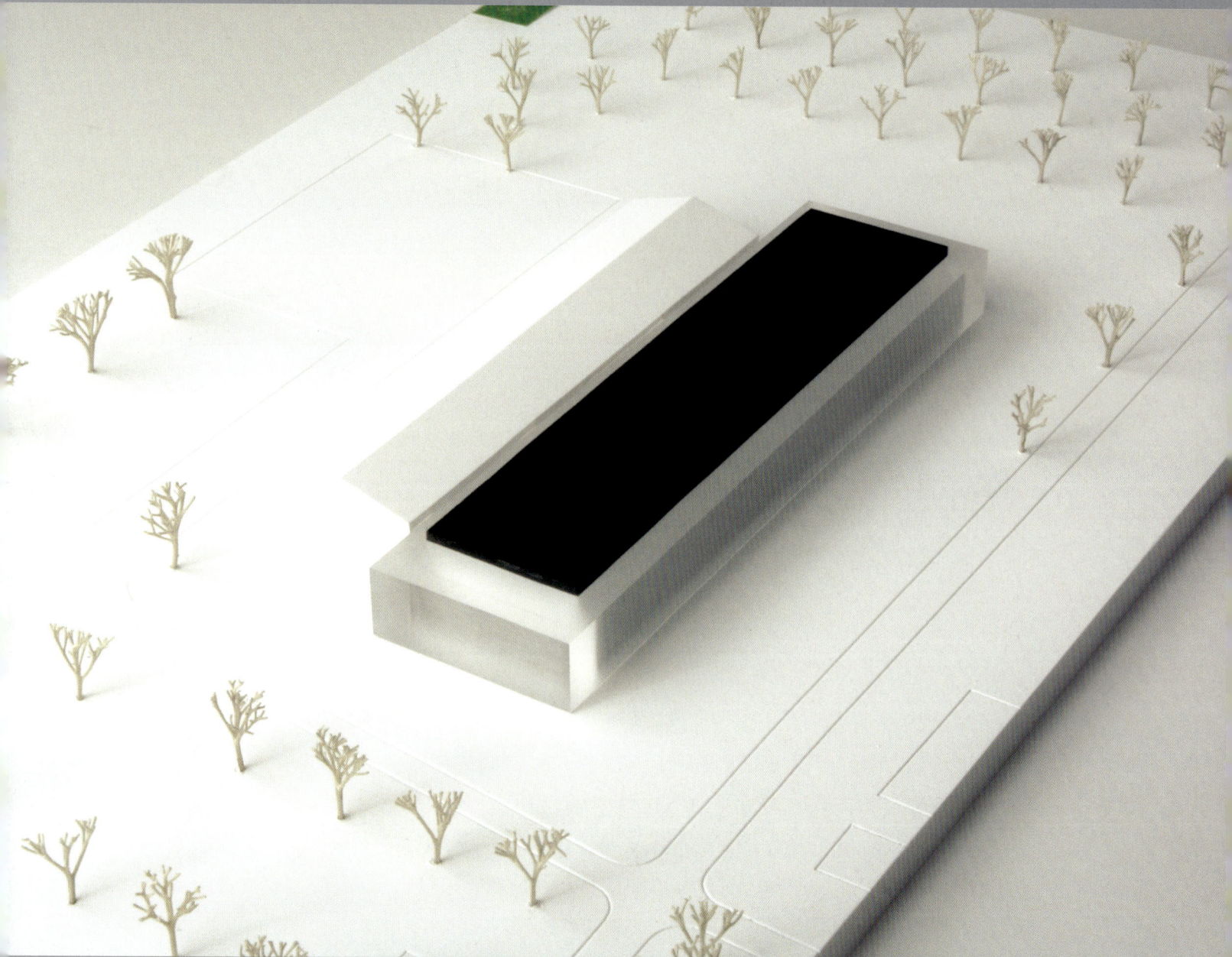

作者 Armand Grüntuch   合作伙伴 Arno Löbbecke, Pascale Busch, Stefan John, Allessio Fossati

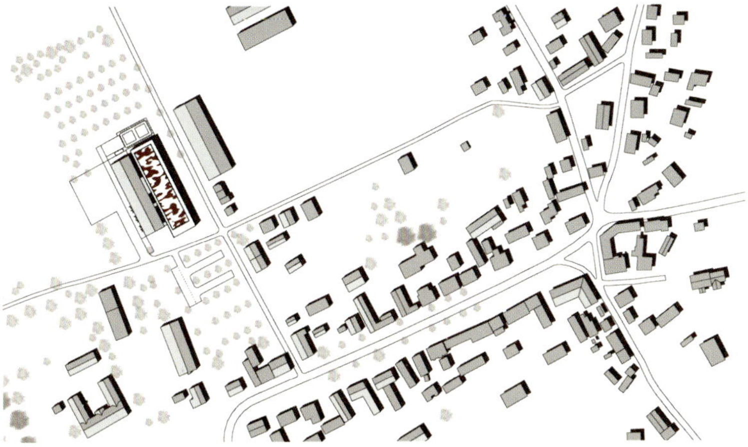

227944

1

Realisierungswettbewerb
Gläserne Molkerei in Müchehofe

Umgebungsplan M 1:2000

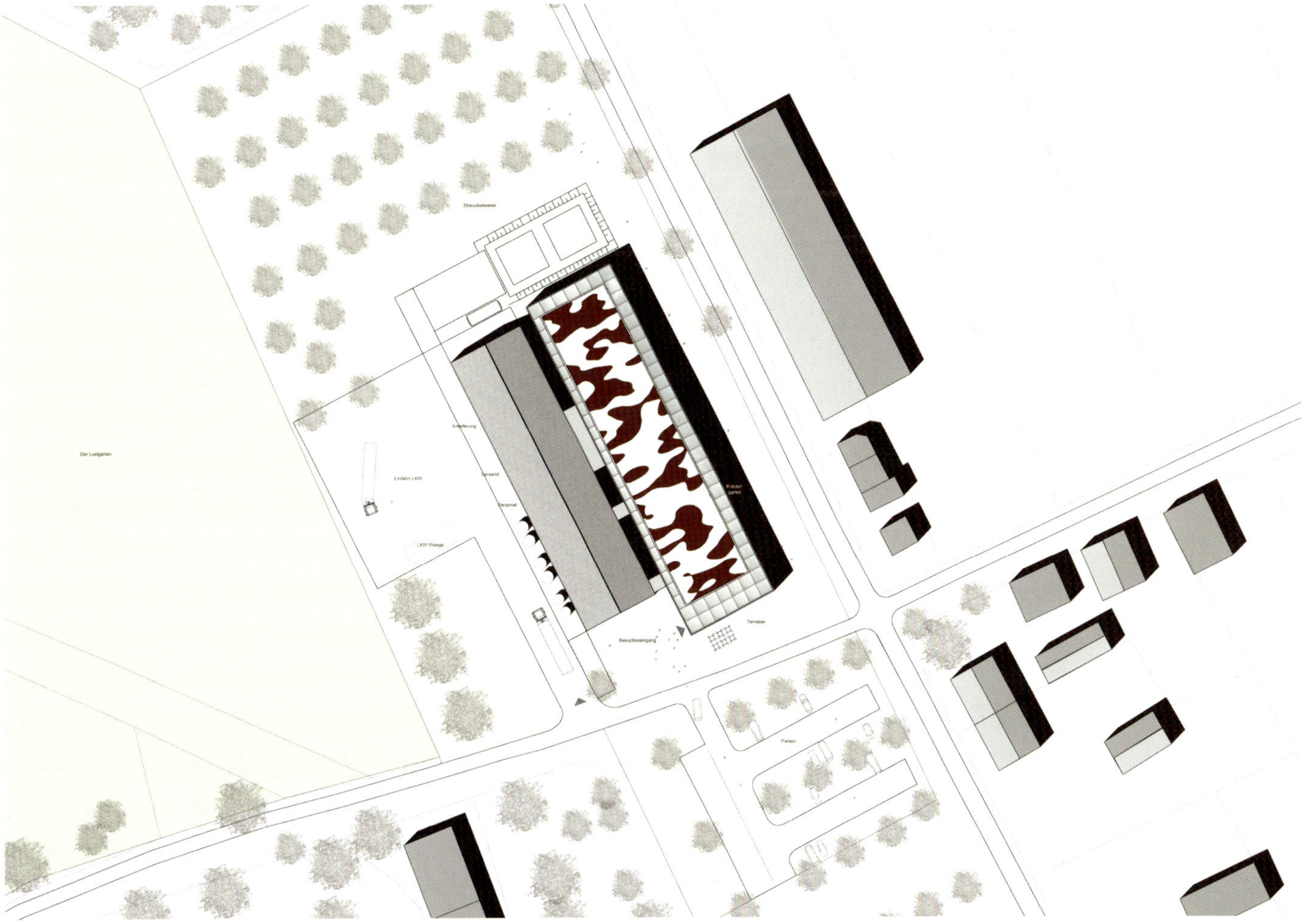

Gerber Architekten

Thomas M

behet bondzio lin arch

Riedberg Campus Commons
in Frankfurt/Main

Riedberg校园公用地
德国美因河畔法兰克福

üller Ivan Reimann

tekten

Riedberg Campus Commons Frankfurt/Main
Riedberg校园公用地 德国美因河畔法兰克福

Restricted interdisciplinatory project competition for architects and engineers preceded by an application procedure
限制严格的建筑师和工程师多科项目竞赛，开始前有申请程序。

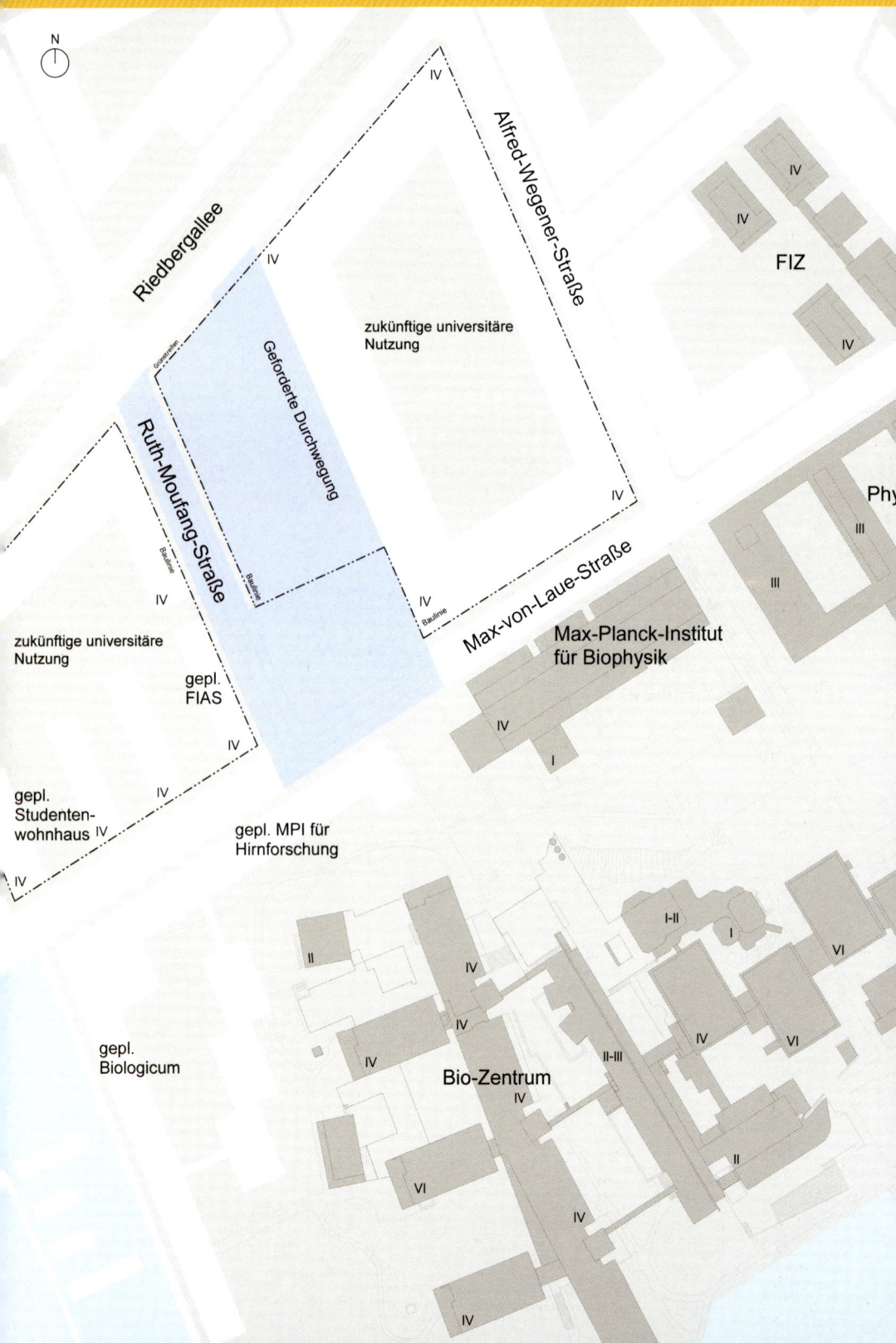

N

Riedbergallee

Alfred-Wegener-Straße

IV

IV

IV

IV

FIZ

IV

IV

zukünftige universitäre Nutzung

Geförderte Durchwegung

Ruth-Moufang-Straße

Baulinie

Grüneburg...

Baulinie

Baulinie

IV

IV

IV

Max-von-Laue-Straße

III

III

Phy...

III

zukünftige universitäre Nutzung

gepl. FIAS

IV

Max-Planck-Institut für Biophysik

gepl. Studenten-wohnhaus

IV

IV

gepl. MPI für Hirnforschung

IV

I

I-II

I

VI

II

IV

IV

IV

gepl. Biologicum

IV

II-III

IV

VI

Bio-Zentrum

IV

VI

II

IV

地点 Frankfurt/Main
时间 06/2006–12/2006
主办方 Ministry of Higher Education, Research, and the Arts, Federal State of Hessen, represented by minister of state Udo Corts, Wiesbaden
参赛者 25 participants
面积 6,000 sq m
竞赛费用 113,000 Euro
专业评奖委员会
Harald Clausen, head of division, Ministry Finance, Federal State of Hessen, Wiesbaden;
Prof. Ansgar Lamott, Stuttgart/Darmstadt;
Prof. Ulrike Lauber, Munich/Berlin;
Prof. Uwe Rotermund, building services engineer, Braunschweig/Münster;
Prof. Kirsten Schemel, Berlin/Münster
专家评奖委员会
Irene Bauerfeind-Roßmann, head of division, Ministry of Arts and Sciences, Federal State of Hessen, Wiesbaden;
Günter Schmitteckert, leading head of division, Ministry of Arts and Sciences, Federal State of Hessen, Wiesbaden
Prof. Dr. Rudolf Steinberg, president, Johann Wolfgang Goethe-Universität Frankfurt/Main;
Prof. Dr. Horst Stöcker, designated vice president , Johann Wolfgang Goethe-Universität, Frankfurt/Main

The Riedberg Campus is situated in Frankfurt's northwest, and intended to house all the science institutions of the university. The spatial proximity of the university buildings to each other and the openness of the campus can be used to strengthen interdisciplinary ties. The task of this competition was the new built construction of the Central Commons. This is essential to the development plan for Riedberg Campus. The site under competition measures approx. 15,000 square metres and is located at the northern edge of the campus, next to downtown Riedberg: Its open-air spaces are to define the venue for entering the campus from there. A part of the task was to design a plaza environment on the southern portion of the site. With approx. 6,000 square metres of useful area, the Central Commons focuses essential supply functions in such a way as to provide the campus with an attractive, well functioning centerpiece. The program of spaces comprises a departmental library, a cafeteria, lecture and seminar facilities plus preparation and storage areas and offices. The competition task took an interdisciplinary approach, including concepts for load-bearing structures and mechanical services as well as architectural design.

Riedberg校园坐落在法兰克福的西北部，大学的全部自然科学机构全部坐落于此。建筑物之间的空间上的接近和校园的开放性能够增加各学科的联系。竞赛任务是中心公用地的重建，该项目对Riedberg校园整个的开发规划非常重要。竞赛地点测量约占1.5万平方米，位于校园的北角，紧挨着Riedberg中心：露天空间作为进入校园的场所，任务之一是在竞赛地点的南部设计一个广场。中心公用地使用面积约6000平方米，强调基本的供给功能，成为校园里引人注意、功能良好的一个中心。空间规划包括一个院系图书馆、咖啡厅、讲座和研讨会设施及准备和储存区和办公区。竞赛任务采用跨学科方法，包括支承结构、力学服务及建筑设计的观念。

Aerial view　鸟瞰图

Competition site　竞赛地点

Competition site　竞赛地点

Skyline of Frankfurt/Main　美因河畔法兰克福的地平线

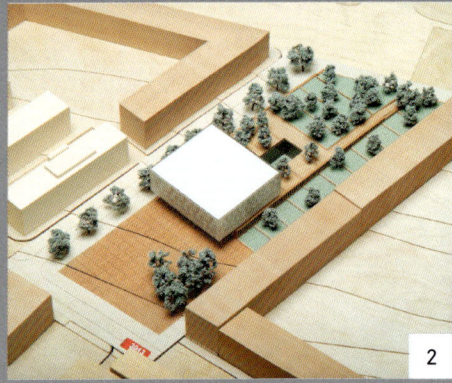

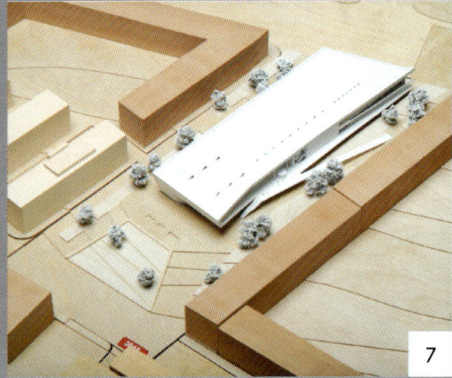

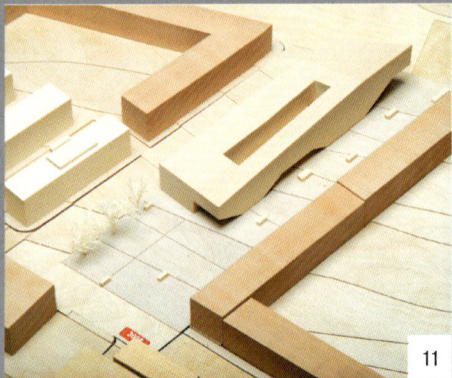

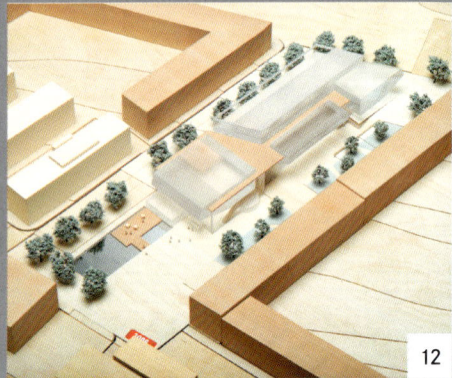

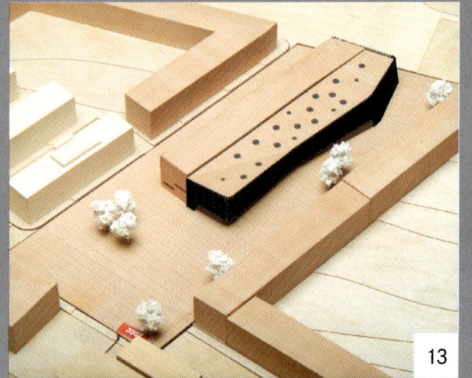

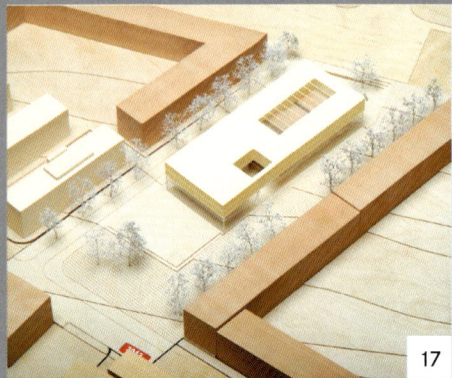

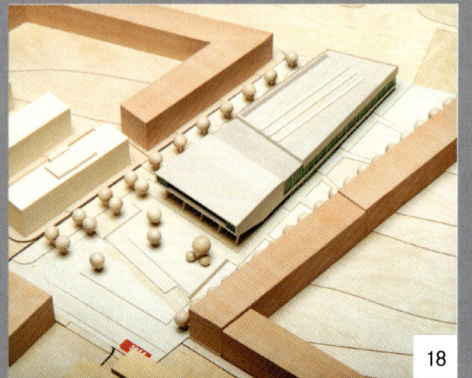

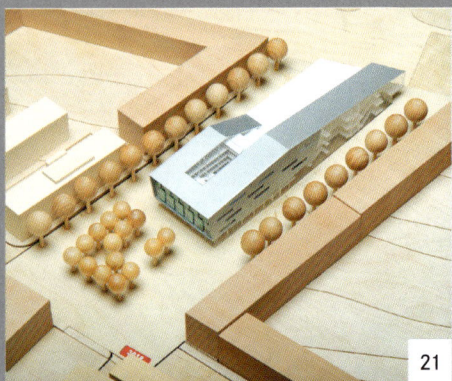

4

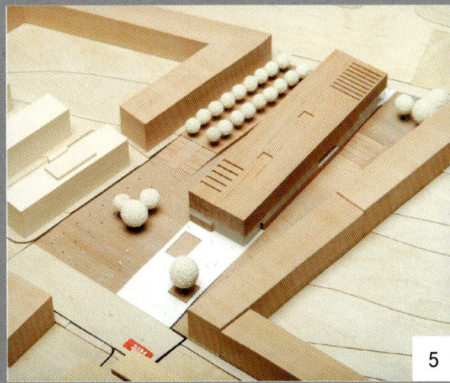

5

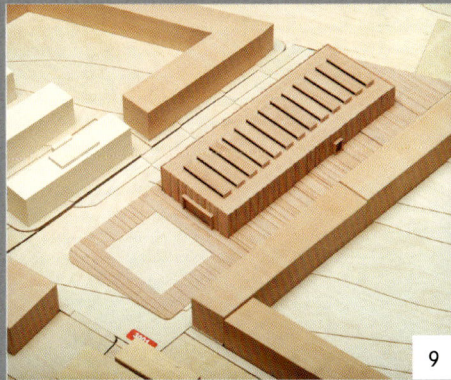

9

10

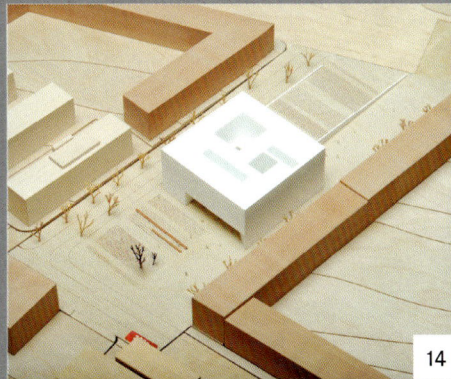

14

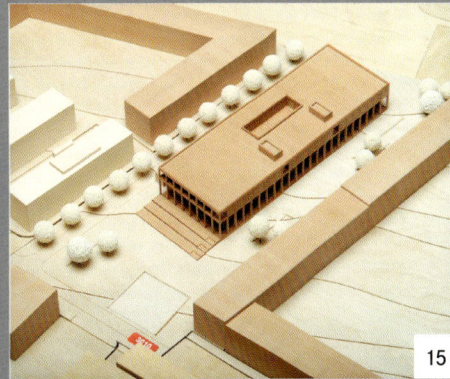

15

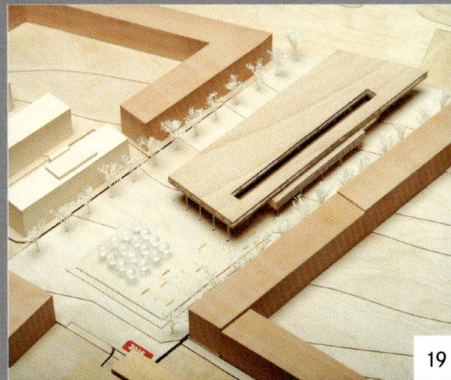

19

20

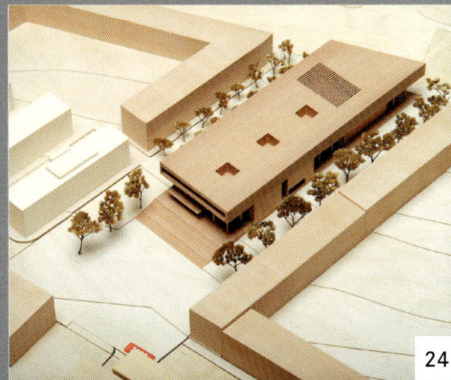

24

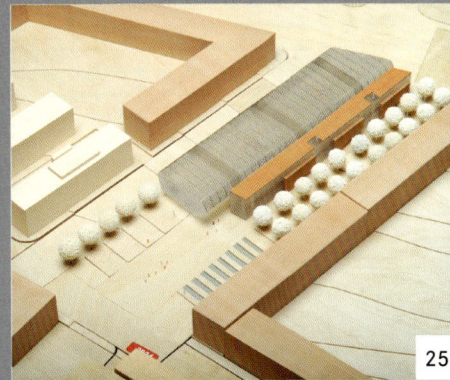

25

Qualified participants
合格的参赛者

1
1st prize   一等奖
Gerber Architekten, Dortmund
2
2nd prize   二等奖
Thomas Müller –
Ivan Reimann, Berlin
3
3rd prize   三等奖
Heinle, Wischer und Partner, Berlin
4
4th prize   四等奖
Ferdinand Heide, Frankfurt / Main
5
5th prize   五等奖
behet bondzio lin architekten, Münster
6
Acquisition   购买
Jockers Architekten, Stuttgart
7
Acquisition   购买
Kissler + Effgen, Wiesbaden
8
Acquisition   购买
kister scheithauer gross, Cologne
9
Prof. Dipl.-Ing. Architekt Benedict
Tonon, Berlin
10
Glass Kramer Löbbert Gesellschaft
von Architekten, Berlin
11
Chestnutt_Niess Architekten, Berlin
12
Kühnl & Schmidt, Karlsruhe
13
sauerbruch hutton, Berlin
14
Henn Architekten, Munich
15
Kleihues + Kleihues, Dülmen-Rorup
16
Wulf & Partner, Stuttgart
17
Nickl & Partner, Munich
18
hks hestermann rommel, Erfurt
19
HASCHER JEHLE, Berlin
20
ReimarHerbst.Architekten, Berlin
21
BMBW Architekten + Partner, Munich
22
b2Architekten Dittmann & Luft, Bonn
23
pfp Architekten, Hamburg
24
Klein & Sänger, Munich
25
KBK Architekten Belz Lutz
Guggenberger, Stuttgart

Gerber Architekten Dortmund
Gerber建筑事务所 多特蒙德

"The configuration and façade design derive from and are in keeping with the conceptual basis and functional assignment of the three built volumes – lecture halls, library, and foyer."

"建筑结构和外观源于并且与三大建筑物——讲座厅、图书馆和休息厅的概念的基础和功能解析协调一致。"

作者 Prof. Eckhard Gerber 合作伙伴 Nils Kummer, Alexandra Kranert, Stefan Lemke, René Albrecht, Martin Pellkofer, Matthias Deilke, Benjamin Sieber, Siegbert Hennecke 专家 building services: Energy Design, Braunschweig; structural analysis: Prof. Pfeifer und Partner, Darmstadt; landscape architecture: Gerber Architekten

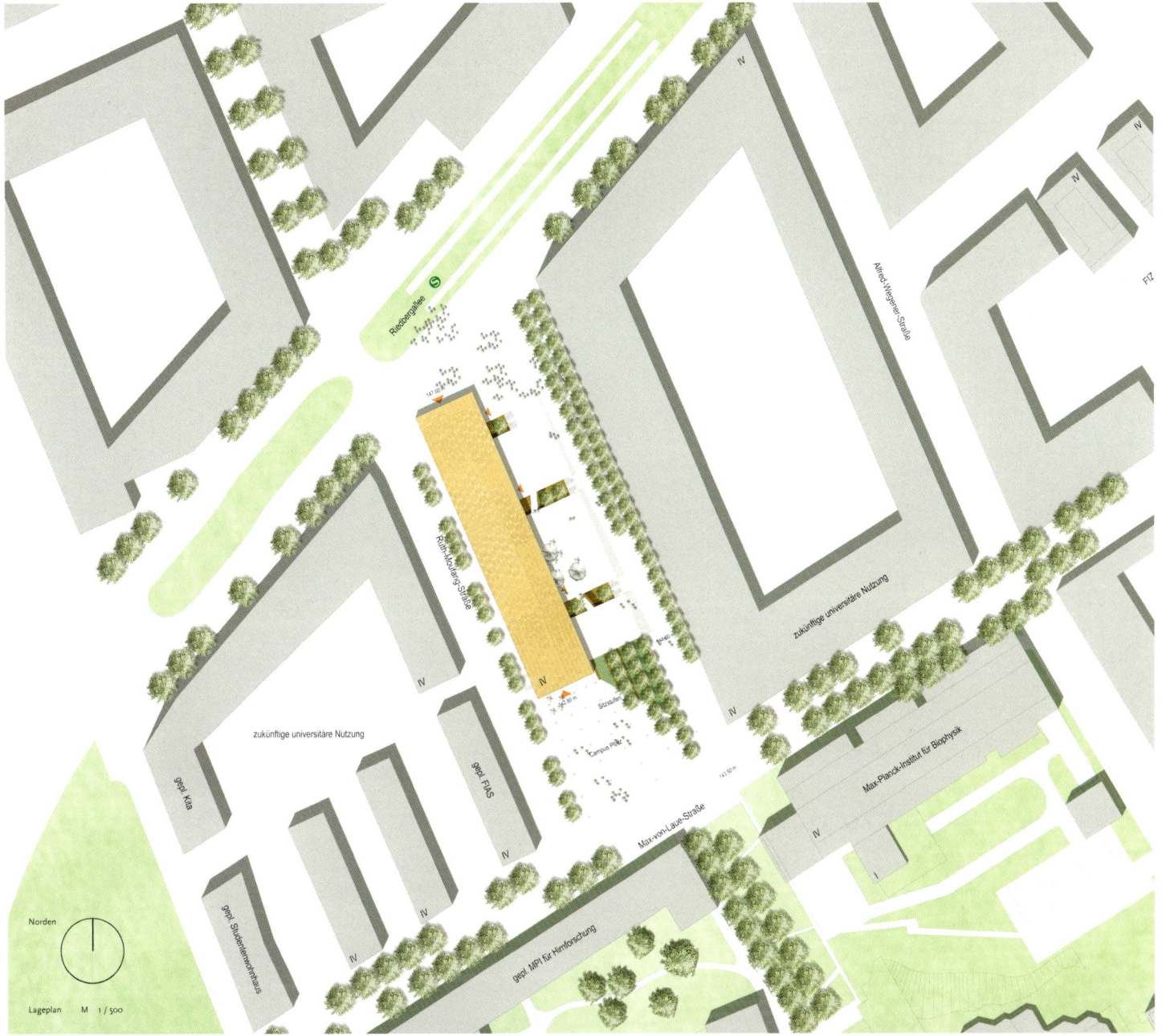

Norden

Lageplan   M 1 / 500

Riedbergallee

Ruth-Moufang-Straße

zukünftige universitäre Nutzu

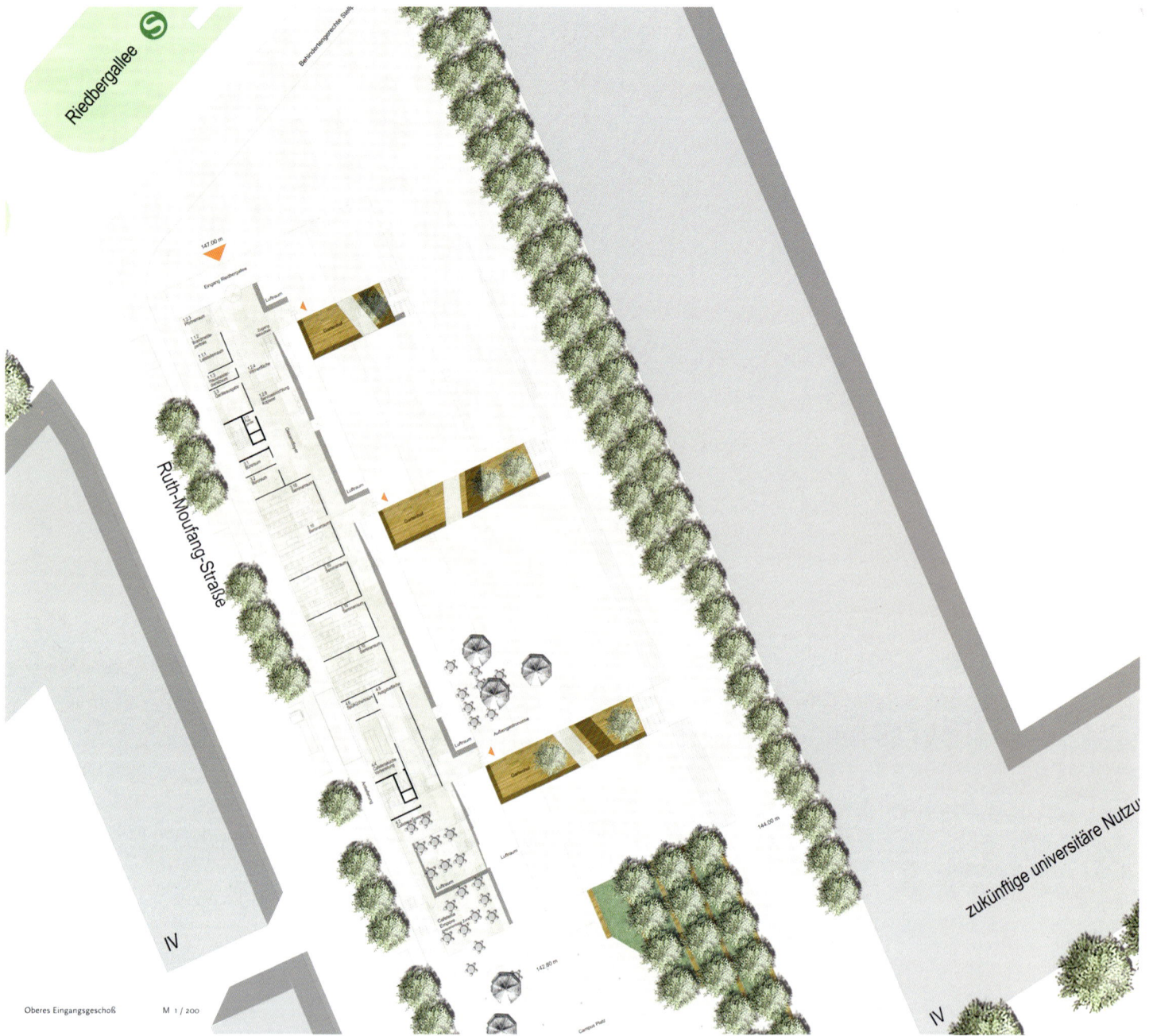

Oberes Eingangsgeschoß   M 1 / 200

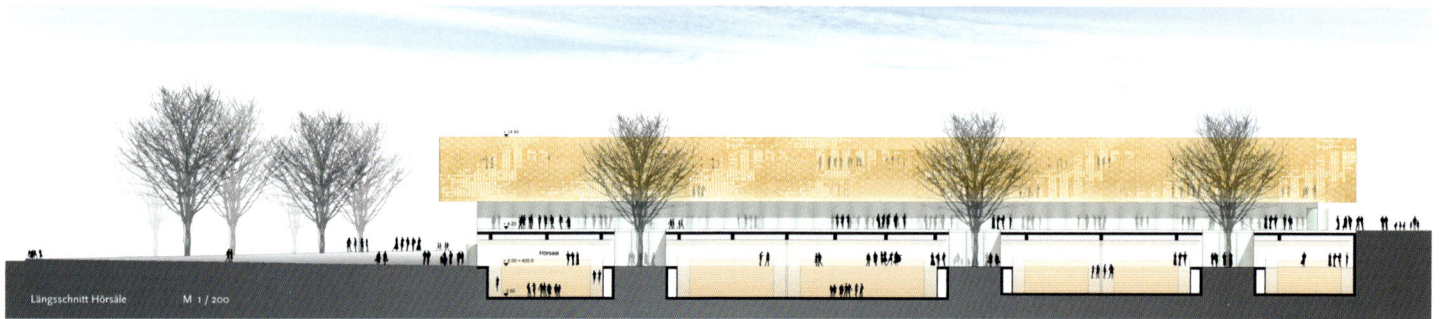

Längsschnitt Hörsäle   M 1 / 200

Ansicht Ruth-Moufang-Straße   M 1 / 200

Detailansicht   M 1/50

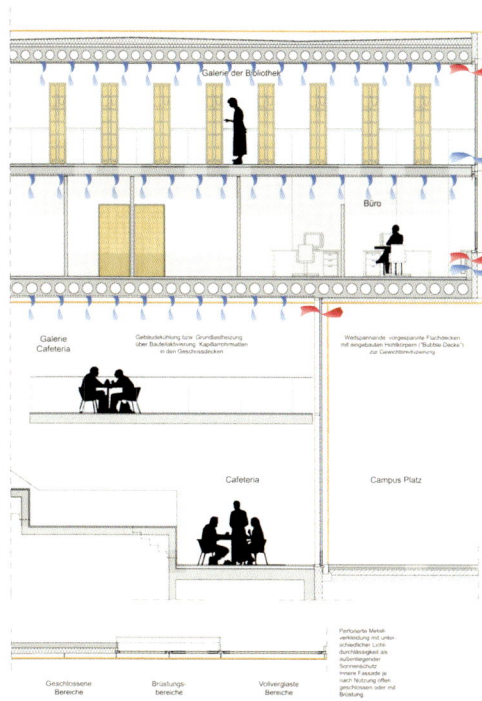

Detailschnitt   M 1/50

Detailgrundriss   M 1/50

Geschlossene Bereiche   Brüstungs- bereiche   Vollverglaste Bereiche

Tragwerk / Systemskizzen

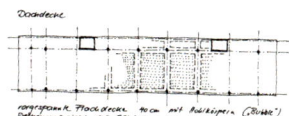

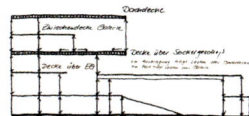

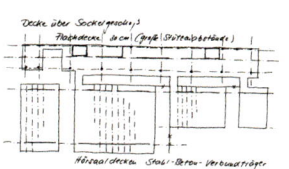

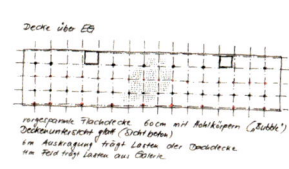

Tragwerkskonzept

Das räumliche Konzept des Entwurfs zeichnet sich durch Offenheit, Großzügigkeit und Transparenz aus.

Das Tragwerk für das Gebäude unterstützt diesen Ansatz durch leistungsfähige Konstruktionen mit großen Spannweiten auf Wänden und wenigen dünnen Stützen.

Energiekonzept / Systemskizzen

Energiekreislauf

Kälterückgewinnung		Lüftung	Dezentrale Lüftung	Bauteil-Aktivierung
	Kompressionskälte			
	Grundwasser			
	Fernwärme			
Bauteil-Aktivierung	Dezentrale Lüftung	Lüftung		Wärmerückgewinnung

Energiekreislauf Kühlung

Kälterückgewinnung		Lüftung	Dezentrale Lüftung	Bauteil-Aktivierung
	Kompressionskälte			
	Grundwasser			
	Fernwärme			
Bauteil-Aktivierung	Dezentrale Lüftung	Lüftung		Wärmerückgewinnung

Energiekreislauf Heizung

Kälterückgewinnung		Lüftung	Dezentrale Lüftung	Bauteil-Aktivierung
	Kompressionskälte			
	Grundwasser			
	Fernwärme			
Bauteil-Aktivierung	Dezentrale Lüftung	Lüftung		Wärmerückgewinnung

Zonierung

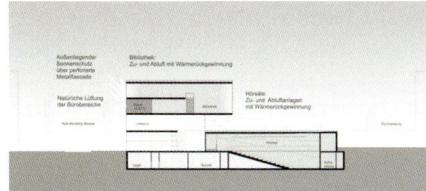

Sommer

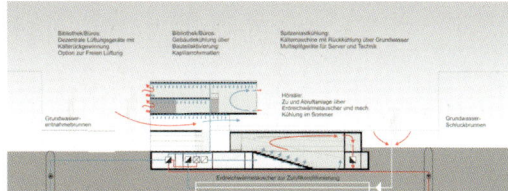

Winter

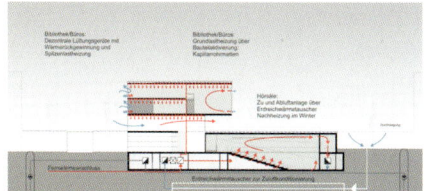

Thomas Müller Ivan Reimann Berlin
Thomas Muller Ivan Reimann建筑事务所 柏林

"We imagine the new Campus Commons not only as a building but also as a sequence of differing public spaces."
"我们认为新校园公用地不仅是一个建筑物，而且也是一连串的不同的公众空间。"

作者 Thomas Müller, Ivan Reimann   合作伙伴 Anna Lemme, Jens Wesche, Marius Förster, Kristina Knapp, Milos Linhard, Thomas Kautsch, Nils Noud, Jana Galicka   专家 building services: IC Ingenieurconsult GmbH, Frankfurt/Main; structural analysis: GSE Ingenieur-Gesellschaft mbH, Berlin; landscape architecture: Jürgen Weidinger Landschaftsarchitekten, Berlin; fire protection: Peter Stanek, Berlin

Perspektive Campusgarten

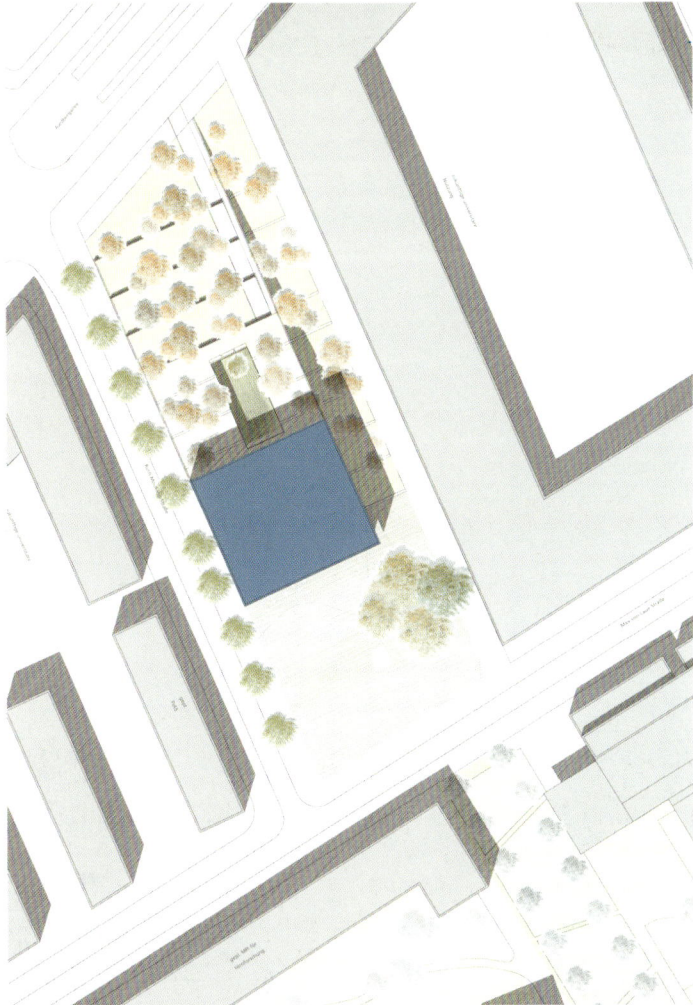
Lageplan M 1:500

Perspektive Campusplatz

Ansicht Süd M 1:200

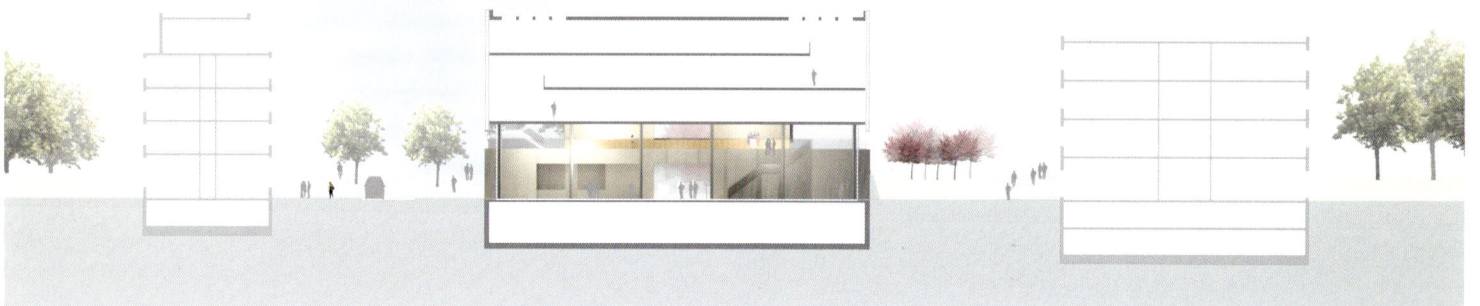
Querschnitt M 1:200

Ansicht Ost M 1:200

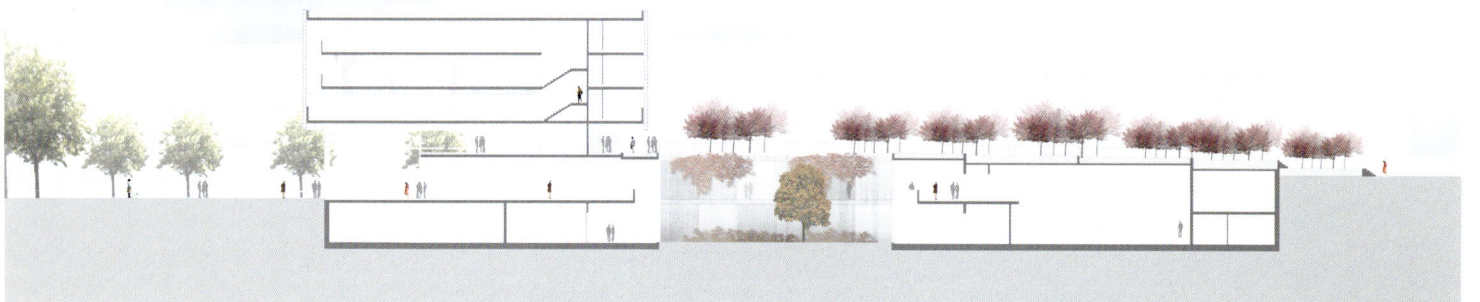
Längsschnitt M 1:200

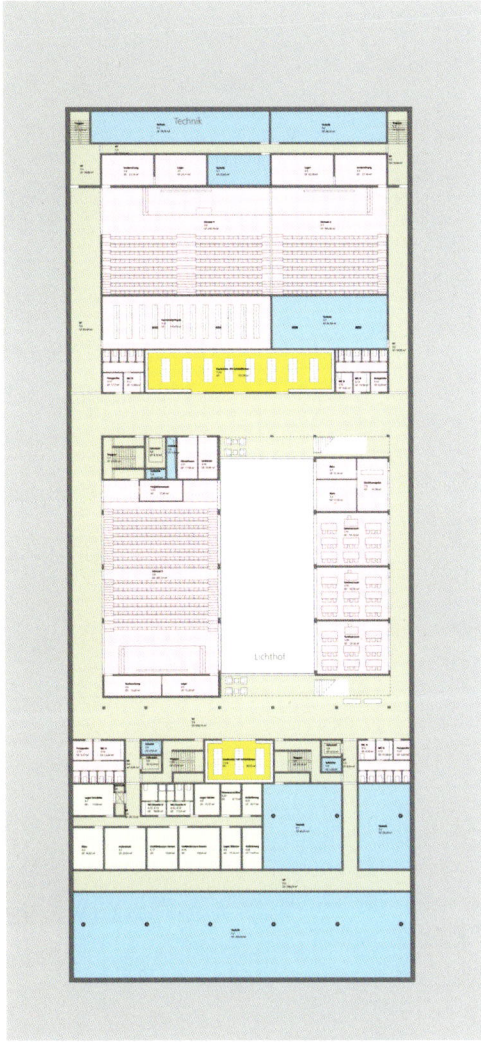
Hörsaalzentrum Sockel M 1:200

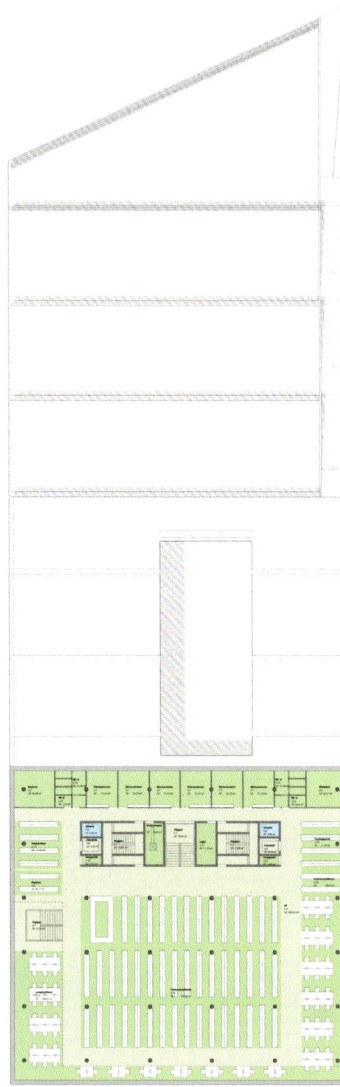
Bibliothek 2.OG M 1:200

Bibliothek 3.OG M 1:200

Bibliothek 4.OG M 1:200

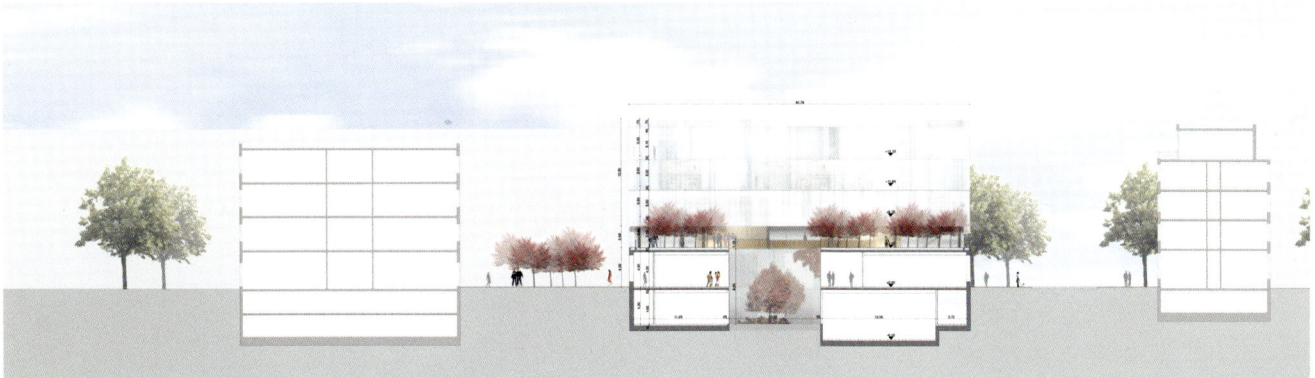
Querschnitt M 1:200

behet bondzio lin architekten Münster

behet bondzio lin建筑事务所 蒙斯特

"It is the primary goal of this concept to find an overall topic of town planning for the whole campus."

"为整个校园找到一个全面的城市规划主题是这个观念的最重要的目标。"

作者 Martin Behet, Roland Bondzio, Yu-Han Michael Lin   合作伙伴 Ulf Düsterhöft, Malte Petersen, Sonja Strickmann, Britta Kasner, Paulo de Aranjo   专家 building services: Ingenieurbüro Nordhorn, Klaus Nordhorn, Münster; collaborator: Thilo Ihle; structural analysis: Ingenieur ARGE HJW, Dr. Jaenisch, Leipzig; collaborator: Mr. Krüger

S-Bahn H.

Eingang Nord

+2,00

+2,00

Zentraler Campusplatz

0,00

Anlieferung

0,00

Passage

+4,00

Anlieferung Küche

+2,00

Foyer

0,00

Eingang Süd

Wasserbecken

0,00

Zentraler Campusplatz

-1,00

Erdgeschoss 2

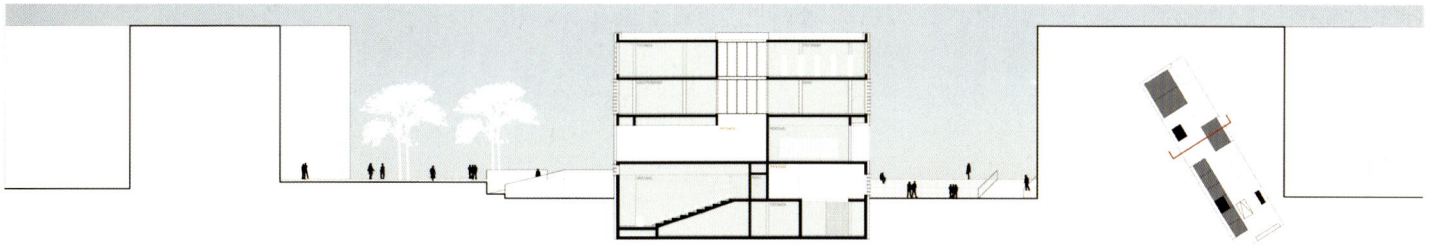

SCHNITT 1:200

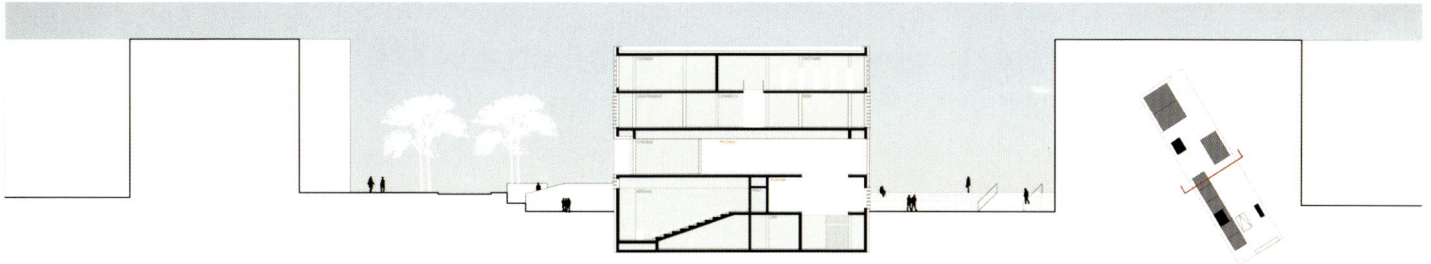

SCHNITT 1:200

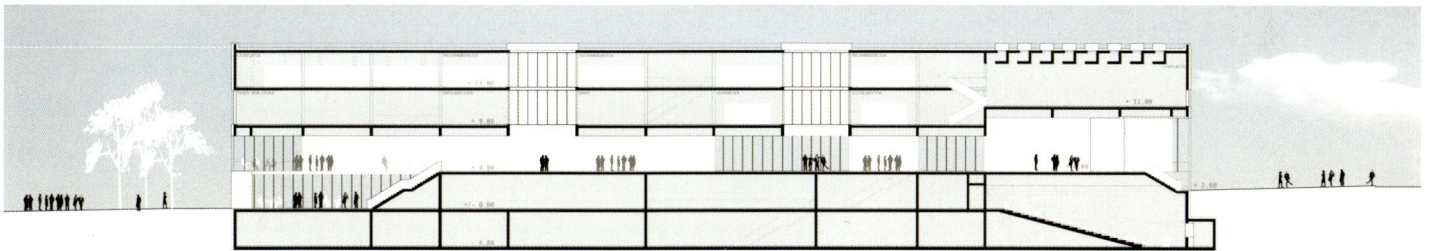

SCHNITT 1:200

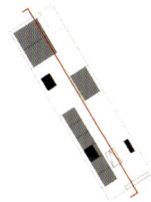

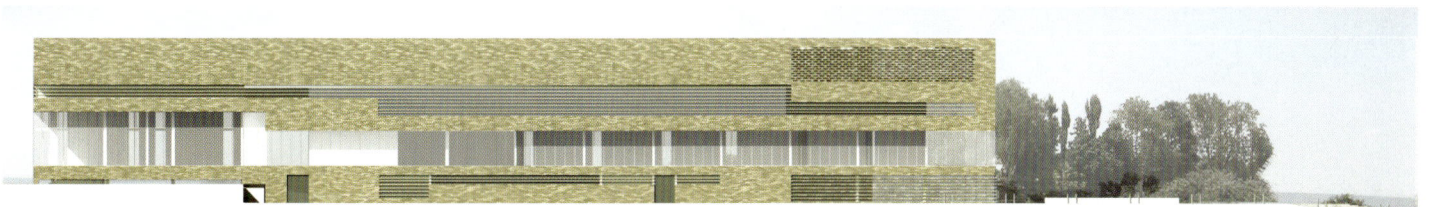

ANSICHT OST 1:200

HASCHER JEHLE

Aue

KSP Engel und Zimme

New KfW Building at the Senckenberganlage in Frankfurt/Main

新KfW大楼

Senckenberganlage，德国美因河畔法兰克福

+Weber +Assoziierte

mann

New KfW Building at the Senckenberganlage Frankfurt/Main

新KfW大楼，Senckenberganlage
德国美因河畔法兰克福

Restricted project competition for general planners preceded by an application procedure

限制严格的一般规划人员的项目竞赛，开始前有申请程序。

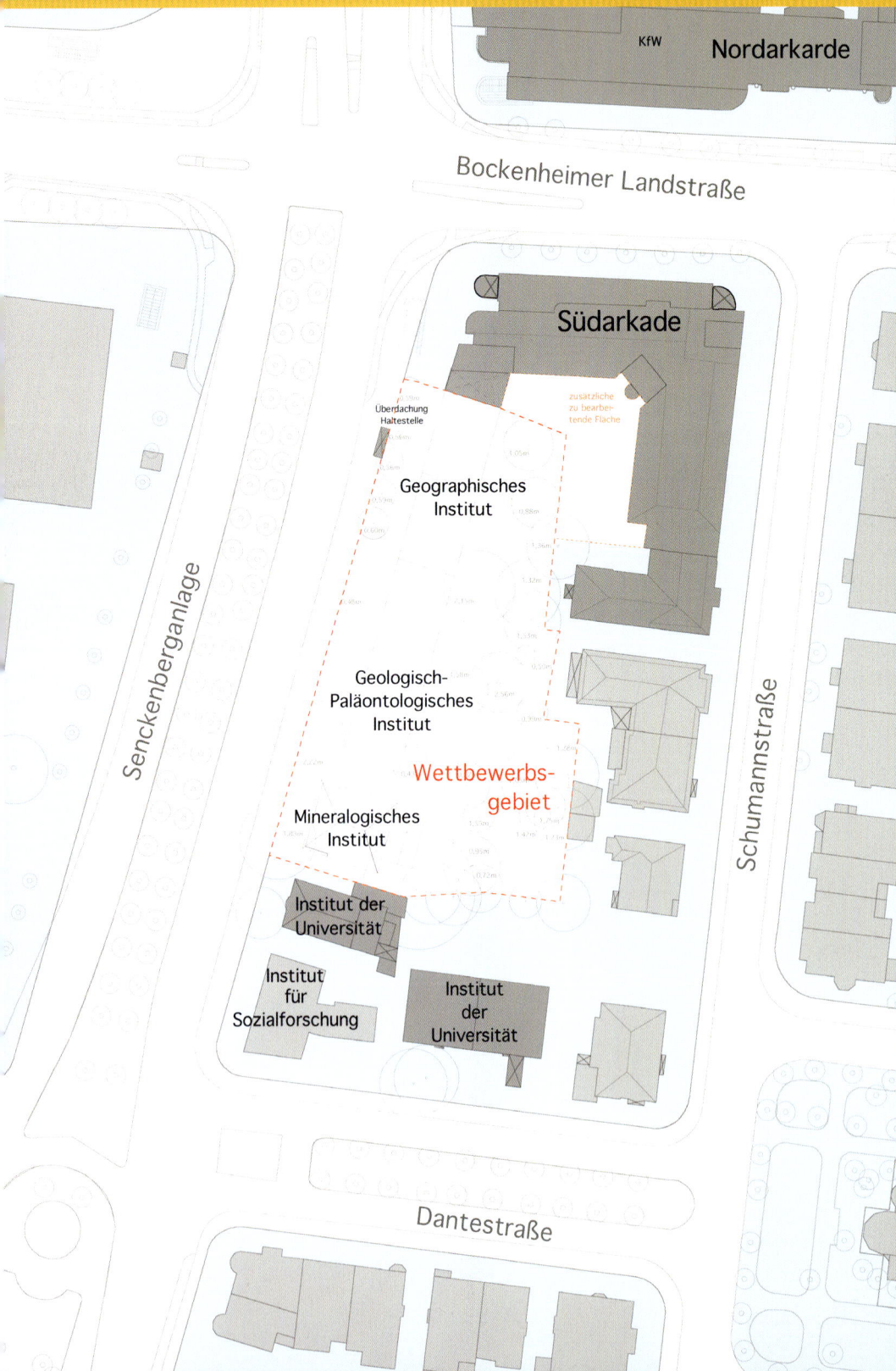

Nordarkarde

KfW

Bockenheimer Landstraße

Südarkade

zusätzliche zu bearbei-tende Fläche

Überdachung Haltestelle

Geographisches Institut

Senckenberganlage

Geologisch-Paläontologisches Institut

Wettbewerbs-gebiet

Schumannstraße

Mineralogisches Institut

Institut der Universität

Institut für Sozialforschung

Institut der Universität

Dantestraße

地点 Frankfurt/Main
时间 10/2005–06/2006
主办方 KfW
参赛者 10
面积 5,400 sq m
竞赛费用 127,000 Euro
专业评奖委员会
Prof. Ulrike Lauber,
Munich/Berlin;
Dieter von Lüpke,
head of city planning office,
Frankfurt/Main;
Prof. Matthias Sauerbruch,
Berlin/Stuttgart;
Alexander Theiss,
Frankfurt/Main
专家评奖委员会
Klaus J. Helms, KfW,
head of city planning office;
department new construction –
design and construction, Frankfurt/Main;
Detlef Leinberger, head of KfW,
Frankfurt/Main;
Hans W. Reich,
chairman of management board KfW,
Frankfurt/Main

The KfW (in full: Kreditanstalt für Wiederaufbau, roughly translatable as reconstruction loan institute) was set up in 1948. As a state-run promotional bank, the KfW made an important contribution to the reconstruction of Germany after World War Two. Subsequently, it was repeatedly assigned new tasks, mainly promoting small and medium-sized businesses and financing aid packages for developing countries. The lot (approx. 5,400 square metres) occupies prime inner-city space at Senckenberganlage, right next to the corporation's longterm headquarters in Frankfurt on the Main. Besides the Senckenberganlage project, the north and south arcade buildings were to be restored and a new structure was to be built at the west arcade. The aim was to increase performance at the site, facilitate workflow and provide more area for use without unduly burdening the neighborhood. The new building on the Senckenberganlage was to connect to the extant south arcade building, without looking like an extension. Rather it was intended to present itself as an independent structure. With a gross floor area of 10,000 square metres (plus two basement levels with space for 100 cars), the new building houses 350 workplaces. The KfW considers flexibility of use a precondition for meeting future needs. This was therefore a key element of the competition task, as well as providing engineering solutions for load-bearing structures and mechanical services.

KfW于1948年建立，为二战后德国重建作出重要贡献。作为一家国立发展促进银行，其工作任务不断增加新的内容，主要是促进中小型企业发展和为发展中国家提供经济援助一揽子计划。竞赛地点位于Senckenberganlage最重要的地点，紧挨着公司在法兰克福的长期总部。除了Senckenberganlage项目以外，还要恢复北面和南面拱形走道的大楼，而且在西面的拱形走道还要建一座新世界建筑物。
竞赛的目的是提高竞赛地点的性能，加速工作流程，在不给周围环境造成负担的前提下提供更多的使用面积。在Senckenberganlage上建的新大楼将连接到南面现有的建筑物，看上去不像是一个延伸部分，而是一个独立的建筑。新大楼总建筑面积共有1万平方米，有350个工作室。KfW认为使用灵活是未来会议需要的一个前提。因此，除为支承结构和机械服务提供工程解决方案以外，使用的灵活性也是竞赛任务的一个关键要素。

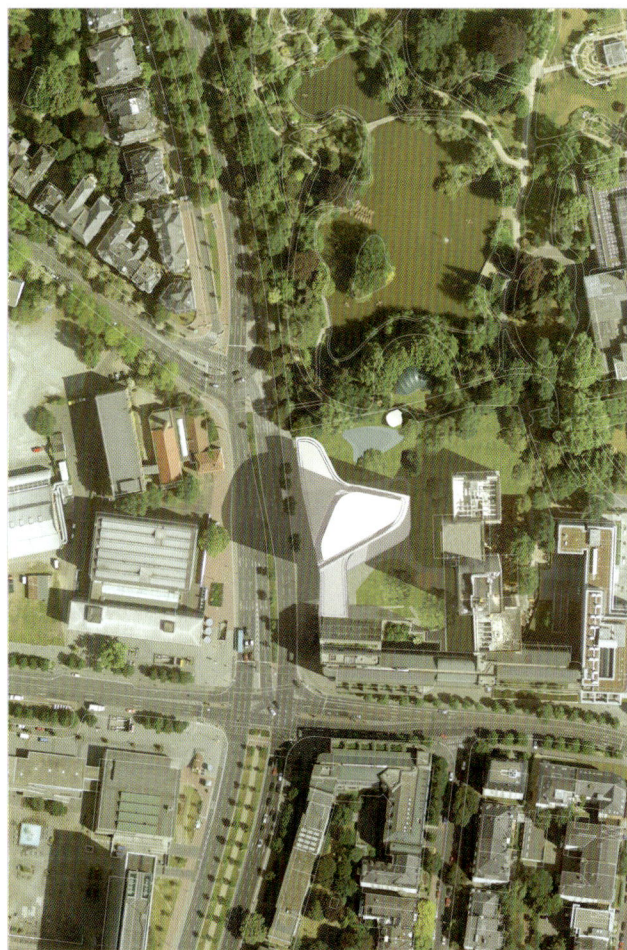
Aerial view　鸟瞰图

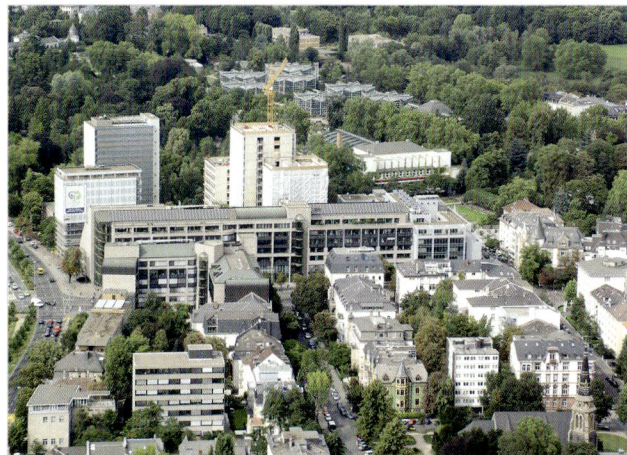
Competition site　竞赛地点

View of the City　城市景色

Qualified participants
合格的参赛者

1
1st prize　一等奖
KSP Engel und Zimmermann
Architekten, Frankfurt/Main

2
2nd prize　二等奖
Auer+Weber+Assoziierte, Stuttgart/
Munich

3
3rd prize　三等奖
HASCHER JEHLE Architektur, Berlin

4
Acquisition　购买
Petzinka Pink Architekten,
Düsseldorf

5
Acquisition　购买
ASP Schweger Assoziierte
Gesamtplanung, Hamburg

6
Further participant　其他参赛者
Ingenhoven Architekten, Düsseldorf

7
Further participant　其他参赛者
BEHNISCH ARCHITEKTEN, Stuttgart

8
Further participant　其他参赛者
schneider + schumacher
Architekturgesellschaft,
Frankfurt/Main

9
Further participant　其他参赛者
struhk architekten
Planungsgesellschaft,
Braunschweig

10
Further participant　其他参赛者
Rhode Kellermann Wawrowsky,
Düsseldorf

KSP Engel und Zimmermann Architekten Frankfurt/Main
KSP Engel und Zimmermann建筑事务所 法兰克福/美因

"The volume of the extant villa is adapted to the volume metrics of the building."
"现有别墅的体积与大楼的体积韵律相适应。"

作者 Jürgen Engel   合作伙伴 Gregor Gutscher, Özgür Ilter, Antonio Vultaggio, Thomas von Girsewald, Anna Stoyanova, Ramona Becker, Silvia Grüning   专家 structural analysis: Ruffert & Partner Ing. GmbH, Limburg; Heinz-Georg Ruffert, Meinhard Rompel, Kay-Uwe Thorn; building services: HTW Hetzel, Tor-Westen + Partner, Düsseldorf; Ralf Tosetto, Sabine Hanel, Olaf Hasse   其他专家 Hegelmann, Dutt + Kist GmbH, Hanno Dutt, Luca Kist; IFFT Institut für Fassadentechnik, Herr Böhm, Karl Otto Schott

[lageplan] 1/500

[körnung]   [motiv villa]   [kommunikative mitte]

Architektonische Intension

Der Ausbau des Hauptsitzes der KfW Bankengruppe entlang der Senckenberganlage nach Süden bietet die Chance ein Ensemble einzigartiger Gebäude zu schaffen. Der vorliegende Entwurf trägt dem Rechnung und verfolgt folgende übergeordnete Ziele:

• Ein Signet der KfW an der Senckenberganlage
• Neue Interpretation des Themas „Villenviertel Westend"
• Flexibel nutzbare, wirtschaftliche Bürobereiche

Historischer Bezug

Interpretation der Villenstruktur
Das Frankfurter Westend wird geprägt durch eine Struktur einzelner Villen und Gebäude. Ein hohes Maß von Grünflächen und ein dichter Baumbestand charakterisieren weiterhin das Areal.
Im Laufe der letzten Jahrzehnte ist diese Struktur durch Großbauten verdrängt worden. Entlang der Bockenheimer Landstraße entstanden eine Reihe Hochhäuser.
Der Mikrostandort wird auch heute noch von der ursprünglichen Idee des Westends geprägt. Der Entwurf nimmt diese Spuren auf und setzt Sie in einer modernen Interpretation der historischen Struktur um.

PLA N01

[perspekive senckenberganlage]

179242

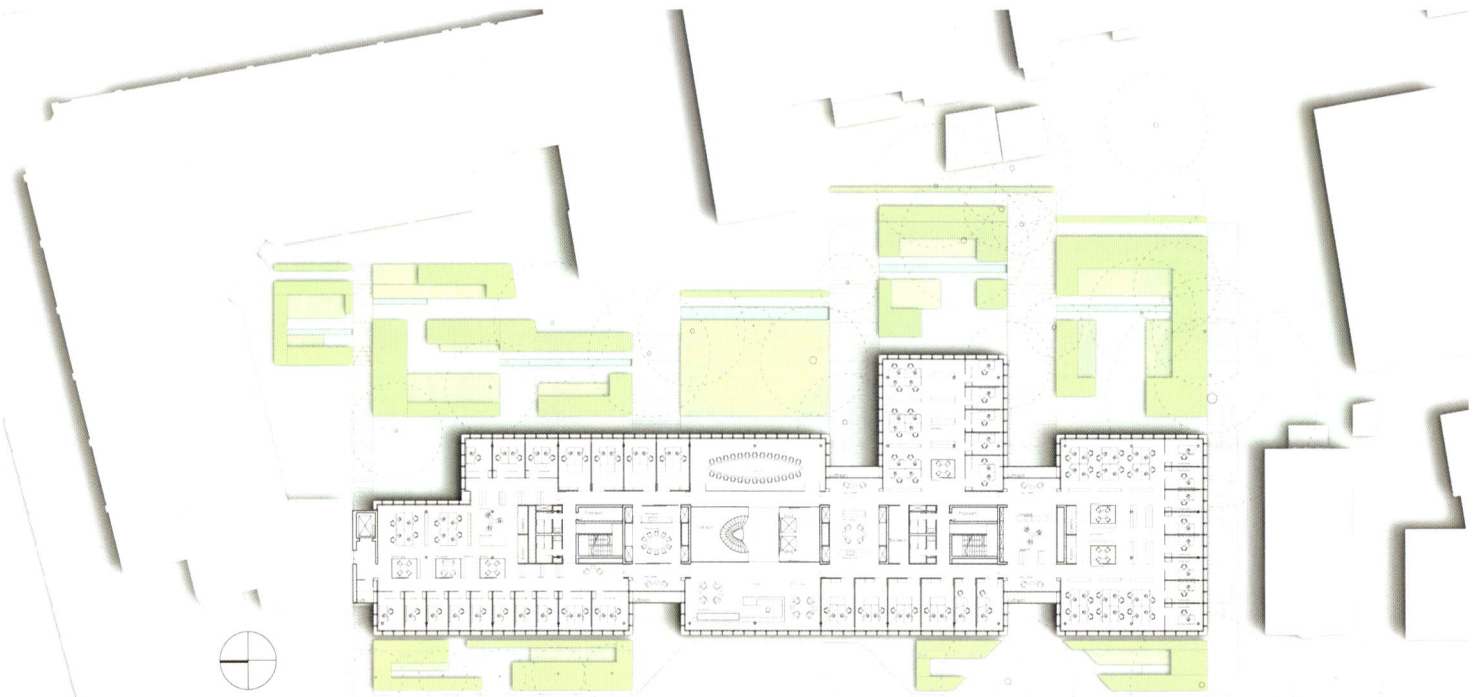

[grundriss alternativmobelierung] 1/200

PLA
N05

[systemschnitt · luftraum]

[demokratische fassade]

Fassadendetail 1/5

[luftungskonzept]

Fassade

Natürlicher Sonnenschutz mit Ausblick
Die Fassaden werden geprägt durch ihre Ost West Ausrichtung. Der außenliegende Sonnenschutz wird durch große, bewegliche Vertikallamellen sichergestellt.

Diese Lamellen erzeugen eine Grundverschattung, ohne die Aussicht maßgeblich zu behindern. Eine zentrale, sonnenstandsabhängige Steuerung kann so durchgeführt werden, ohne dass die einzelne Mitarbeiter gestört wird. Der Anteil der Fassade, der dem direkten Sonnenlicht ausgesetzt ist wird auf ein Minimum reduziert (Restsonneneinstrahlung im Grenzfall Sommersonnenwende gegen 18.00Uhr).
Die nun noch einfallende Reststrahlung wird durch eine Sonnenschutzverglasung plus innenliegendem Blendschutz absorbiert. Der Blendschutz wird als Schiebeelement konzipiert und kann individuell der Sonne und dem Sonnenschutz nachgeführt werden (auch hier wäre eine Automation denkbar).
Ein hohes Maß an indirektem Tageslicht gelangt so noch in die Bürobereiche und reduziert den Kunstlichtanteil bei der Beleuchtung.

Materialität

Hochwertige Oberflächen entsprechen dem Anspruch der KfW
Die Struktur des Sonnenschutzes prägt das Erscheinungsbild des Gebäudes. Die Lamellen werden entsprechend der Ausrichtung zur Sonne mir zwei unterschiedlichen Oberflächen ausgestattet.
Zur Sonne, bzw. Außenseite hin, wird eine etwa 4cm starke Natursteinschicht eingelassen. Entsprechend der Gliederung des Gebäudes werden leicht differierende Oberflächen verwendet.

Die zum Gebäude zeigende Seite wird, wie die gesamte Lamellenkonstruktion in Aluminium ausgeführt. Die Oberfläche in Eloxal erzeugt eine helle Fläche mit hohem Reflexionsgrad für Tageslicht an den Arbeitsplätzen.

Je nach Öffnungsgrad der Lamelle erhält das Gebäude so eine eher steinerne oder metallische Erscheinung. Die horizontalen Halteprofile der Lamellen werden in Chromstahl ausgeführt. Die warm glänzende Oberfläche schafft einen dezenten Kontrast zu den Materialien der Lamelle.
Die gewählten Oberflächen geben dem Neubau der KfW ein zeitloses Design, dass sich sowohl in die städtebauliche Umgebung, als auch in das Ensemble der KfW einfügt.

Arbeitswelten

Flexibel nutzbarer, transparenter Dreibund
Die Arbeitsbereiche liegen, mit gleichberechtigten Ausblickmöglichkeiten, entlang der Fassade. Die Kontur des Gebäudes ist auf eine möglichst große Abwicklungslänge hin konzipiert, um eine größtmögliche belichtete Fläche zu erzeugen.
Die Innenzone, der kommunikativen Mitte, des Gebäudes liegen, neben der Infrastruktur, den Sanitärbereichen und der Erschließung auch Besprechungsräume mit Teeküche, Meetingpoints, sowie flexibel nutzbare Bereiche. Diese Zonen sind den Einschnitten zugeordnet, um eine der Nutzung entsprechende natürliche Belichtung zu gewährleisten.
Über die Mittelzone hinweg sind offene Wegbeziehungen zwischen den Abteilungen und offene Möblierungen möglich.

[luftungskonzept]

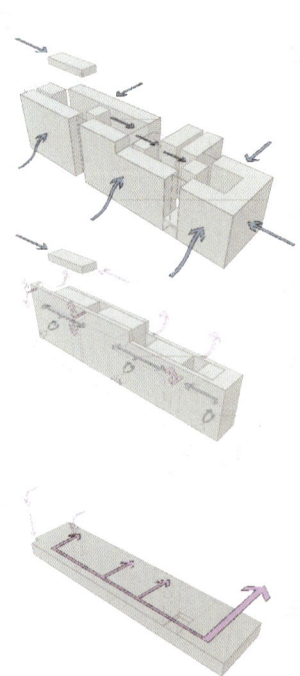

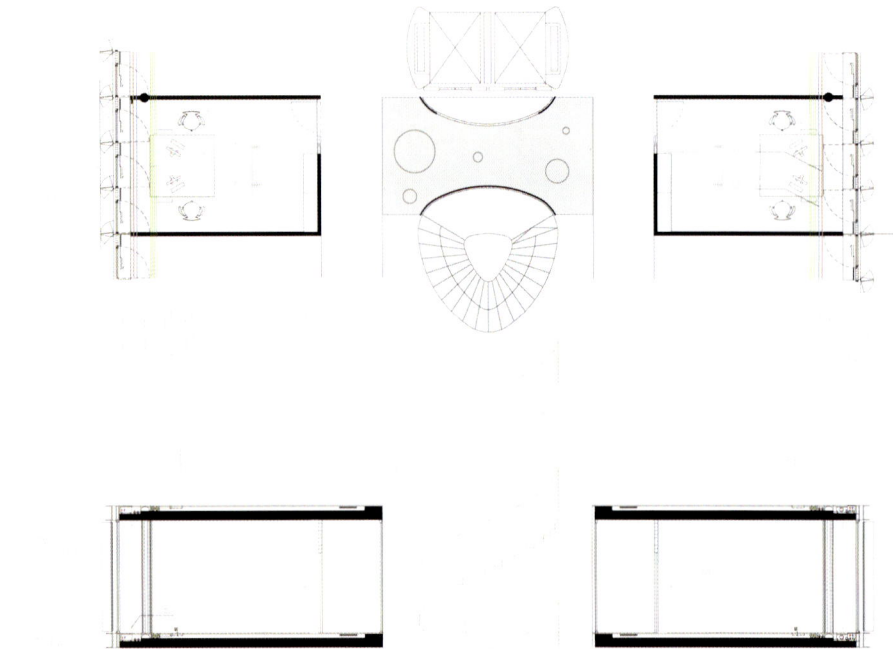

[querschnitt]

PLA N06

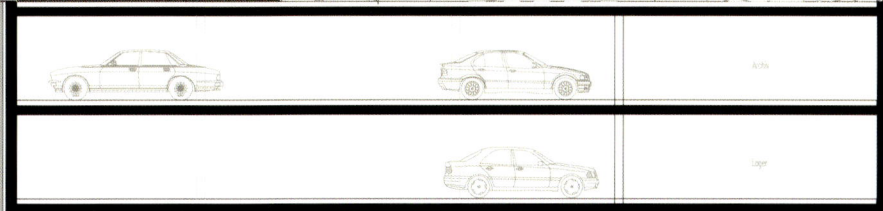

[innenraumperspektive]

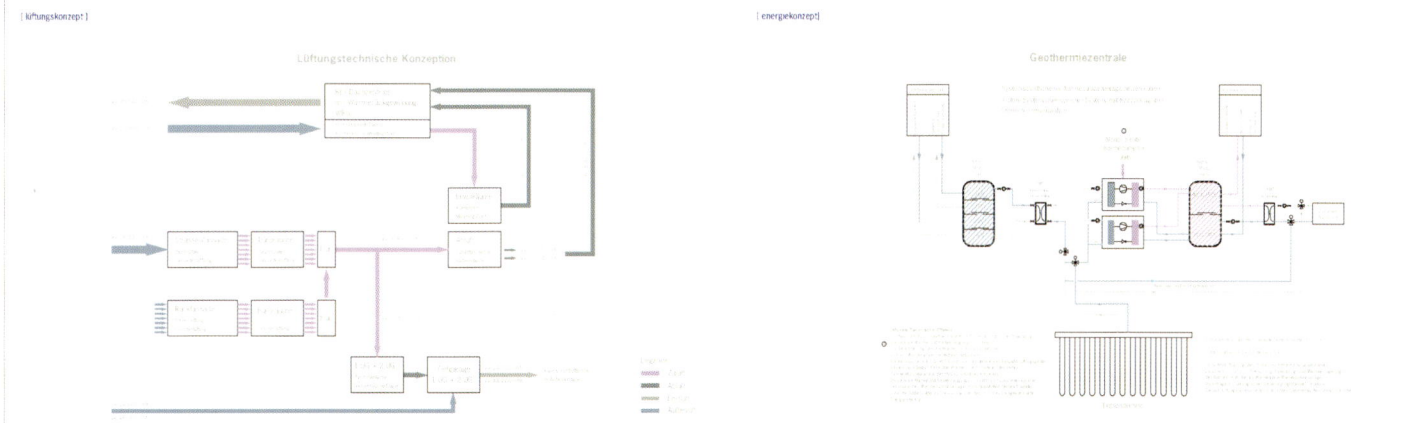

[lüftungskonzept]

Lüftungstechnische Konzeption

[energiekonzept]

Geothermiezentrale

Auer + Weber + Assoziierte Stuttgart/Munich
Auer+ Weber+Assoziierte建筑事务所 斯图加特/慕尼黑

"What is intended is an integrative, while autonomous building that responds in a pliant, easy-to-get-along-with manner to the various requirements."

"我们的目的是综合性的、然而也是自治的大楼，柔韧地、随和地对不同的要求作出反应。"

作者 Auer+Weber+Assoziierte, Stuttgart/Munich   合作伙伴 Achim Söding (associate), Henrike Schlinke, Karsten Schuch, Rainer Oertelt, Daniel Hänelt, Jan Huettel, Marianne Strauss   专家 structural analysis: Pfefferkorn Ingenieure, Stuttgart; building services: Zibell, Willner + Partner, Munich   其他专家 Jörg Stötzer, landscape architect, Waldkirch

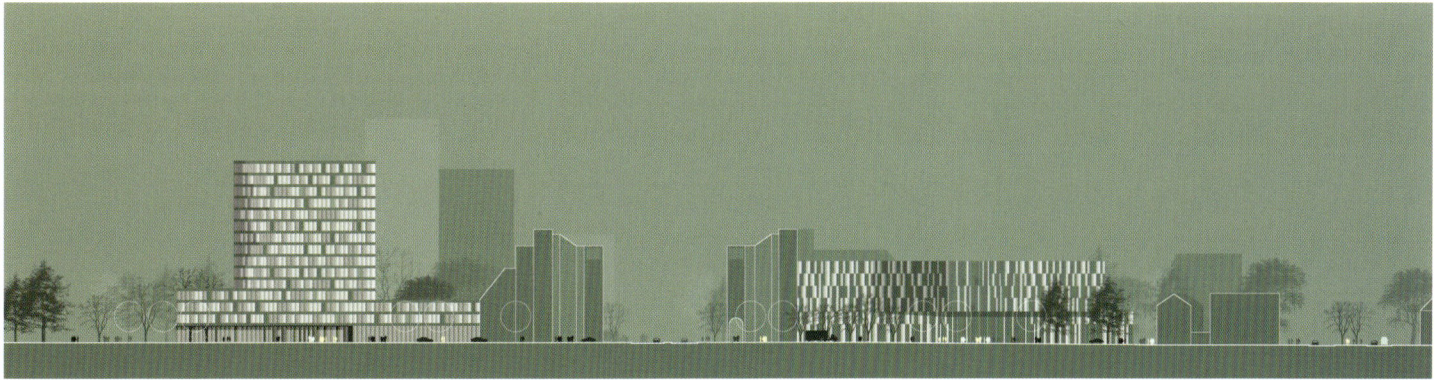

HASCHER JEHLE Architektur Berlin
HASCHER JEHLE建筑事务所 柏林

"Individual 'villas' with a depth effect blend into the texture of the surroundings. Glazed halls set back from the building line connect these 'villas' and turn them into a functional unit."

"有深邃效果的别墅个体融入周围环境的质地中，从建筑线中后移的镶有玻璃的大厅把这些'别墅'连接起来组成一个功能体。"

作者 Prof. R. Hascher, Prof. S. Jehle   合作伙伴 Fleur Keller, Michael Meier, Moritz Müller-Werther, Florian Sell   专家
structural analysis: RPB Rückert GmbH Planer + Berater, Berlin; building services: SCHOLZE Ingenieurgesellschaft mbH, Berlin
Weitere Fachberater hutterreimann Landschaftsarchitektur, Berlin

Perspektive Senckenberganlage

New KfW Building at the Senckenberganlage, Frankfurt/Main

Bär, Stadelmann, Stöc

netzwer

Burger Rudacs Archite

Thoma

SCHULT

Gerber Architekten

er

agps

architekten

University and State Library
in Darmstadt

达姆施塔特大学和州立图书馆
德国

ten

Müller Ivan Reimann

S FRANK ARCHITEKTEN

University and State Library Darmstadt
达姆施塔特大学和州立图书馆 德国

Restricted two-stage project competition preceded by an application procedure
限制严格的项目竞赛，分为两个阶段，开始前有申请程序。

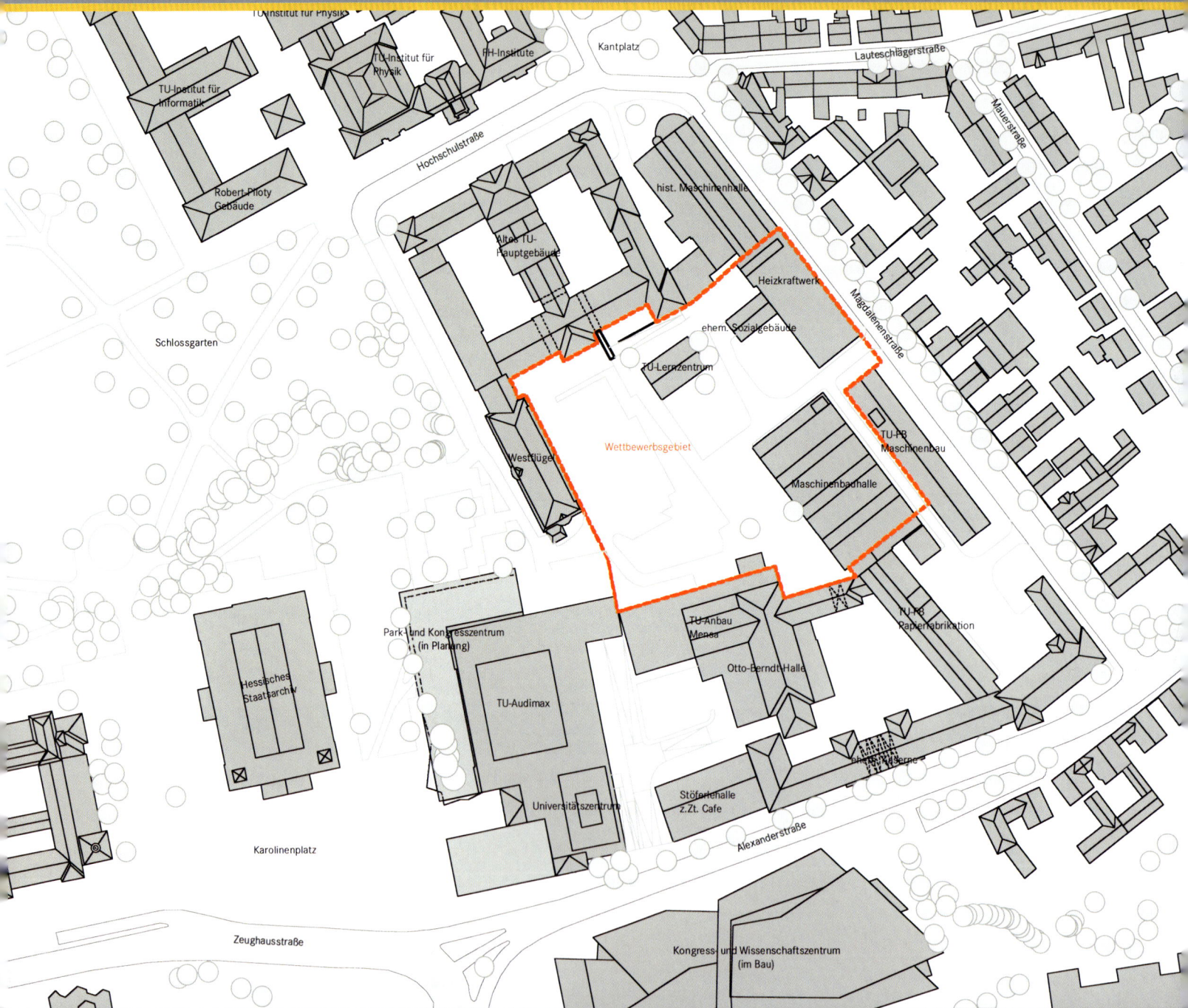

Map labels:
TU-Institut für Physik · PH-Institute · Kantplatz · Lauteschlägerstraße · Mauerstraße · TU-Institut für Physik · TU-Institut für Informatik · Hochschulstraße · hist. Maschinenhalle · Robert-Piloty Gebäude · Altes TU-Hauptgebäude · Heizkraftwerk · Magdalenenstraße · Schlossgarten · ehem. Sozialgebäude · TU-Lernzentrum · Wettbewerbsgebiet · Westflügel · TU-FB Maschinenbau · Maschinenbauhalle · Park- und Kongresszentrum (in Planung) · Hessisches Staatsarchiv · TU-Audimax · TU-Anbau Mensa · Otto-Berndt-Halle · N.FB Papierfabrikation · Universitätszentrum · Stöferlehalle z.Zt. Cafe · Karolinenplatz · Alexanderstraße · Zeughausstraße · Kongress- und Wissenschaftszentrum (im Bau)

地点 Darmstadt   时间 06/2005–12/2005   主办方 TU Darmstadt   参赛者 1st stage: 55; 2nd stage: 14   面积 19,000 sq m
竞赛费用 172,000 Euro   专业评奖委员会 Prof. Hilde Barz-Malfatti, Weimar; Prof. Rebecca Chestnutt, Berlin; Prof. Markus Gasser, Darmstadt/ Zurich; Prof. Wolfgang Lorch, Darmstadt/Saarbrücken; Prof. Matthias Sauerbruch, Berlin/Stuttgart   专家评奖委员会 Dr. Hans-Georg Nolte-Fischer, director ULB Darmstadt; Günter Schmitteckert, ministry of higher education, research and the arts, Federal State of Hessen, Wiesbaden; Dieter Wenzel, local council, City of Darmstadt; Prof. Dr.-Ing. Johann-Dietrich Wörner, president TU Darmstadt

It sees itself as a modern hybrid library: As its first function, it will be an academic service center for information retrieval, a place for study and work, for communication and human encounter. The second, more public function is that of a state library whose importance, in view of its highly valued historic collection, surpasses regional boundaries. The project site is located at TUD's long-established premises, adjacent to the palace and Darmstadt's city center. It is framed around an inner courtyard which is listed as a heritage site and borders on the "old suburb", likewise a listed site. The area lies in between the TUD's historic main building and the university commons which dates from the 1970s. The main task set to the architects was to find sensitive and original ways to integrate the new structure into the context of a highly eclectic group of buildings, some of them are listed, and thus create a new overall architectural scenario. The program of spaces includes various reading rooms, administrative areas, event venues, magazines, and workshops.

规划兴建的新大学和州立图书馆是一个现代的混合图书馆。首先，它是一个学术服务中心，可以检索信息、工作学习、交往沟通。其次，州立图书馆由于其历史收藏珍贵，其重要性已远远超出地区的界限。项目地点位于达姆施塔特的技术大学历史悠久的校园，紧临宫殿和市中心。项目在邻近"老郊区"的一个被列为历史地段的一个内庭旁边进行，位于达姆施塔特的技术大学的具有历史意义的主楼和始建于70年代的大学公用地之间。竞赛的主要任务是找到敏感的、新颖的方式使新建筑物与一些非常折中的建筑群浑然一体。空间规划包括各种阅览室、管理区、活动场馆、杂志室和工作室。

Aerial view　鸟瞰图

Competition site　竞赛地点

Competition site　竞赛地点

View of the City　城市景色

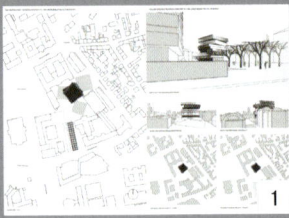

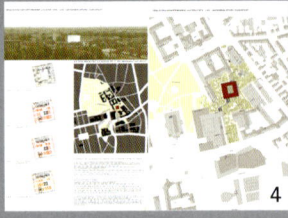
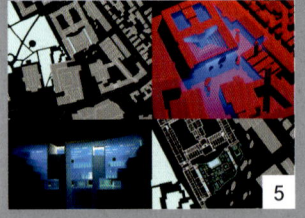

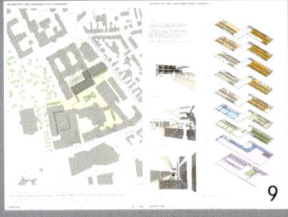

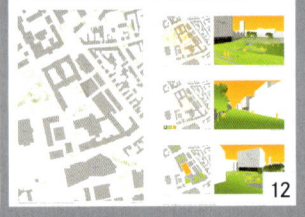
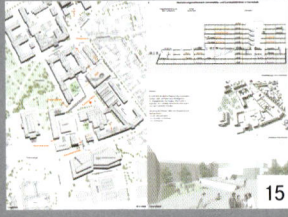

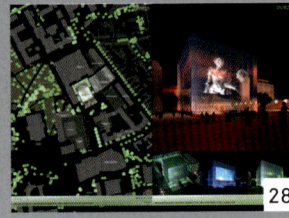
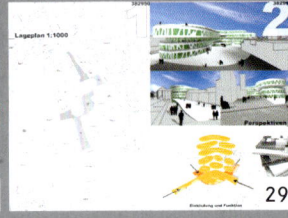

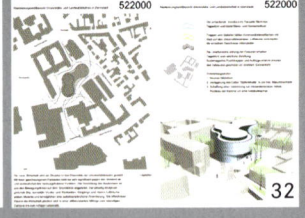
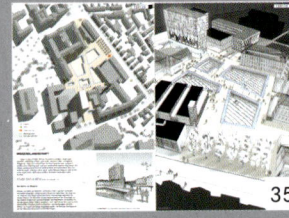
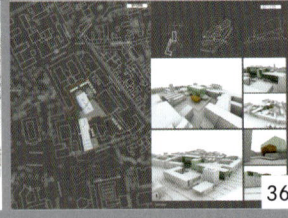

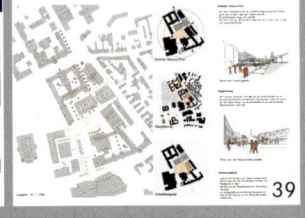
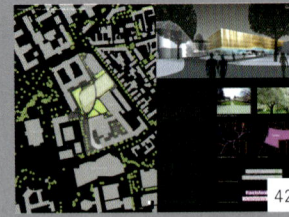
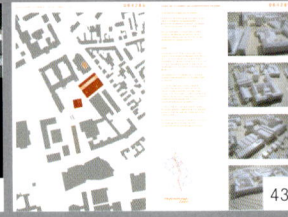

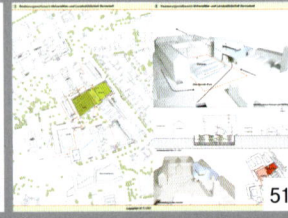
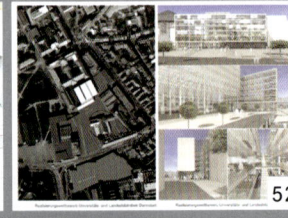

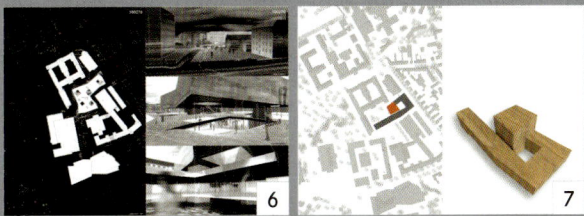

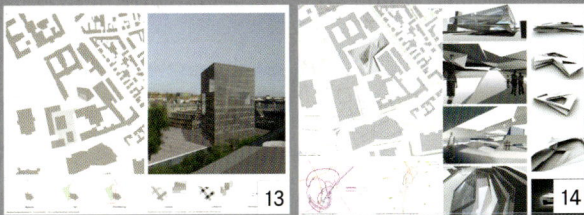

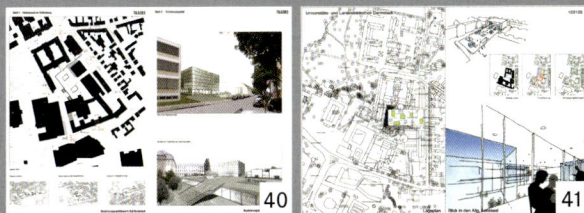

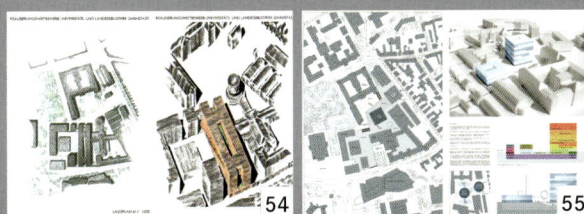

Qualified participants 1st stage 第一阶段合格参赛者
1 Peter Kulka Architektur, Cologne **2** ff-Architekten, Berlin **3** Auer+Weber+Assoziierte, Stuttgart/Munich **4** Gerber Architekten, Dortmund **5** SCHULTES FRANK ARCHITEKTEN, Berlin **6** Kirsten Schemel Architekten BDA, Berlin **7** Bär, Stadelmann, Stöcker Architekten BDA, Nuremberg **8** KSP Engel und Zimmermann, Frankfur/Main **9** Thomas Müller Ivan Reimann, Berlin **10** Burger Rudacs Architekten, Munich **11** netzwerkarchitekten, Darmstadt **12** ASTOC Architects & Planners, Cologne **13** agps architecture, Zurich **14** Plasma Studio, London

Not qualified participants 1st stage 第一阶段不合格参赛者
15 Hoechstetter und Partner, Darmstadt **16** Baumschlager-Eberle Ziviltechniker GmbH, Lochau **17** Ferdinand Heide Architekt, Frankfurt/Main **18** Ortner + Ortner Baukunst, Berlin **19** Goldfinger A. Roloff Ruffing Sill, Hamburg **20** Pahl + Weber-Pahl Architekten, Darmstadt **21** Du Besset-Lyon Architectes, Paris **22** Heckmann - Jung Freie Architekten, Stuttgart **23** Abelmann Vielain Pock Architekten, Berlin **24** HASCHER JEHLE Architektur, Berlin **25** Prof. Jörg Friedrich, PFP Architekten, Braunschweig **26** Schneider + Sendelbach Architektengesellschaft, Braunschweig **27** Gatermann + Schossig Architekten Generalplaner, Cologne **28** gmp - Architekten von Gerkan, Marg und Partner, Hamburg **29** Blauraum Architekten, Hamburg **30** HG Merz Architekten, Berlin **31** Henning Larsen Architects, Kopenhagen **32** K+P Architekten und Stadtplaner GmbH – Koch Drohn Schneider Voigt, Munich **33** Van Den Valentyn - Architektur, Cologne **34** Scheuring und Partner Architekten, Cologne **35** Architectenbureau Micha de Haas, Amsterdam **36** Keith Williams Architects, London **37** ASP Schweger Assoziierte Gesamtplanung GmbH, Hamburg **38** AssmannSalomon, Berlin **39** Gössler Architekten, Berlin **40** Bernhardt + Partner, Darmstadt **41** Schuster Architekten, Düsseldorf **42** Benthem Crouwel GmbH, Aachen **43** Herzog + Partner, Munich **44** Braunfels Architekten, Berlin **45** AS&P - Albert Speer und Partner GmbH, Frankfurt/Main **46** waechter+waechter architekten, Darmstadt **47** Max Dudler, Berlin **48** feuerstein + gerken, Munich **49** Anin·Jeromin·Fitilidis & Partner Architekten & Ingenieure, Düsseldorf **50** Lederer+Ragnarsdóttir+Oei, Stuttgart **51** Univ.Prof.Arch.DI Klaus Kada, Graz **52** Opus Architekten, Darmstadt **53** Bez+Kock Architekten, Stuttgart **54** Architekturbüro Böhm **55** Knoche Architekten, Stuttgart

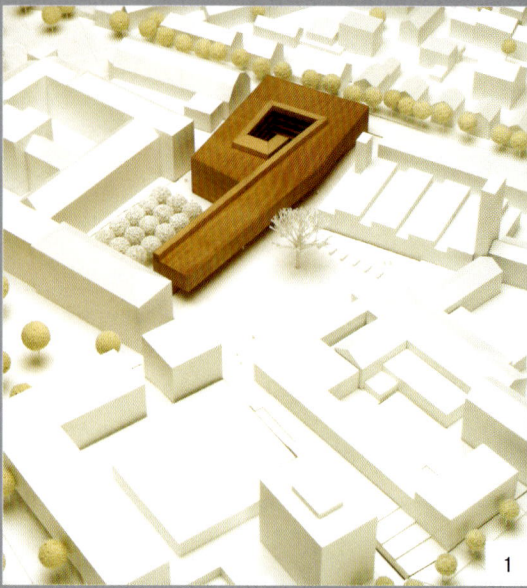

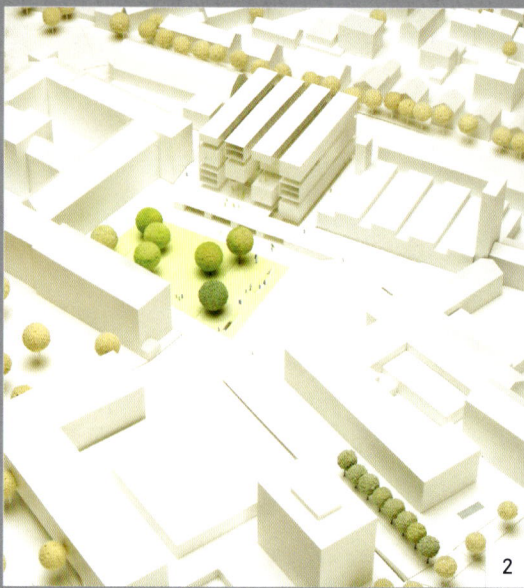

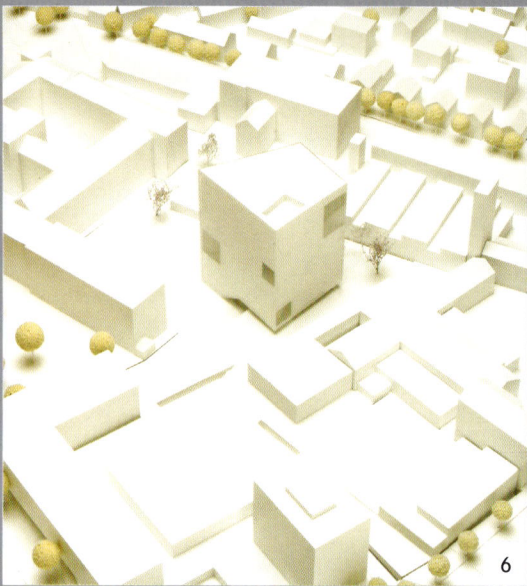

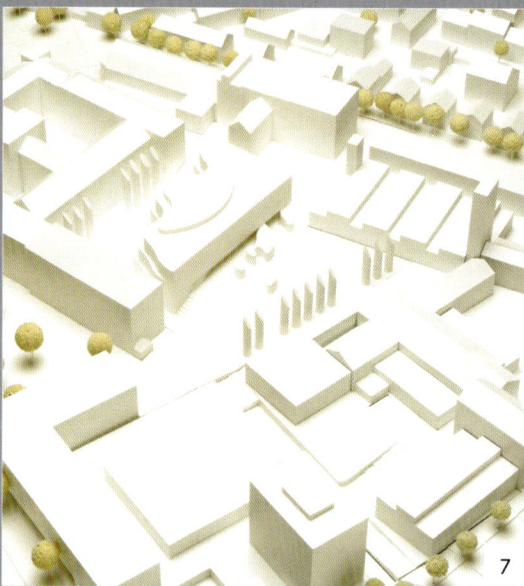

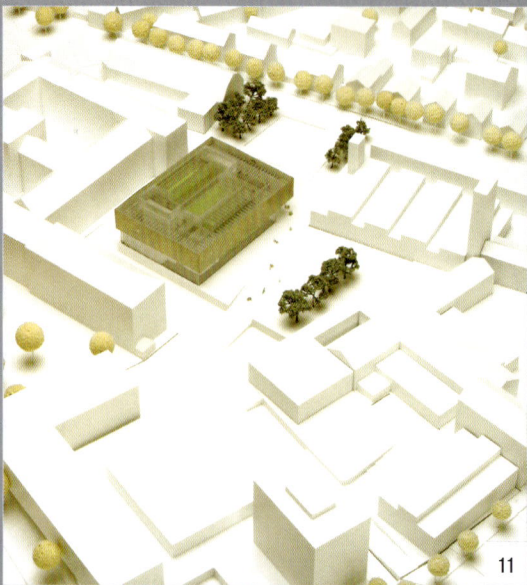

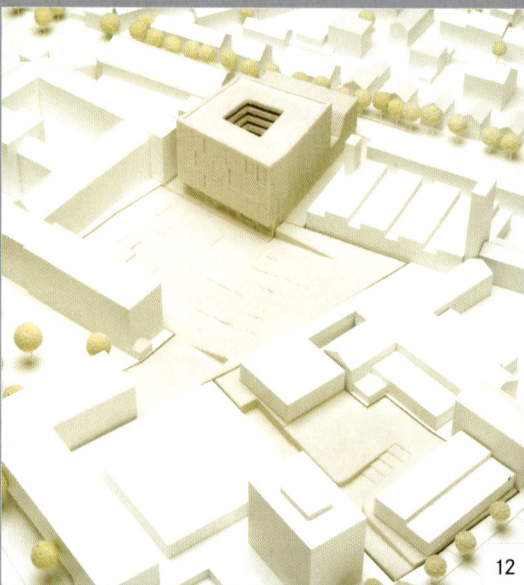

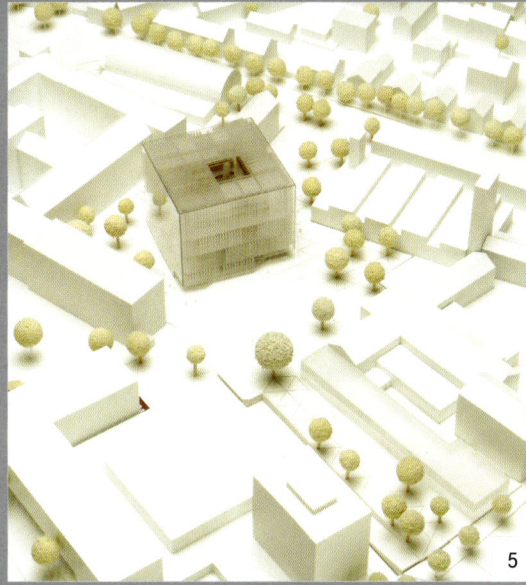

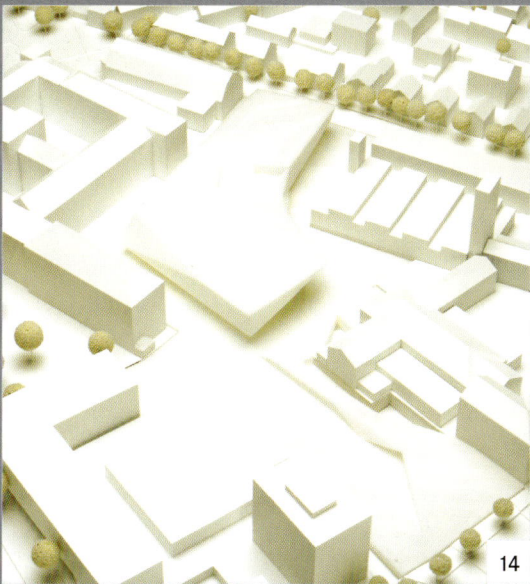

Participants 2nd stage
第二阶段参赛者

1
1st prize 一等奖
Bär, Stadelmann, Stöcker
Architekten, Nuremberg

2
2nd prize 二等奖
netzwerkarchitekten, Darmstadt

3
3rd prize 三等奖
Burger Rudacs Architekten, Munich

4
3rd prize 三等奖
Thomas Müller Ivan Reimann, Berlin

5
5th prize 五等奖
Gerber Architekten, Dortmund

6
1st acquisition 购买一等奖
agps, Zurich

7
Acquisition 购买
SCHULTES FRANK ARCHITEKTEN,
Berlin

8
Acquisition 购买
Plasma Studio, London

9
2nd round 第二轮
Peter Kulka Architektur, Cologne

10
2nd round 第二轮
ASTOC Architects & Planners,
Cologne

11
2nd round 第二轮
Kirsten Schemel Architekten, Berlin

12
1st round 第一轮
Auer+Weber+Assoziierte, Stuttgart/
Munich

13
1st round 第一轮
ff-Architekten, Berlin

14
1st round 第一轮
KSP Engel und Zimmermann
Architekten, Frankfurt/Main

Bär, Stadelmann, Stöcker Architekten Nuremberg
Bar, Stadelmann, Stocker建筑事务所 纽伦堡

"The design is marked by the idea of space creation, interweaving, and identification, and by the respect for the university's heterogeneous environs."

"该设计以其空间创造、相互交织和等同的思想，也以其对大学周围各异环境的尊重而引人注目。"

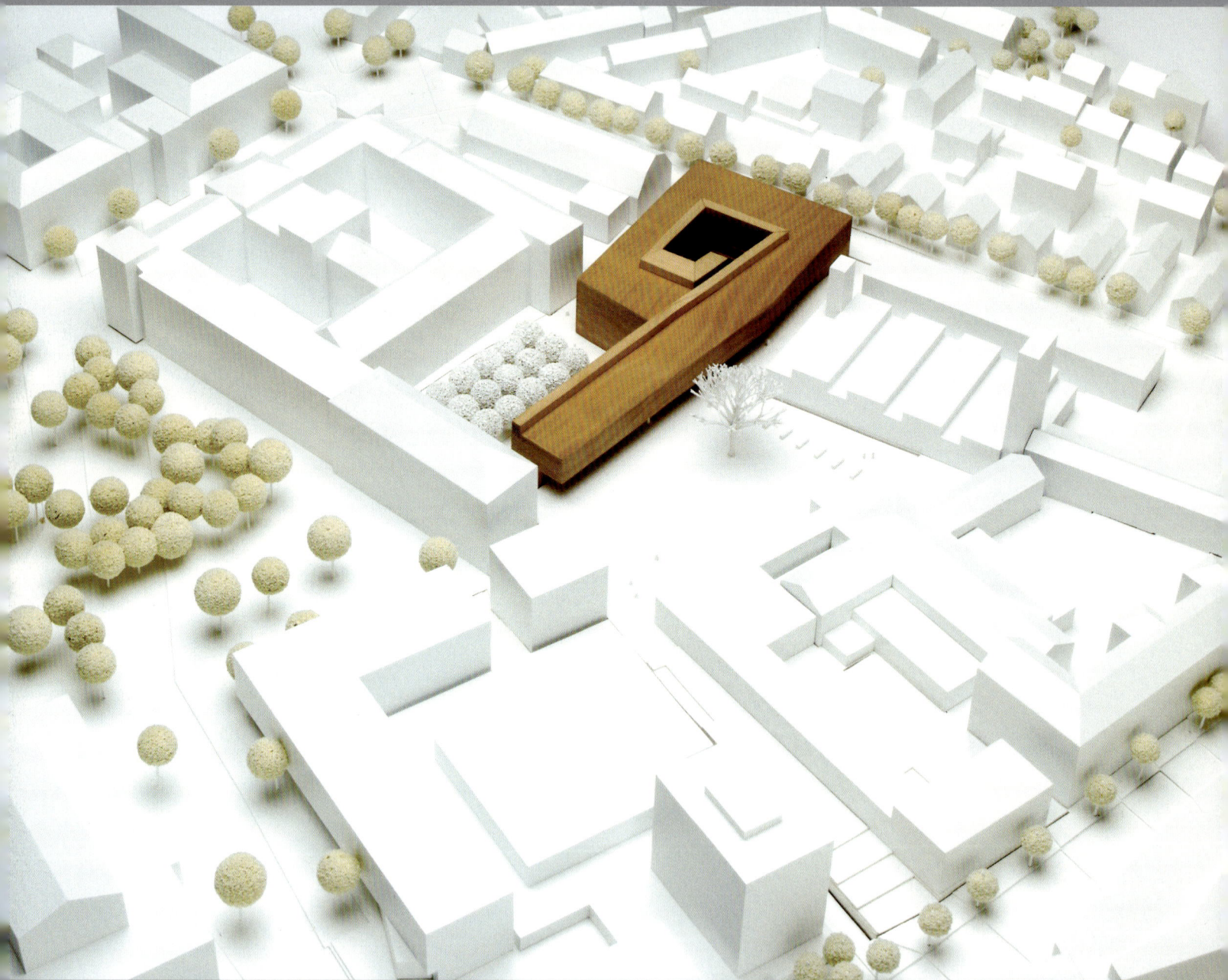

作者 Friedrich Bär   合作伙伴 Anja Vogl   专家 building services: Ingenieurbüro Hausladen GmbH, Josef Bauer, Kirchheim

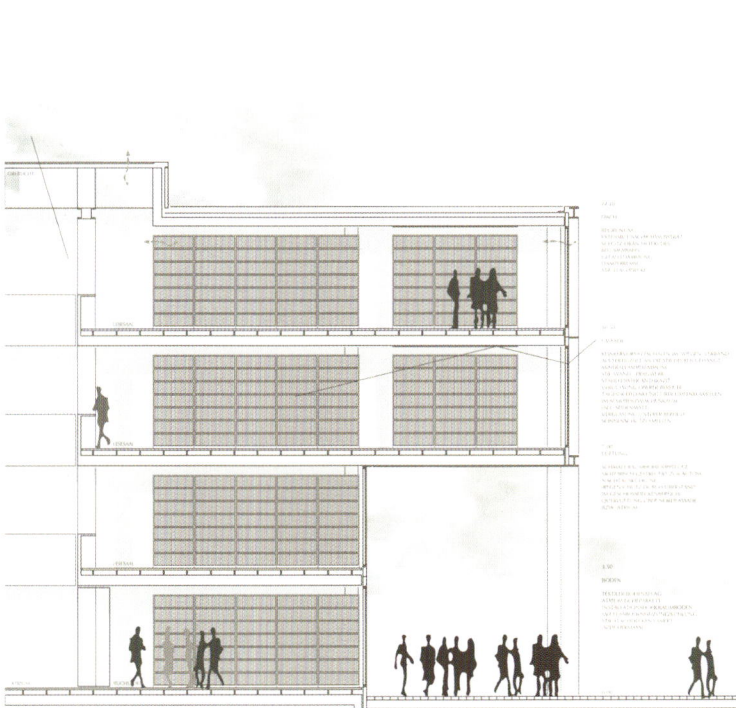

Detail 1:50

Schnitt + Klimakonzept

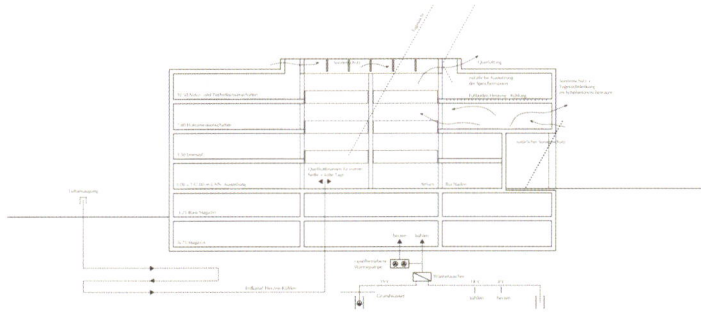

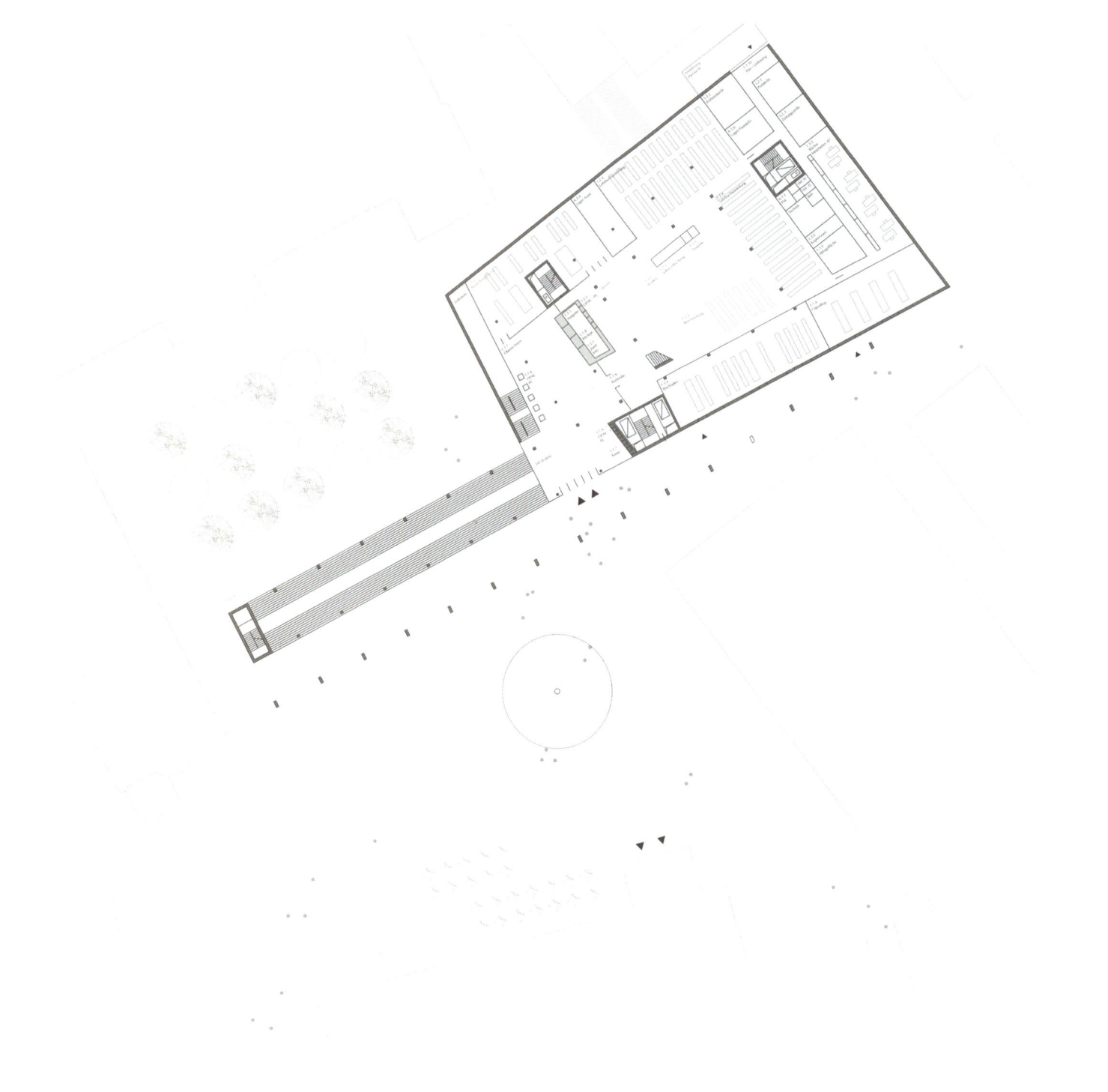

Erdgschoss 1:200

3

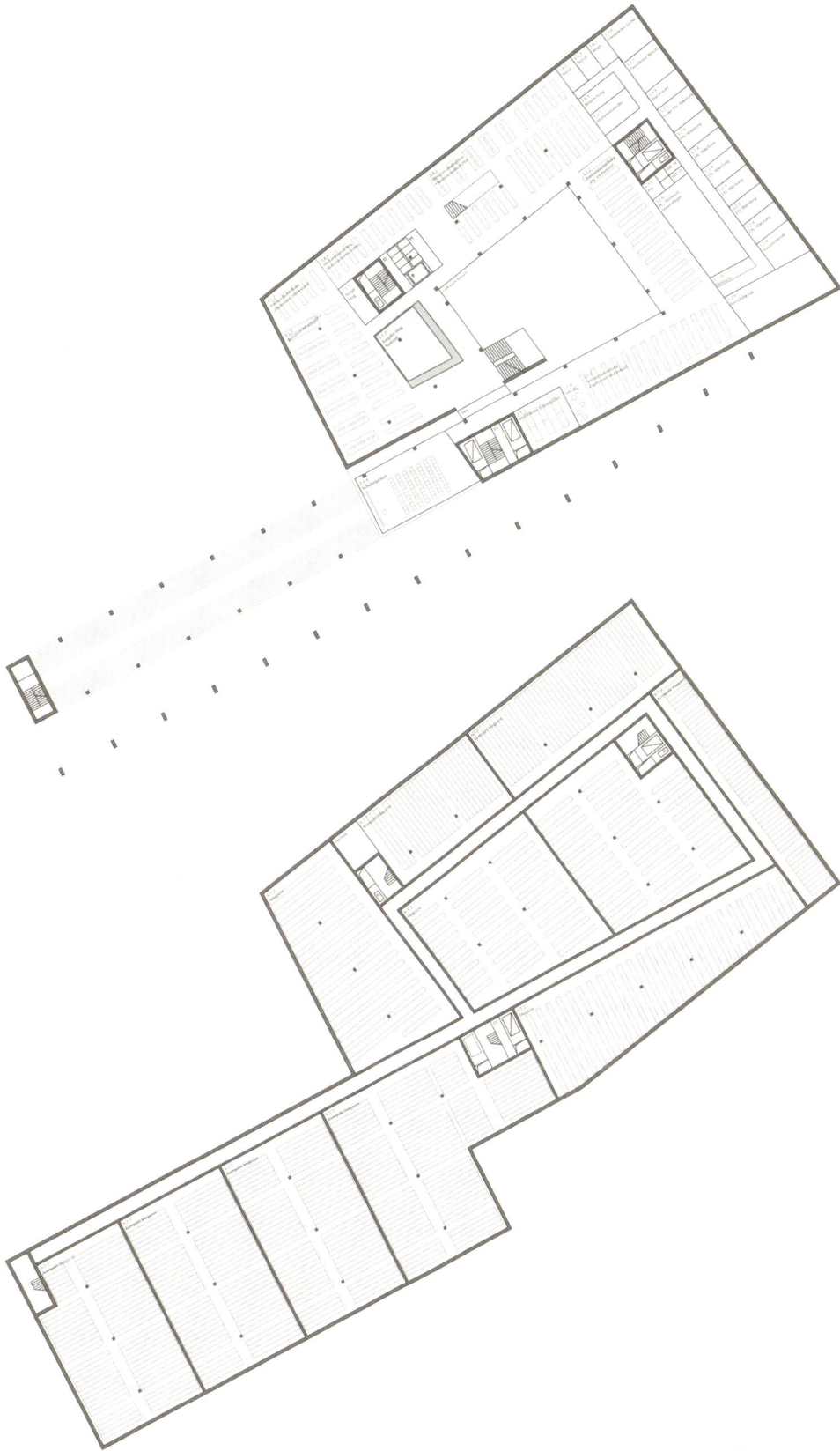

1. Obergeschoss 1:200

UG 1:200

netzwerkarchitekten Darmstadt
Netzwerk建筑事务所 达姆施塔特

"The structure of the new university and state library will be located between Magdalenen Straße and the block's interior, thus freeing space for an ample campus."

"新大学和州立图书馆建筑将位于Magdalenen StraBe和区域内部之间，从而为校园释放了更多的空间。"

作者 Thilo Höhne, Karim Scharabi, Philipp Schiffer, Jochen Schuh, Marcus Schwieger, Oliver Witan 合作伙伴 Petra Lenschow, Jeremias Lorch , Marvin King 专家 structural analysis: Bollinger und Grohmann, Frankfurt/Main; building services: Platzer Ingenieure, Bad Nauheim; landscape architects: Club L94, Cologne

Detailschnitt

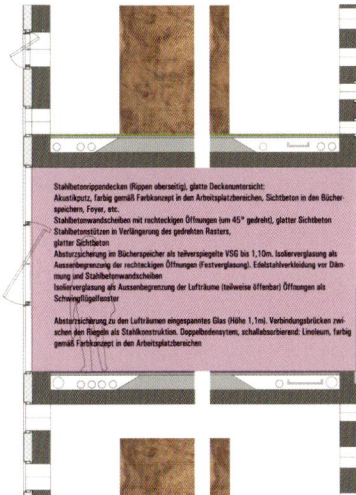

Stahlbetonrippendecken (Rippen oberseitig), glatte Deckenuntersicht:
Akustikputz, farbig gemäß Farbkonzept in den Arbeitsplatzbereichen, Sichtbeton in den Bücher-
speichern, Foyer, etc.
Stahlbetonwandscheiben mit rechteckigen Öffnungen (um 45° gedreht), glatter Sichtbeton
Stahlbetonstützen in Verlängerung des gedrehten Rasters,
glatter Sichtbeton
Absturzsicherung im Bücherspeicher als teilverspiegelte VSG bis 1,10m. Isolierverglasung als
Aussenbegrenzung der rechteckigen Öffnungen (Festverglasung). Edelstahlverkleidung vor Däm-
mung und Stahlbetonwandscheiben
Isolierverglasung als Aussenbegrenzung der Lufträume (teilweise öffenbar) Öffnungen als
Schwingflügelfenster

Absturzsicherung zu den Lufträumen eingespanntes Glas (Höhe 1,1m). Verbindungsbrücken zwischen den Riegeln als Stahlkonstruktion. Doppelbodensystem, schallabsorbierend: Linoleum, farbig
gemäß Farbkonzept in den Arbeitsplatzbereichen

Technik

Allgemein

Die kompakte Gebäudestruktur hat ein A/V-Verhältnis von ca. 0,3. Sie öffnet
sich mit seinen Lesebereichen im Wesentlichen gegen Nordwest und schließt
sich zu den wärmebelasteten Himmelrichtungen Nordost bis Südwest. Bei
Außentemperaturen oberhalb + 17 °C werden Raumlufttechnische Anlagen
für den Außenzonenbereich (bis ca. 8 m Tiefe) abgeschaltet um diese natürlich
zu belüften und unterstützend durch wasser-basierte Kühldecken zu kühlen
um einen angenehme Raumzustände zu erreichen. Bei Außentemperaturen, die ein
Beheizen des Gebäudes nach sich ziehen, soll das Gebäude komplett geschlossen
werden.
Durch den Eintrag von Luft über eine Lüftungsanlage kann in hohem Maße eine
Wärmerückgewinnung erfolgen. Die Magazinbereiche werden im Wesentlichen
hermetisch abgeschlossen und mit einem ca. 2fachen Luftwechsel klimatisiert,
um sowohl Temperaturschwankungen wie auch Feuchteschwankungen in sehr
geringen Bandbreiten zu erreichen.

Lüftungskonzept:

Die Zuluftzuführung erfolgt über geschlitzte Doppelbodenstrukturen
(Druckboden), so dass sich eine Quellüftung einstellt und mit minimalen
Luftmengen gearbeitet werden kann. In den Randbereichen des Gebäudes
werden Unterflurkonvektoren installiert, die die Glasflächen hinsichtlich ihrer
Oberflächentemperaturen abdecken und den Wärmeverlust ausgleichen.

Durch die besondere Struktur des Gebäudes und die offenen Raumbereiche
kann die Abluft und die Entrauchung über das gleiche Abluftsystem aus dem
Gebäude abgeführt werden das mit einem regenerativen Wärmerückgewinnungs-
Systemen ausgestattet ist. Zur Reduzierung des Kälteenergiebedarfs sind sie als
Desorptionsanlagen ausgebildet (Nutzung adiabat abgekühlter Abluftströme).

Leselandschaft

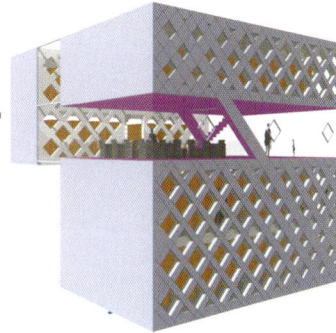

Buchbereiche

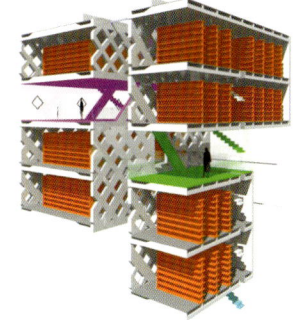

städtebauliche Ergänzung

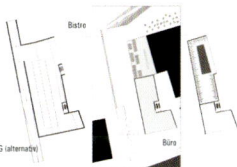

Lüftung /Entrauchung

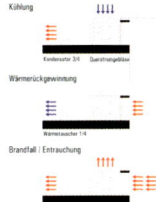

Licht

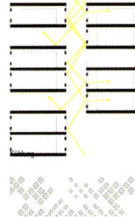

Schall

Erschließung

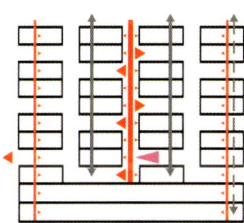

Buchförderanlage

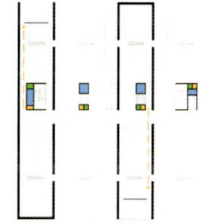

Westansicht

Erläuerungstext

Städtebauliche Einbindung

Der Baukörper für die neue Universitäts- und Landesbibliothek Darmstadt wird im Baufeld zwischen Magdalenenstraße und Blockinnenbereich platziert, so dass Raum für einen großzügigen Campus im Inneren entsteht. Zur Magdalenenstraße hin wird die Bibliothek aus den Gebäudefluchten der angrenzenden Bebauung über eine platzartigen Aufweitung zurückgesetzt. Hierdurch wird einerseits der Höhenentwicklung des Gebäudes im Verhältnis zum angrenzenden Quartier Rechnung getragen und andererseits die Durchwegung und Anbindung des Campus für den Fuss- und Radverkehr gestärkt. Die historische Maschinenhalle wird in ihrer Solitärcharakter thematisiert und erhält eine neue Kopfsituation zur Bibliothek hin. Das Gebäude stapelt sich mit seinen Bücherboxen und offenen Lesedecks in die Höhe und gibt dem neuen Universitätscampus sein Gesicht.

städtebauliche Einbindung

Der Campus wird als Kreuzungs- und Treffpunkt entwickelt, an dem sternförmig alle wesentlichen Fußwegeverbindungen der Hochschulumgebung anknüpfen, bzw. an dem die beiden wesentlichen Niveaus - Herrengarten und Plattform/ Uni - miteinander verflochten werden:

Von der Stadtseite her wird die Bestandsebene des Universitätszentrums großzügig über Sitzstufen an das obere Niveau von Mensa, Maschinenbaugebäude, Haupteingang Bibliothek und ehemaligem Hauptgebäude der Uni angeknüpft.

Vom Kongreßzentrum her kommend bereitet ein kleiner Platz das Entrée zur baumbegleiteten Wegeverbindung zum Campus. Die gegenwärtigen Gebäudeteile von Cafeteria und anschließenden Büroräumen werden durch einen neuen, gestreckten Gebäudeflügel entlang dieser Durchwegung ersetzt. Da in dieser Konfiguration nun die Anlieferung der Mensa und Cafeteria ostseitig des neuen Gebäudeflügels und somit getrennt von der Rampe zur TG der Hochschule organisiert ist, kann die TG- Zufahrt in weiten Teilen abgedeckt werden. Die Hörsäle erhalten großzügige Oberlichter.

Das Foyer des Audimax, wie auch der Herrengarten bzw. die Hochschulstraße, sind nach wie vor an die untere Ebene des Campus angeschlossen. Über die neue Mitte mit baumbestandener `Leselounge` erreicht man das Sockelgeschoss des ehemaligen Hauptgebäudes und den unteren Eingangsbereich der Bibliothek an dem sich die öffentlichen Funktionen wie Cafeteria, Buchhandlung und Copyshop bündeln. Eine großzügige Treppenanlage mit diagonaler Rampe führt barrierefrei auf das obere Niveau zur Mensa hinauf.

Vom Kantplatz her führt eine gestreckte Rampe zwischen ehemaligem Hauptgebäude und Maschinenhalle zum Campus. An der Südwestecke des ehemaligen Hauptgebäudes, also in unmittelbarer Nähe des Haupteingangs der Bibliothek, wird an das 1. Obergeschoss angeschlossen, so dass die Bestandsbrücke entfallen kann und das Hauptgebäude in seiner gegenwärtigen Organisation kaum angepasst werden muss.

Strukturprinzip

Speicher

Ordnung

Schnittstelle

Verknüpfung

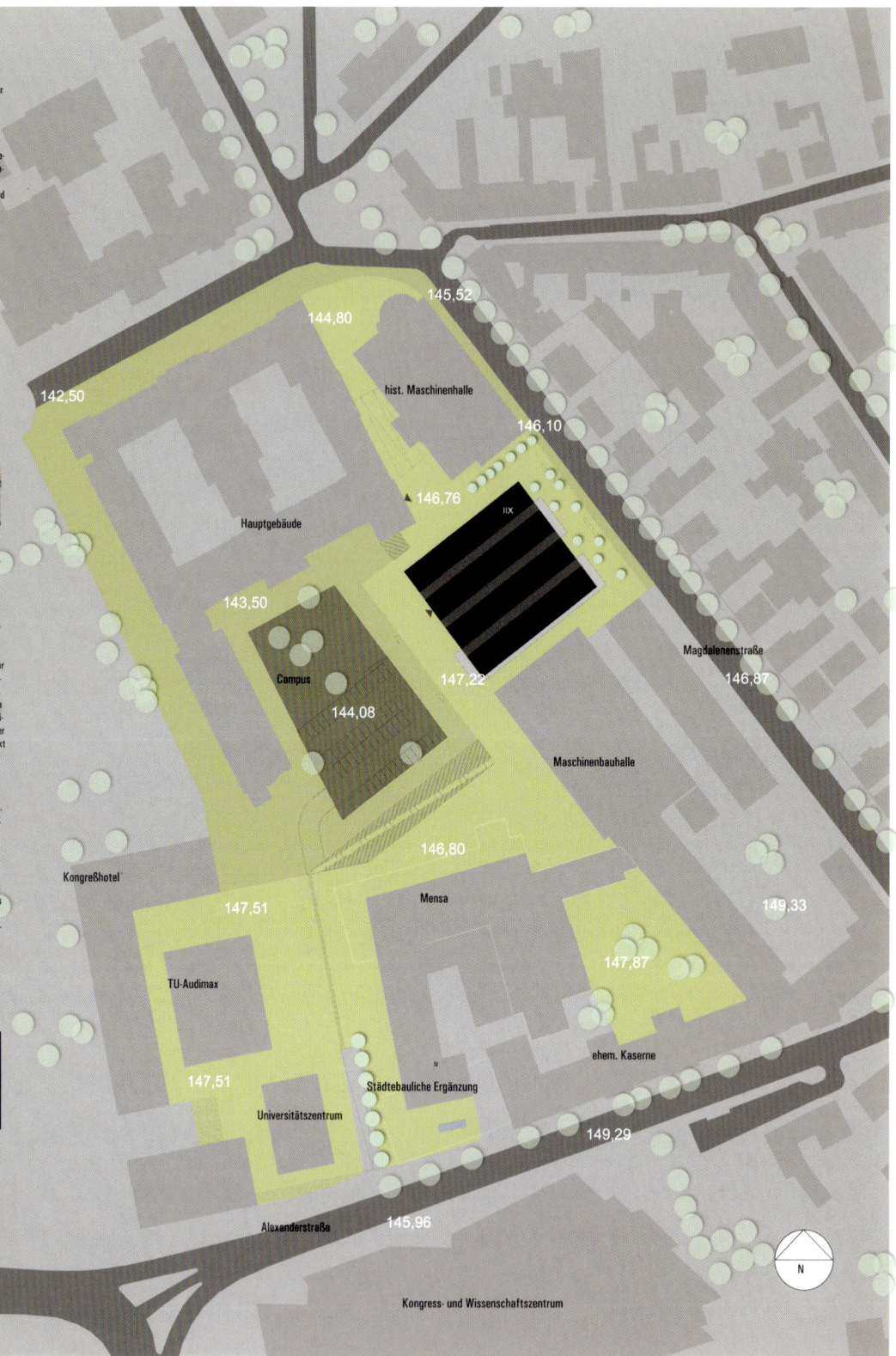

Lageplan M. 1:500

Ansicht Süd M. 1:500

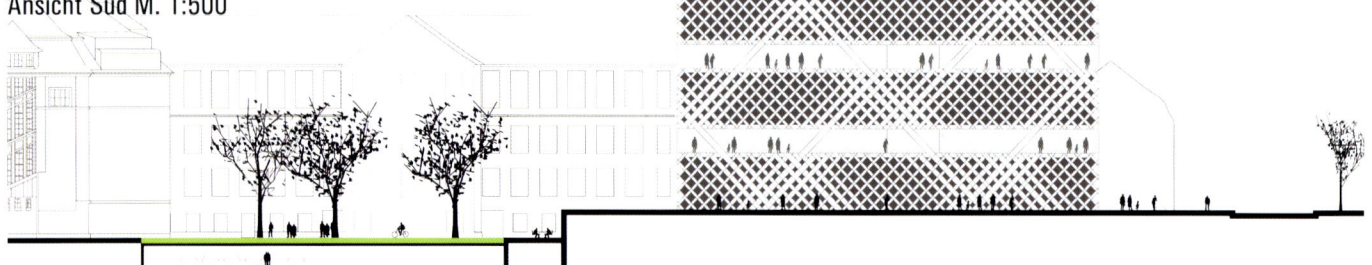

Blatt 1

Grundrisse M. 1:200

OG 5

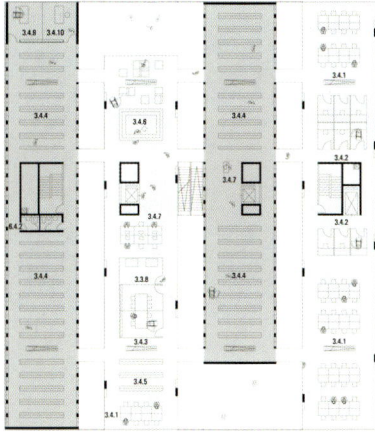

OG 6

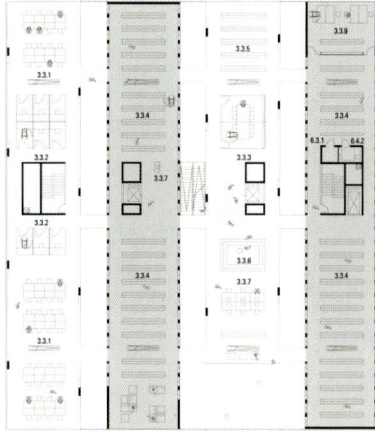

OG 7

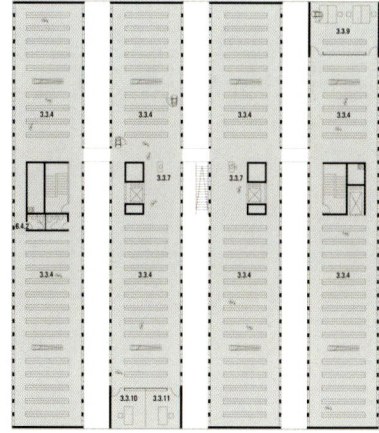

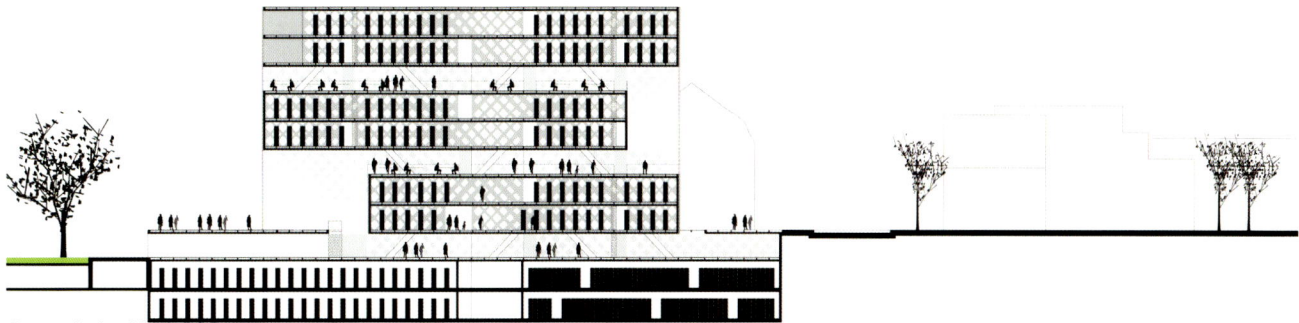

Querschnitt M. 1:200

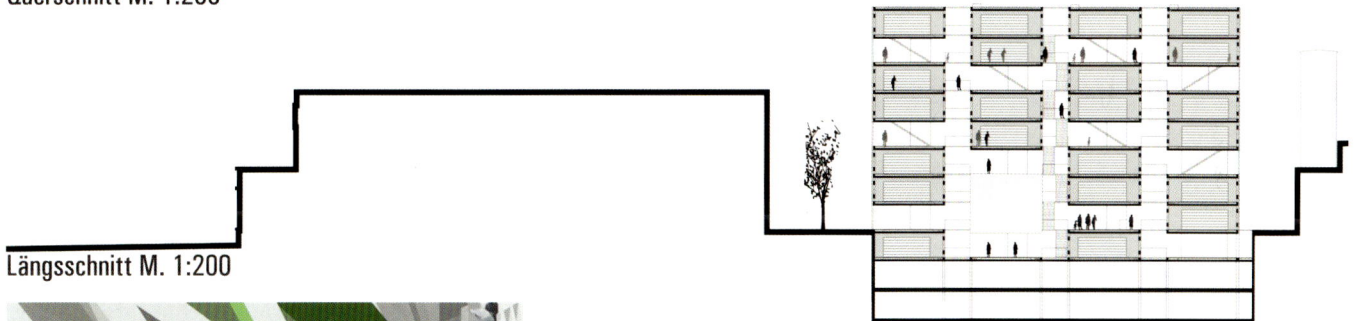

Längsschnitt M. 1:200

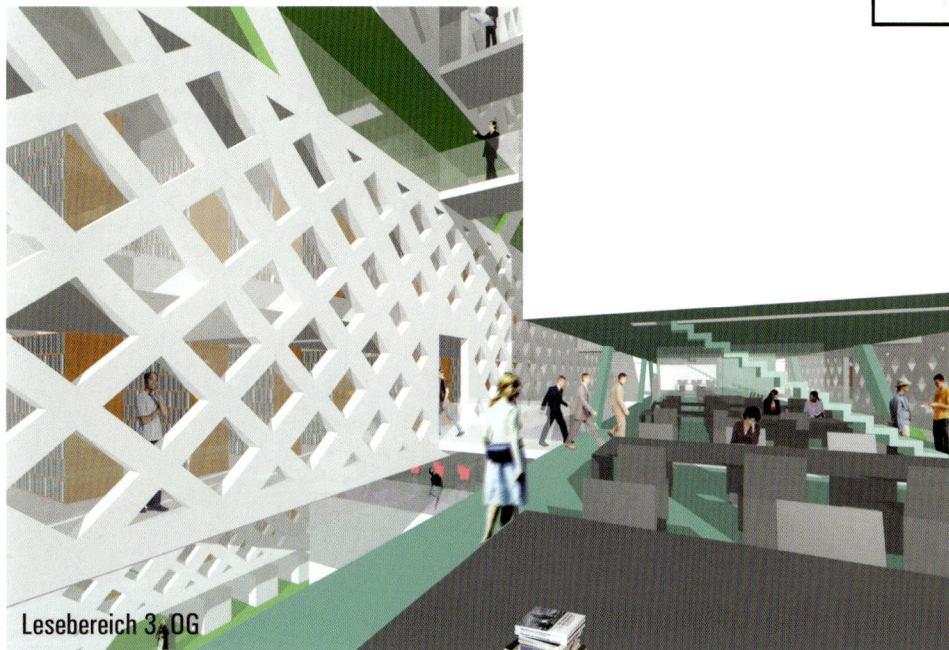

Lesebereich 3. OG

Blatt 5

Burger Rudacs Architekten Munich
Burger Rudacs建筑事务所 慕尼黑

"The new building for Darmstadt's university and state library is conceived as a large, four-storey, east-west-oriented bar."
"达姆施塔特的大学和州立图书馆新大楼被看成是一个大的、四层楼的、东西朝向的栅栏。"

作者 Stefan Burger, Birgit Rudacs   专家 landscape architects: Lohrer Hochrein landscape architects BDLA, Munich; Team Pawlowski, Ingenieurbüro im Bauwesen, Munich; building services: Schreiber Ingenieure GmbH, Ulm

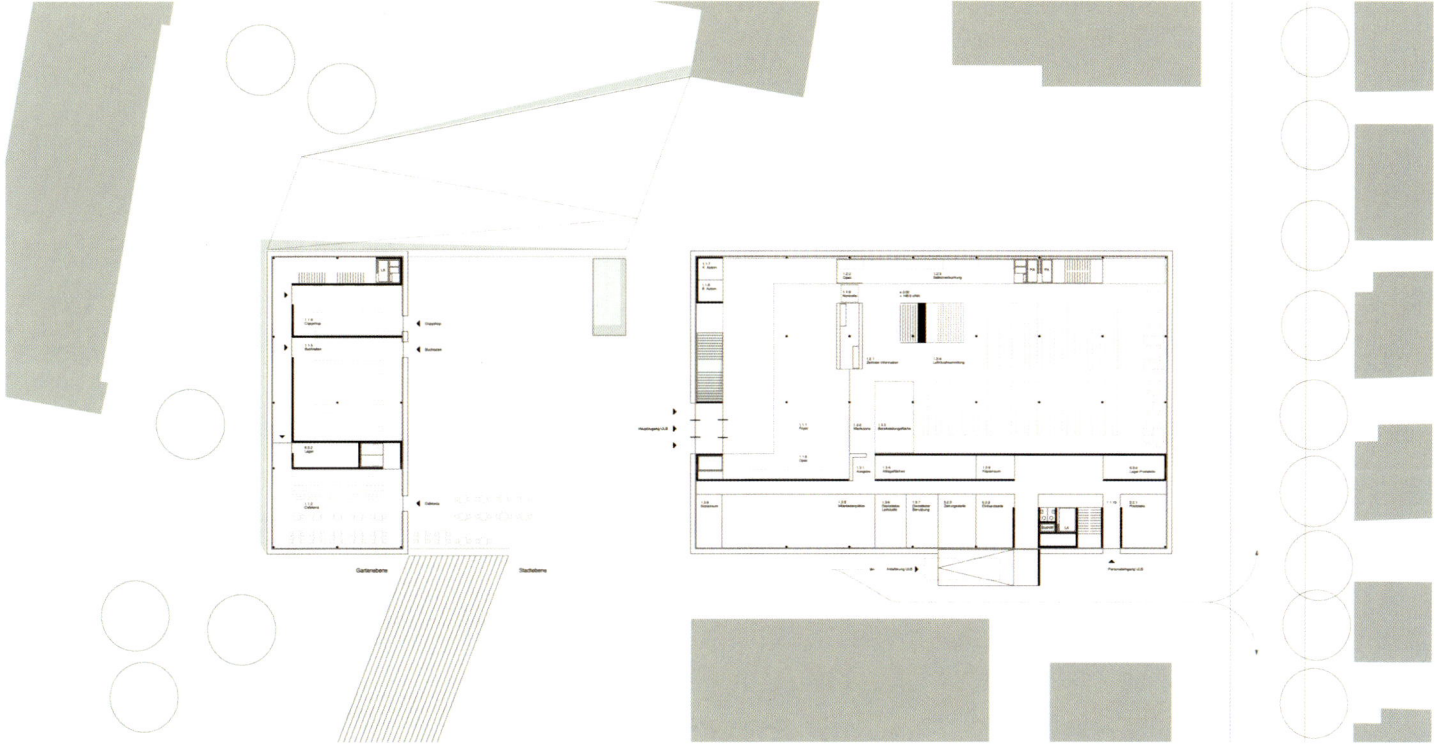

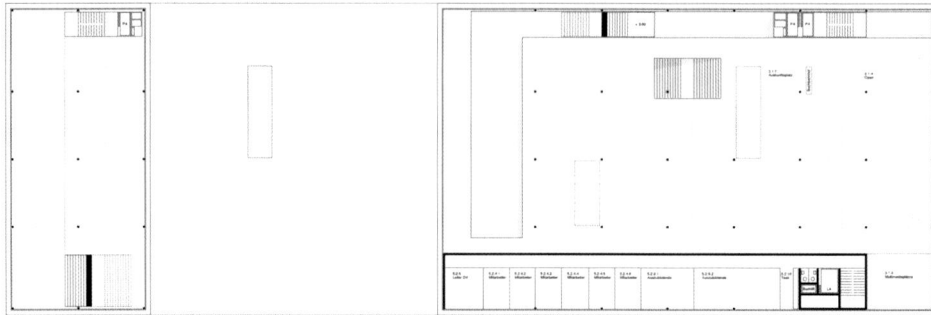

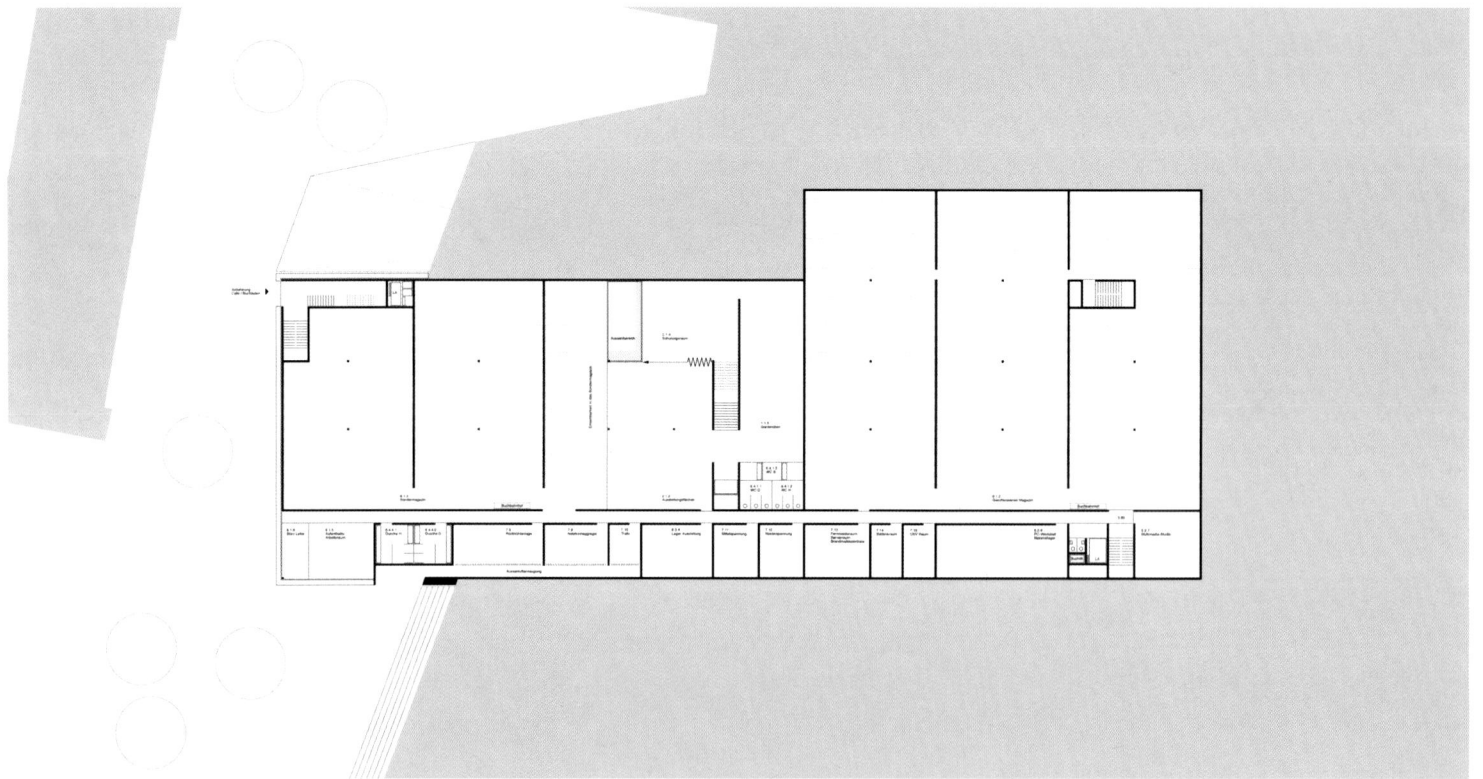

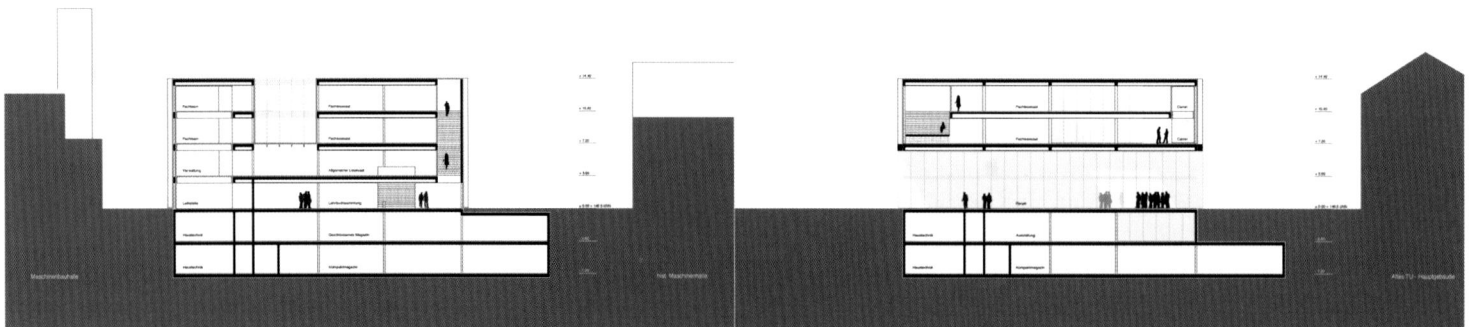

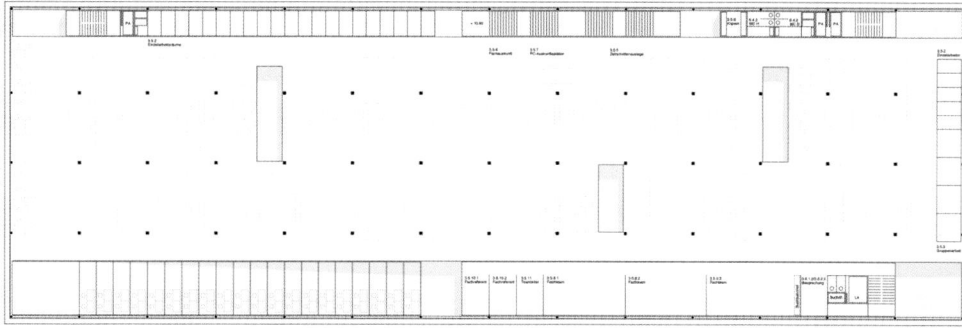

3. obergeschoss 1/200
detailansicht 1/40

Thomas Müller Ivan Reimann Berlin
Thomas Muller Ivan Reimann建筑事务所 柏林

"The new library integrates smoothly with the extant built environment and complements latent spaces and circulation routes."

"新图书馆流畅地与现有建筑环境融合一起，补充潜在的空间和区内交通线。"

作者 Ivan Reimann   合作伙伴 Erik Frenzel, Ferdinand Oswald, Edna Lührs, Thomas Möckel, Jens Böttcher   专家
landscape architecture: Jürgen Weidinger; fire protection: Büro Stanek; further: Alhäuser + König, H. Dunschmann; energy:
Transsolar GmbH, H. Auer

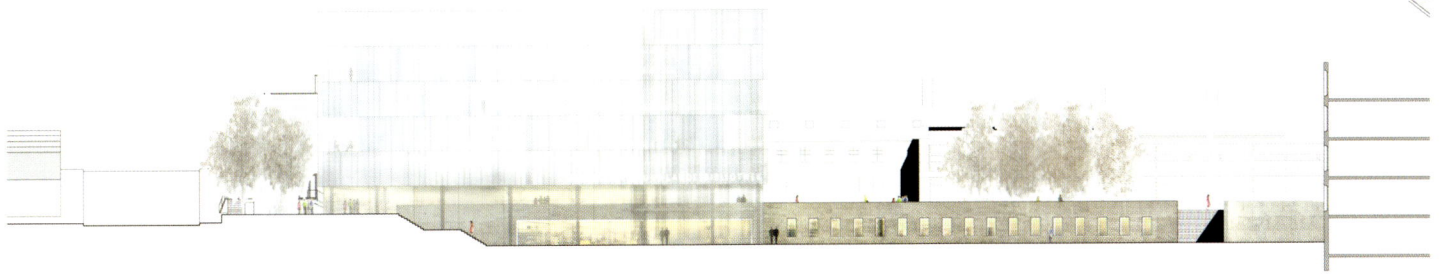

Ansicht Nord M 1:200

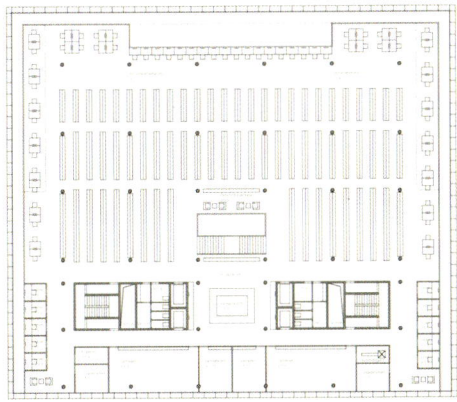

Technikwissenschaften 4.OG M 1:200

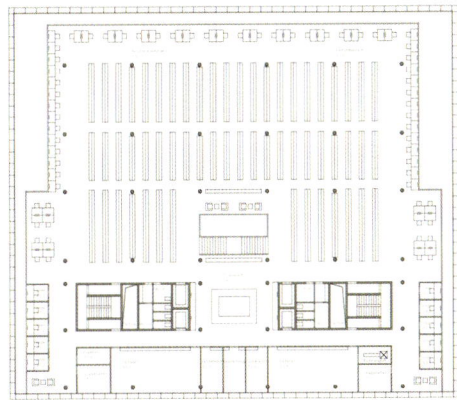

Technikwissenschaften 5.OG M 1:200

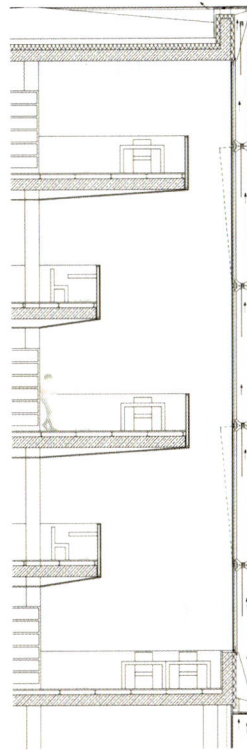

Fassadenausschnitt M 1:50

Gerber Architekten Dortmund
Gerber建筑事务所 多特蒙德

"From a built volume reduced to the most basic geometric form, the cube, the energy of a central building would radiate into the heterogeneous environs."

"从一个被缩减为最基本的立体形式——立方体的建筑物中，中心大楼的能量向四周多样化的环境中辐射。"

作者 Prof. Eckhard Gerber   合作伙伴及建筑系在校生 Sandra Kroll, Marius Puppendahl, Siegbert Hennecke, Van Hei Nyguen, Manuela Perz, Karsten Liebner, Benjamin Siebner, Lilian Panek, Vanda Govedarica; modelmaking: Henrik Hilsbos, Alexandra Kranert
专家 building services: DS-Plan AG, Hr. Moesle, Stuttgart; structural analysis: OSD, Prof. Kloft, Darmstadt; open space planning: Kienle Planungsgesellschaft Freiraum und Städtebau mbH, Stuttgart

BLICK AUF DIE STA

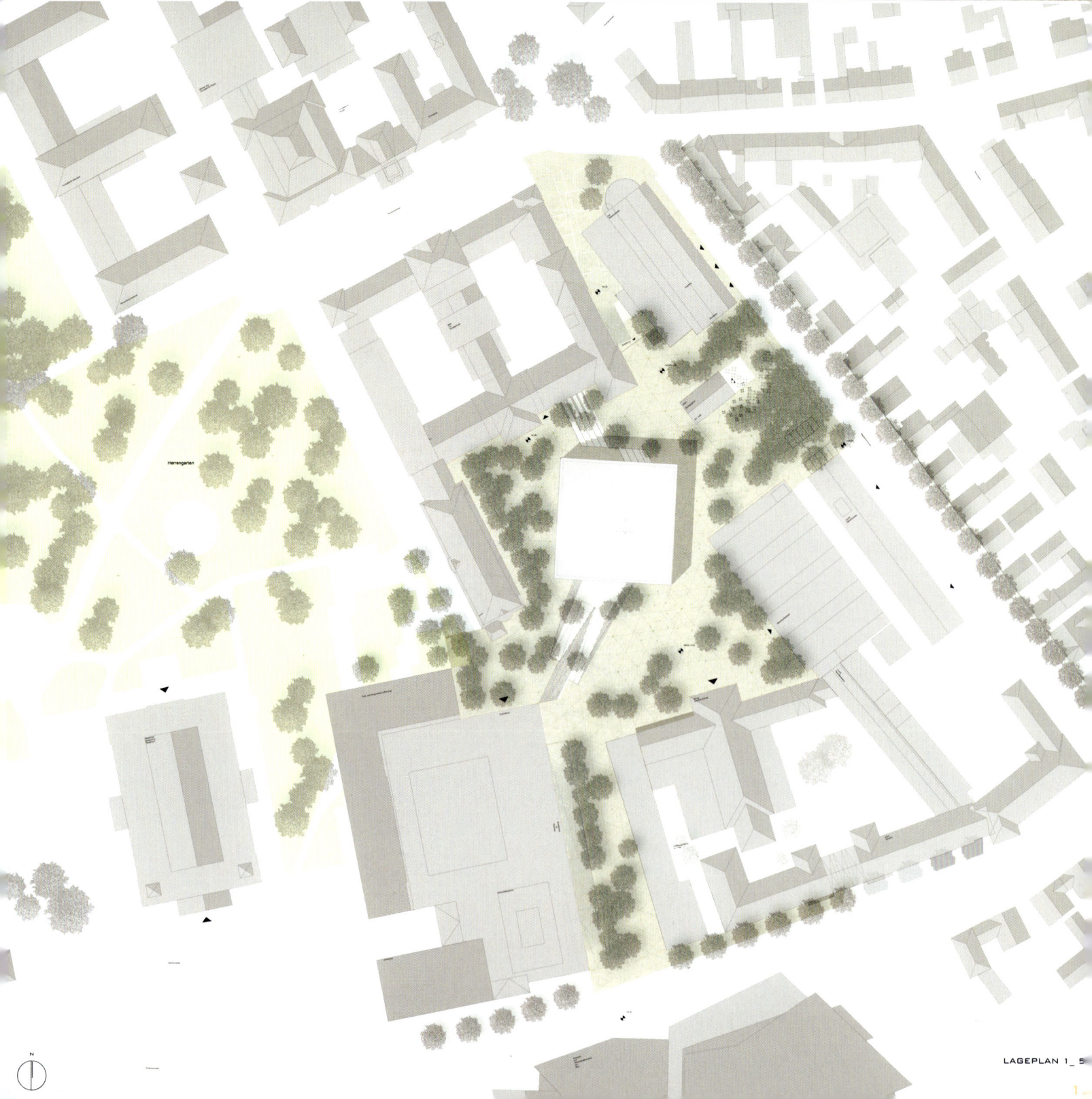

Herrengarten

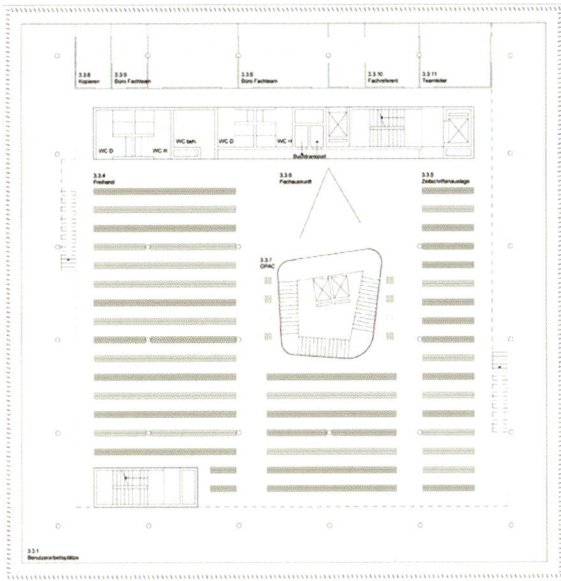

GRUNDRISS EBENE 03 1 _ 200

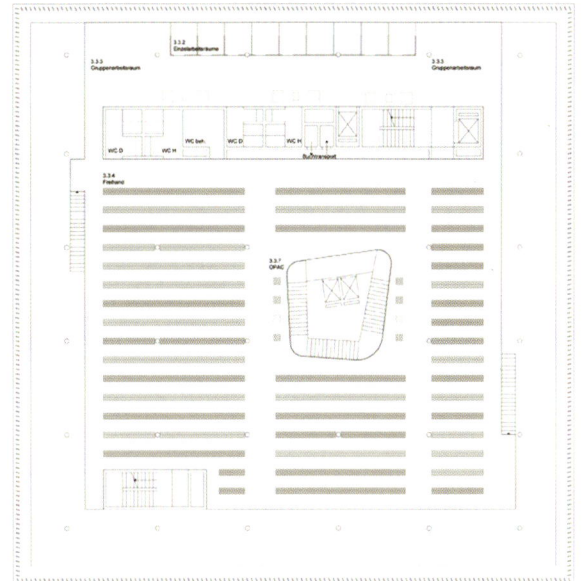

GRUNDRISS EBENE 04 1 _ 200

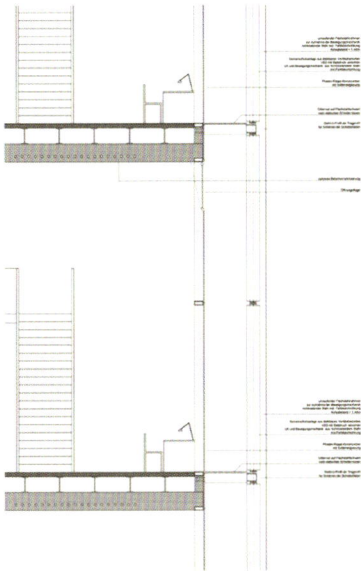

FASSADENDETAIL 1_50

LESESAALBEREICH

377

ANSICHT NORD 1_200

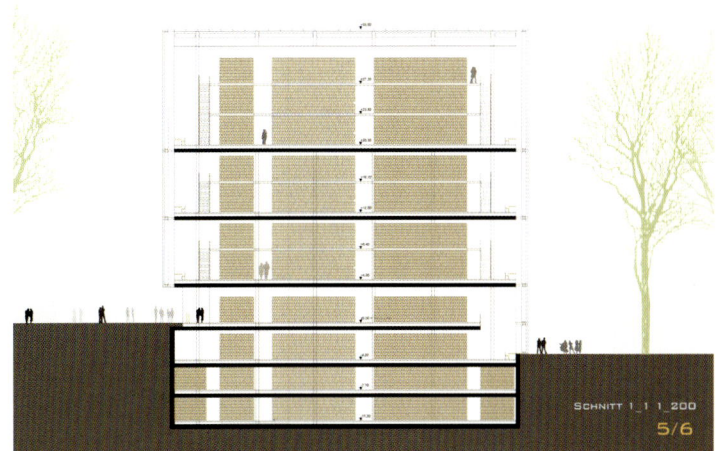

SCHNITT 1_1 1_200

5/6

agps Zurich
agps建筑事务所 慕尼黑

"To see and to be visible."
"看见，也可见。"

作者 Hanspeter Oester, Dr. Marc Angélil, Reto Pfenninger   合作伙伴 Denise Ulrich, Phil Steffen, Roger Naegeli, Thomas Summermatter, Andreas Kopp, Nelson Tam   专家 APT Ingenieure GmbH, Andreas Lutz, Zurich; Amstein+Walthert AG, Adrian Altenburger, Zurich

02

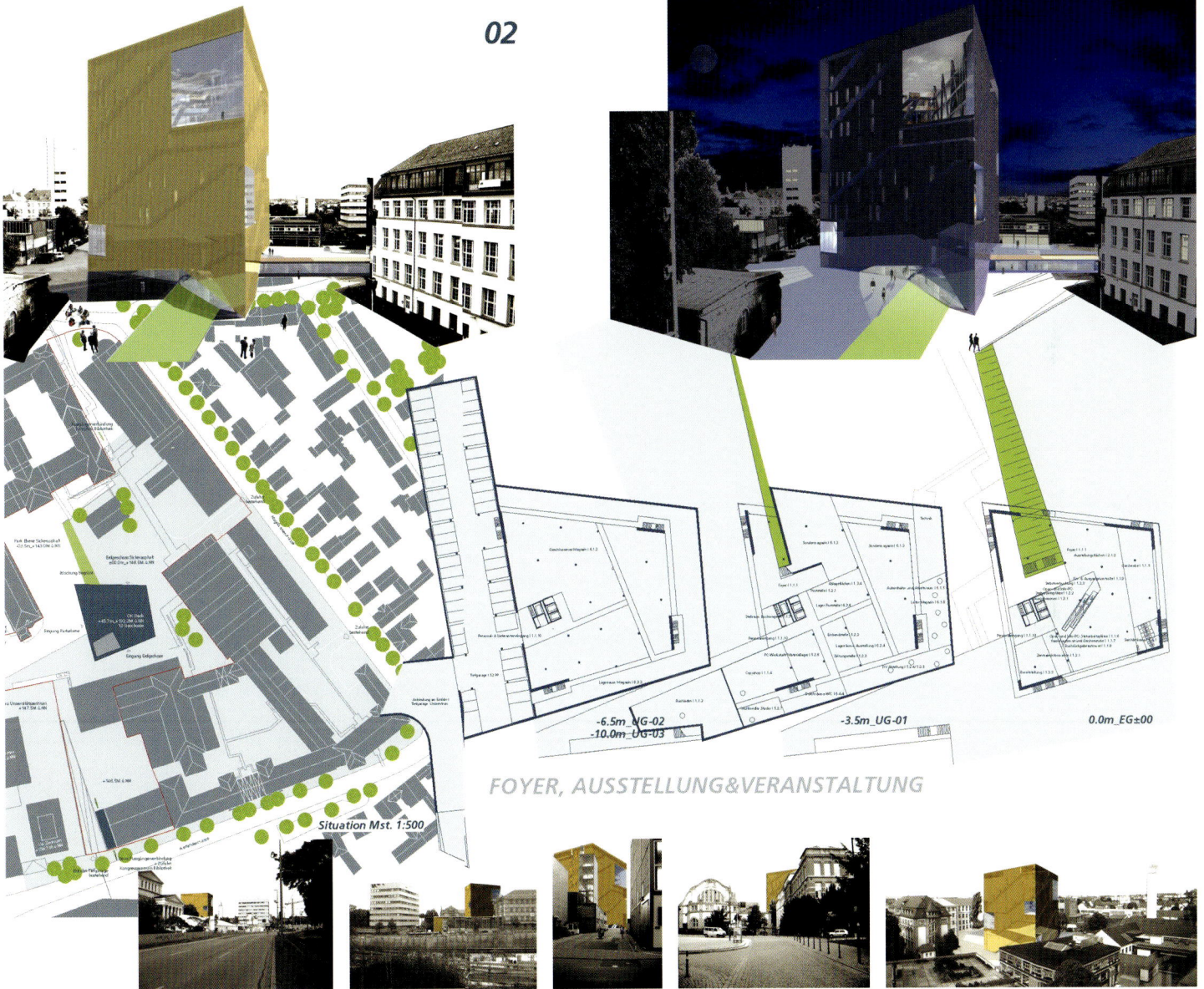

-6.5m_UG-02
-10.0m_UG-03

-3.5m_UG-01

0.0m_EG±00

FOYER, AUSSTELLUNG&VERANSTALTUNG

Situation Mst. 1:500

N

147258

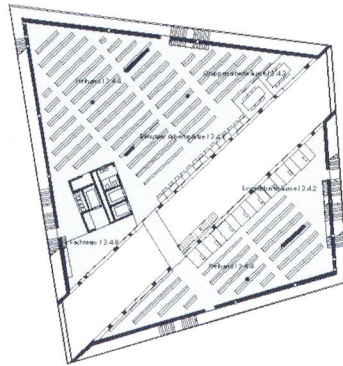

24.5m_OG+07

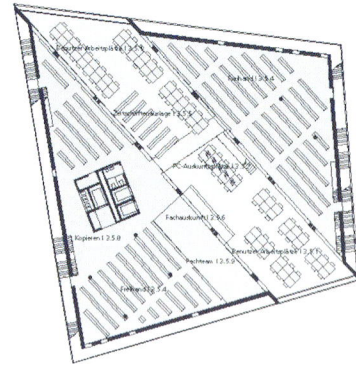

28.0m_OG+08

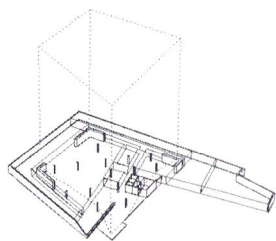

Untere Eingangshalle

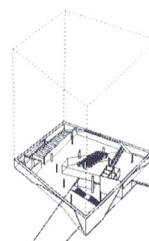

Obere Eingangshalle

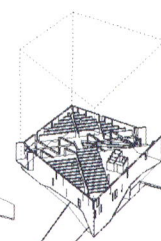

Allgemeine Bibliothek

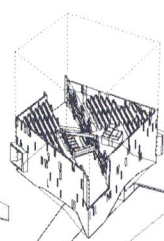

Lesesaal Recht&Wirtsch

Universitäts- und Landesbibliothek Darmstadt   Realisierungswettbewerb 2. Phase   November 2005

OG+09

35.0m_OG+10

38.5m_OG+11

42.0m_OG+12

Saal Humanwissenschaften   Lesesaal Natur&Technik   Dachgeschoss

N

0   10   25

Fassadenschnitt Mst. 1:50

45.7m_OK Dach

Cafeteria

42.0m_OG+12

Lesesaal

28.0m_OG+08

Bibliothek

24.5m_OG+07

Bibliothek

21.0m_OG+06

Bibliothek

Schulungsraum 12.1.4

Technik

Besprechungs & Sozialraum 1 2.6.1/2.6.2

Lager Cafeteria 16.2.2.2

Cafeteria 11.1.2

Terrasse

10.5m_OG+03

Bibliothek

7.0m_OG+02

Bibliothek

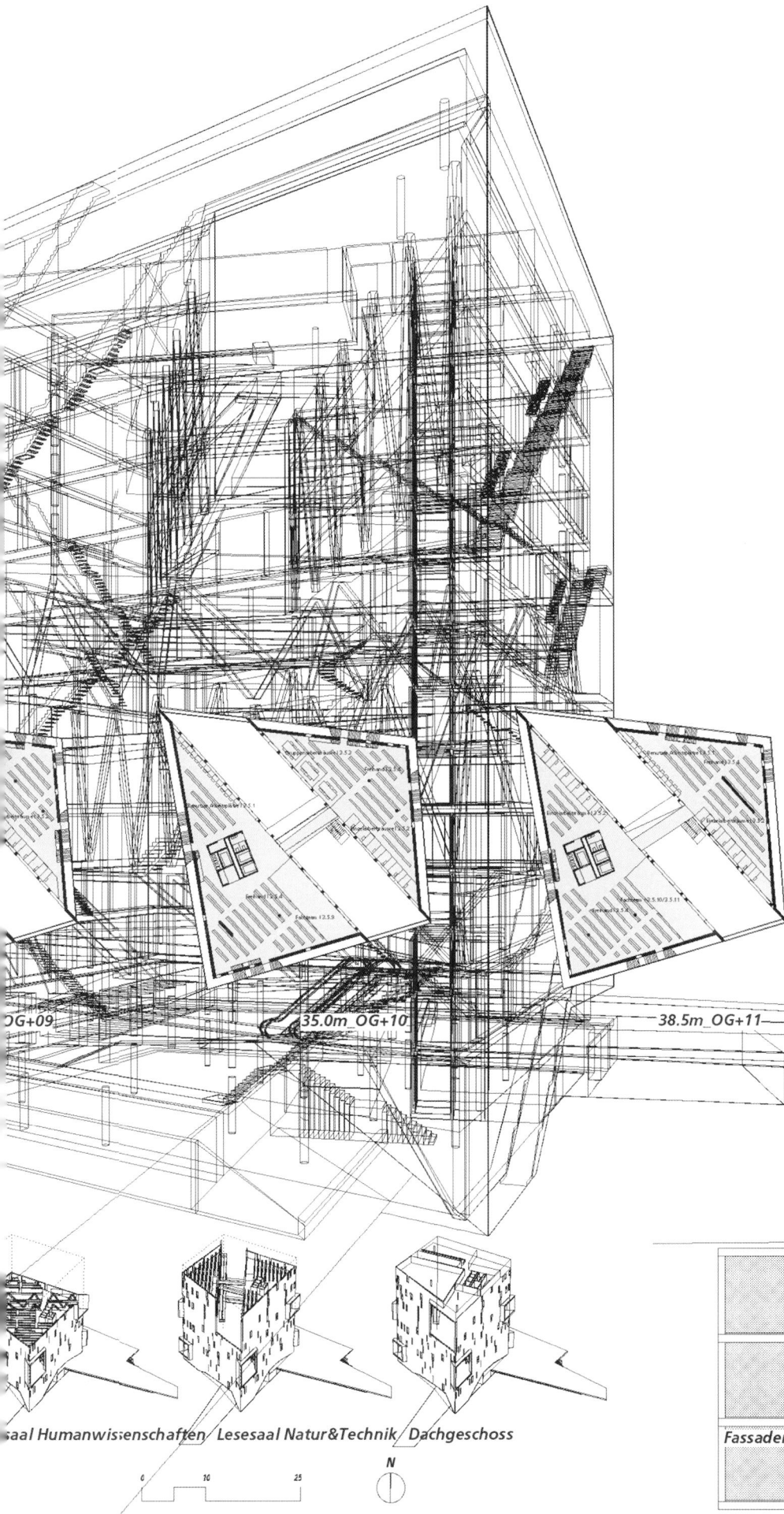

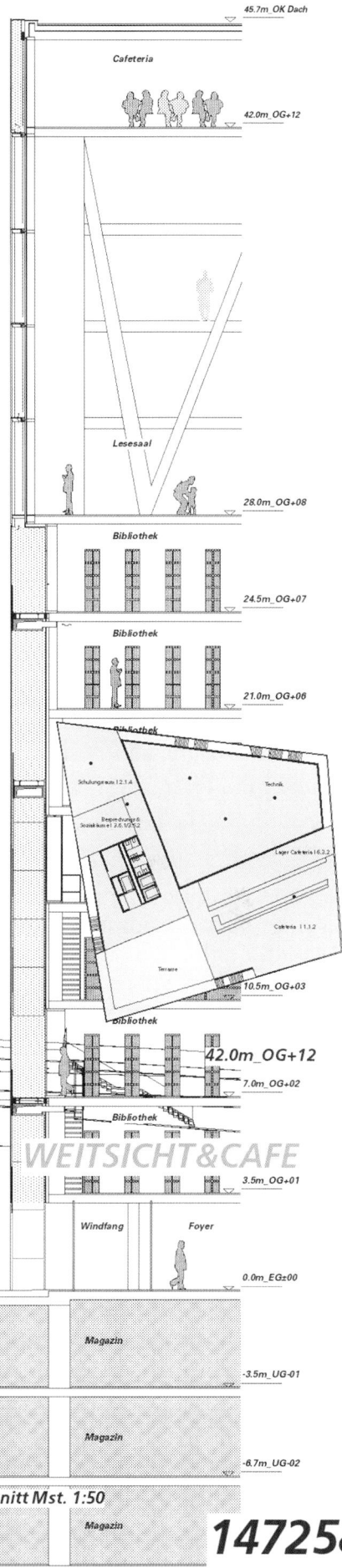

WEITSICHT&CAFE

3.5m_OG+01

Windfang   Foyer

0.0m_EG±00

Magazin

-3.5m_UG-01

Magazin

-6.7m_UG-02

Magazin

147258

SCHULTES FRANK ARCHITEKTEN Berlin
SCHULTES FRANK建筑事务所 柏林

"'Order is' – Louis Kahn is right, and even if he were wrong, here, in the chaos around the old TUD building, order is definitely the planner's foremost obligation."

"'秩序使然'——路易斯康是正确的，即使在达姆施塔特老楼里他错了，秩序无疑是规划人员最重要的义务。"

作者 Axel Schultes, Charlotte Frank   合作伙伴及建筑系在校生 Fritz Lobeck, Andreas Schuldes   专家 building services: HL-Technik GmbH, Prf. Dr. Klaus Daniels, Munich

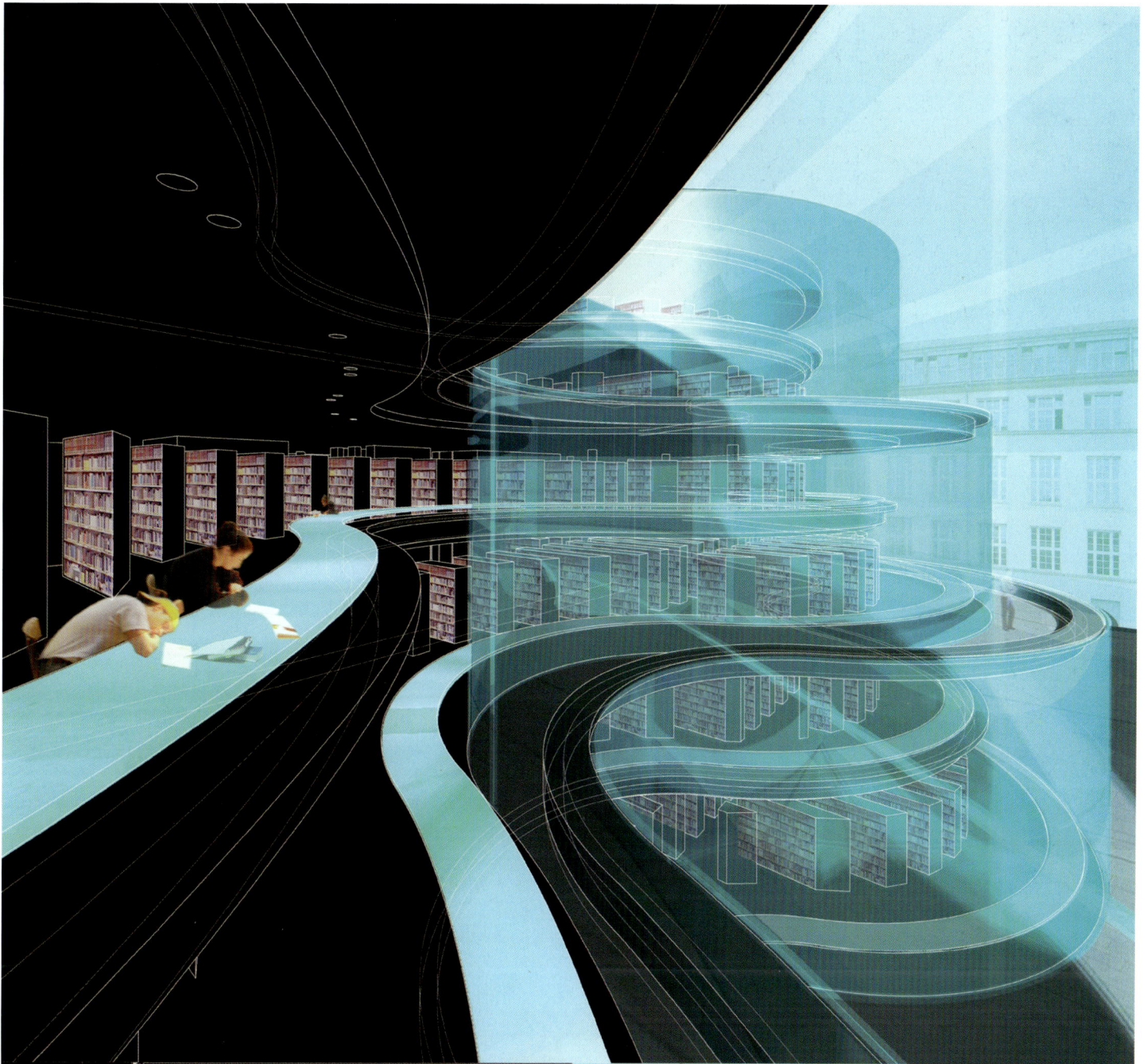

Fassade Schnitt, Ansicht 1:20

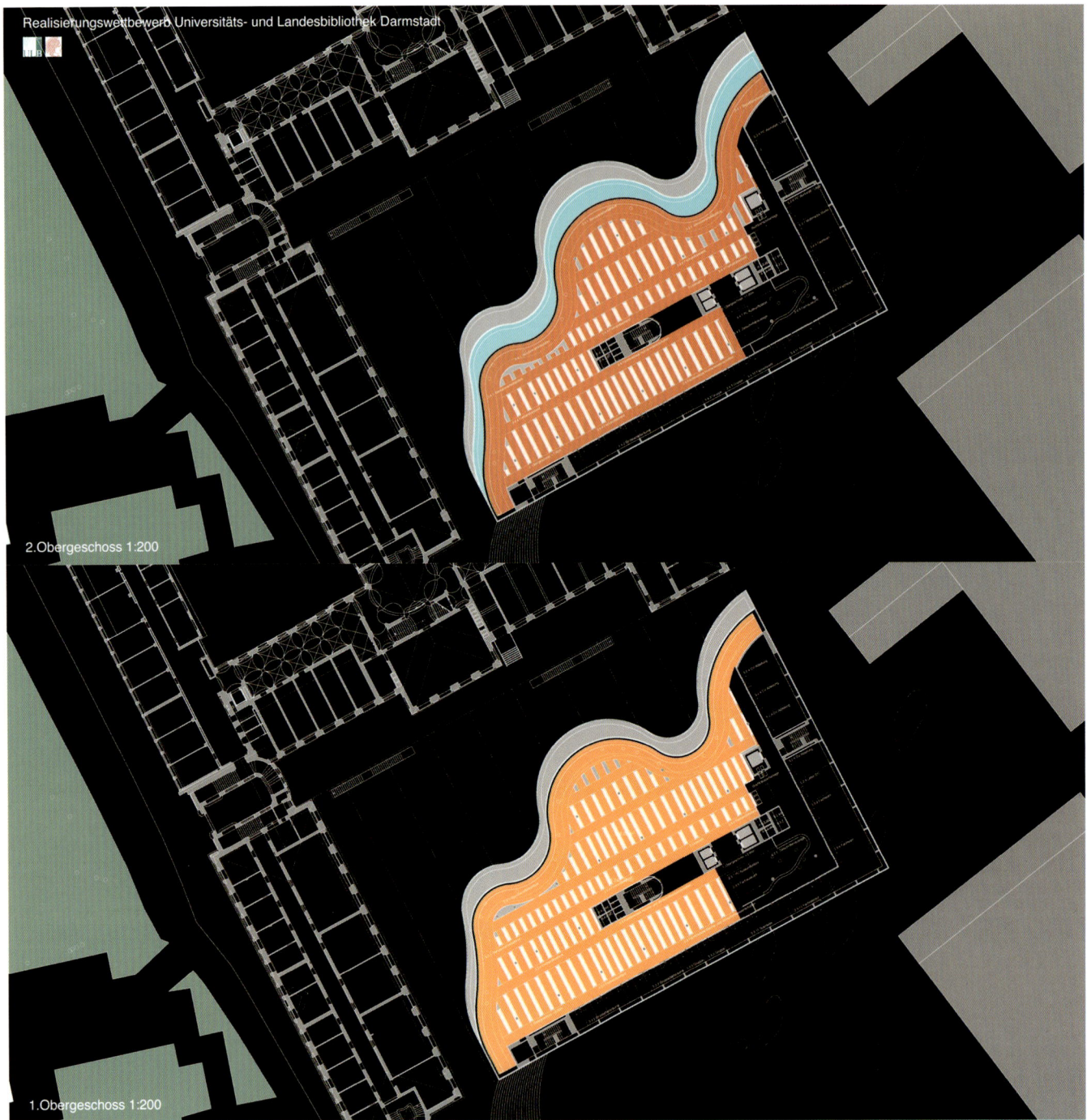

2.Obergeschoss 1:200

1.Obergeschoss 1:200

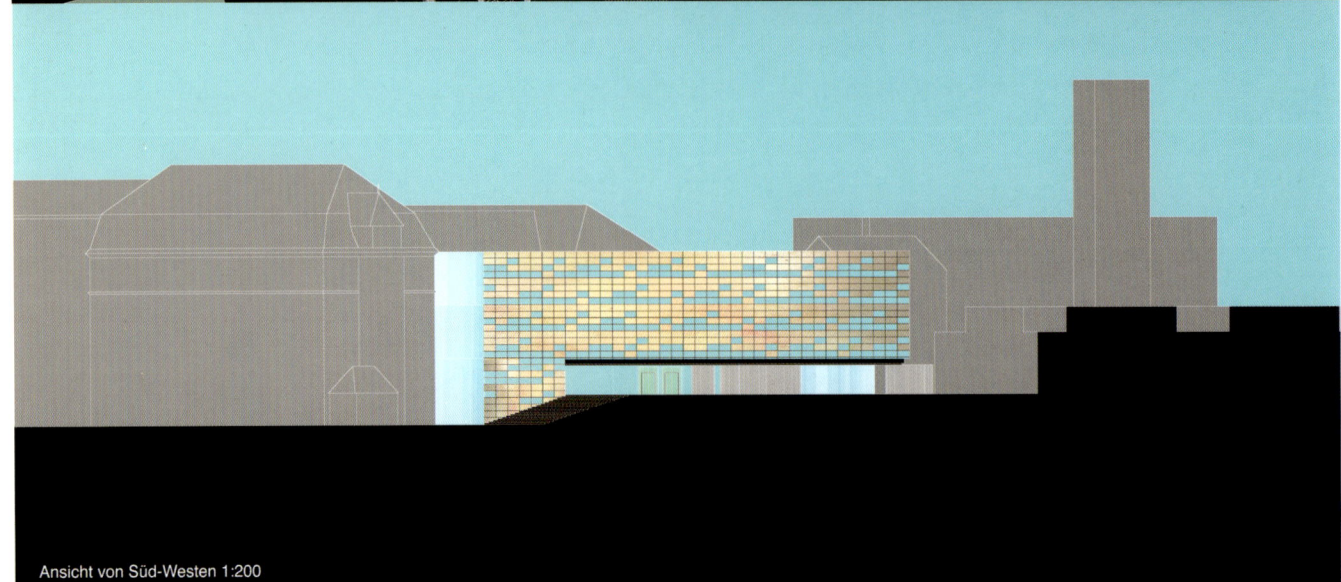

Ansicht von Süd-Westen 1:200

Lageplan 1:500

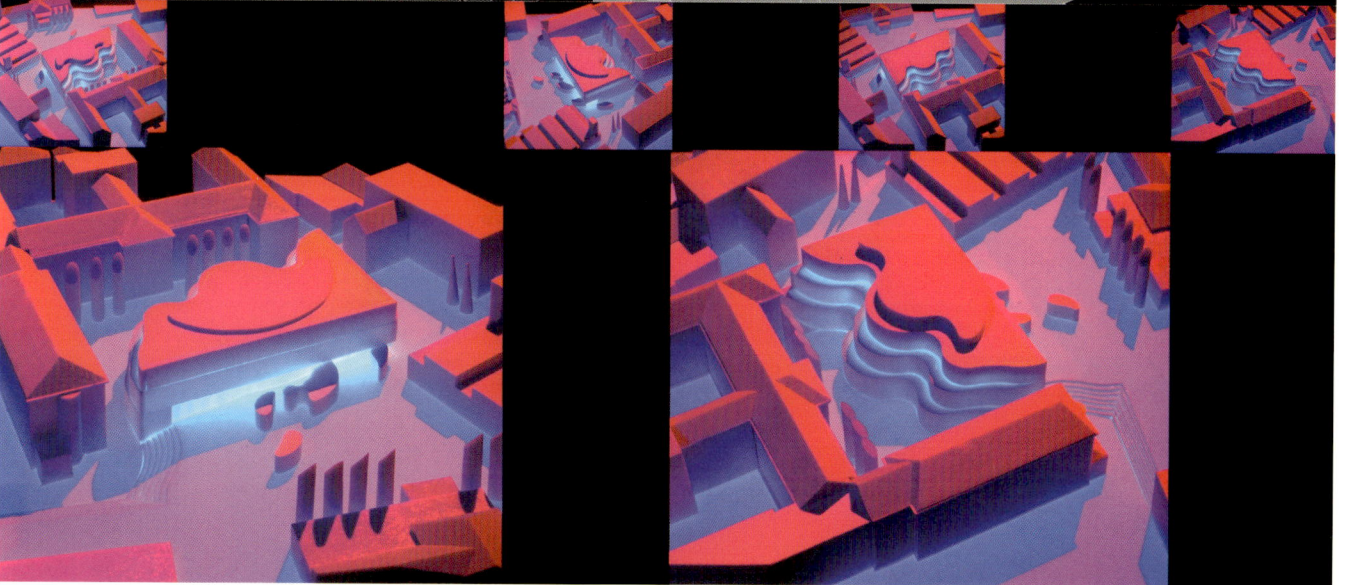

LOVE architecture and

Martin Mechs
Architekturbüro U

thread coll

urbanism

Southbank Project in Stellenbosch

南部海岸项目，南非斯泰伦博斯

with

Tischler

ctive & normaldesign

Southbank Project Stellenbosch, South Africa
南部海岸项目 南非斯泰伦博斯

Open, two-stage project competition
公开项目竞赛，分为两个阶段。

Competition site

地点 Stellenbosch, South Africa   时间 07/2006-09/2006   主办方 Spier Holdings   参赛者 1st stage: 96; 2nd stage: 6
面积100 ha   竞赛费用 225,000 USD   专业评奖委员会 Adrian Enthoven, Tanner Methvin, Spier Holdings, South Africa; Ikem Stanley Okoye, Ni-
geria/USA; Luyanda Mphalwa, South Africa; Michael Keniger, Australia; Mike Rainbow, UK; Salah Hassan, curator and
art historian, Sudan/USA; Anne Lacaton, France

As yet, Africa has no single institution representing the culture of the entire continent under one roof. When the sponsors of this competition, old-established wine-makers in South Africa's western cape province, decided to build "Southbank", a new residential development, they also set themselves the ambitious goal of making the "Africa Centre" its centerpiece. The obvious plan was to run the procedure for this project as an open competition; the management was entrusted to [phase eins]. and the University of Witwatersrand, Johannesburg.

The plan called for a usable area of 150,000 square metres, mainly for various forms of residential uses, on an 80-hectare site embedded in a valley landscape. The Africa Centre proper was to cover approx. 11,000 square metres of the usable area. Furthermore there would be a guesthouse, retail, sports facilities, school, nursery school, etc. The project committed itself to a high standard: A vibrant blend of residential, artistic and cultural functions, a meeting place for traditional and contemporary art and culture, an amalgam of museum and culture centre set in a community where artists can reside temporarily or permanently. To view the result, please visit *www.southbank.co.za*.

至今非洲还没有一个建筑物能够代表整个非洲大陆的文化的。当竞赛主办方，南非西部开普省的一家历史悠久的葡萄酒制造商决定建起一个新的居住区"南部海岸"的时候，他们决心把"非洲中心"做成一个中心。确定项目竞赛采取公开的形式，管理协调工作委[phase eins].公司和约翰内斯堡金山大学进行。

该项目规划需要在藏于山谷风景中一块80公顷的地上留出15万平方米的使用面积，主要用作各种住宅使用。非洲中心拟占1.1万平方米的使用面积。此外，还会建一家宾馆、商店、运动设施、学校、幼儿园等。项目要求极高，是一个集住宅、艺术和文化功能为一体的充满活力的建筑，是一个传统和现代的艺术和文化的碰撞处，也是一个置身风景如画社区里的博物馆和文化中心的混合物。若要浏览项目结果，请访问网站www.wouthbank.co.za

Aerial view　鸟瞰图

Competition site　竞赛地点

H. Prins, A. Lacaton, L. Mphalwa, S. Hassan, M. Rainbow, I.S. Okoye

Southbank Project, Stellenbosch

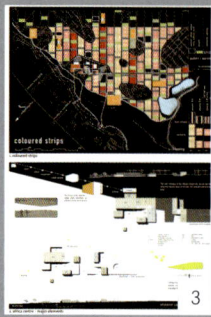

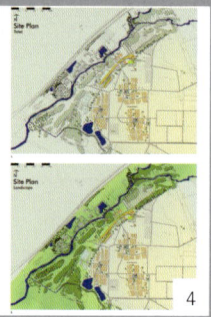

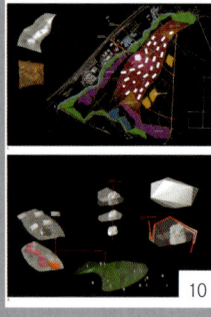

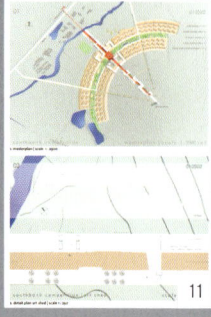

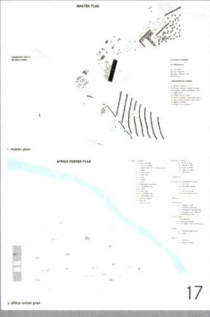

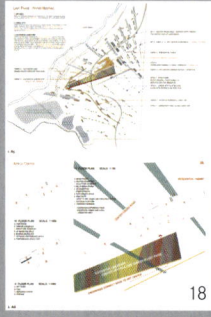

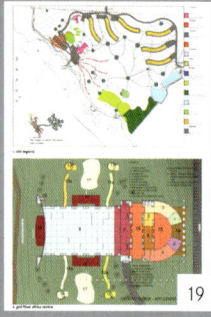

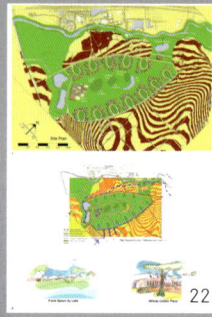

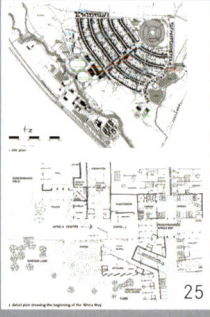

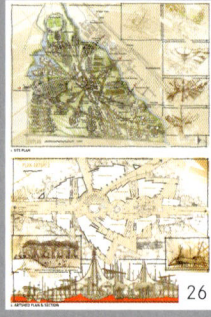

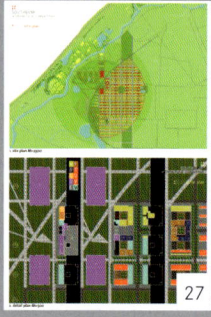

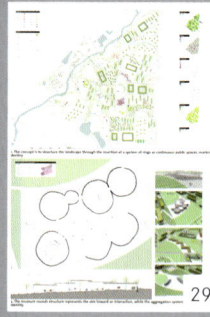

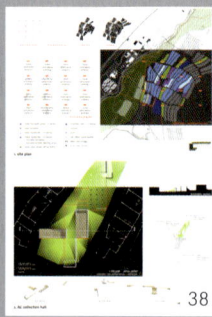

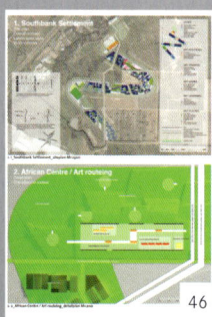

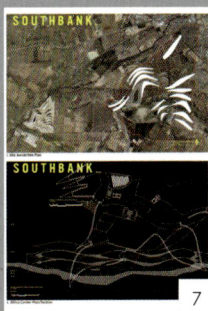

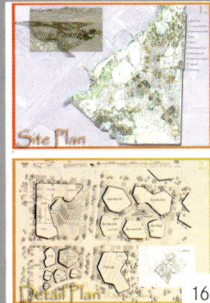

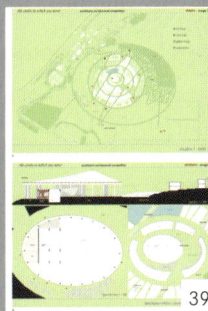

Participants 1st stage 第一阶段参赛者

1 constructconcept, Berlin, Germany 2 LOVE architecture and urbanism, Graz, Austria 3 Martin Mechs with Architekturbüro Uli Tischler, Graz, Austria 4 Gerber Architekten, Dortmund, Germany 5 MATSUOKASATOSHITA MURAYUKI, Tokyo, Japan 6 1j2b architects, Fribourg, Switzerland 7 Qua'Virarch, Chicago, Illinois 8 MODIStudio_Associati, Campobasso, Italy 9 GreenhilLi Design PTE LTD, Singapore 10 SHEPPARD ROBSON LTD, London, UK 11 gaudlarchitekten, Dessau, Germany 12 Studio 3 Architects/Planners, Hout Bay, South Africa 13 James Atkinson, Edinburgh, UK 14 Technum NV, Hasselt, Belgium 15 Eugen Ulirsch, Zurich, Switzerland 16 Red Landscape Architects (Pty) Ltd, Pretoria, South Africa 17 Christophe Hutin, Bordeaux, France 18 Hsin-Hung Tsao, Astoria, New York 19 r. van wezel and associates, Joburg, South Africa 20 Bollati architects, Montevideo, Uruguay 21 Mashabane Rose Associates cc; Johannesburg, South Africa 22 Dalhousie University, West Pennant, Canada 23 Luca De Gol, Hugo Castaneda, Alvaro Corredor, Helsinki, Finland 24 The Workplace Development Firm CC, Port Elizabeth, South Africa 25 Harber & Associates, Durban, South Africa 26 AJ Architects, Kapstadt, South Africa 27 Louis Krüger, Adelfia, Italy 28 Trace and associates, Johannesburg, South Africa 29 A+P architettura, Rom, Italy 30 Nightingale Associates, Kapstadt, South Africa 31 Edina Osmanovic, Mannheim, Germany 32 Total Design + Associates, Castries, Saint Lucia 33 STUDIO EGRET WEST, London, UK 34 Andrade Morettin, Sao Paulo, Brazil 35 Verzone Woods Architectes, Rougemont, Switzerland 36 Santos Prescott & Associates, San Francico, USA 37 Junya Toda Architect & Associates, Osaka, Japan 38 normaldesign/thread collective, Brooklyn, New York 39 rabaschus und rosenthal, Dresden, Germany 40 Dodi Moss Srl, Milan, Italy 41 sitengineering srl, Vigevano, Italy 42 SLAG, Firenze, Italy 43 Frédéric Haesevoets, Brüssel, Belgium 44 the idom group, London, UK 45 Magdalena Szczypka, Tychy, Poland 46 Worldlab, Århus, Denmark 47 Masauso Branda, Harare, Zimbabwe 48 Architecton Design Studio, Harare, Zimbabwe

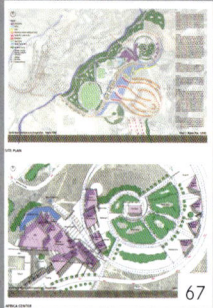

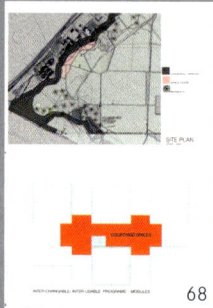

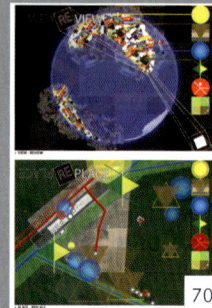

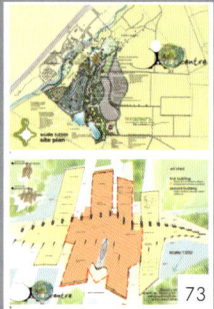

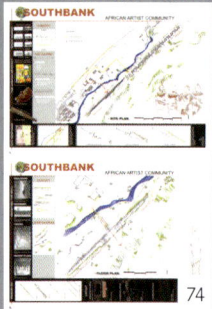

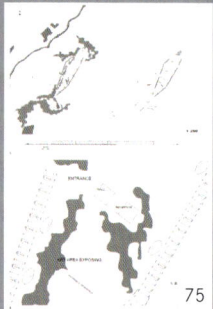

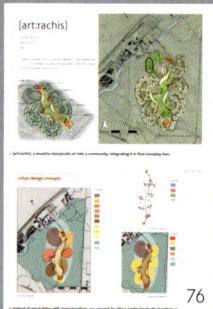

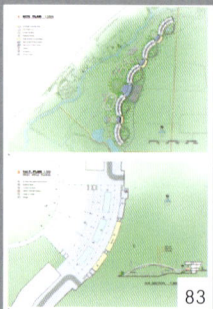

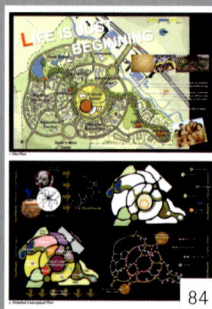

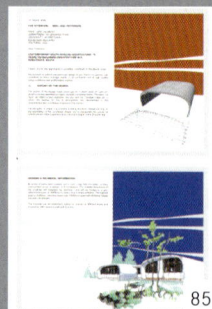

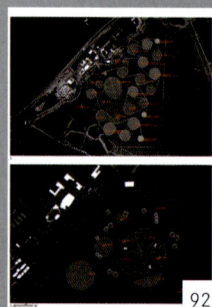

55

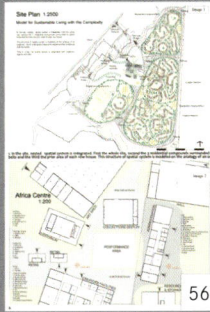

56

63

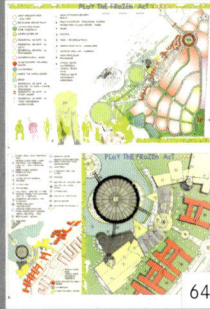

64

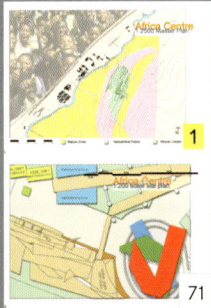

71

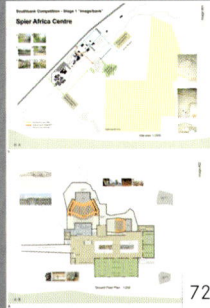

72

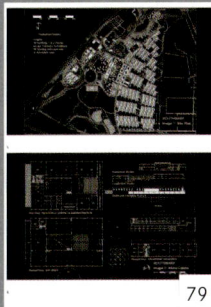

79

86

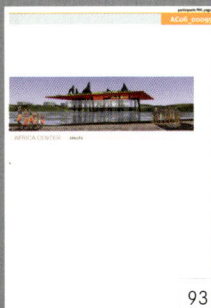

93

49 buero blickpunkt, Berlin, Germany **50** konyk architecture pc, Brookly, New York **51** GROUP A, Rotterdamm, Netherlands **52** Takayuki Kamei, Fukuoka, Japan **53** motsepe architects, Johannesburg, South Africa **54** Aliso Odinyo, UNIVERSITY OF NAIROBI – architecture department, Nairobi, Kenya **55** Antarctica Group, Melbourne, Australia **56** Daisuke Matsushita, Kyoto University, Japan **57** François Machado, Poilhes, France **58** Atelier Sanna Lahti, Helsinki, Finland **59** Albonico Sack Mzumara Architects & Urban Designers, Johannesburg, South Africa **60** GLAM+; Johannesburg, South Africa **61** arcmatic, Johannesburg, South Africa **62** studio83, Jakarta, Indonesia **63** MGF ARCHITEKTEN GMBH, Stuttgart, Germany **64** BH Architects, Pretoria, South Africa **65** Pieter Jooste, Joannesburg, South Africa **66** Wesley Hindmarch, Hobart, Australia **67** Bentley Faulmann Architects, Yellowknife, Canada **68** Habib Chanzi, Melbourne, Australia **69** Patchara Jumpangen, Nonthaburi, Singapore **70** SCAD, Savannah, South Africa **71** Savannah college of Art and Design, Savannah, South Africa **72** Smit & Serfontein Architects Pty (Ltd), Santa Carina, Brazil **73** umar.pareja architecture, Alexandria, South Africa **74** Gary White and Associates, Pretoria, South Africa **75** Juan Carlos Moreno Coriat, Santiago de Cali, South Africa **76** studioMAS, Johannesburg, South Africa **77** Tim Wippich, Hannover, Germany **78** Manuela Perz, Munich **79** D Hirschman, Architects, Kapstadt, South Africa **80** Wiswedel Architects, Düsseldorf, Germany **81** Wolf & Wolf Architects, Kapstadt, South Africa **82** Patrick Kathuli, Mwingi, Kenya **83** Zhou Zhenxiong; Shenzhen, China **84** Realrich Sjarief, London, UK, UK **85** Philip Kilian, Pretoria, South Africa **86** Thomas C Falck Architects, Worcester, UK **87** CMSP Arquitetura + Design, Florianópolis, Brazil **88** ecoedge architecture, Richmond, Australia **89** Valentin Oleynik, St. Petersburg, Russia **90** Quagga Group, Kapstadt, South Africa **91** Sean Mahoney, Kapstadt, South Africa **92** Kohlmayer Oberst Architekten, Stuttgart, Germany **93** Mohsen Sane'i, Tehran, Iran

thread collective & normaldesign Brooklyn, USA
Thread collective & normaldesign设计工作室   美国布鲁克林

"The collage integrate existing objects, materials and fragments, creating new forms, facilitating a dialog between parts, and revealing unexpected meanings; it is this concept that spatially organises and links the three elements of the South-bank design: residential fabric, landscape and the Africa Centre."

"拼贴把现有物体、材料和片段合为一体，创造出新形式，利于部分之间的对话，也表示意想不到的意义；正是这种概念把南部海岸项目设计的三个要素，即居住建筑物、景观和非洲中心从空间上组织起来。"

作者 thread collective: Elliott Maltby, Mark Mancuso und Gita Nandan; normaldesign: Matthias Neumann   建筑系在校生 Nazia Aftab   专家 modelmaking: Michael Caton + Jerome Barbu; renderings: John Watson; green consultant: Lauren Gropper, solar + wind calculations: solar e systems

Southbank Project, Stellenbosch

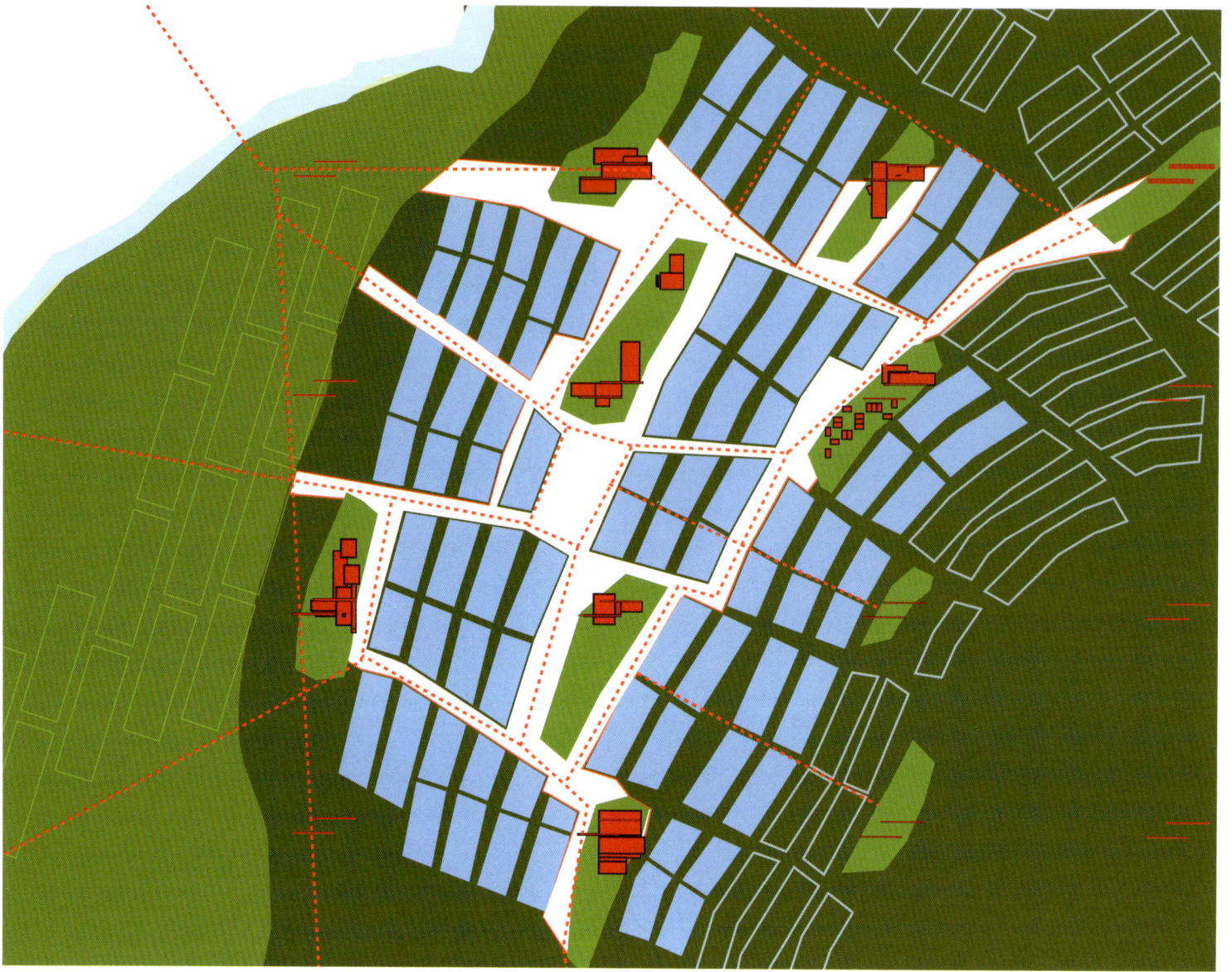

sustainability diagram

in contrast the africa center, the residential fabric is subtle in its display of sustainability, while maintaining the same degree of energy efficiency and water conservation

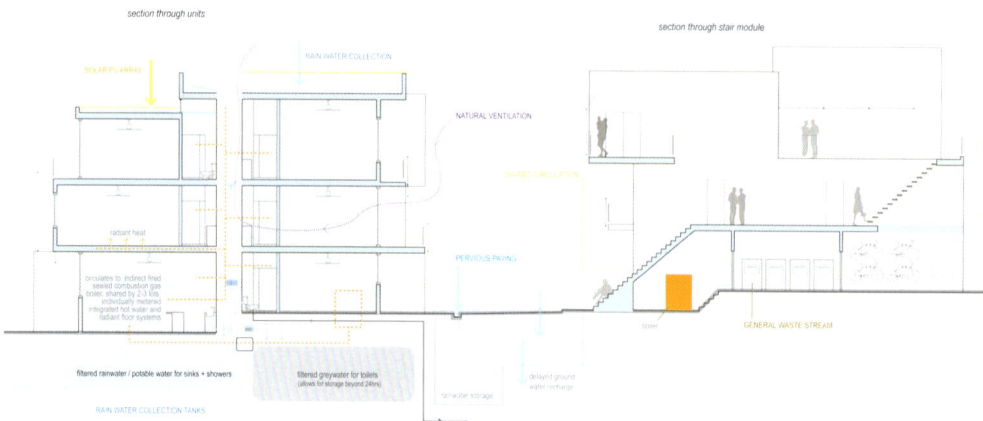

section through units

section through stair module

RAIN WATER COLLECTION

SOLAR PV ARRAY

NATURAL VENTILATION

radiant heat

PERVIOUS PAVING

circulates to indirect fired
sealed combustion gas
boiler, shared by 2-3 lots,
individually metered
integrated hot water and
radiant floor systems

GENERAL WASTE STREAM

filtered rainwater / potable water for sinks + showers

filtered greywater for toilets
(allows for storage beyond 24hrs)

RAIN WATER COLLECTION TANKS

delayed ground
water recharge

general water saving measures

low-flow showerheads and tap aerators installed throughout
no bath tubs
toilets use recycled grey water, not low flush to ensure proper
water levels at treatment dam
energy efficient dishwashers
accessible rooftops have small water barrel for plant irrigation

general energy efficient measures

domestic solar hot water heaters
fans, natural ventilation and solar orientation eliminate need
for air conditioners
radiant heat installed throughout
all lighting hardwired compact flourescent lights
solar orientation minimizes solar heat gain, yet maximizes
natural light
adobe bricks have strong thermal mass properties

LOVE architecture and urbanism Graz

LOVE建筑及城市规划事务所 格拉茨

"Each of the built volumes is wing-shaped, thus creating a feeling of innumerous wings sweeping the vast natural land-scape. Together, they form an urban pattern reflecting the expanse and vigor of the landscape."

"每个建筑物外形都很像翅膀，因此创造出无数翅膀飞过广袤自然景观的感觉。他们一起构成一种城市模式，反映景观的广阔和活力。"

作者 LOVE architecture and urbanism   合作伙伴 DI Gerald Brencic, Iulius Popa   专家 HKLS planer: La TEC GmbH Nfg Keg; landscape architecture: Koala; building physics: Müller; structural analysis: Ingenieurbüro Petschnigg

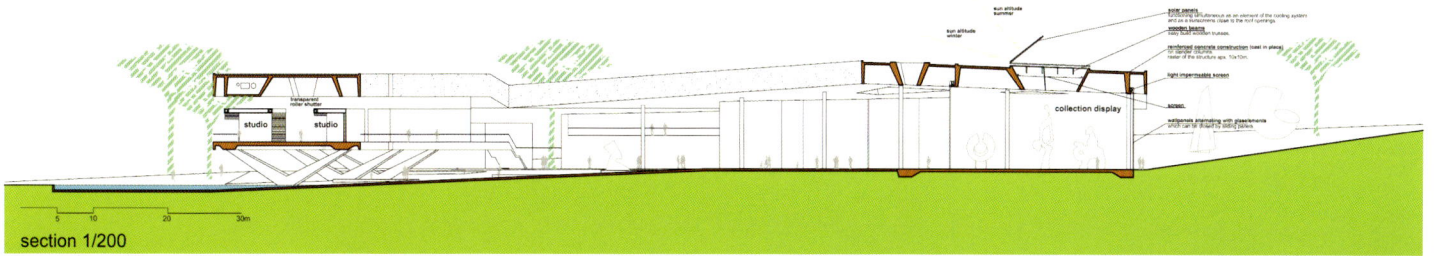

section 1/200

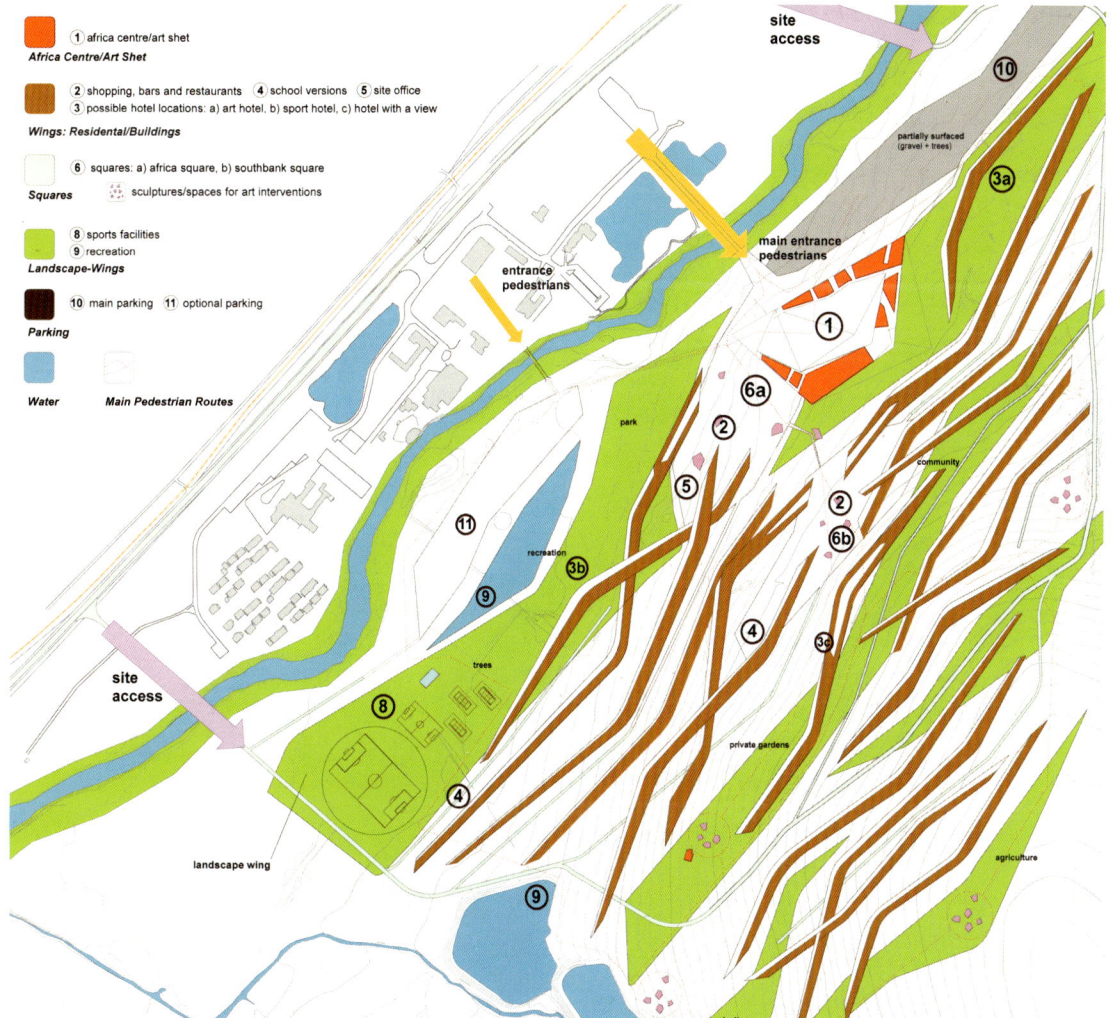

① africa centre/art shet
Africa Centre/Art Shet

② shopping, bars and restaurants ④ school versions ⑤ site office
③ possible hotel locations: a) art hotel, b) sport hotel, c) hotel with a view
Wings: Residental/Buildings

⑥ squares: a) africa square, b) southbank square
Squares sculptures/spaces for art interventions

⑧ sports facilities
⑨ recreation
Landscape-Wings

⑩ main parking ⑪ optional parking
Parking

Water **Main Pedestrian Routes**

Martin Mechs with Architekturbüro Uli Tischler Graz
Martin Mechs with Architekturburo Uli Tischler建筑事务所 格拉茨

"The 'colored strips', while signaling the acceptance of existing differences, are moved toghether closely to allow reciprocal perception 'across the alley' – and the potential for communication."

"'彩色纸条'在标志对存在差异的认可的同时被移近，允许'穿过小巷'这一互惠和沟通潜力的看法。"

作者 Martin Mechs with Architekturbüro Uli Tischler   合作伙伴 consultant: Felicitas Konecny; phase 1: Wolfgang Isopp, Christina Kimmerle; phase 2: Markus Hopferwieser, Christoph Wiesmayr, Herwig Marx, Jakob Winkler (renderings)   专家 landscape architecture: Thomas Proksch "Land in Sicht" Vienna; energy- and water supply; sustainability: Arge Energie AEE INTEC (Charlotta Isaksson, Christian Platzer); modelmaking: Patrick Klammer

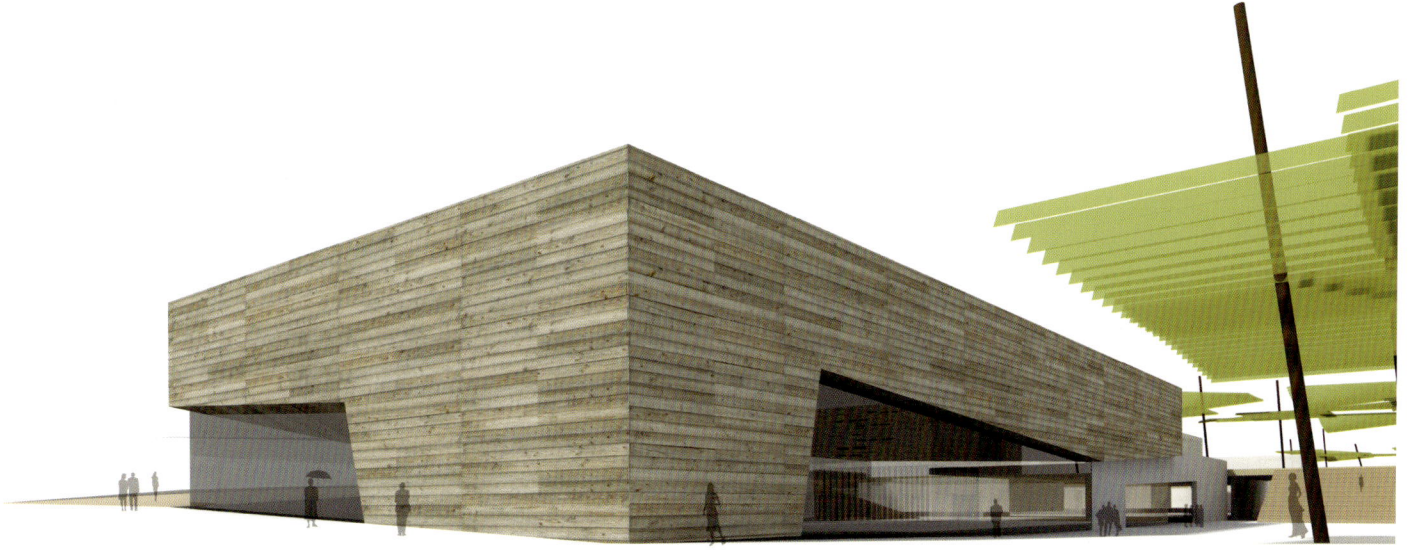

Martin Mechs with Architekturbüro Uli Tischler, Graz

Southbank Project, Stellenbosch

Ateliers Lion archited

Jafar Tukan Architects

Kisho Kurokawa

L

SCHULTES FRANK AR

von Gerkan, Marg un

tes urbanistes

Administration Complex in Tripoli
行政办公建筑区，利比亚的黎波里

rchitect & associates

on Wohlhage Wernik

HITEKTEN

l Partner

aha Hadid Architects

Administration Complex Tripoli
行政办公建筑区，利比亚的黎波里

Project competition for architects and city planners preceded by an application procedure
建筑师和城市规划人员的项目竞赛，开始前有申请程序。

Agricultural Land

Proposed Third Ring Road

Airport Highway

Forest

Proposed Train Station

Railway Track

Islamic World Call Society

Proposed Metro Line

Forest

G.M.M.R. Water Pipeline

Competition Site

Building
Forest
Agricultural Land

地点 Tripoli   时间 12/2006-6/2007   主办方 ODAC   参赛者 8   面积 615,000 sq m   竞赛费用 300,000 USD
专业评奖委员会 Craig Dykers, Oslo/New York; Guido Hager, Zurich; Prof. Rodolfo Machado, Boston; Prof. Matthias Sauerbruch, Berlin/London; Prof. Peter Zlonicky, chairman, Munich   其他专业评奖委员会 Prof. Bruno Sauer, Valencia
专家评奖委员会 Ali I. Dabaiba, managing director, organisation for development of administrative centers, Tripoli; Dr Mostafa Al Mezughi, chairman of general corporation for housing and infrastructure, Tripoli; Anwar A. Sassi, chairman of Brega & Ras Lanuf higher committee, Tripoli; Dr Ali Shebani, chairman of national consulting bureau, Tripoli   其他专家评奖委员会 Mohsen M. Ben Halim, national project coordinator, national consulting bureau, Tripoli; Dr. Ahmed M. Shembesh, director general, Libyan national center for standardisation and metrology, Tripoli

Aerial view　鸟瞰图

The governmental complex which will be the hub of a new business and administrative district in the years to come, will be situated approx. 7 kilometres south of the city center. Next to the site is where the highway to the international airport crosses Tripoli's third main ring road. The first task was to draw up a master plan (including landscaping and transportation grids) for a spacious open quarter that Libyans can identify with, complete with high-performing internal and external access routes, well connected to adjoining green spaces and other districts the city plans to develop in the future. Next, the project's most important buildings were to be designed in a style that would reconcile the regional vernacular (e.g. manual construction) with the visual language of contemporary architecture and at the same time address climatic challenges. The overall program called for a gross floor area of some 70,000 square metres including as key elements the parliament building, a conference palace with room for 1,500 delegates, and the buildings for the secretariat of the general people's committee, and the coordinating council of the people's leaderships. These are Libya's highest-ranking political functions after that of the top leader-ship. Another element of the program was a fivestar, 540-bedroom hotel where guests of the state and, during parliamentary sessions, the people's representatives will be lodged. Lastly, it called for designing buildings to house 20 government departments. In keeping with the comprehensive and ambitious nature of the program, a procedure to select candidates internationally preceded the actual competition which offered generous prizes to the winners.

Competition site　竞赛地点

Competition site　竞赛地点

该行政办公建筑区位于市中心南部7公里的地方。项目地点旁边是国际机场高速公司与的黎波里三环路的交叉处。竞赛任务之一是起草一份总平面图。其次，项目的重要建筑物的设计风格应该与该区的本土特点和谐一致，运用现代建筑的视觉艺术语言，同时也要适合当地气候。

整个项目规划需要约7万平方米的建筑总面积，包括议会大楼、能容纳1500名代表的会议厅、人民委员会秘书处和人民领导权协调委员会的办公大楼。规划的另一内容是一家有540间客房的五星级酒店。最后，项目还要求设计20家政府部门的办公大楼。

该项目综合性强、气势恢宏，因此在实际竞赛前有一个在全球范围内选择参赛候选人的程序，竞赛向获胜者提供丰厚的奖品。

View of the city　城市景色

1

2

5

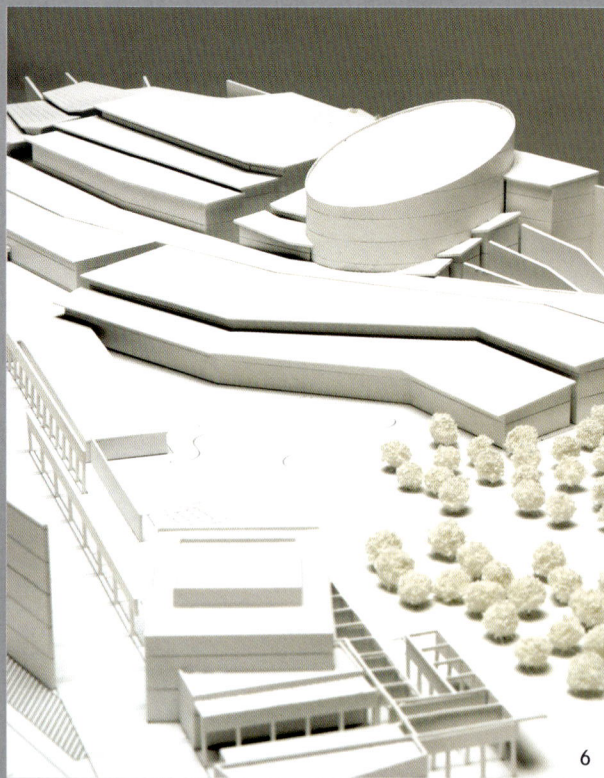

6

3

4

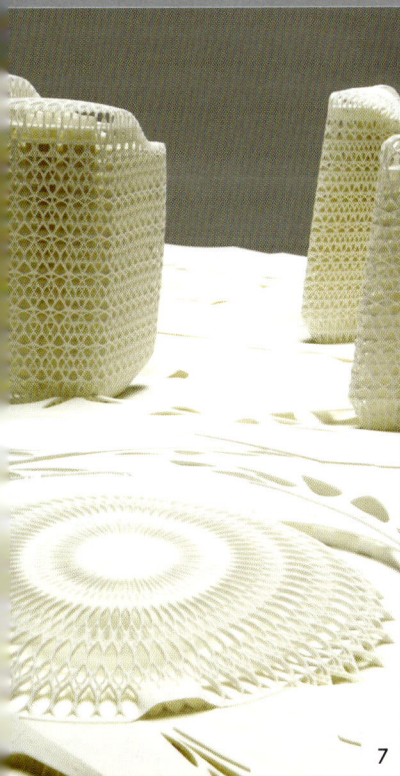

7

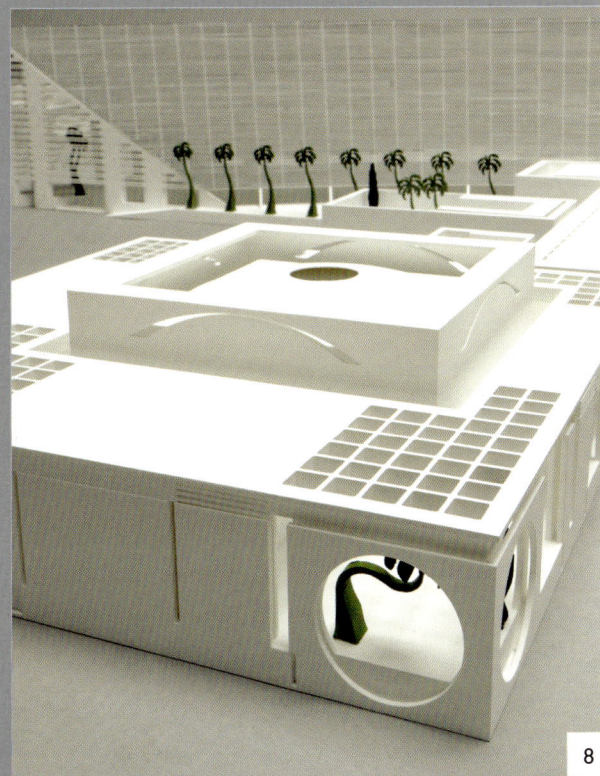

8

Qualified participants
合格的参赛者

1
1st prize　一等奖
Léon Wohlhage Wernik Architekten,
Berlin

2
2nd prize　二等奖
Burckhardt + Partner, Zurich

3
3rd prize　三等奖
von Gerkan, Marg und Partner
Architekten, Hamburg

4
4th prize　四等奖
Ateliers Lion architectes urbanistes,
Paris

5
Further participant　其他参赛者
Kisho Kurokawa architect &
associates, Tokyo

6
Further participant　其他参赛者
Consolidated CE – Jafar Tukan
Architects, Amman

7
Further participant　其他参赛者
Zaha Hadid Architects, London

8
Further participant　其他参赛者
SCHULTES FRANK ARCHITEKTEN,
Berlin

Léon Wohlhage Wernik Architekten Berlin
Leon Wohlhage Wernik建筑事务所 柏林

"Tripoli Greens – fulfills the Organisation for Development of Administration Centers' desire for a bold and visionary symbol of modern Libya. Our concept is iconic, unique and it will create an identity for the site that will be recognisable from afar."

"的黎波里绿色实现了行政办公中心开发建设现代利比亚一个大胆的、有远见的标志的愿望。我们的概念是标志性的、独特的，从远处就能认出的竞赛地点的身份标志。"

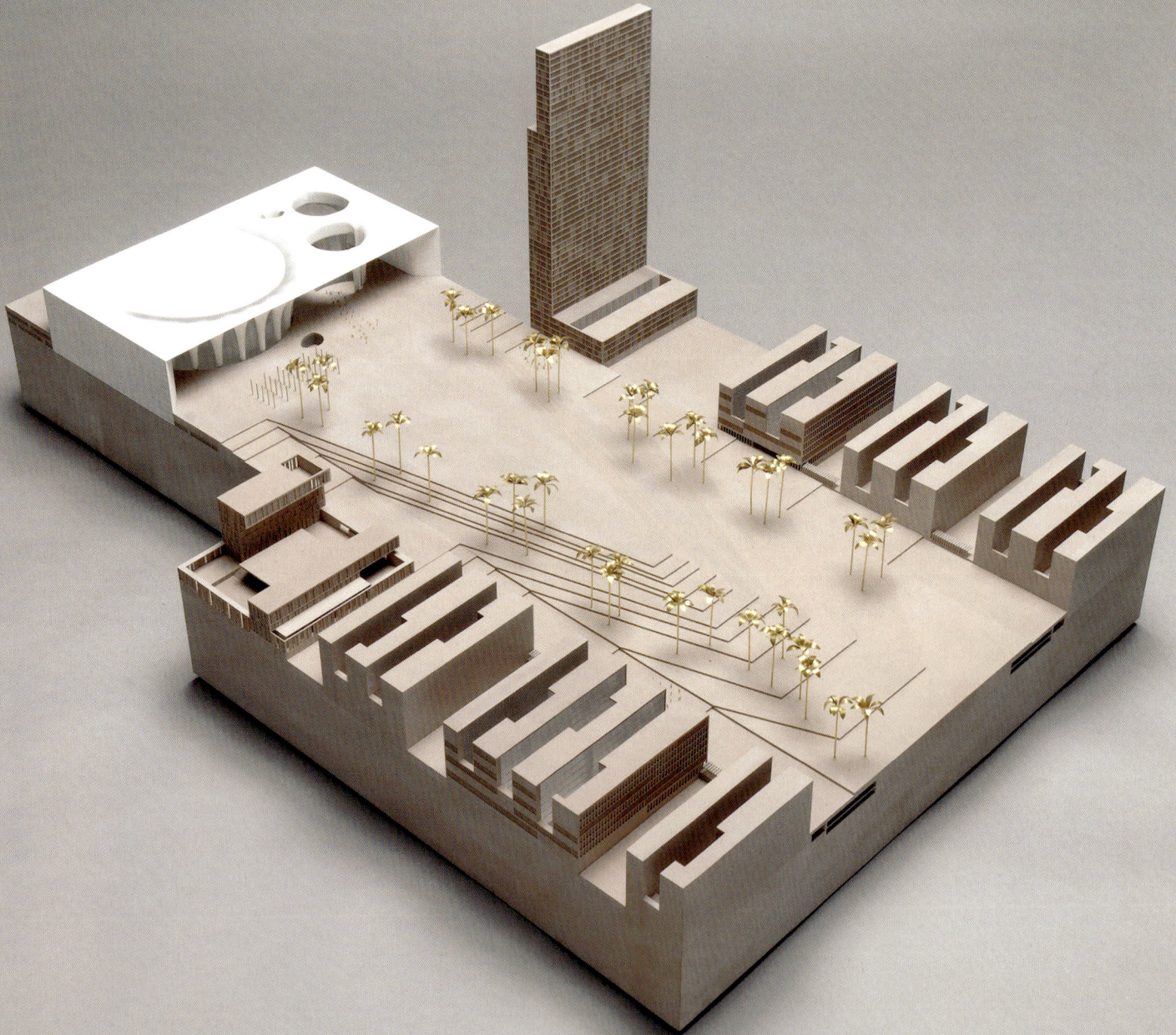

作者 Hilde Léon, Konrad Wohlhage (†), Siegfried Wernik   合作伙伴 Klaus-Tilman Fritzsche, Sebastian Lippok, Julius von Holst, Marius Mensing, Florian Dreher, Tim Lindner, Adrian König, Hans-Josef Lankes, Gerrit Neumann, Jutta Kliesch, Thiele Nickau, Miriam Göllner
专家 structural analysis: Werner Sobek Ingenieure GmbH, Stuttgart; sustainability: Happold Ingenieurbüro GmbH, Berlin; landscape planning: ST raum a. Landscape architecture; real estate identity: MetaDesign AG, Berlin

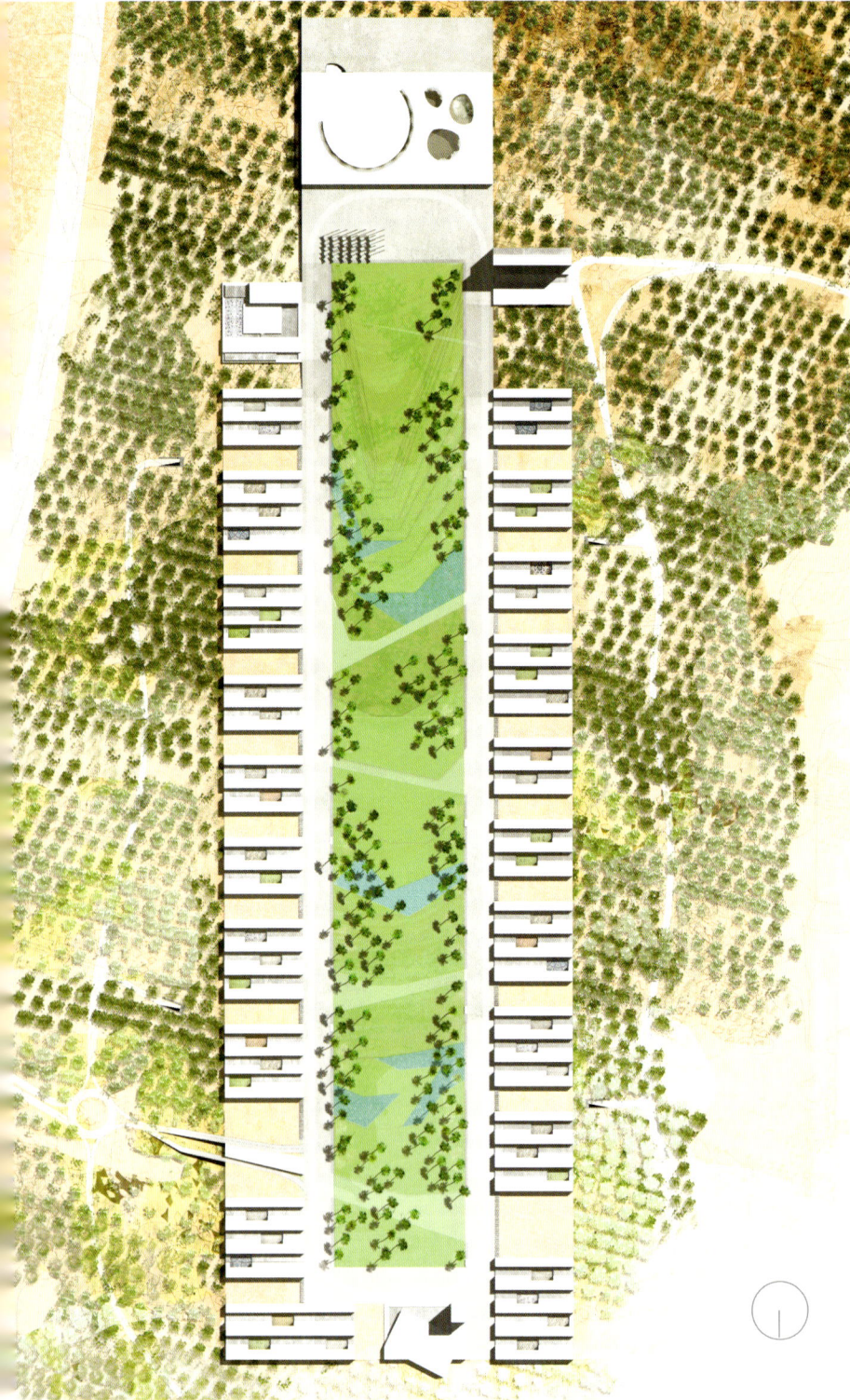

411

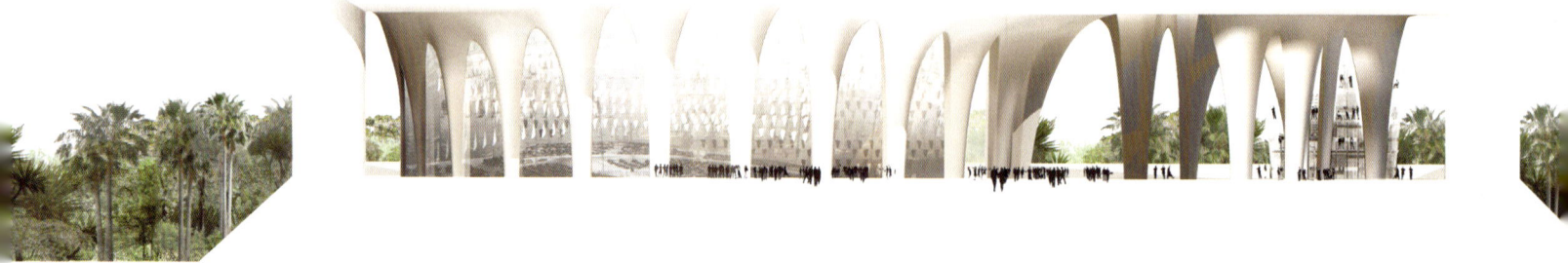

The followed vision is that the building illustrates the political culture of Libya in a sculptural form. Representatives of the people meet under its big roof to discuss and shape the future of the country. The roof provides protection from the sun while allowing cool breezes to waft through. The atmosphere is that of a great open foyer allowing one to feel rather than see the expansive volume of the generous Main Conference Hall of the Conference Palace through its translucent membranes. The special design of the load-bearing structures through their sculptural moulding creates an uplifting effect. The mix of varying volumes, courtyards and roof openings dramatises the interplay of space, light and shadow under the roof. Several expansive staircases, escalators and elevators lead one from the foyer in to the lower level of the plinth.

All required functions of the administration, including the press centre, are housed in the sides of the building, the giant pillars supporting the roof. At the moment, the roof is reserved for technical and construction purposes, with an option for accommodating other special functions. The goal was to propose and develop a concept based on a very strong and central core.

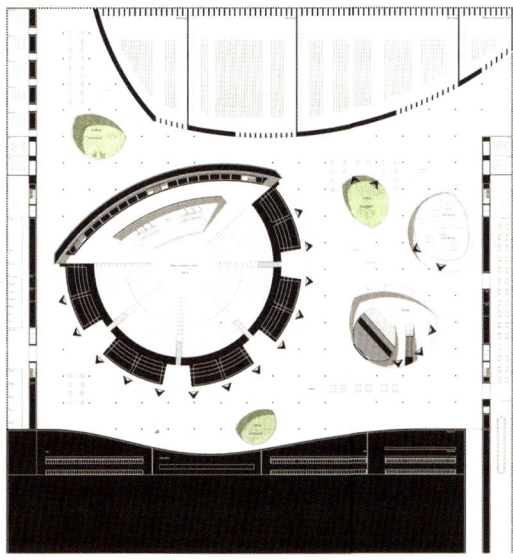

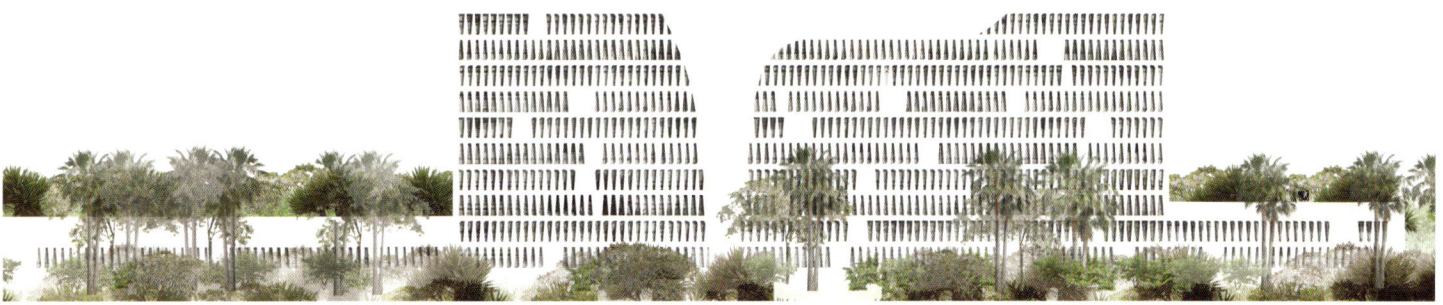

The
Administration
District

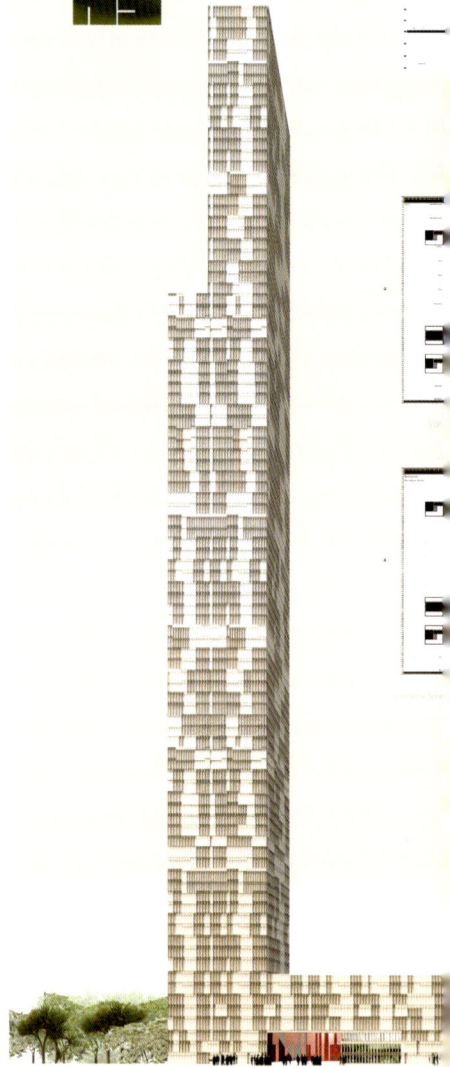

GENERAL PEOPLE'S COMMITTEE'S BUILDINGS AND VIP HOTEL

floor plans, conceptual plan views and sections

The programme calls for three varying sizes of ministries - small, medium and large. One basic type was developed allowing for the rational and economic use of space for offices on the upper floors, with special functions being accommodated on the ground floor or even the first floor. The requirements for natural daylight, ventilation, and shade as well as a strong architectural expression determined the design. The east-west oriented long narrow stretches within the individual building blocks provide shade to neighbouring entities while allowing a cool breeze to flow. These are interspersed by courtyards in differing position. They cut into the building sometimes going down to the first floor thus adding character to the ground floor. The façade is perforated with recesses to reduce the heat from the southern light.

The VIP Hotel takes on a sculptural form like the other buildings. The recessed part of the tower starts above the wellness floor, which runs along its entire length while providing outside access and higher views. Rooms are generously proportioned with luxury private baths. The plinth level rooms have been designed and arranged to cut out any disturbances. There is a direct and secure connection to the Conference Palace via the plinth level. Two separate entry ways cater to the public spaces of the hotel. The deep three-dimensional relief cladding of the hotel's façade diffuses the light evoking an image of a sculptured wall.

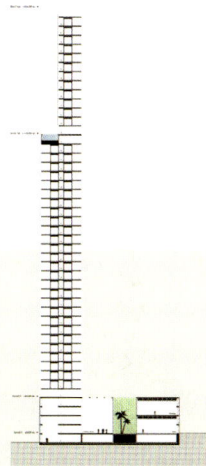

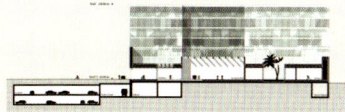

section aa, 1:500   section a-a, 1:500   section b-b, 1:500   north ele...

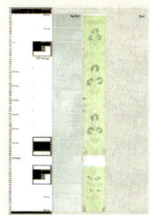

Floor level 5, 1:500
VIP Lounge / Conference rooms

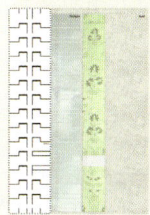

Floor levels 6-25, 1:500
Double bed rooms

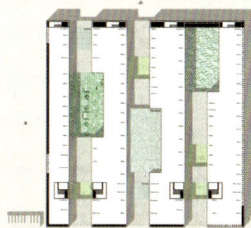

Floor level 3-6, 1:500   Floor level 14, 1:500   Floor level 14, 1:500

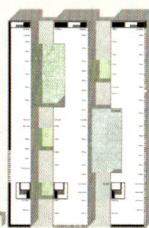

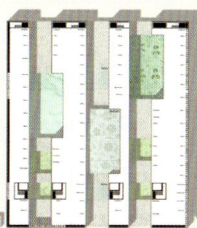

ground floor - entrance level, 1:500
GPC Building, large

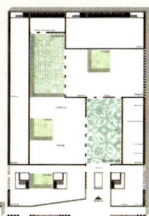

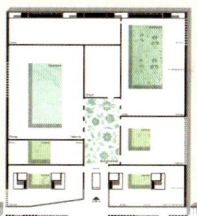

ground floor - entrance level, 1:500
GPC Building, medium

ground floor - entrance level, 1:500
GPC Building, small

square elevation, 1:200
GPC Building - large

square elevation, 1:200
GPC Building - medium

Burckhardt + Partner Zurich

Burckhardt + Partner建筑事务所 慕尼黑

"The new Government district for the state of Libya is placed in a garden. It distinguishes itself in its form and arrangement from the surrounding urban pattern."

"利比亚的新政府区位于一个花园里，在形式和安排上与周围城市模式不同。"

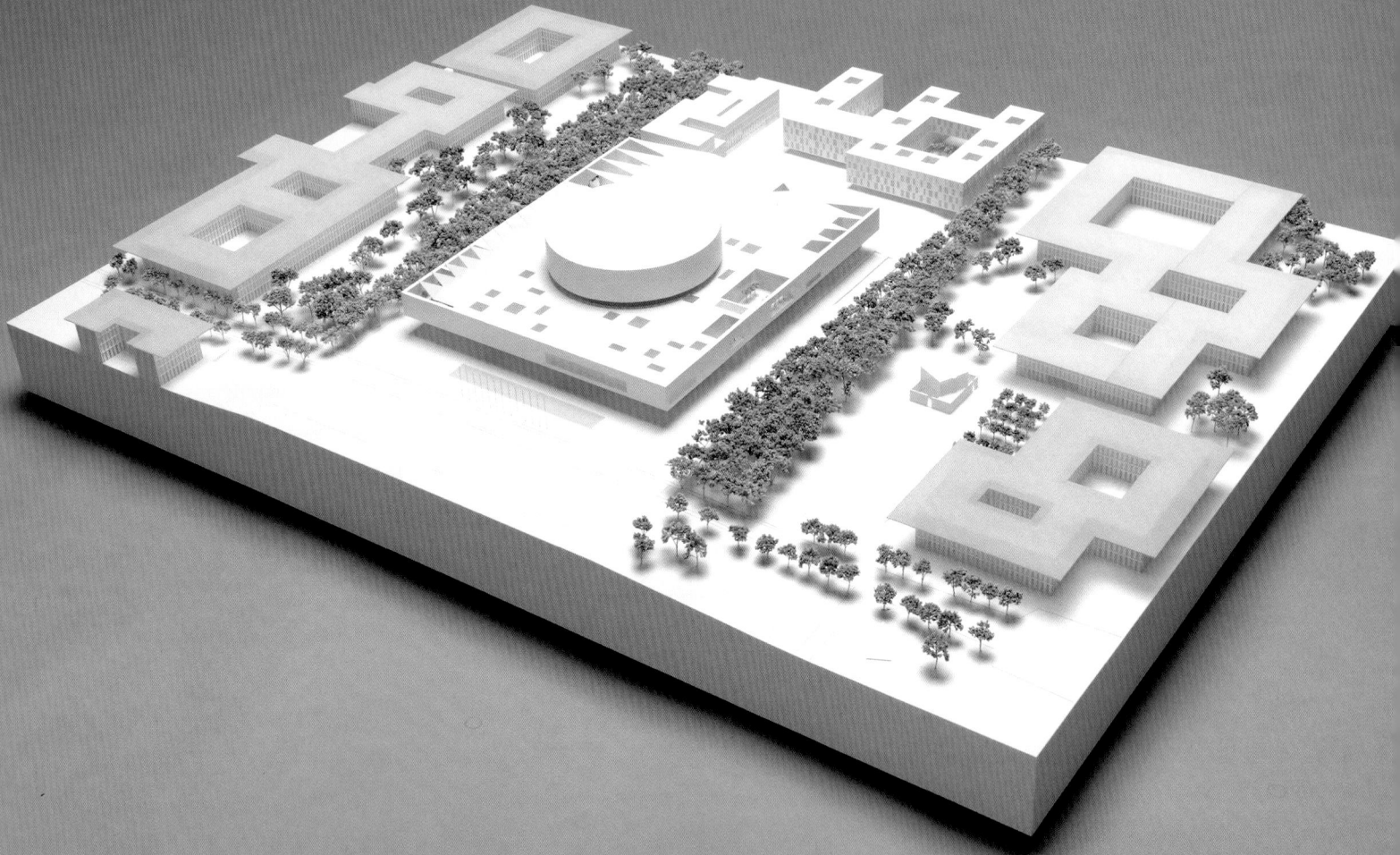

作者 Mathis Simon Tinner   专家 landscape architecture: Vogt Landschaftsarchitekten, Zurich; mechanical & MVAC: HL-Technik AG, Zurich; traffic: Ernst Basler + Partner, Zurich; renderings: Raumgleiter GmbH, Zurich

Administration Complex, Tripoli

Urban Strategies

Garden Landscape

The new Government district for the state of Libya is placed in a garden. It distinguishes itself in its form and arrangement from the surrounding urban pattern. On one hand it creates a green island in the outskirts of Tripoli. On the other hand the new neighbourhood connects with its surrounding by means of pedestrian and bicycle paths that pick up existing street patterns and form a continues pattern on the site and beyond. The national forest on the south end of the site will merge into the garden landscape and will become an integral part of the new administration complex.

Urban Pattern

The urban pattern proposed for the site has the potential to extend beyond the actual competition site. In a first phase the different ministries embrace the government palace, the VIP hotel and the office of the general secretary in a protective ring. A second phase allows for an extension along the ring road. If desired even across the freeway. The existing apartment block on the North West end of the site becomes an integral part of the garden. Current uses on the site such as the ostrich farm and the camel farm can be integrated and will remain.

Height Development

The new government headquarter consists of mainly low buildings with a maximum height of four stories. Except the government palace distinguishes itself in height from the surrounding structures. Its shape is visible and recognisable from far away. Be it during the day or in the night. Great importance is given to the visibility from the airport highway.

The Alleys

Both sides of the Airport Highway will have landscaped flat fills that create a different spatial experience along the airport highway. They also serve as acoustical barriers and are made out of the excavation from the buildings on the site. The ring road and the streets to the existing neighbourhood are expressed by the use of clearly delineated alleys of trees.

The Park

The Urban Plazas

The Ancillary Uses

Spread throughout the park and placed along the pedestrian paths are numerous public uses. Sports facilities invite everyone to use the park. Restaurants, café and playgrounds animate and vitalize the park. The ostrich farm and the camel herd can remain in its current place. The concept proposes to split the mosques in different entities that can be erected near the program parts they are needed for.

The Courts

Social Sustainability

Urban Scale

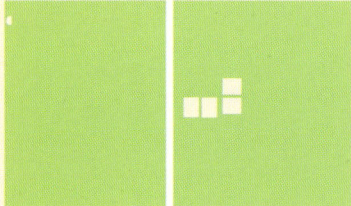

The garden landscape ➡ The green island in the urban fabric

Pedestrian and bicycle paths ➡ The web to the existing surrounding

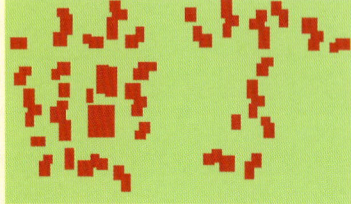

The low rise buildings ➡ Generous open spaces at a human scale

Urban public plazas ➡ The link and the identity

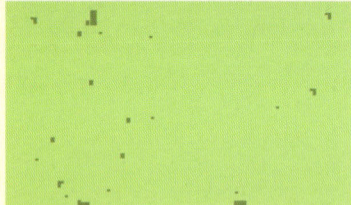

The outdoor facilities ➡ An invitation to the garden for everyone

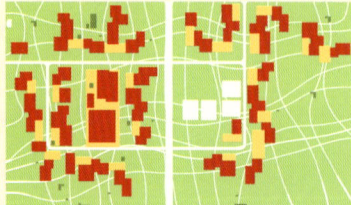

The garden landscape ➡ The green island in the urban fabric

Building Scale

The committee buildings, the Government Palace and the VIP Hotel are based on courtyard schemes. In contrast to the generous public garden, the inner landscaped courts offer more private places to enjoy. A series of common uses are organised around the courtyards on ground level of the committee buildings. Lounge areas and communication spots are spread throughout the regular office floors.

- 🟥 Entrance hall
- ⬜ Common uses
- 🟩 Landscaped courtyards

Environmental Sustainability

Garden Landscape

The urban pattern of the buildings allows for a large area to be landscaped. Only one third of the entire site is occupied with built structures, urban plazas or streets. The entire project is developed to avoid heavy excavation. The excavated material will be re- used on site to form the acoustical "land art" barriers on both sides of the airport highway. Trees and plants will be local.

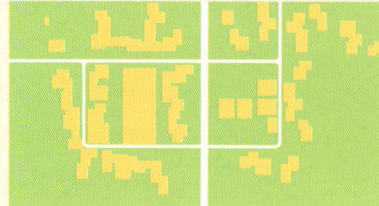

- 🟨 Built area = 27%
- 🟩 Green area = 73%

Facades

All facades are developed to avoid the extensive use of air conditioning. Natural fixed, shading elements in the horizontal direction and in the vertical direction (in the façade) will provide sufficient shading. Users will have the option to either employ natural ventilation or air conditioning. All buildings are in light colour to minimize heat gain through the façade.
The façade design and the building mass is optimized to minimize "heat loss" in the winter months and to reduce cooling loads in the summer months. All buildings will have

Committee buildings and the Secretariat of the General People's Congress

The façade is based on a prefabricated, insulated concrete panel that alternates with operable windows. The form and the arrangement of the panels provide sufficient shading and allows for optimum natural lighting for the interior.

VIP Hotel

The façade of the VIP Hotel is based on a prefabricated panelling system that will provide sufficient shading. In contrast to the committee buildings the outer skin is more open to allow for views into the garden landscape and the courtyards.

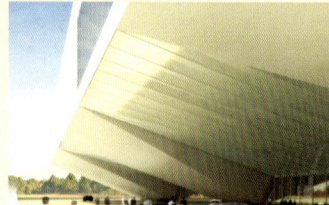

Conference Palace

The construction of the Government Palace is based on a structural steel grid system with prefabricated concrete panels. Natural light for the meeting spaces within the Palace will be mainly achieved with a system of Northern lights on the roof.

Economical Sustainability

Phasing

The urban scheme and the general layout of the buildings allows for a high flexibility for the construction and the organization of the Administration Complex. In a first phase the core of the complex, consisting of the Government Palace, the VIP Hotel and the Secretariat of the General People's Congress can be built. The positioning of the ring road allows for further phases of the complex to be added to the scheme as needed.

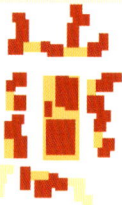

Saving Energy

The building envelope and the building mass are designed to minimize "heat loss" in the winter months and to reduce cooling loads in the summer months. All buildings will have sufficient insulation and will be equipped with fixed shading devices where necessary.

Cooling

The buildings will be cooled with air-conditioning systems (input of fresh air) and static cooling systems. The static cooling systems are calculated with totally 32.34 MW. The air-handling units need a total cooling capacity of 12.06 MW.
The cold water of all chillers will be served on the basis of 12/16 °C. Each building has its own cooling exchange system to serve cold water for the cold of the air-conditioning systems (12/16 °C) or cooling ceilings (16/20 °C). The buildings will be cooled only at outside temperatures over 24 °C. Otherwise natural ventilation provides for a pleasant climate. In order to reduce the gas or oil consumption to run boilers or EPS, a vacuum pipe collector system should be installed on the surface of the water tanks or elsewhere appropriate.

Electrical power supply

2 electrical supply nets will serve the building complex. A second electrical power supply will be served by gas or oil-driven motors with generators. To reduce the electrical power, run by the public network or EPS-system, a photovoltaic system (integrated in the shading elements on the roofs or in the landscape) can serve a peak power supply.

Boiler System

The boiler system with a total capacity of 17.85 MW will serve a maximum heating capacity of 7.55 MW in winter. The total heat loss of all buildings is calculated with 6.1 MW, hot water for all bathrooms (especially hotel and lavatories 0.65 MW) and hot water for all kitchens (0.8 MW).

Fresh Water and Waste Water

The calculation for the buildings shows that only 30 % of the complete water consumption will need to have drinking water quality. A biological cleaning system with water tanks, which will also be served by storm water or rainwater, can be used for flushing toilets. The water tanks can also serve sprinkler systems which might be needed in different parts of the Administration district.

Principles Energy - Systems — Total cool.cap. 40.500 kW$_{TH}$

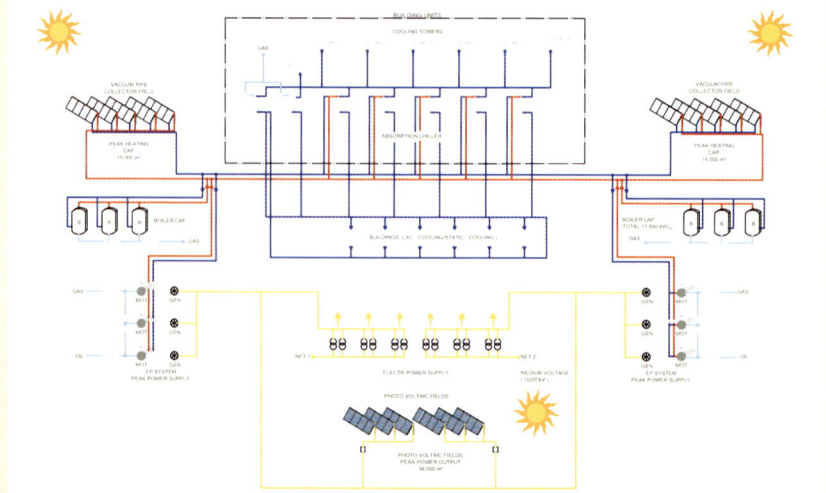

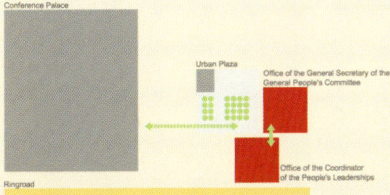

Concept

The Office of the Coordinator of the People's Leadership is based upon a standard courtyard structure. It shares an urban plaza with the Office of the General Secretary of the General People's Committee. The program is organized on two floors around a landscaped courtyard. Two additional floors allow for an extension within the same building.

The Office of the Coordinator of the People's Leaderships is placed in immediate proximity to the Office of the Secretary of the General People's Committee and the Conference Palace.

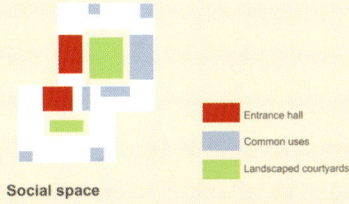

Conference Palace

Urban Plaza

Office of the General Secretary of the General People's Committee

Ringroad

Office of the Coordinator of the People's Leaderships

Social space

Entrance hall	
Common uses	
Landscaped courtyards	

South	10m
East / West	7.5m
North	5m

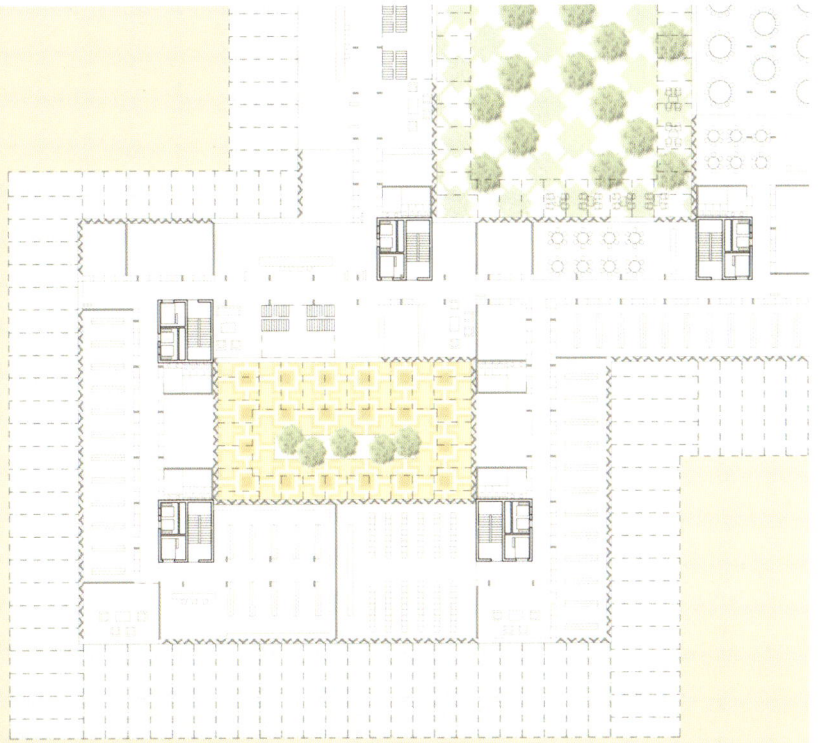

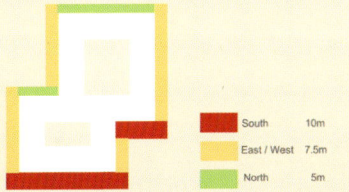

Roof Shading Concept

The fixed roof shading devices are designed with a permanent, perforated membrane that duplicates the "shadow of natural leaves" by using a geometric, repetitive pattern. The facades are developed to avoid the extensive use of air conditioning. Natural fixed, shading elements in the horizontal direction and in the vertical direction (in the facade) will provide sufficient shading. Users will have the option to either employ natural ventilation or use air conditioning.

Courtyard

The interior court of the Offices of the Coordinator of the People's Leaderships is part of the overall landscape scheme for the courtyards. A pattern of geometrically arranged tiles of different colours will contrast with a canopy of trees that will offer additional natural shading.

von Gerkan, Marg und Partner Architekten Hamburg
von Gerkan, Marg und Partner建筑事务所 汉堡

"A beautiful landscape with a rich natural scenery will be created as a park for the new government district."
"为新政府区创造出一个像花园一样的美丽的自然风景的景观建筑。"

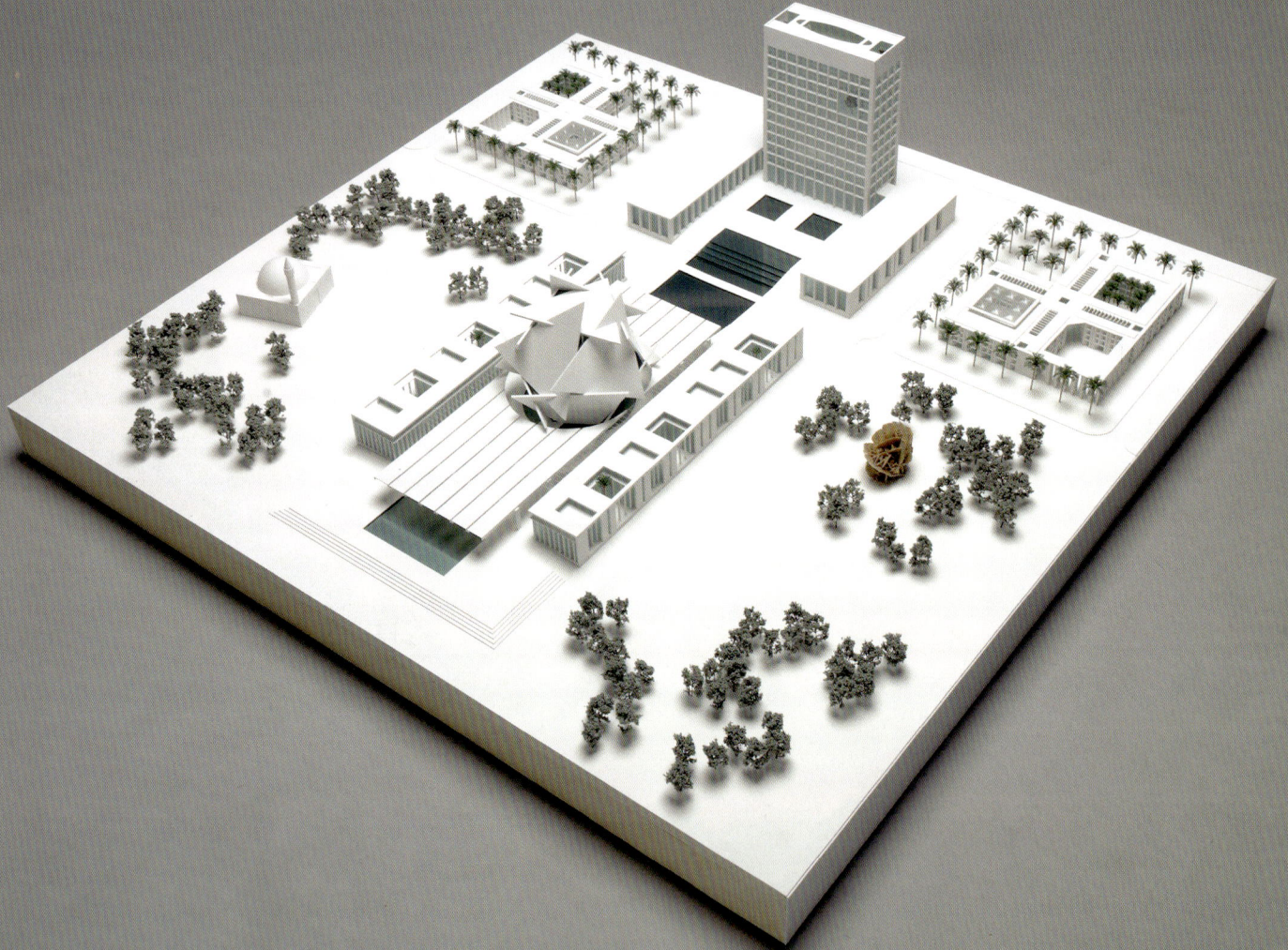

作者 Meinhard von Gerkan, Jürgen Hillmer   专家 landscape architecture: Breimann Bruun, Hamburg;
energy planning: Transsolar, Stuttgart

Administration Complex, Tripoli

von Gerkan, Marg und Partner Architekten, Hamburg

Description Administration Complex in Tripoli

The area chosen for the development of the Administration Complex is located at the Airport Highway south of the proposed Third Ring Road. A beautiful landscape with a rich natural scenery will be created as a central park for the new Government District. The iconic building is located in the center part of the park in order to create a powerfull visual ascer-tainability.

The concept is based on 7 positions

Structural Order
As well as every living existence a proposed health park also depends on principles of structural order. Therefore each structural element has to be defined in relation to the whole. The principle has to structure the building volumes and the uses, plan and elevation have to be derived out of the same order. Analogue to the grammar of language the structural order has to take over a rational steering-function.

The overruling concerns of structural order are:
1. Serenity of architectural form
2. Hierarchy
3. Visual ascertainability
4. Orientation in space

Individuality
In traveling over the world, we experience locations, which stick to our memory and others, which mingle with other pictures very fast. The more specific and explicit the appearance of a certain location is, the more our memory will be touched by it. Most of the contemporary city planning is characterized by ran-domness and commutability. There are only a few plannings, which are equipped with a unique and specific identity.
The objective for the new health industrial park should be to develop the specific identity out of the building-task and regional situation.

Grid
The building- and infrastructure are due to the ap-proach of antique Greek cities based on a grid. The traffic-grid serves with several categories of facilities.
1. Walkways
2. Bicycle lanes
3. Streets for cars and buses
4. Mixed used areas for pedestrians, cyclists, delivery and the fire brigade.

The grid offers alternations of narrow alleys and wide squares which are also intersections in the network. Different themes of landscape design should create a distinct character for each area.

Public space
One main issue is the defined quality of public space. The design of street-objects and sections, as well as the early realisation of parks and plazas is as important as the layout and design of the buildings for the success of the development.

Balance of determination and freedom
The premise for a lively Health Park is a balance of determination and freedom of architectonic articulation. The determination is given through the definition of an urban structure with public space, plots and building highs based on a measure. Within this system of structural order the freedom of finding an individual form for each building has to be kept.

Simplicity
The command of simplicity does not mean primitive, banal or unimaginatively. Meant is simplicity in the sense of plausible, self-evident and clear. Architec-tural solutions which take on arbitrary forms and need complex structures to be interesting are less suitable to meet the requirements of the new Software Park, neither in form nor in content.

Balance of variety and unity
The uncomfortableness of our environment is very often based on an excess of sameness, which is perceived as monotony or an excess of variety, which is registered as chaos. How important the balance of variety and unity is can be emphasized by analysing architectural history. The recipe for the highly appreciated fabric of ancient cities is exactly rooted in this balance. It is necessary to analyse these examples to transform the principles into the future, but under no circumstances, to copy them.

The above mentioned principles are used to create a set of different geometrically defined areas which contrast the natural landscape. Each area responds to the specific function and condition of its location.

Landscape

The setting of the architecture within the city, bordering a greened area, allows this green to surround and float through the new building complex. The entire site is embedded in a forested area of palm trees and eucalyptus, continuing the existing plant-ings to its south.

When driving towards this new governmental city you have to pass through this thick green belt. Once you reached the first ring road, the green opens up for the outer ring of governmental buildings. The spaces between the buildings are planted with palm trees. They grow up between the parking lots, from one floor below. They spread out in an even grid, according the architectural grid. The recipe for the highly function as plaza trees as well as street trees. The inner ring road surrounds the central park of the site.

The par-iament house is the central part of this park. It sits in the middle axis on a slightly raised stone plaza. A large water basin marks its middle axis. The surrounding park works with the structures and patterns of the surrounding landscape. Paths divide the area into amorphous shapes. Wherever the cross each other a small plaza opens up and creates spaces to rest, for pavilions, water plays, etc. On a second layer trees stand loosely on the entire park. The surrounding landscape interfaces this newly created open space.

The inner courtyards of the buildings are places to take a break and enjoy the fresh air under the translucent fabric shade roofs. Additional to these roofs, some palm trees provide a second roof layer. They "dance" playful around a central round pool that sits in the centre of each courtyard, directly under the circular opening in the roof structure. The entire surface is paved in a light colored stone, with paths that run towards the pool in the centre. The theme of the central park is interpreted in these much smaller open spaces.

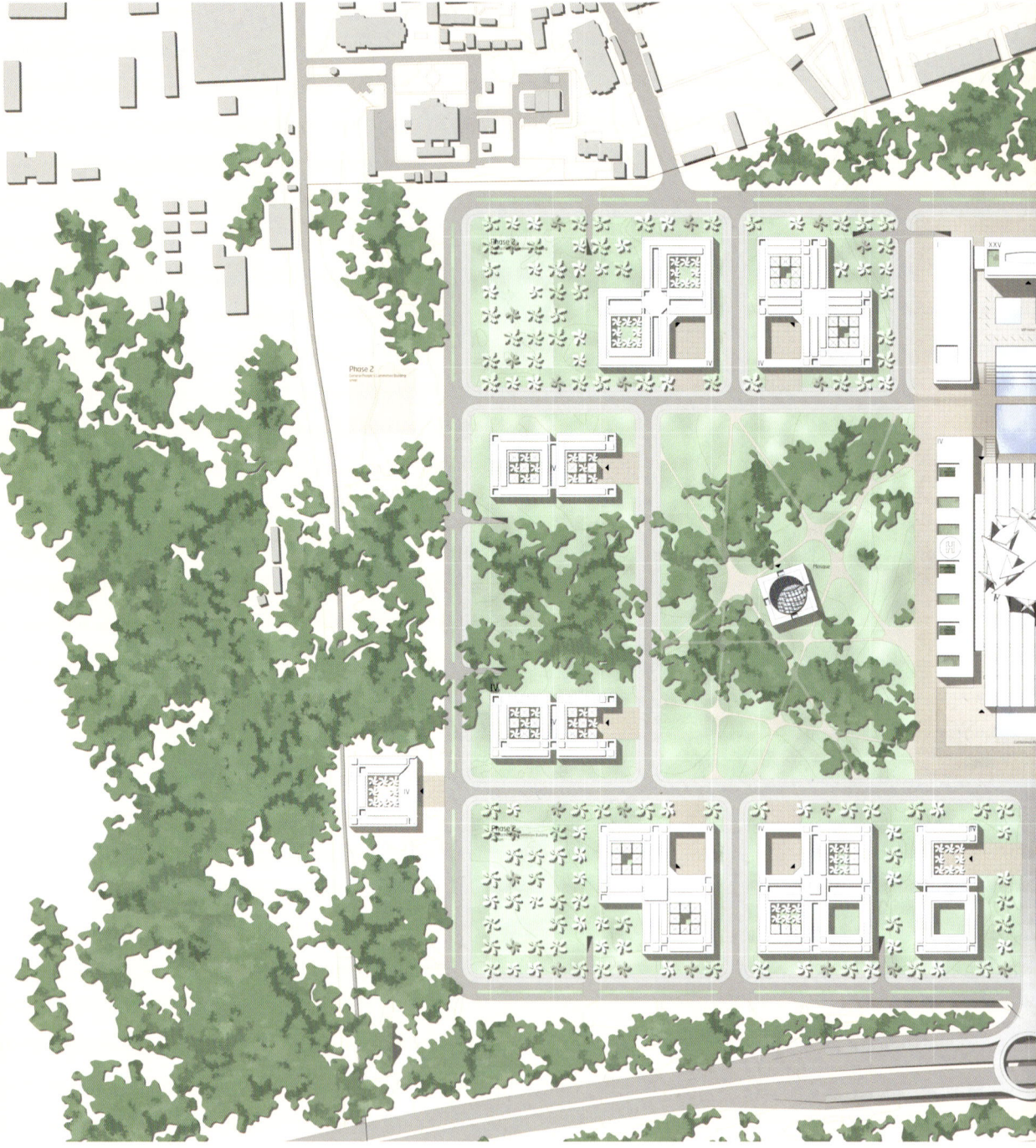

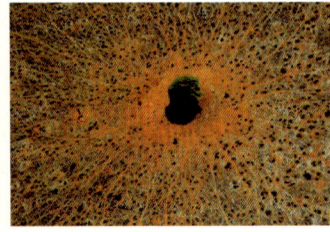

site plan . scale 1:1000   0   40   100   200

administration complex in tripoli

site plan . scale 1:1000   0   40   100   200

administration complex in tripoli

IV

Phase 2

Phase 2

distribution

traffic

parking

pipe plan

site plan . scale 1:1000 |0 |40 |100 |200

administration complex in tripoli

von Gerkan, Marg und Partner Architekten, Hamburg

Administration Complex, Tripoli

conference palace and secretariat of the general people's congress and other facilities

administration complex in tripoli

Ateliers Lion architectes urbanistes Paris
Ateliers Lion建筑及城市规划事务所 巴黎

"... what can be more important in the symbolic issue than confronting the political institution with the Libyan land and its climatic contrasts?"

"在象征问题中，什么可能比面对在利比亚土地上气候对比强烈的政府办事机构更重要的事情呢？"

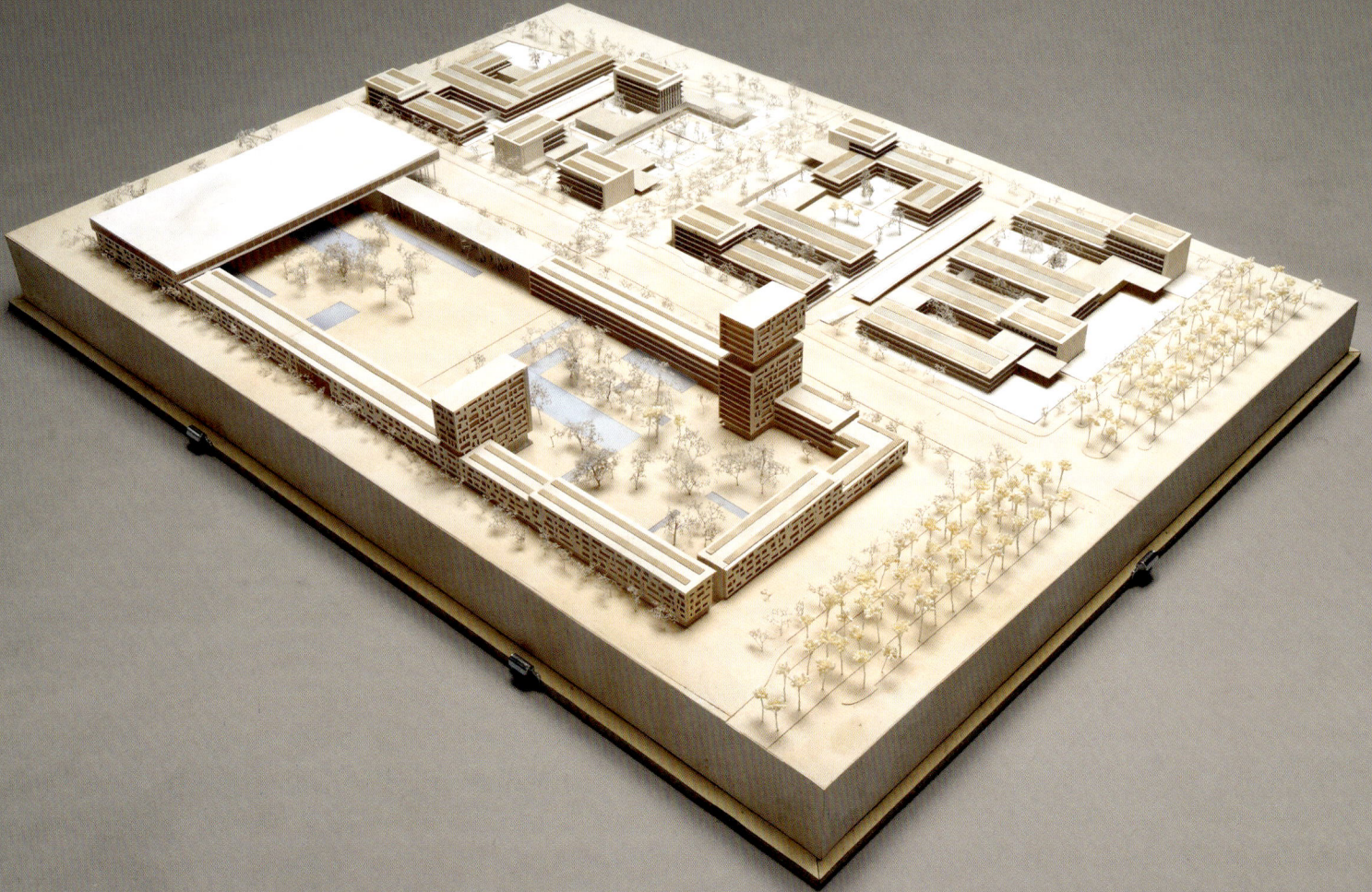

作者 Yves Lion, Francois Leclercq, Claire Piguet   合作伙伴 Delonne Léonard, Kim Hyon Seok, Le Minh Triet, Mahajan Reena, Ramone Laurianne, Ré Christelle   自由建筑师 Nicolas Laisné, architect, Christophe Roussel, architect   专家 energy planning: Transsolar, Thomas Auer, Stuttgart; traffic: Citec, Philippe Gasser, Geneva

South facade

Conference palace & Secretariat of the General people's Congress West facade

Main conference hall

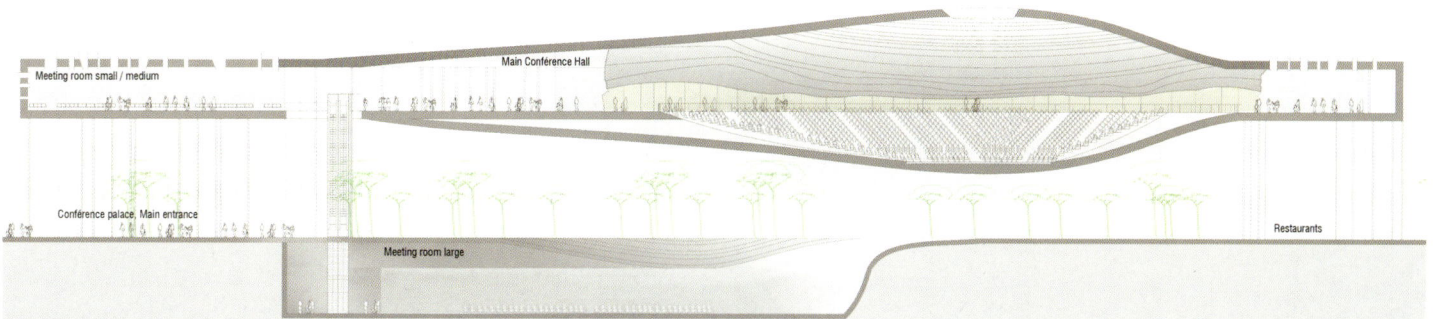

Meeting room small / medium

Main Conférence Hall

Conférence palace, Main entrance

Meeting room large

Restaurants

Longitudinal section   1:200

SITE PLAN

N

0 10   50   100   200

1 : 1000

Ateliers Lion architectes urbanistes, Paris

Administration Complex, Tripoli

Kisho Kurokawa architect & associates Tokyo

Kisho Kurokawa建筑事务所 东京

"Iconic central government dome, introduction of a cultural hub, linking these two main elements by the 'urban axis'"

"通过'城市轴'把标志性的中央政府圆顶和一个文化区这两个主要元素联结起来。"

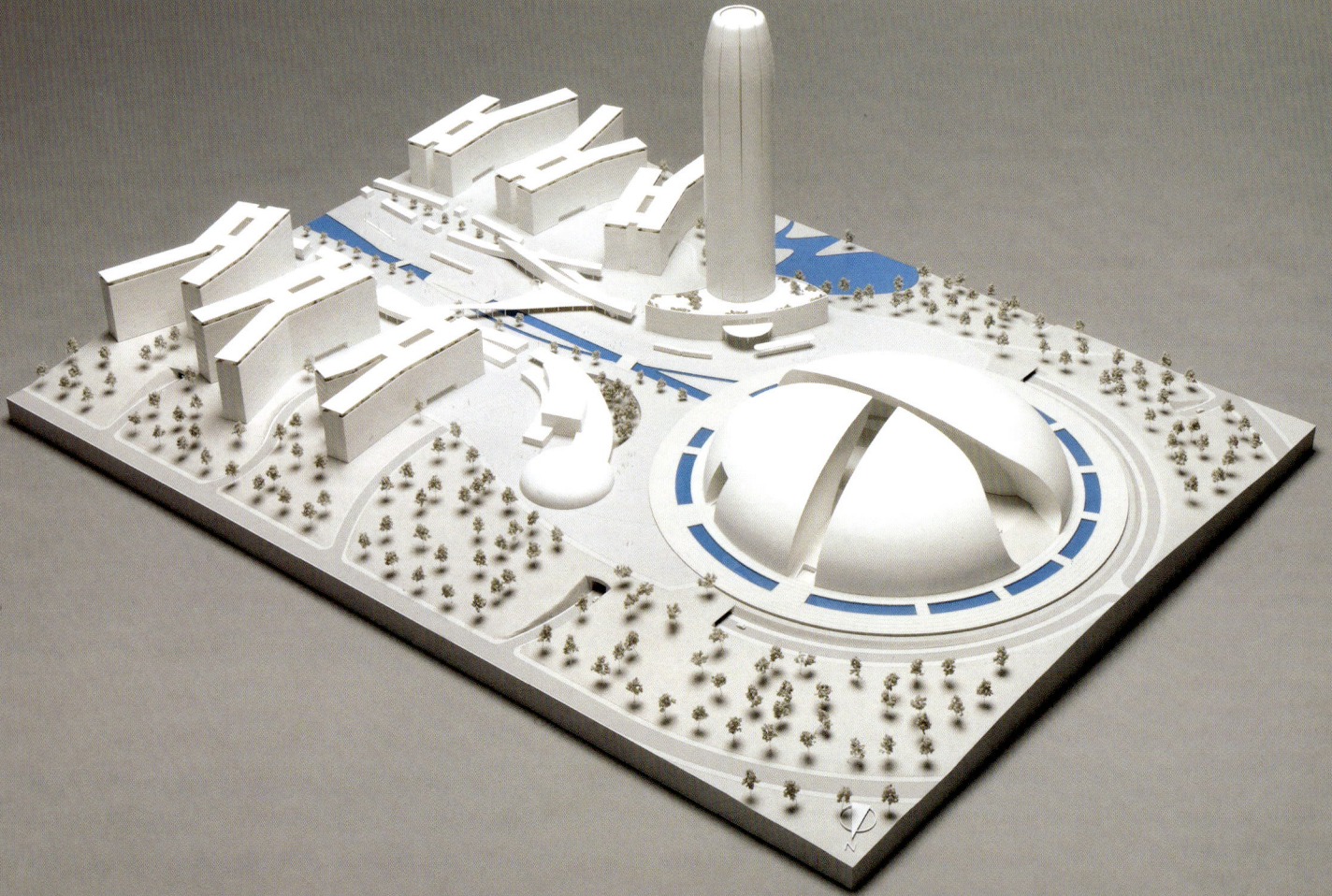

作者 Kisho Kurokawa (†)

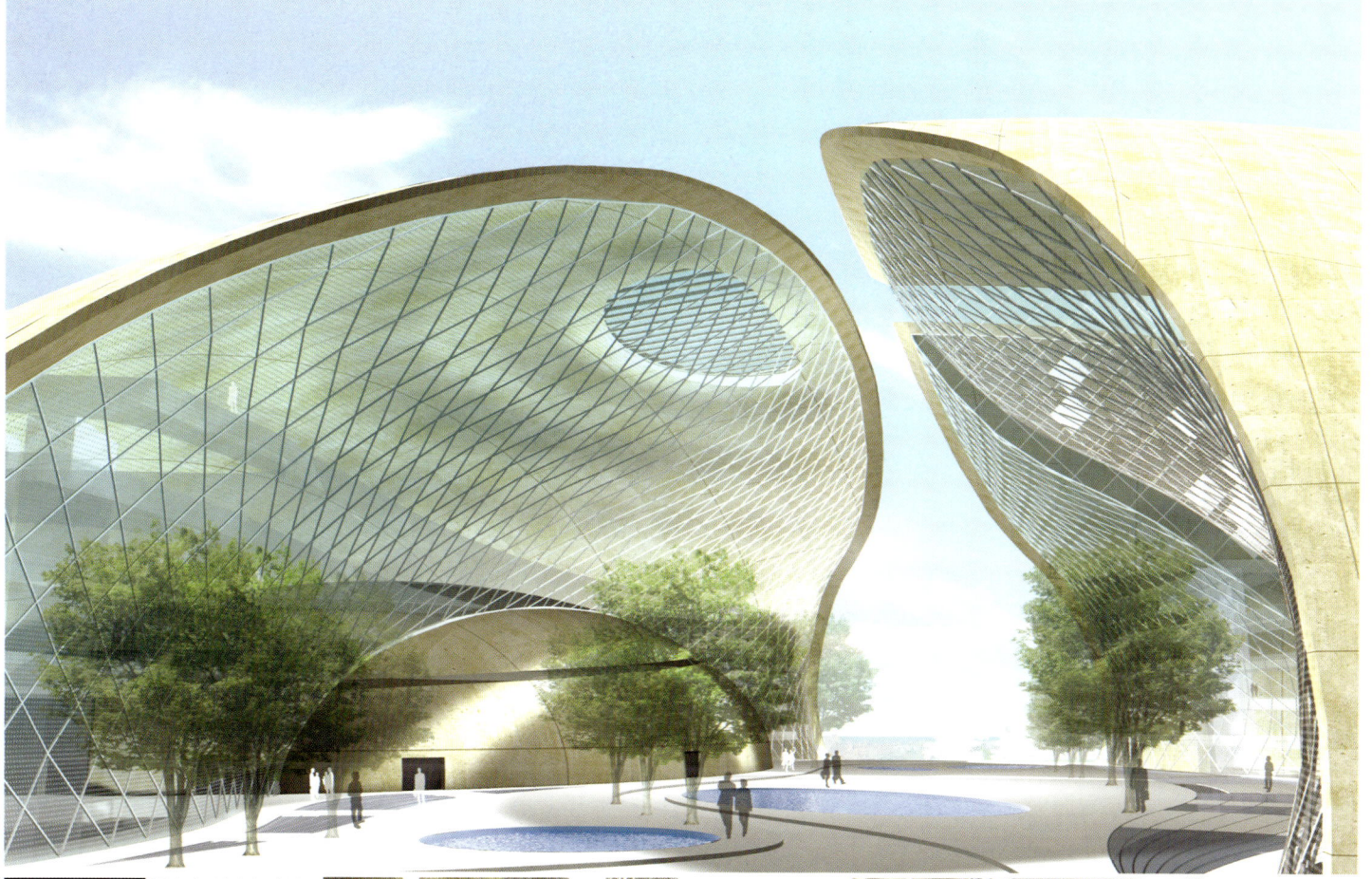

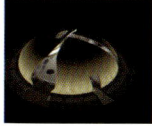   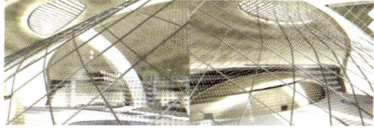 

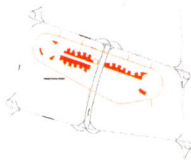

ROADS

GREEN BUFFER

WATER SYSTEM

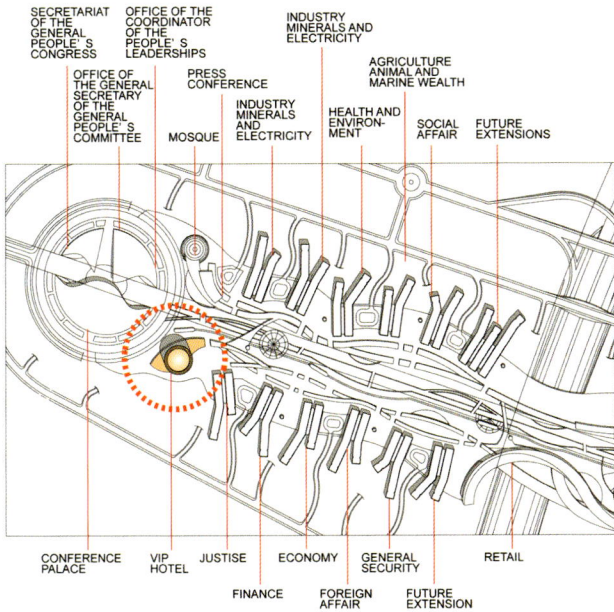

Connection to Adjacent Buildings
Scale1:6000

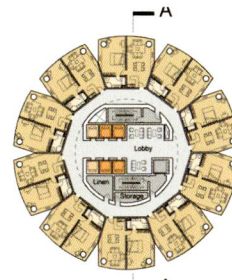

SECRETARIAT OF THE GENERAL PEOPLE'S CONGRESS

OFFICE OF THE GENERAL SECRETARY OF THE GENERAL PEOPLE'S COMMITTEE

OFFICE OF THE COORDINATOR OF THE PEOPLE'S LEADERSHIPS

PRESS CONFERENCE

MOSQUE

INDUSTRY MINERALS AND ELECTRICITY

INDUSTRY MINERALS AND ELECTRICITY

HEALTH AND ENVIRON-MENT

AGRICULTURE ANIMAL AND MARINE WEALTH

SOCIAL AFFAIR

FUTURE EXTENSIONS

CONFERENCE PALACE

VIP HOTEL

JUSTISE

FINANCE

ECONOMY

FOREIGN AFFAIR

GENERAL SECURITY

FUTURE EXTENSION

RETAIL

Sketches

Typical Plan
Boarding Suites   Scale1:600

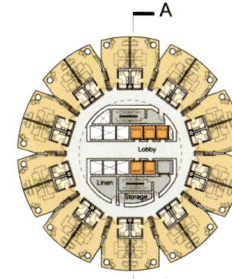

Lobby

Linen

Storage

Typical Plan
Double Bed Rooms   Scale1:600

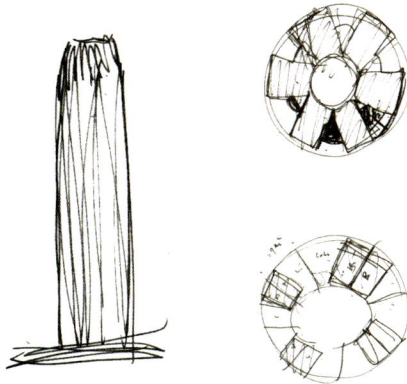

Lobby

Linen

Storage

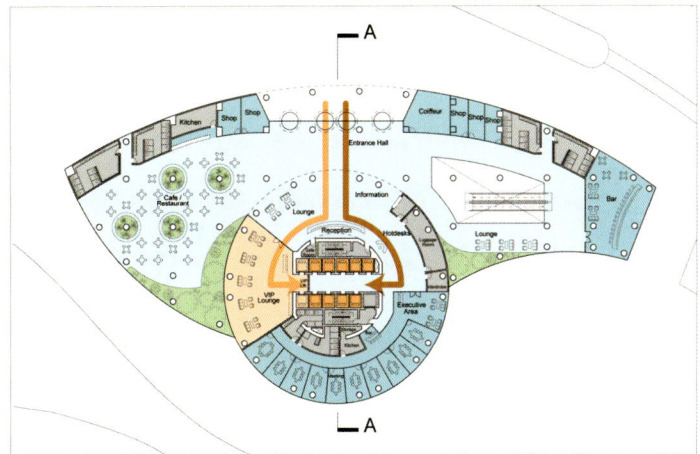

Ground Floor Plan
Hotel Entrance Lobby   Scale1:600

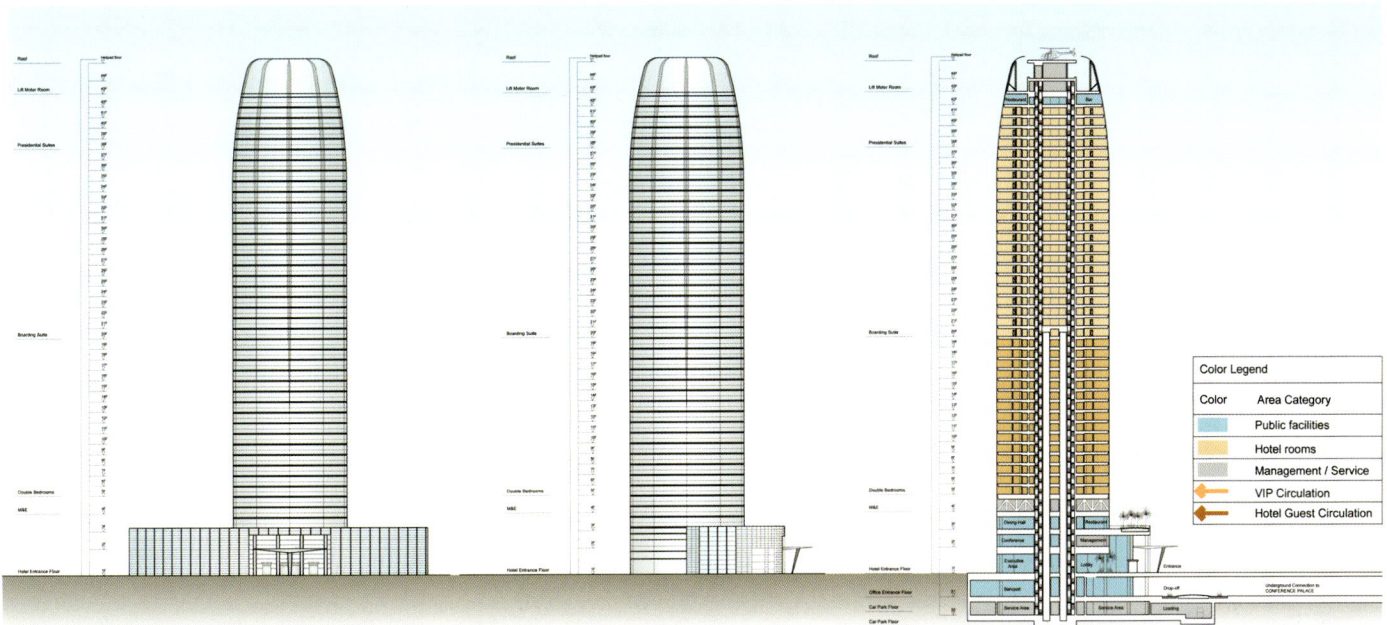

Kitchen

Shop

Coiffeur   Shop

Entrance Hall

Cafe / Restaurant

Lounge

Information

Reception

Bar

VIP Lounge

Executive Area

Lounge

Front Elevation
Scale 1:1200

Side Elevation
Scale 1:1200

Section A-A
Scale 1:1200

Color Legend	
Color	Area Category
	Public facilities
	Hotel rooms
	Management / Service
	VIP Circulation
	Hotel Guest Circulation

Consolidated Consultants Engineering & Environment Jafar Tukan Architects Amman

工程环境综合咨询公司和Jafar Tukan建筑事务所 安曼

"Rhythm of the city flows into the site and collides with nature."

"城市节奏流入竞赛地点，与自然产生碰撞。"

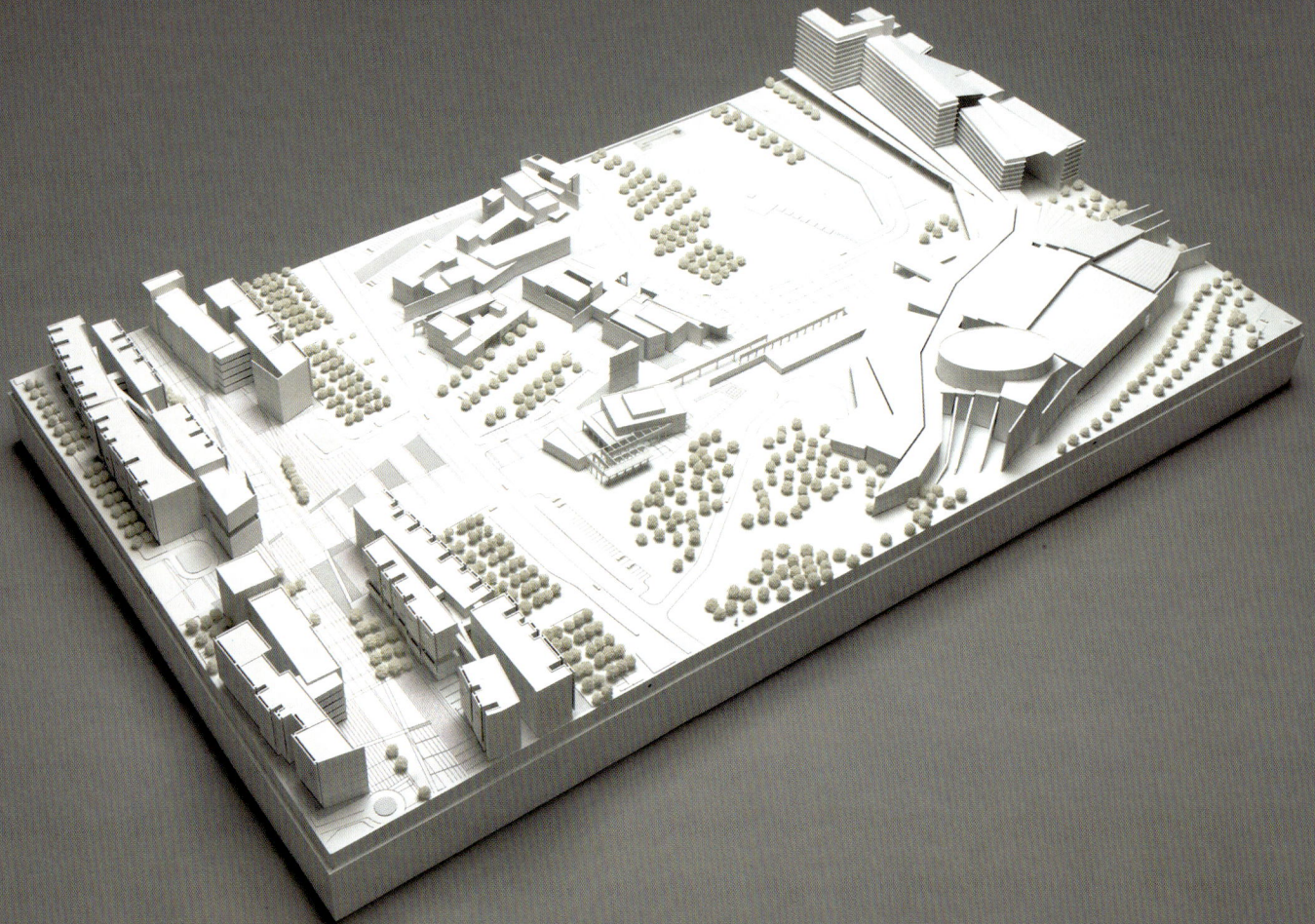

作者 Jafar Tukan, Yasser Darwish, Mohammad Abbas   合作伙伴 Tareq Ghanam, Saba Innab, Andaleeb Bizreh, Hadeel Hamad, Jumana Hamadani, Ahmad Seyam, Nemeh Mansour   专家 traffic: Consolidated Consultants Engineering & Environment –Jafar Tukan – Architects, Issam Bilbesi, Amman

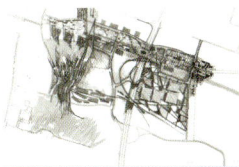

Libya is embarking on new era where massive construction will be one of its major traits. Large projects will most likely take place outside the city along the highways. Sprawling is undesirable pattern of growth and will strip Tripoli of its unique spatial qualities.

Such large projects are vital to the city and can contribute a lot to its character if planned properly. We believe that the location of the Governmental HQ on this site is one step in the right direction but should be inscribed as part of a global vision of the city. The site is located on a pivot area right between the city and the countryside. The latter is very present due to the relatively well conserved forest. Our strategy consists of locating future large projects along the third ring road. These projects can extend all the way east to the old airport thus creating a well defined zone with a transitional typology. This zone is beneficial in two ways; first, it has the merit of preserving and encouraging the densification of the existing city. Second, it acts as a solid base for future developments that are to take place beyond it. This zone is hybrid between city and nature; its genesis is the forest; its heart is the hill. Rhythm of the city flows into the site and collide with nature.

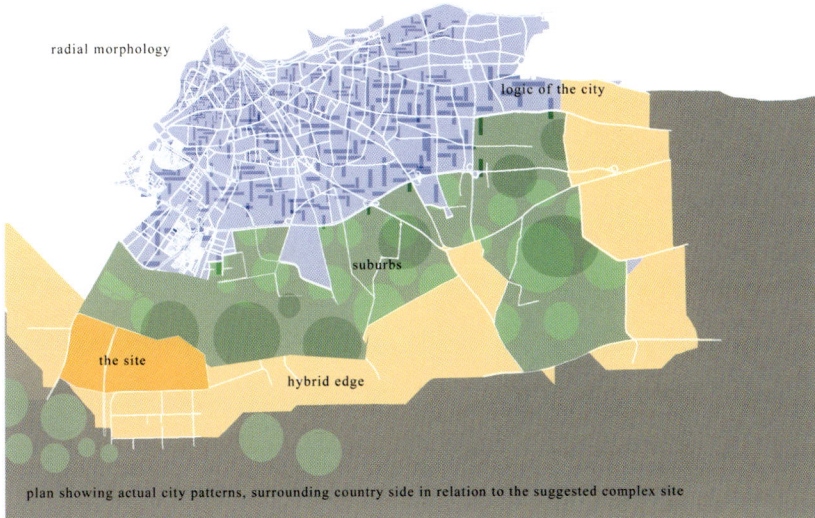

radial morphology

logic of the city

suburbs

the site

hybrid edge

plan showing actual city patterns, surrounding country side in relation to the suggested complex site

existing

sprawl

sprawl vs. planned growth

densification

channeling interaction

defining the edge

filtering the expansion

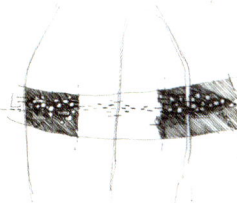

the edge, hybrid pivot

vip hotel and conference center

collision of green and city

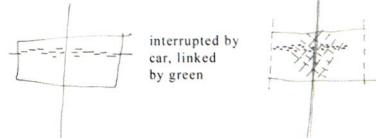

interrupted by car, linked by green

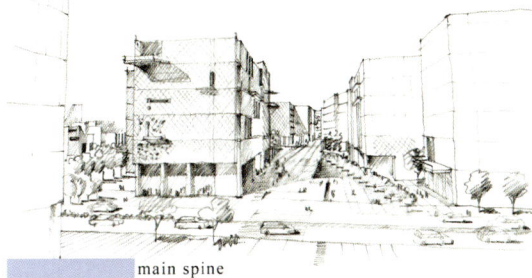

main spine

New Hybrid Typology, responding to ground situation

city

Linear structures, spine

organic structures

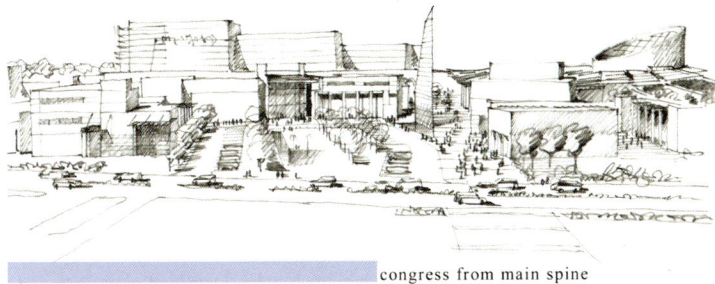

congress from main spine

view from airport highway

forest

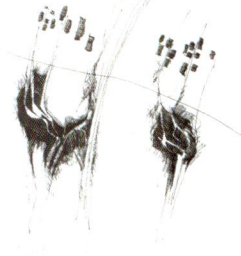

The General people's committees are configured in a linear arrangement producing a main spine that cuts through the site longitudinally. The spine is interrupted by the existing highway but connected through the emerging green platforms climbing up to hover over just above the highway. Besides "linking" the two sides of the spine, the green hill also defines the hierarchy created in the planning of the committees by opening up from the political to the more public as we move from extreme west to extreme east. As we move along the axis we see a clear defined spine intersecting with another flow coming from The Grand Plaza. This defined spine gradually climbs the green hill up opening up to a more fluid configuration where the logic of the organic structure intersects with the city logic. The committees in this part hold the more open to the public role, for example; culture, agriculture, etc...in a more interactive way and unfold into supporting s   e   r   v   i   c   e   s   .

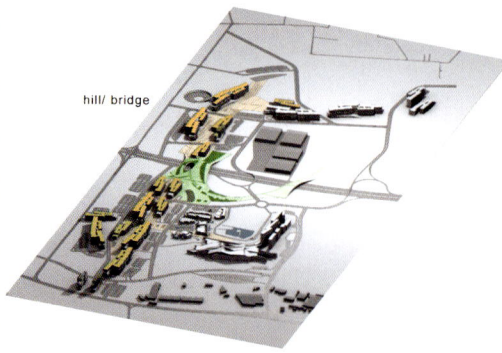

hill/ bridge

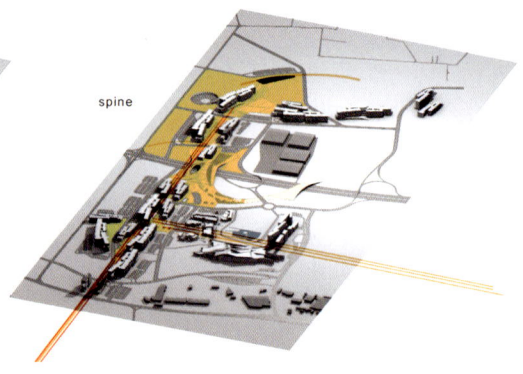

spine

Administration Complex, Tripoli

Consolidated CE – Jafar Tukan Architects, Amman

Zaha Hadid Architects London
Zaha Hadid建筑事务所 伦敦

"The new seminal government buildings will be situated in a national botanic garden representing the geographic of Libya."
"有创意的新政府大楼将坐落在一个表现利比亚地理特点的国家植物花园里。"

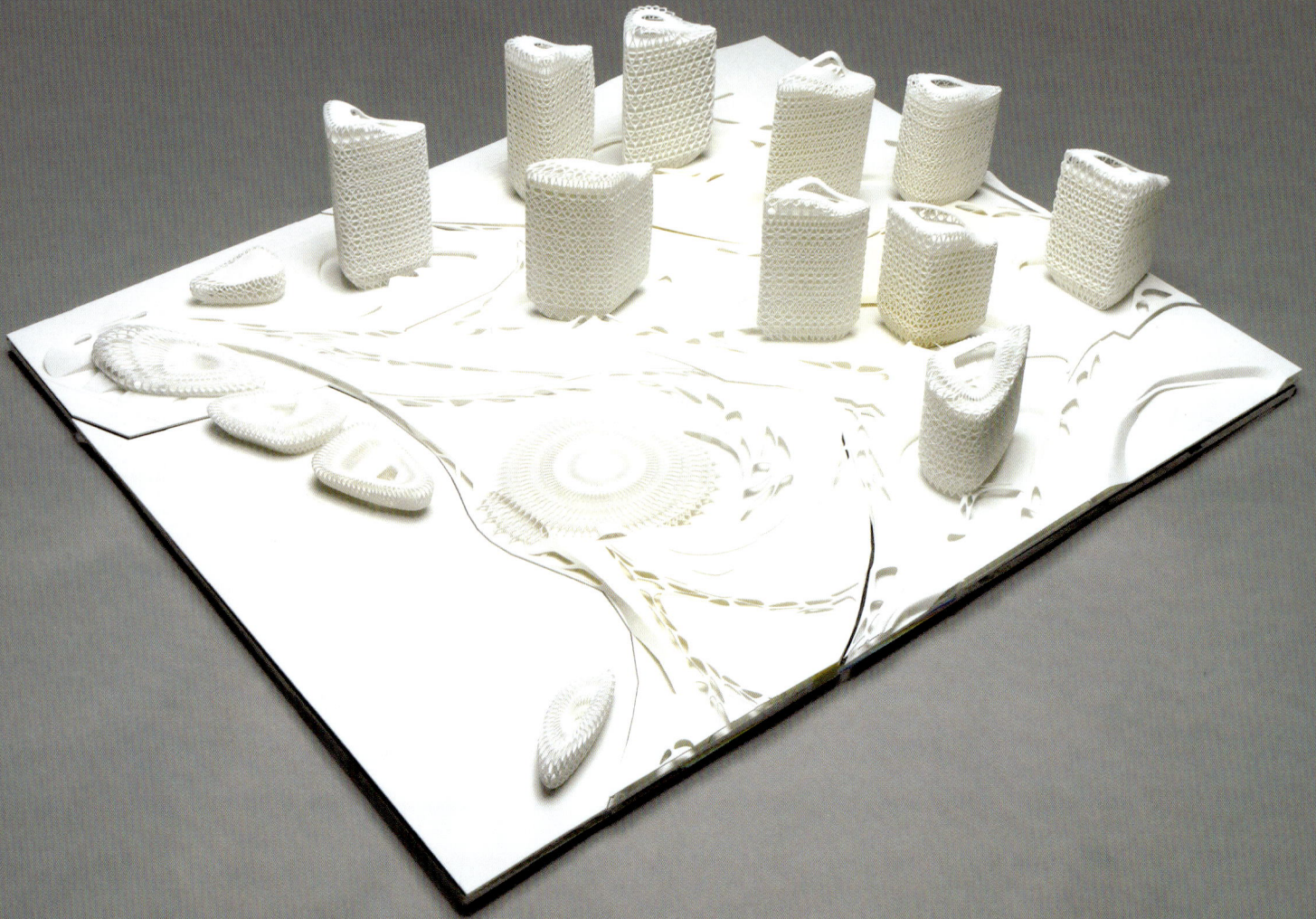

作者 Zaha Hadid, Patrick Schumacher   合作伙伴及建筑系在校生 Helmut Kinzler, Ebru Simsek, Lars Teichmann, Enrico Kleinke, Brian Dale, Lauren Barclay, Deniz Manisali, Dadatsi Dominiki, Emily Chang, Dimitris Akritopoulos, Hala Sheikh, Kelly Lee, Lillie Liu, Oznur Erboga, Pavlidou Eleni, Shih-Chin Wu, Shiqi Li, Nantapon Junngurn, Sylvia Georgiadou, Theodora Ntatsopoulou, Eirini Fountoulaki, Claus Voigtmann   专家 structural analysis: ARUP Engineering, Keith Jones, London; landscape architecture: Gross.Max, Eelco Hooftman, London; traffic: ARUP Engineering, Tim Gabbitas, London

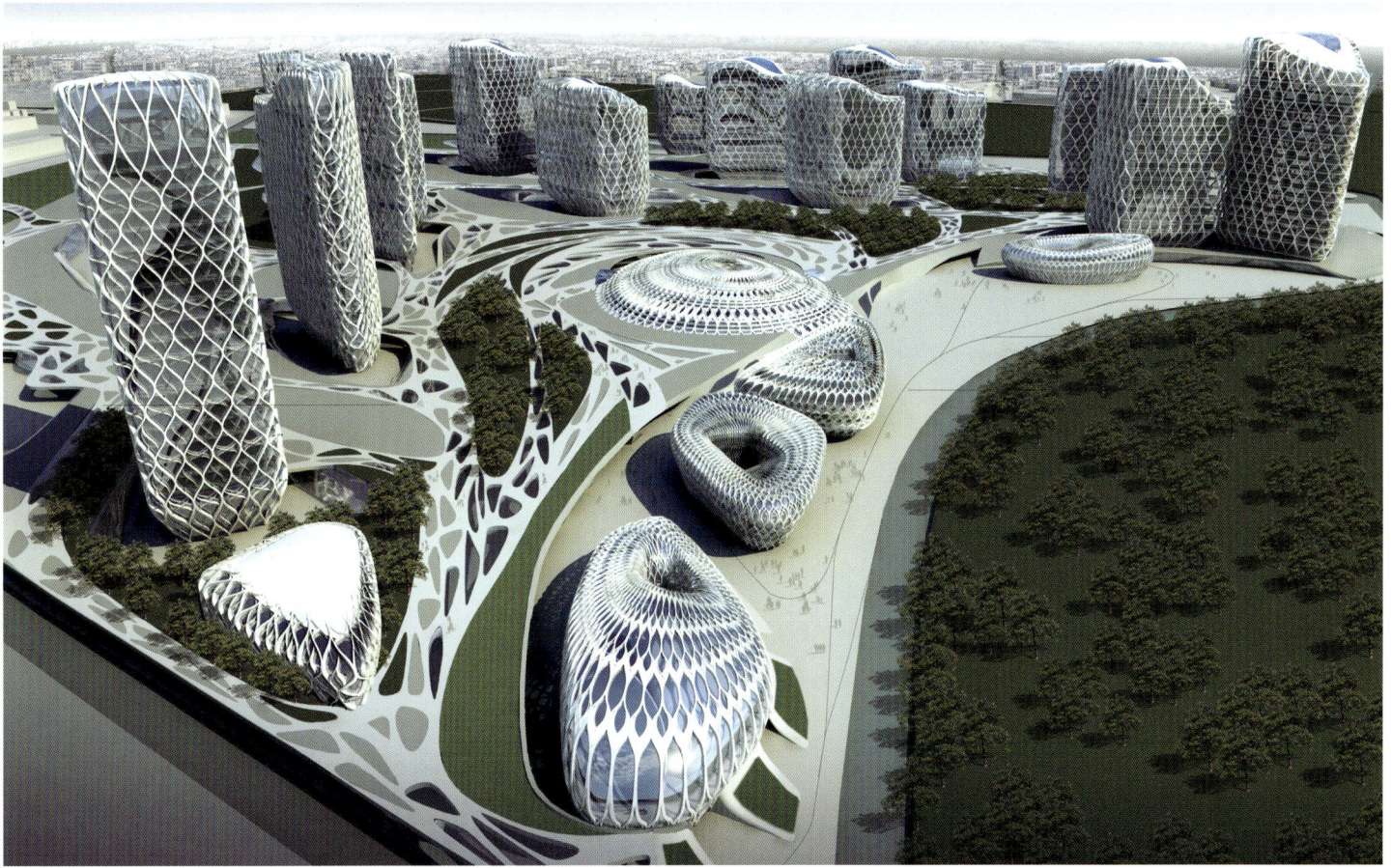

MASTER PERSPECTIVE VIEW TO THE NORTH

Landscape Concept

The new seminal governmental buildings will be situated in a unique 200 hectare National Botanic Garden representing the distinct geographic regions in Libya. The National Botanic Garden will become a striking symbol and representation of the country as a whole. The three main geographical regions in Libya which will be represented are Tripolitania, Cyrenaica and Fezzan. These regions consist of a variety of landscape typologies ranging from coast, desert and mountain. The gardens will have a public and recreational dimension and will become a public resource through conservation, education and scientific research representing the face of modern Libya for the 21st century. The dynamic configuration of the governmental quarter will be extended into the site and across the motorway by means of an undulating, free flowing, botanic spine whose shape and forms are inspired by various geomorphologic landscape formations. A circuit of curvilinear walkways provides connectivity, linkages and circulation loops. Emphasis is placed on the dynamics of ecosystems rather than to horticultural classification using native species and representing plant communities and associative habitats. The two patches of existing woodland are extended and incorporated into a botanical forest. Importantly the whole site forms part of a wider green structure and ecological corridor. The woodland can provide a landscape framework for eventual future expansion of built fabric within a coherent whole.

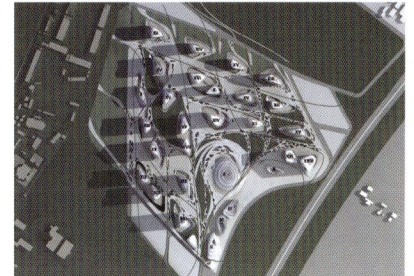

MASTER PLAN WITH NUMBER OF FLOORS

MASTER PERSPECTIVE VIEW TO THE SOUTH

LANDSCAPE CROSS SECTION

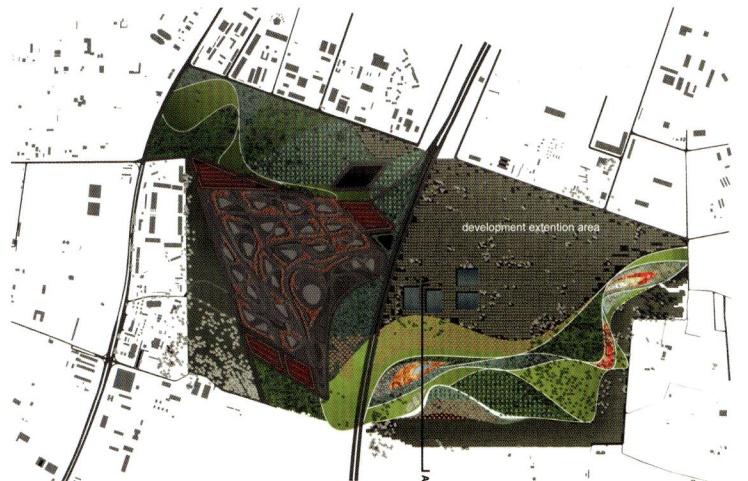

BIRD'S EYE VIEW OF OVERALL MASTER PERSPECTIVE

development extention area

OVERALL LANDSCAPE DEVELOPMENT

MASTER PERSPECTIVE

ADMINISTRATION COMPLEX IN TRIPOLI
SHEET NO.1

CONGRESS PALACE VIEW

CONGRESS HALL INTERIOR VIEW

CONFERENCE PALACE COURTYARD VIEW

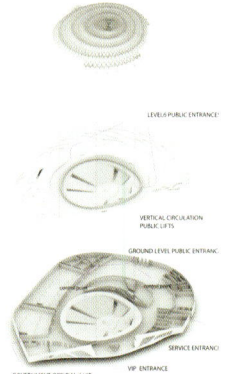

LEVEL 6 PUBLIC ENTRANCE

VERTICAL CIRCULATION PUBLIC LIFTS

GROUND LEVEL PUBLIC ENTRANCE

SERVICE ENTRANCE

VIP ENTRANCE

GOVERNMENT OFFICIAL & VIP ENTRANCE

CONGRESS HALL ACCESS DIAGRAM

ENTRANCE_PUBLIC AREAS
DINING AND KITCHEN
CORE
OFFICES
MEETING ROOMS
BOARD ROOMS
MAIN CONFERENCE HALL
GENERAL BUILDING SERVICES

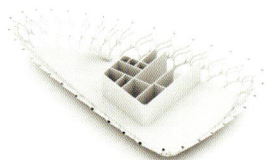

PROGRAM DIAGRAM

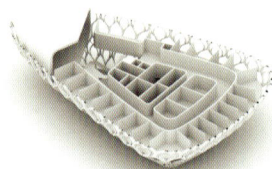

GROUND FLOOR

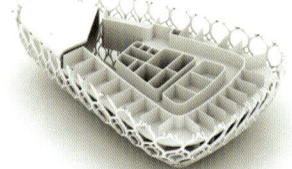

FIRST FLOOR

SECOND FLOOR

PROGRAM USAGE DIAGRAM

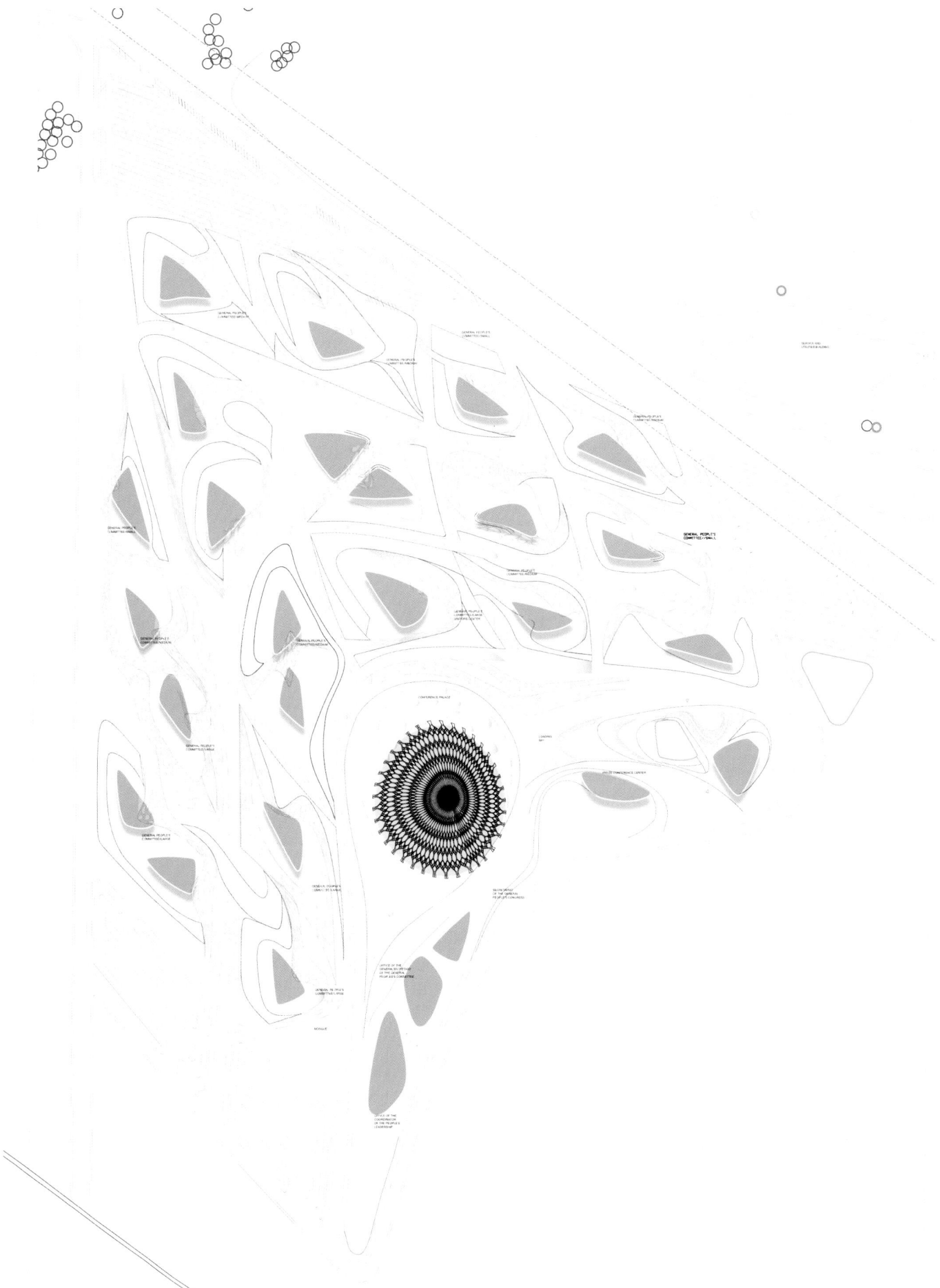

SITE PLAN   LEVEL +6.00 m

ADMINISTRATION COMPLEX IN TRIPOLI
SHEET NO.3b

SCHULTES FRANK ARCHITEKTEN Berlin
SCHULTES FRANK建筑事务所 柏林

"A 'megaform as urban landscape', the village at the foot of the mountain, a symbiosis of opposites, may well serve the need for an icon of the Libyan constitution, may well serve as a symbol of Libya's turn to openness, inside and to the world abroad."

"一个'作为城市景观的巨型建筑',山脚下的村庄,对立面的互生现象很好地为利比亚政府标志建筑的需要服务,也成为利比来内部和对外开放的标志。"

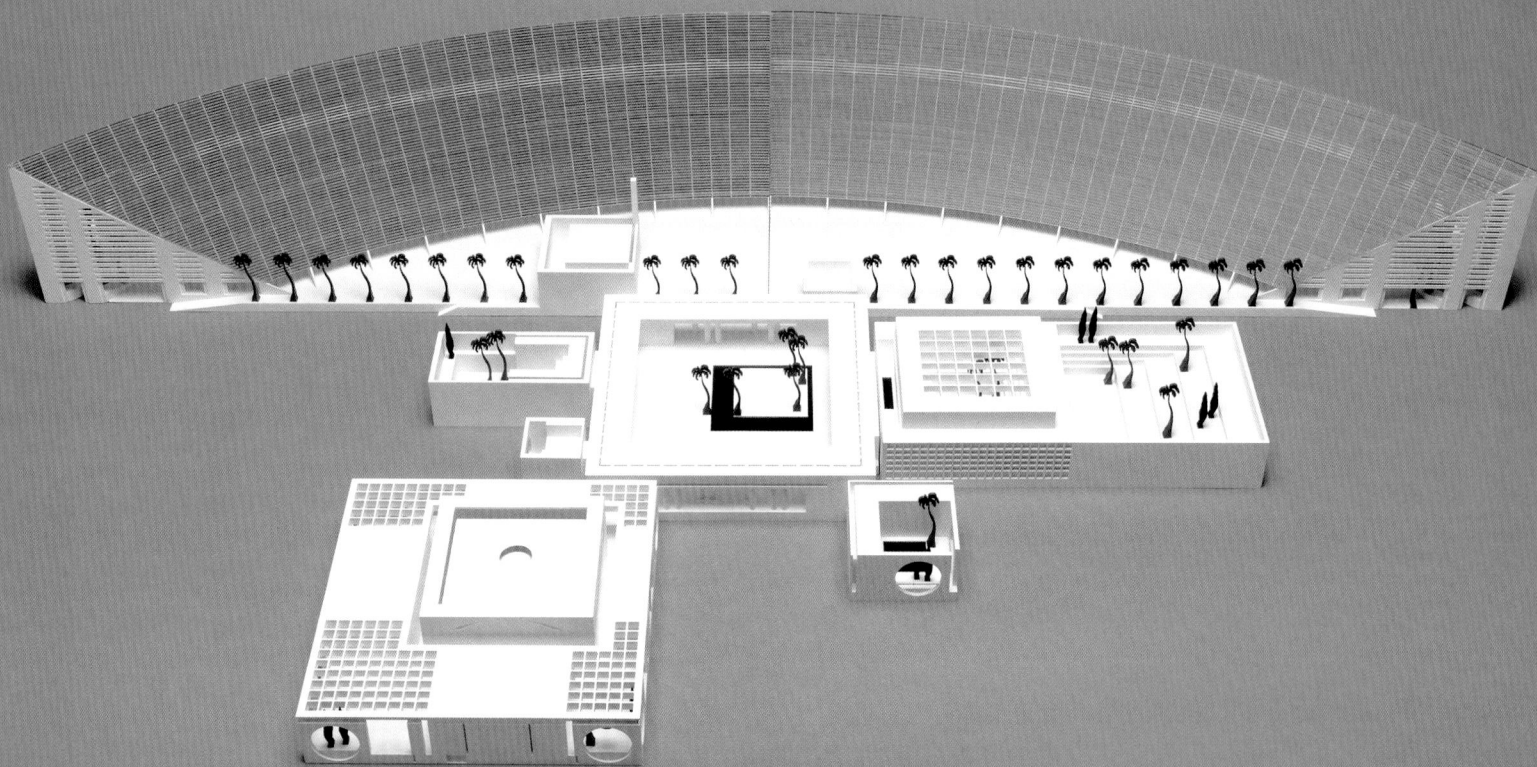

作者 Axel Schultes, Charlotte Frank   合作伙伴及建筑系在校生 Monika Bauer, Fritz Lobeck, Sören Timm, Christian Laabs, Frithjof Kahl   专家 landscape architecture: Büro Thomas, Kirsten Thomas, Berlin; building services engineering: HL Consulting and Shareholder Comp. Ltd., Klaus Daniels, Munich; traffic planning/infrastructure: GRI, The Company for Traffic Infrastructure Regionalisation and Infrastructure, Bodo Fuhrmann, Berlin

Administration Complex Tripoli

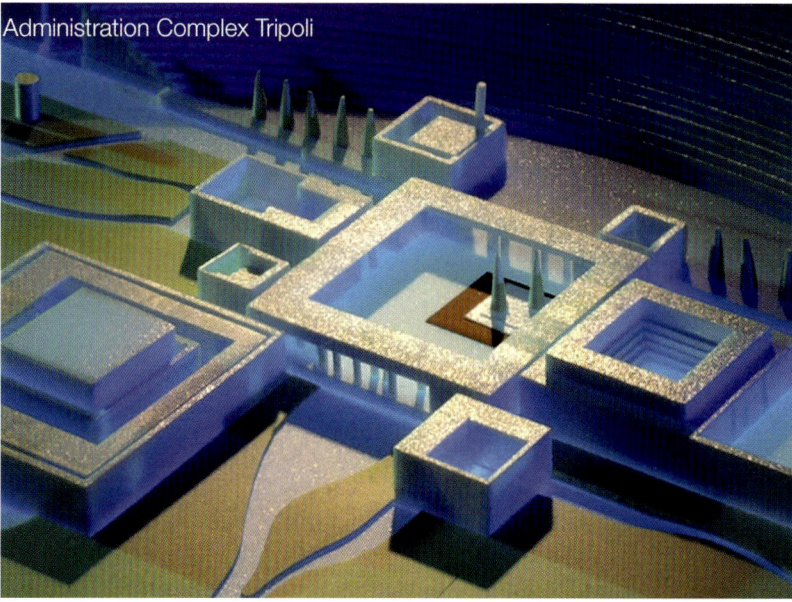

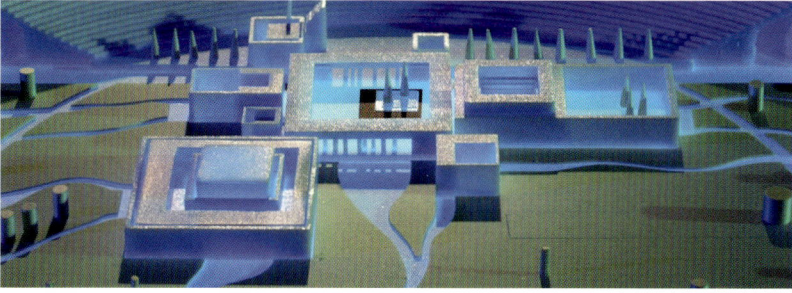

Administration Complex Tripoli

Administration Complex Tripoli

Administration Complex Tripoli

445

SCHULTES FRANK ARCHITEKTEN, Berlin

Administration Complex, Tripoli

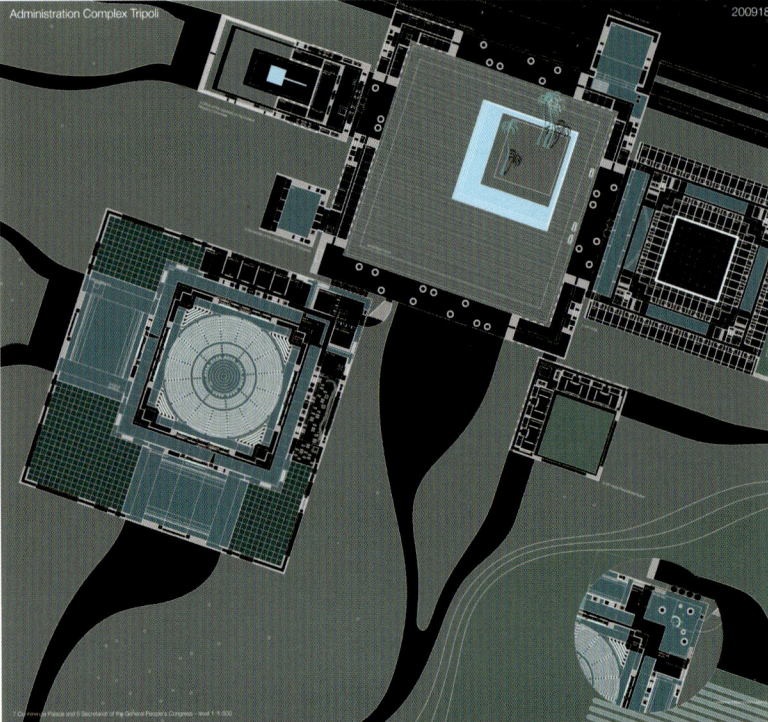

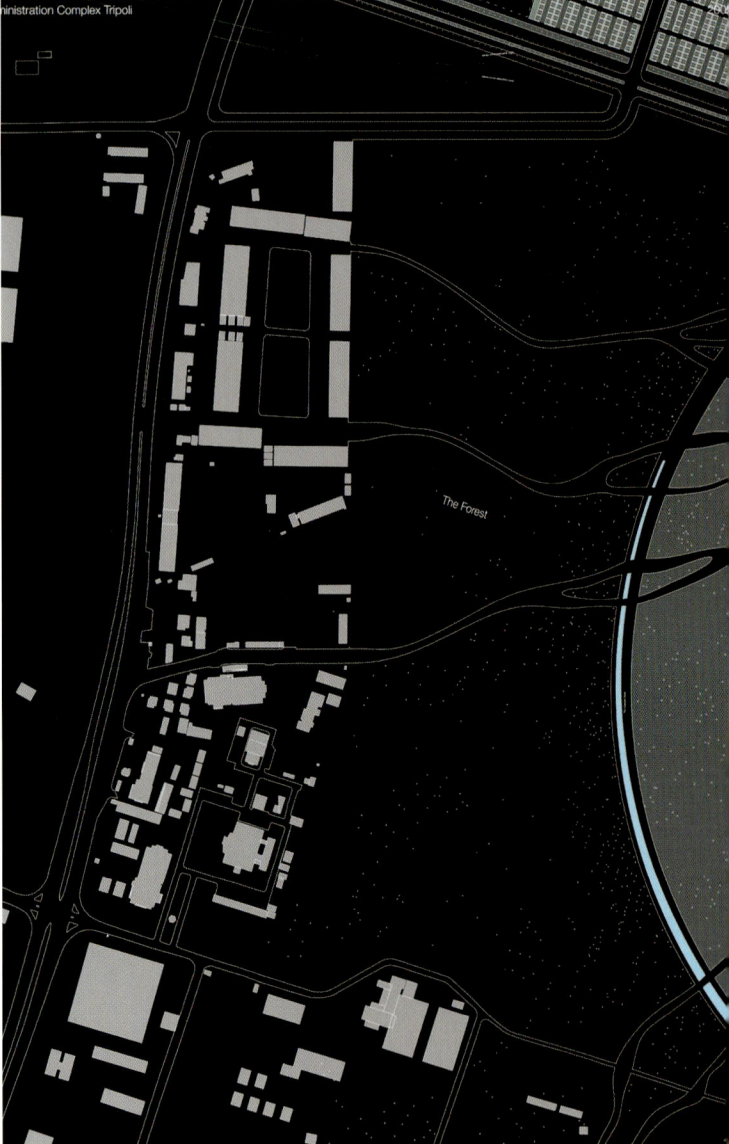

The Forest

The Committees Crescent

Jamahiriya Forum

The Pleasure Ground

The Park

The Forest

Administration Complex, Tripoli

Aedas

K

COOP HIMMELB(L)AU

Harry Seidler and

Peia As

Massimiliano Fuksas

Tower at Suk Al Thalath Al Gadeem in Tripoli

Suk Al Thalath Al Gadeem高楼 的黎波里

eihues + Kleihues

Associates

sociati

Tower at Suk Al Thalath Al Gadeem Tripoli
Suk Al Thalath Al Gadeem高楼 的黎波里

Restricted project competition preceded by an application procedure
限制严格的项目竞赛，开始前有申请程序。

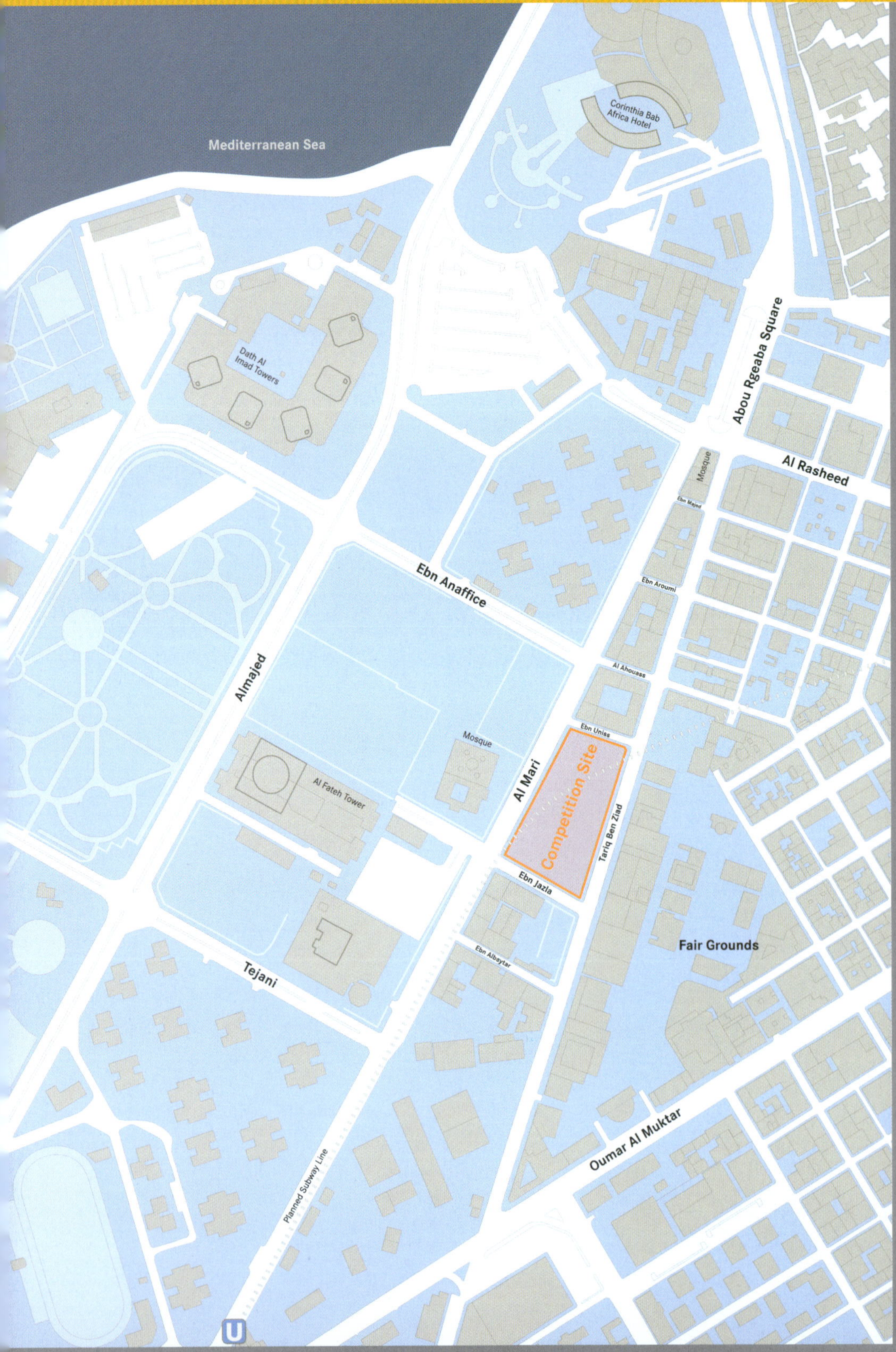

Mediterranean Sea

Corinthia Bab Africa Hotel

Dath Al Imad Towers

Abou Rgeaba Square

Mosque

Al Rasheed

Ebn Majed

Ebn Anaffice

Ebn Aroumi

Almajed

Al Ahouass

Mosque

Al Fateh Tower

Ebn Unias

Al Mari

Competition Site

Tariq Ben Ziad

Ebn Jazia

Tejani

Ebn Alhaytar

Fair Grounds

Oumar Al Muktar

Planned Subway Line

地点 Tripoli
时间 09/2007–01/2008
主办方 Libya Africa Investment
Portfolio (LAP)
参赛者 6
面积 8,200 sq m
竞赛费用 100,000 USD
专业评奖委员会
Faisal Khalil Al-Bannani, Tripoli;
Donald Bates, Melbourne/London;
Prof. Hilde Léon, Berlin;
Jafar Tukan, Amman
专家评奖委员会
Manfredi Anello, Dublin
Technical jurors
Bashir Saleh,
chairman of LAP, Tripoli;
Dr. Ali Shebani,
chairman of NUH, Tripoli;
Dr. Mustafa Mezughi,
chairman of NCB, Tripoli;
Ahmed Bashir Saad,
director "projects and
investment", LAP;
Mohsen M. Ben Halim
"technical affairs"-coordinator,
NUH, Tripoli

Tripoli has a history of ancient building traditions and sweeping political events. It has always been a gateway to the Sahara. But years of political and economic isolation have distorted the city's image overseas: Tripoli became one of "those not-so-nice places" in travel guides. Now Tripoli reopens to the world One of the few inner-city development areas is situated around the Suk Al Thalath Al Gadeem. Today it rather looks like a small industry zone. This area, between the old town, the Italian quarter, and the coast, will become a densely built-up center marked by high-rises and housing offices, residential and retail uses. The task set for the competition was to design a high-rise ensemble (105,000 square metres of gross floor area) on a site of 8,000 square metres, for a deluxe hotel, apartments and offices. As the flagship of the project, the hotel is to be a distinctive landmark, and, public space included, the ensemble is to become the center of the neighborhood. An important element of the task was to interpret local building traditions and create a sustainable response to climatic conditions, so as to give the building an unmistakably Libyan appearance.

的黎波里建筑历史悠久。它一直是撒哈拉沙漠的门户。但是多年政治和经济上的与世隔绝歪曲了的黎波里的国际形象，使它成为旅游指南中不受欢迎的地方之一。现在，的黎波里重新向世界开放。内城开发地区之一位于Suk Al Thalath Al Gadeem。这个区域现在像是一个小型工业区，将会成为一个建筑密集的中心，有许多高层标志性的建筑和办公楼、住宅区和商店。竞赛任务是在一个8000平方米的竞赛地点上设计一个高层综合性大楼，内设豪华酒店、公寓和写字间。作为项目的旗舰，酒店将成为一个显著的地标。包括公共空间，综合大楼将成为该区的中心。任务的一个重要方面是诠释当地建筑传统，对当地气候条件作出一种可持续性的反应，赋予建筑明显的利比亚外观。

Aerial view　鸟瞰图

Competition site　竞赛地点

Competition site　竞赛地点

Roman triumphal arch in Tripoli　的黎波里的罗马凯旋门

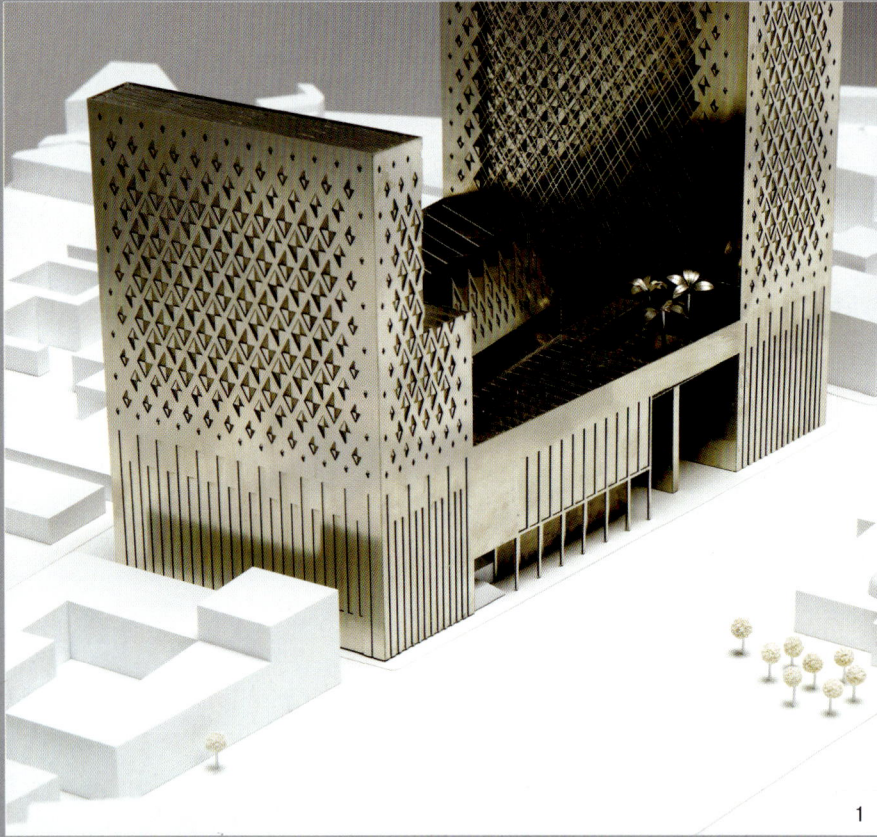

1

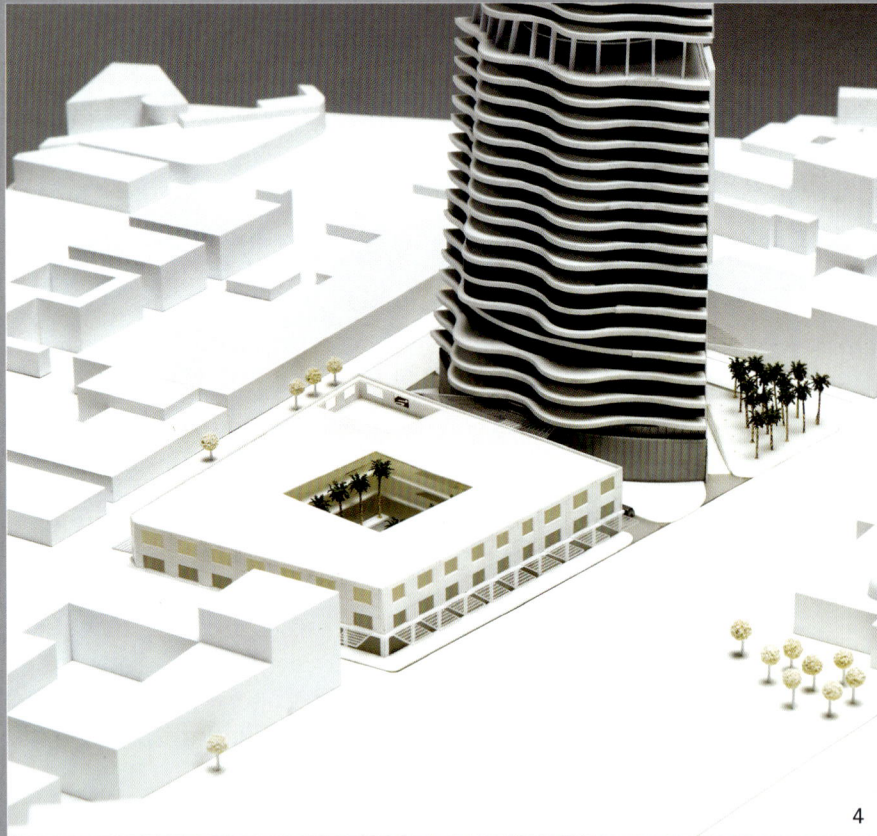

4

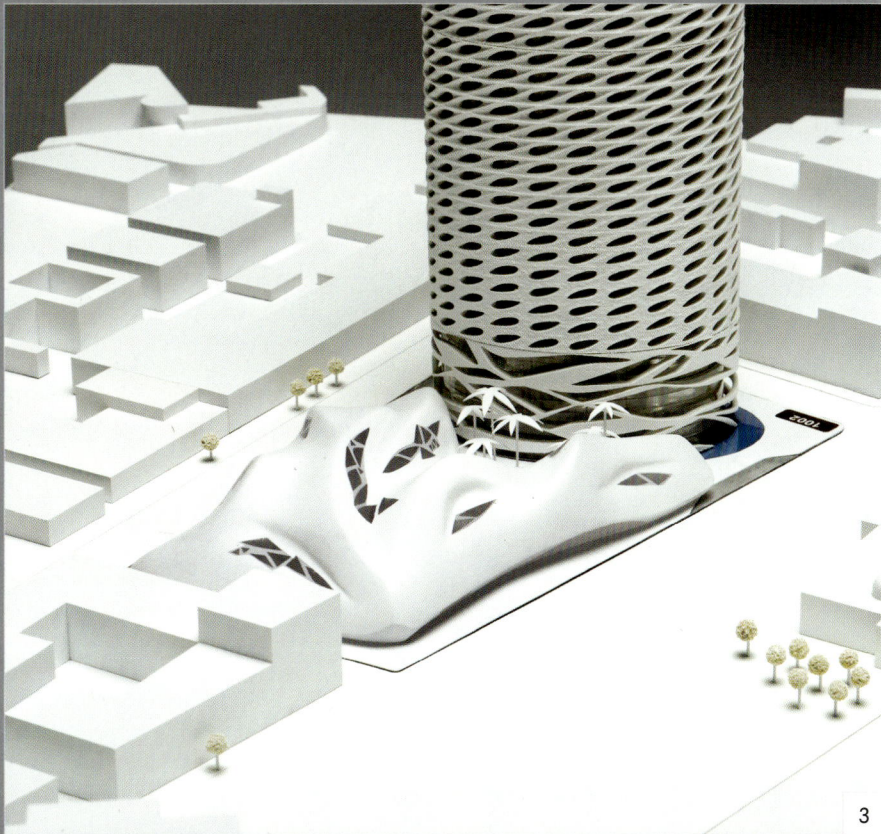

Tower at Suk Al Thalath Al Gadeem, Tripoli

Qualified competitors
合格的参赛者

1
1st prize　一等奖
Kleihues + Kleihues Gesellschaft von
Architekten mbH, Berlin

2
3rd prize　三等奖
Peia Associati,
Milan

3
3rd prize　三等奖
Massimiliano Fuksas architetto,
Rome

4
Further participant　其他参赛者
Harry Seidler and Associates,
Sydney

5
Further participant　其他参赛者
COOP HIMMELB(L)AU,
Prix/Dreibholz & Partner, Vienna

6
Further participant　其他参赛者
Aedas, Dubai

Kleihues + Kleihues Gesellschaft von Architekten mbH Berlin
Kleihues + Kleihues Gesellschaft von Architekten mbH建筑事务所 柏林

"The 'tower of falling water' is a modern interpretation of the Arabic and Libyan court house."
" '瀑布' 高楼是对阿拉伯和利比亚法院的一种现代诠释。"

作者 Jan Kleihues　合作伙伴及建筑系在校生 Götz Kern, Anna Liesicke, Alice Berresheim, Roland Block, Daniel Horn, Philipp Zora, Sonja Grötzebach, Philipp Buschmeyer, Veronika Weber, Susanne, Thesen, Marc Helbach　自由建筑师 Marc Hensel
专家 structural analysis: HMI Hartwich Mertens Ingenieure, Berlin; fire protection: Büro für Brandschutz (Prein), Wuppertal; landscape architecture: ST Raum A Gesellschaft von Landschaftsarchitekten mbH, Berlin; façade engineering: Erich Mosbacher Beratungs- und Planungsgesellschaft für Fassadentechnik mbH, Friedrichshafen

Kleihues + Kleihues Gesellschaft von Architekten mbH, Berlin

Tower at Suk Al Thalath Al Gadeem, Tripoli

Mediterranean Sea

Ahou Rgada Square

Al Rasheed

Ebn Anafice

Al Matii

Zaid Ben Zaid

Office Tower

Hotel

Al Fateh Tower

Oumar Al Muktar

+255.70*
+230.25
+217.25

+100.25

+83.84*

+52.61

+29.40

+0.00

Kleihues + Kleihues Gesellschaft von Architekten mbH, Berlin

Tower at Suk Al Thalath Al Gadeem, Tripoli

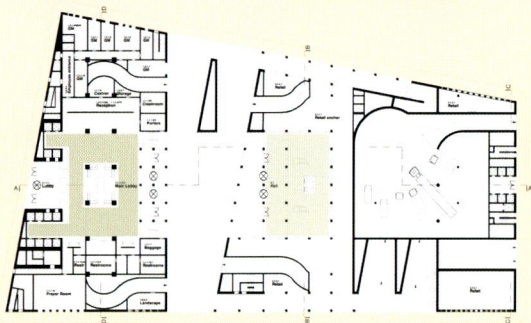

± 0,00 m / Groundfloor and -2,00 m / Delivery

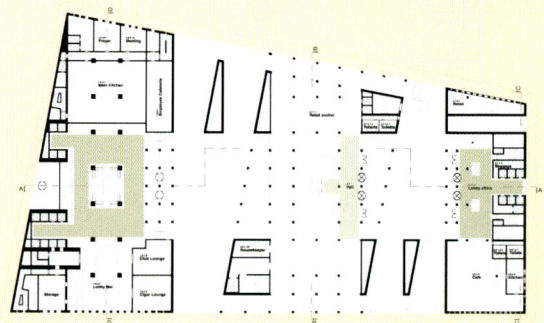

+ 3,50 m / 1st floor

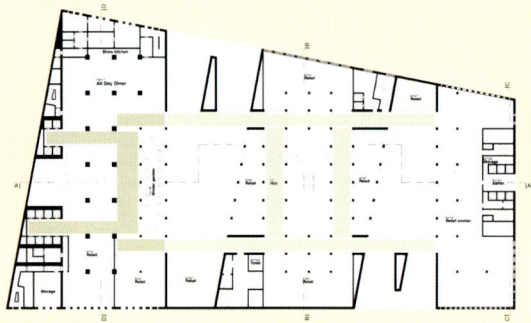

+ 8,50 m / 2nd floor

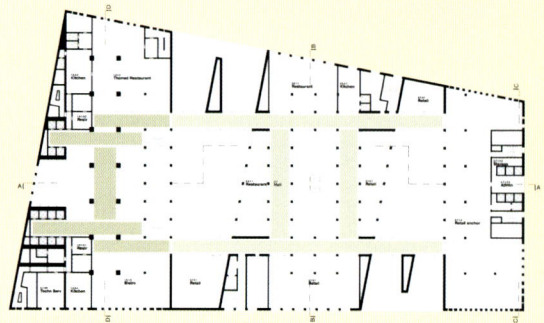

+ 13,50 m / 3rd floor

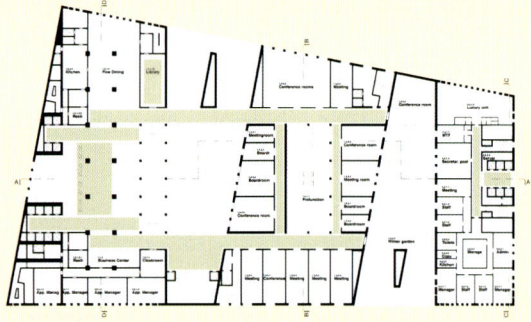

+ 18,50 m / 4th floor

+ 23,50 m / 5th floor

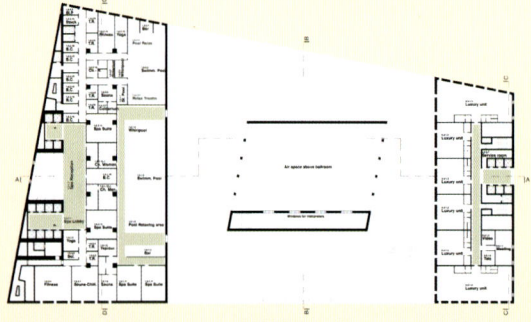

+ 28,50 m / 6th floor

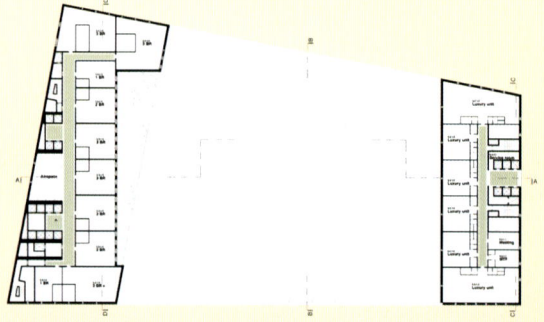

+ 35,00 m / 7th floor

Tower at Suk Al Gadeem, Tripoli, Libya

Elevation of the facade, scale 1 : 50

Detail section A, scale 1 : 50

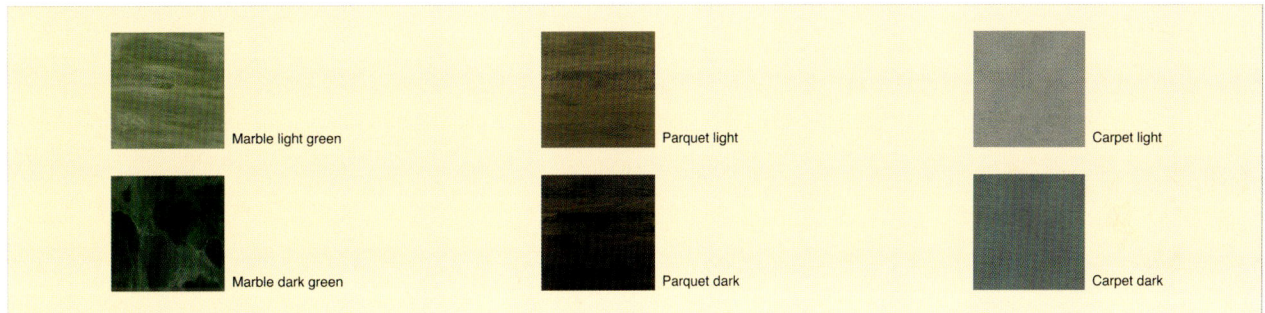

Marble light green   Parquet light   Carpet light

Marble dark green   Parquet dark   Carpet dark

Material and furniture proposal

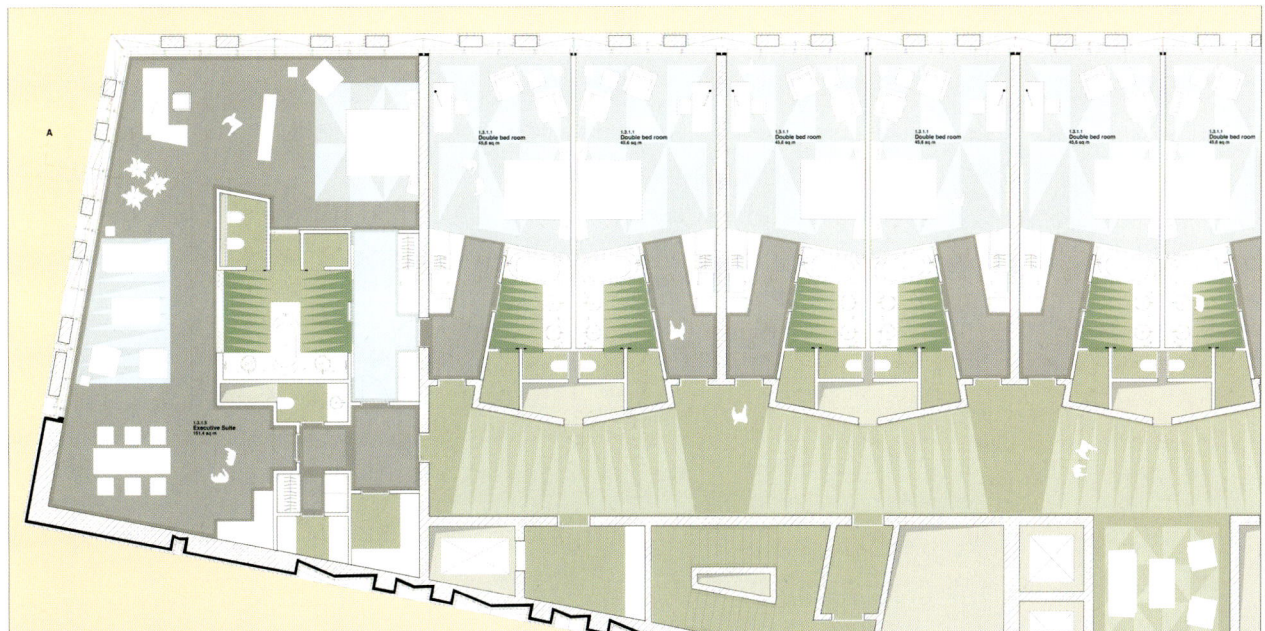

Hotel rooms, scale 1 : 50

Sheet 12

459

Section B - B

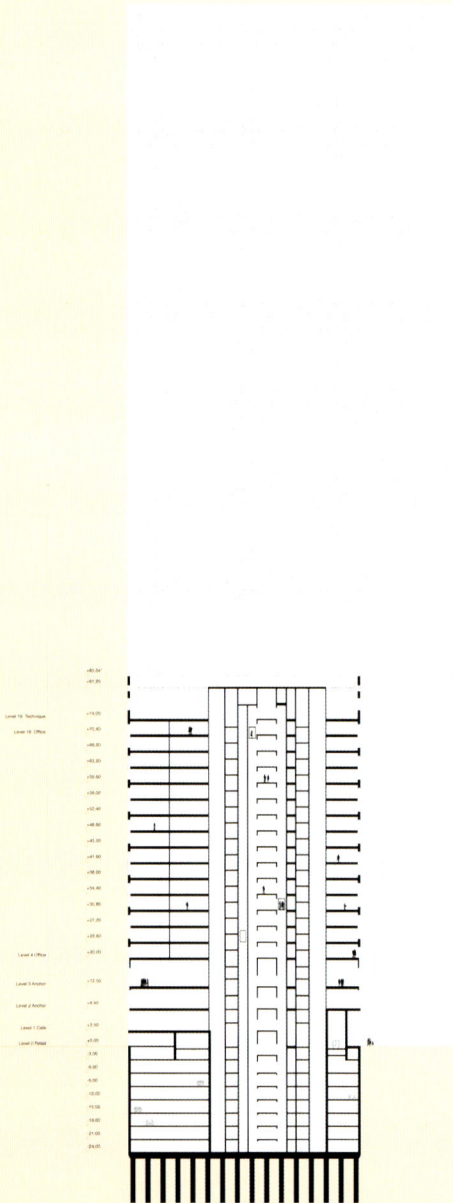

Section C - C

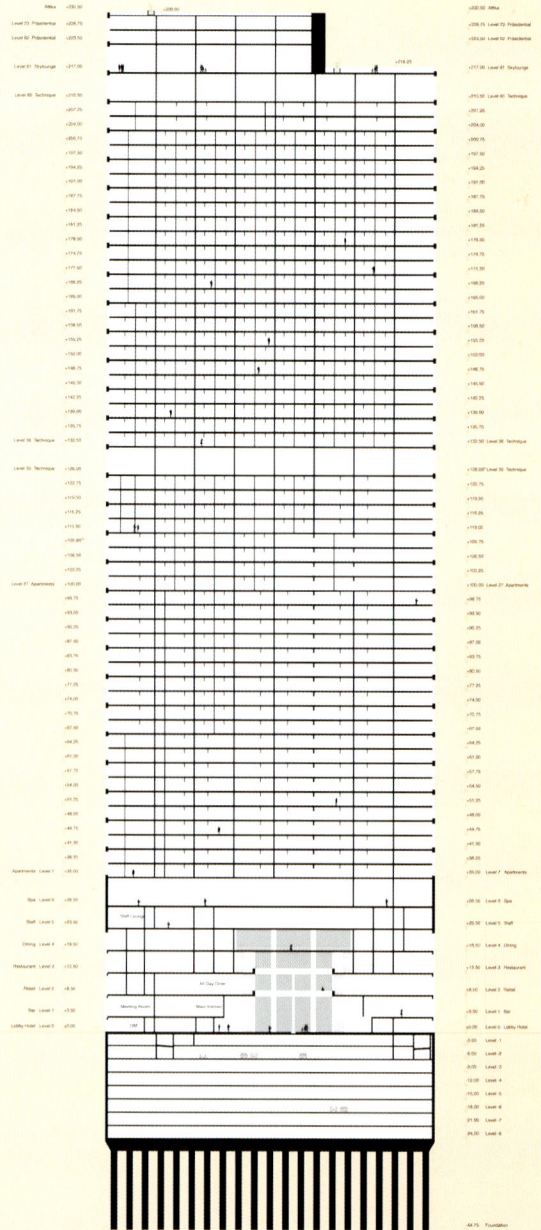

Section D - D

Tower at Suk Al Gadeem, Tripoli, Libya

Standard Hotel Guest Room

Lobby Hotel

Lobby Hotel with Reception Desk and Falling Water

Elevation on Al Mari Street

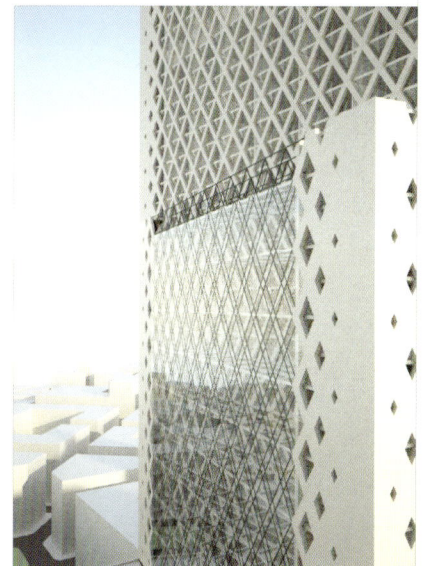

Elevation with Wintergarden   Sheet 13

Peia Associati Milan
Peia Associati建筑事务所 米兰

"The main concept is based on the formal abstraction of the palm tree trunk shape as the most recognisable metaphorical meaning of the African identity and unity. A perfect balance between strength and sinuosity."

"主要概念建立在棕榈树干是非洲身份和团结最易辨识的比喻意义的形式抽象上。力量和弯曲的完美平衡。"

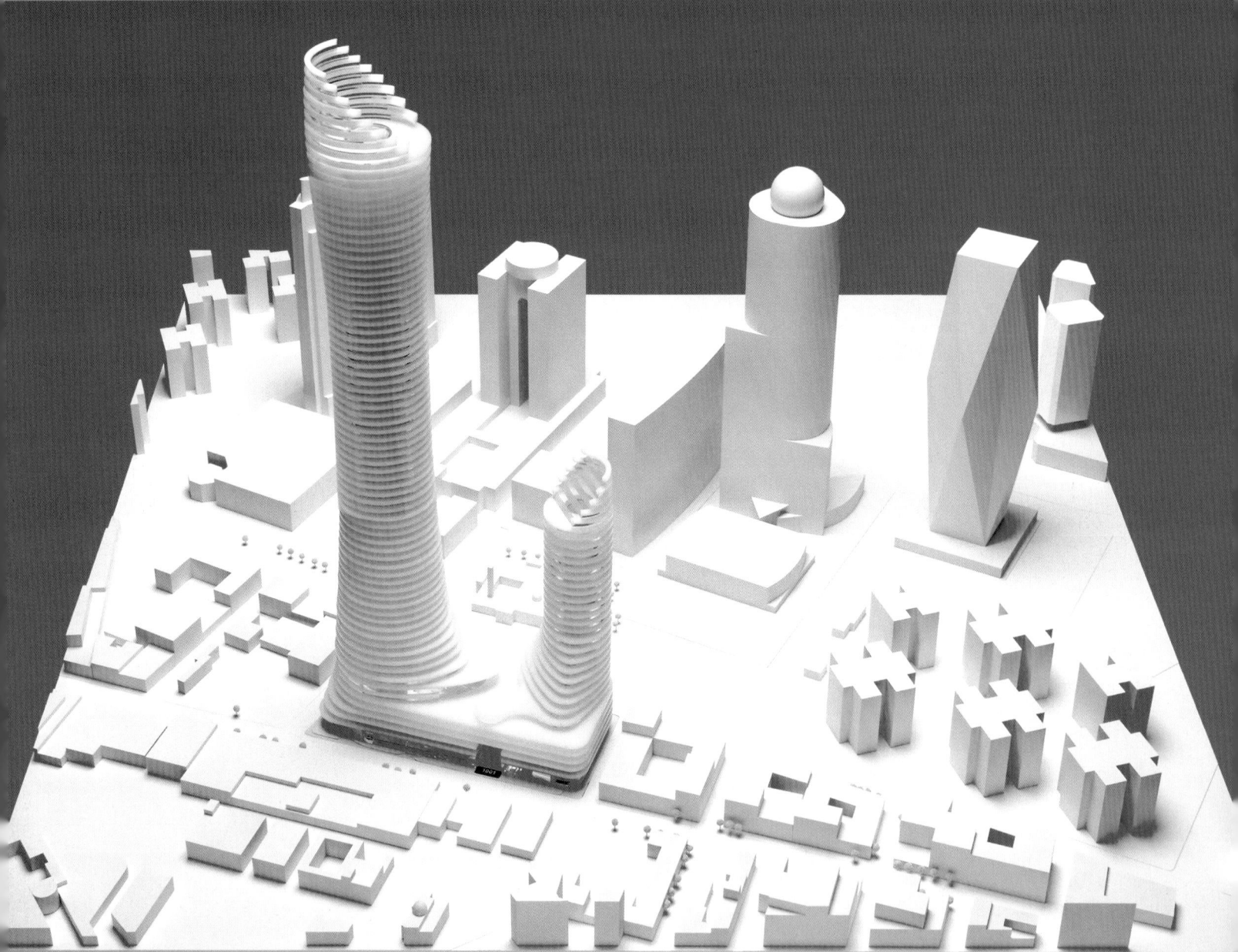

作者 Giampero Peia   合作伙伴及建筑系在校生 Luca Bonazzoli, Massimo de Mauro, Alessandro, Garzaro, Johnny Hugnot, Mayumi Kishimoto, Andrea Martelu, Lorenzo Merloni, Marta Nasazzi, Matteo, Nicotra, Francesca Patti, Anna Pavoni, Michaela, Ricciotti, Luigi Vaciago   专家 building services: Hani Awni Hawamdeh; AEB Engineering, Doha, Qatar; structural analysis: Amr el Khady, AEB Engineering Doha, Qatar; architekt: Cenon D. Ditan Jr., AEB Engineering, Doha, Qatar

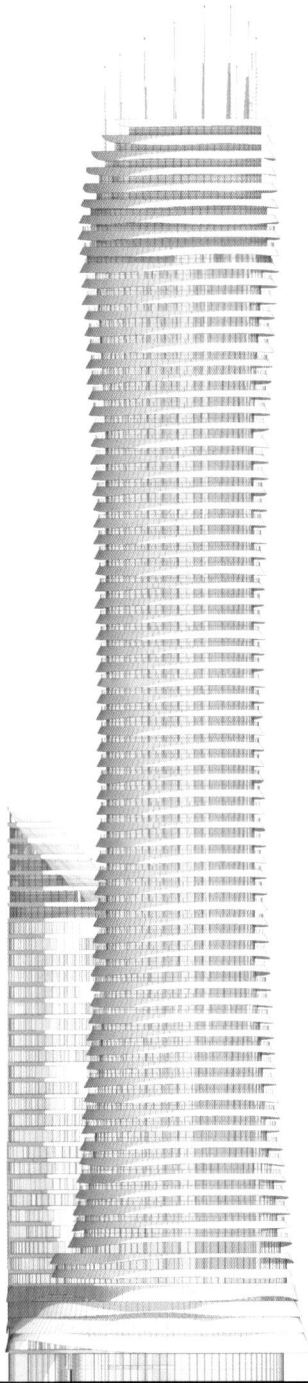

Ebn Jazla elevation
(South/south-east elevation)

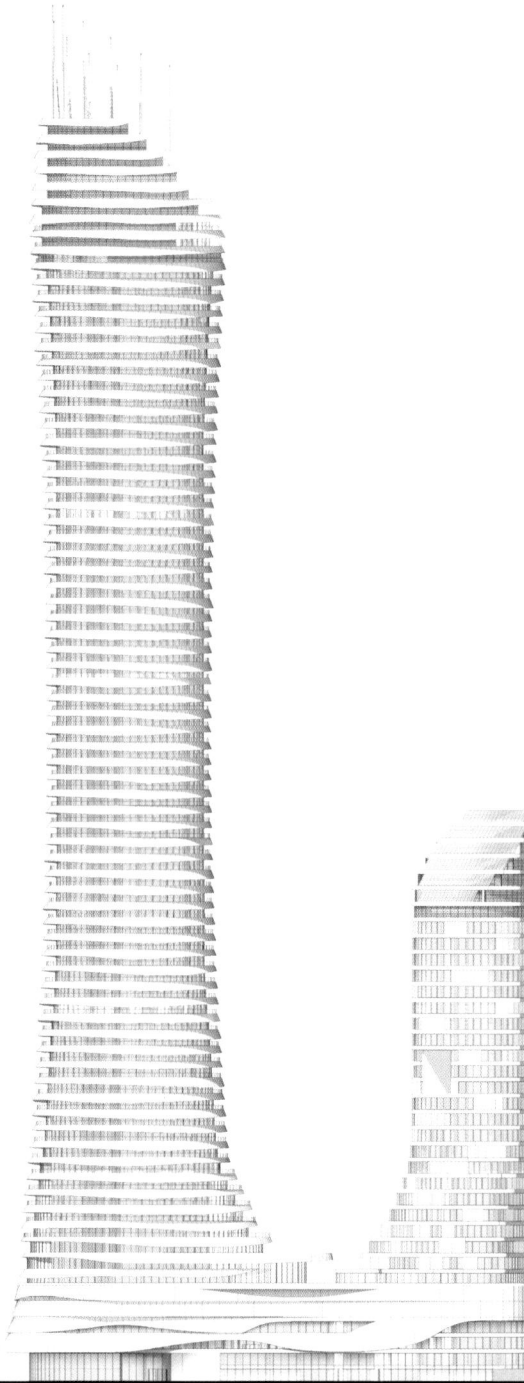

Tariq Ben Ziad elevation
(East/south-east elevation)

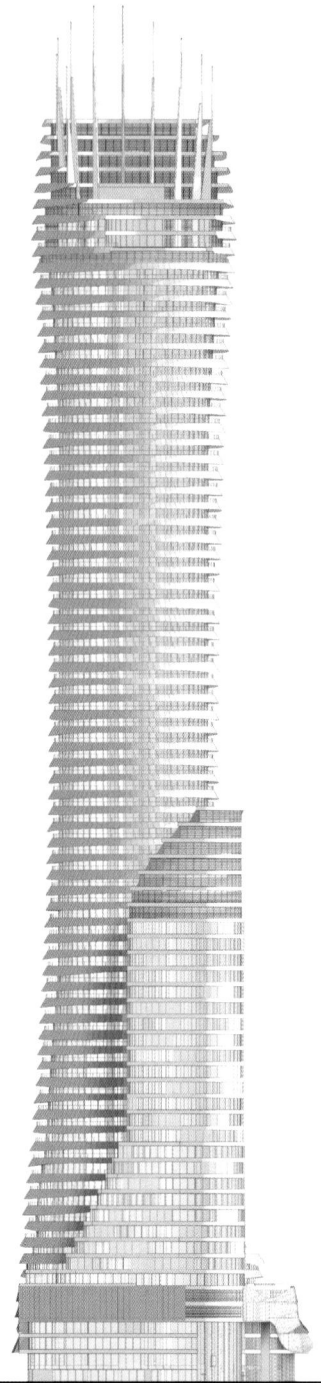

Ebn Uniss elevation
(North/north-east elevation)

North
Makkah

Mosque

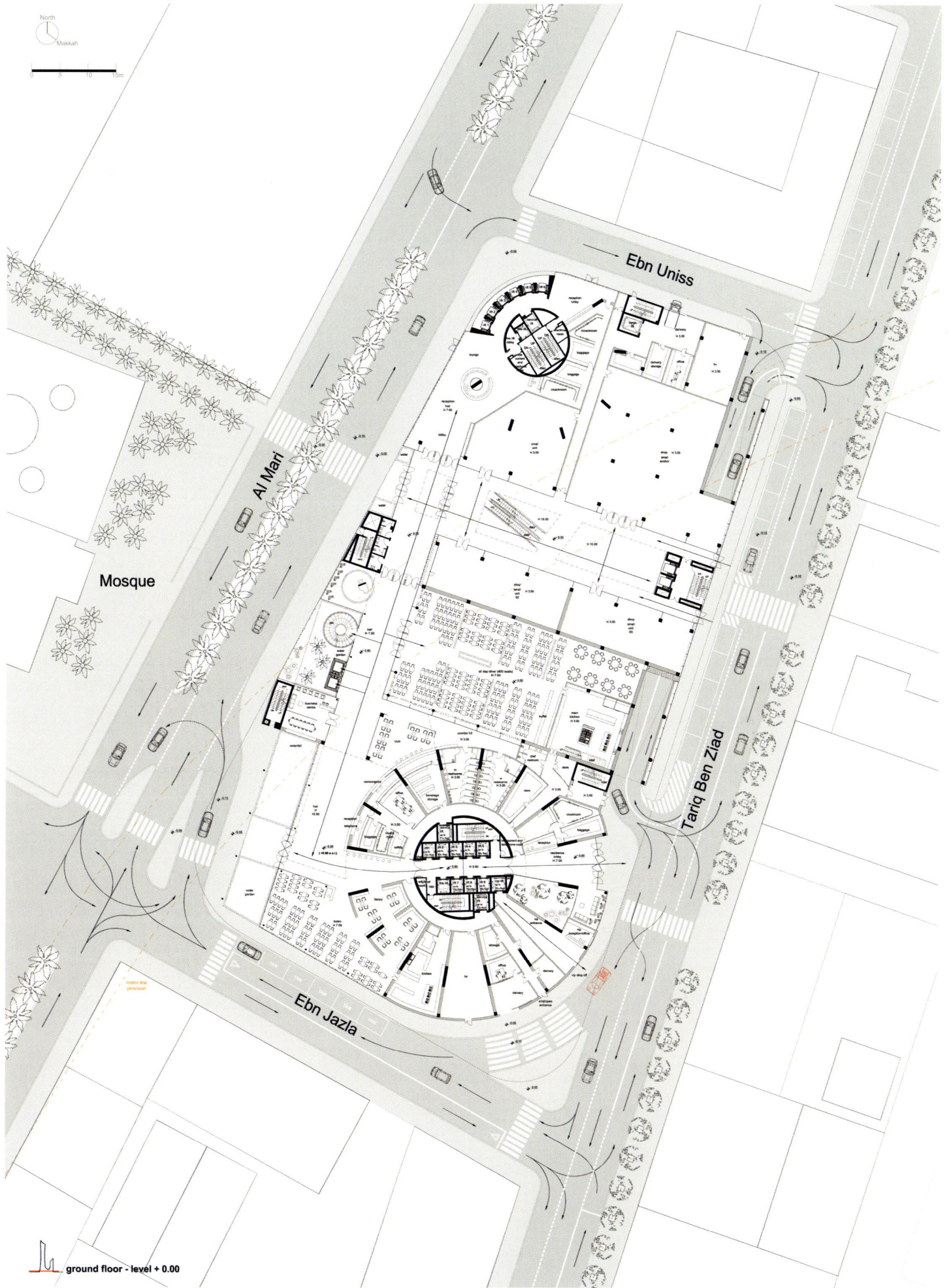

North
Makkah

0 5 10 15m

Ebn Uniss

Al Mari

Mosque

Tariq Ben Ziad

Ebn Jazla

ground floor - level + 0.00

Peia Associati, Milan

465

Tower at Suk Al Thalath Al Gadeem, Tripoli

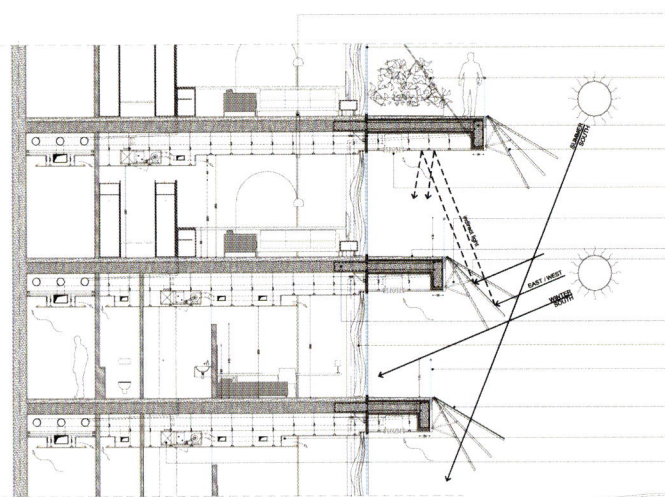

detailed longitudinal section

detailed cross section

finishing 8 mm ceramic porcelain tiles 8cm screed
8 mm impact-sound insulation, 30 cm lightweight
reinforced concrete, 60-120 cm service space
12+12 mm plasterboard ceiling

double glazing (4+4/12 argon cavity/5+5) and low
emission (face2) internal coating

stainless steel tie bar 30mm to support
cantilever slab over 3 meter

air conditioning silent duct

concrete external ring beam 300x800mm

aluminium sheet with micro perforated decoration
thickness 2mm and maximum lenght 2400mm
angle 60° - east and west side

aluminium sand trap louver to eliminate
sand admittance in sandy weather (TYP)
and for individual fresh air intake

aluminium profile pivoting bracket 40mm

resin floor finishing: 30mm max screed and 30mm
extruded polistirene thermal insulation

angle 45° - south east / south side

concrete radial beam 500x800mm

finishing and screed 125mm
acoustic insulation 10mm, concrete slab 300mm

concrete ring beam 1000x300mm

silk curtain and blackout curtain and double railing

stainless steel handrail 60mm diam
balustrade 12+12 with tempered extraclear safety
glass and heat strengthened glass (2mm PVB)

angle 30° - south side
eventual photovoltaic panel on south side

cyma reversa with recessed light and air diffuser
linear grid around the bed plasterboard 15 mm)

HAUC / AC with motorized damp to control
the quantity of admitted fresh air (control by BMS)

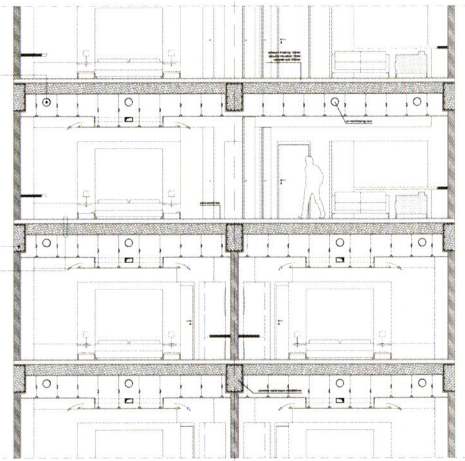

teak wooden timber slats

swinging panel to close
view toward outside
assure privacy

trasparent glazing

recessed carpet

lock

ALL BEDROOMS AND BATHROOMS HAVE THE SAME
ENLARGED PANORAMIC VIEW

inspection panels for M.E.P. shafts in the corridor

sand stone tiles 60x60

white italian marble (calacatta)

teak wood panels

red persian marble

sand stone tiles 60x60

1. presidential suites II - sqm 365,0 - n°2 - from 66 th to 68 th
2. double bed rooms - sqm 42,5 - n°247 - from 39 th to 55 th
3. junior suites - sqm 63,3 - n°36 - from 56 th to 59 th
4. suites - sqm 86,5 - n°25 - from 60 th to 62 th
5. executive suites - sqm 108,5 - n°16 - from 63 th to 65 th

hotel rooms floor plan

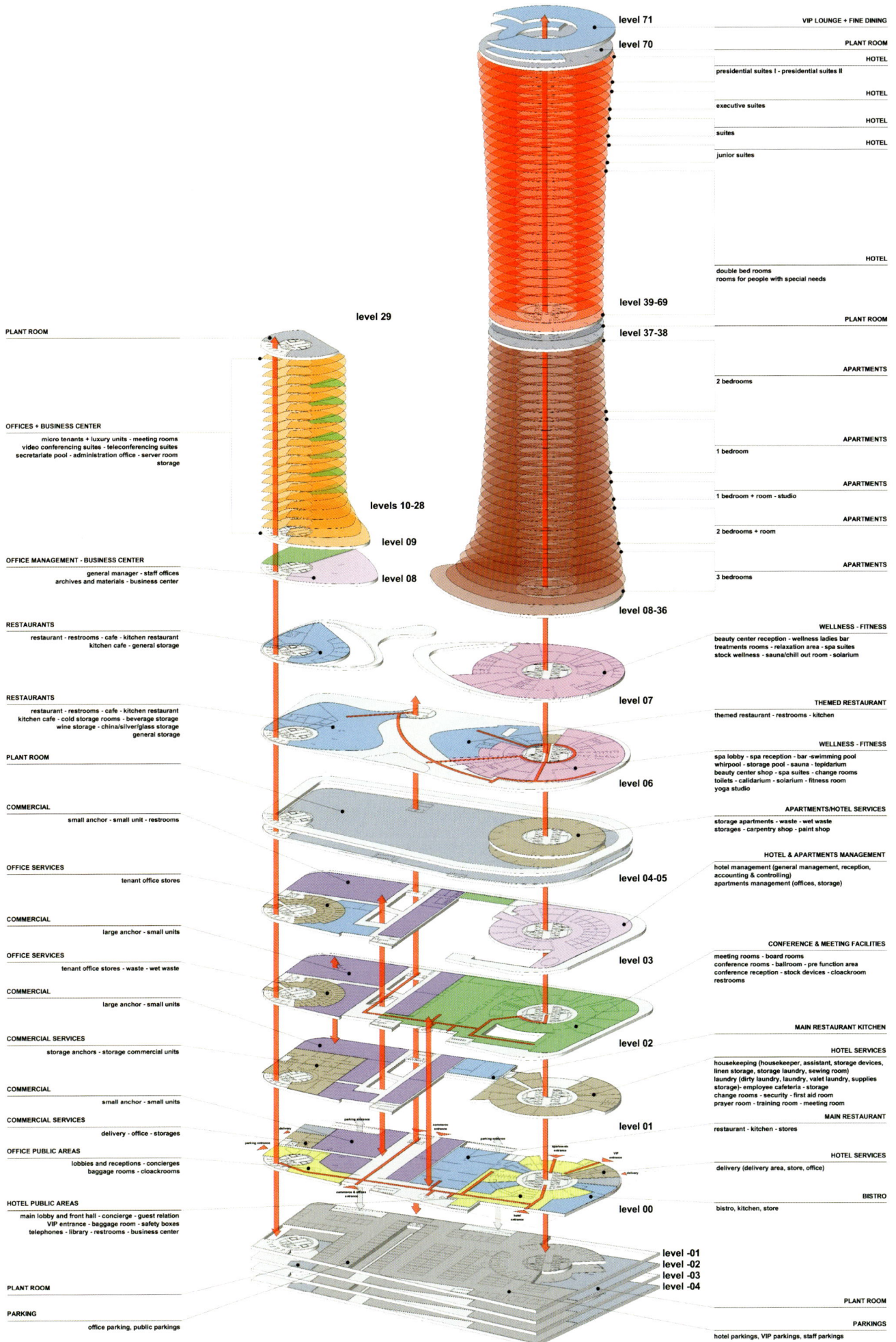

PLANT ROOM

OFFICES + BUSINESS CENTER
micro tenants + luxury units - meeting rooms
video conferencing suites - teleconferencing suites
secretariate pool - administration office - server room
storage

OFFICE MANAGEMENT - BUSINESS CENTER
general manager - staff offices
archives and materials - business center

RESTAURANTS
restaurant - restrooms - cafe - kitchen restaurant
kitchen cafe - general storage

RESTAURANTS
restaurant - restrooms - cafe - kitchen restaurant
kitchen cafe - cold storage rooms - beverage storage
wine storage - china/silver/glass storage
general storage

PLANT ROOM

COMMERCIAL
small anchor - small unit - restrooms

OFFICE SERVICES
tenant office stores

COMMERCIAL
large anchor - small units

OFFICE SERVICES
tenant office stores - waste - wet waste

COMMERCIAL
large anchor - small units

COMMERCIAL SERVICES
storage anchors - storage commercial units

COMMERCIAL
small anchor - small units

COMMERCIAL SERVICES
delivery - office - storages

OFFICE PUBLIC AREAS
lobbies and receptions - concierges
baggage rooms - cloackrooms

HOTEL PUBLIC AREAS
main lobby and front hall - concierge - guest relation
VIP entrance - baggage room - safety boxes
telephones - library - restrooms - business center

PLANT ROOM

PARKING
office parking, public parkings

level 29

levels 10-28

level 09

level 08

level 39-69

level 37-38

level 08-36

level 07

level 06

level 04-05

level 03

level 02

level 01

level 00

level -01
level -02
level -03
level -04

VIP LOUNGE + FINE DINING

PLANT ROOM

HOTEL
presidential suites I - presidential suites II

HOTEL
executive suites

HOTEL
suites

HOTEL
junior suites

HOTEL
double bed rooms
rooms for people with special needs

PLANT ROOM

APARTMENTS
2 bedrooms

APARTMENTS
1 bedroom

APARTMENTS
1 bedroom + room - studio

APARTMENTS
2 bedrooms + room

APARTMENTS
3 bedrooms

WELLNESS - FITNESS
beauty center reception - wellness ladies bar
treatments rooms - relaxation area - spa suites
stock wellness - sauna/chill out room - solarium

THEMED RESTAURANT
themed restaurant - restrooms - kitchen

WELLNESS - FITNESS
spa lobby - spa reception - bar -swimming pool
whirpool - storage pool - sauna - tepidarium
beauty center shop - spa suites - change rooms
toilets - calidarium - solarium - fitness room
yoga studio

APARTMENTS/HOTEL SERVICES
storage apartments - waste - wet waste
storages - carpentry shop - paint shop

HOTEL & APARTMENTS MANAGEMENT
hotel management (general management, reception,
accounting & controlling)
apartments management (offices, storage)

CONFERENCE & MEETING FACILITIES
meeting rooms - board rooms
conference rooms - ballroom - pre function area
conference reception - stock devices - cloackroom
restrooms

MAIN RESTAURANT KITCHEN

HOTEL SERVICES
housekeeping (housekeeper, assistant, storage devices,
linen storage, storage laundry, sewing room)
laundry (dirty laundry, laundry, valet laundry, supplies
storage)- employee cafeteria - storage
change rooms - security - first aid room
prayer room - training room - meeting room

MAIN RESTAURANT
restaurant - kitchen - stores

HOTEL SERVICES
delivery (delivery area, store, office)

BISTRO
bistro, kitchen, store

PLANT ROOM

PARKINGS
hotel parkings, VIP parkings, staff parkings

Massimiliano Fuksas architetto Rome
Massimiliano Fuksas建筑事务所 罗马

"... a dramatic and elegant tower that can become a symbol of both the country's proud and deep heritage and of its position as a progressive cultural leader in the 21st century."

"一个引人注目的和优美文雅的塔楼可以成为国家引人为豪的深厚传统的象征，同时也是作为21世纪先进文化的领袖。"

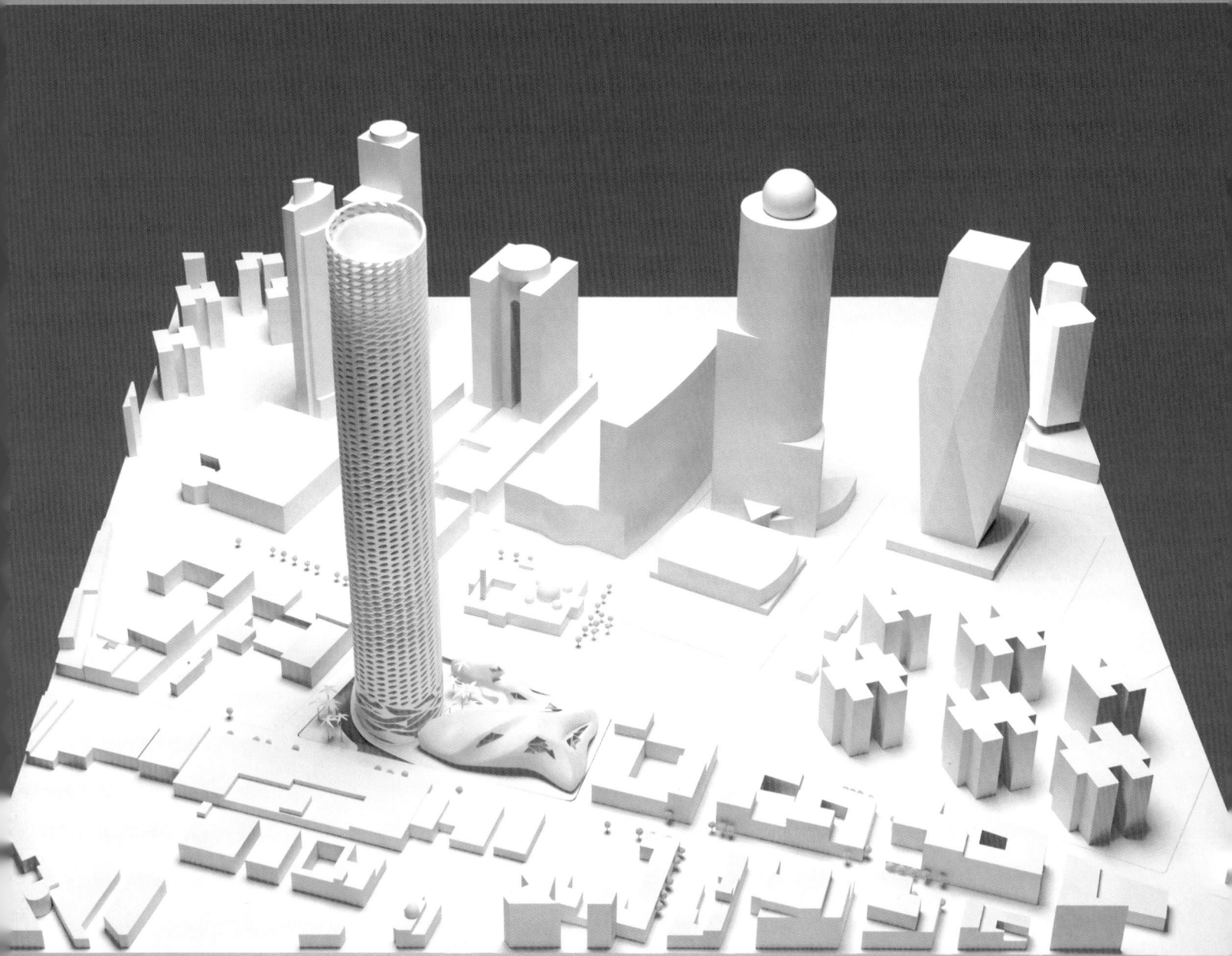

作者 Massimiliano Fuksas, Doriana Mandrelli 合作伙伴及建筑系在校生 Grazia Patruno, Serena Mignatti, Joshua Mackley, Tommaso Villa, Nicola Cabiati, Luca Vernocchi 专家 structural analysis: Knippers und Helbig Beratende Ingenieure, Stuttgart

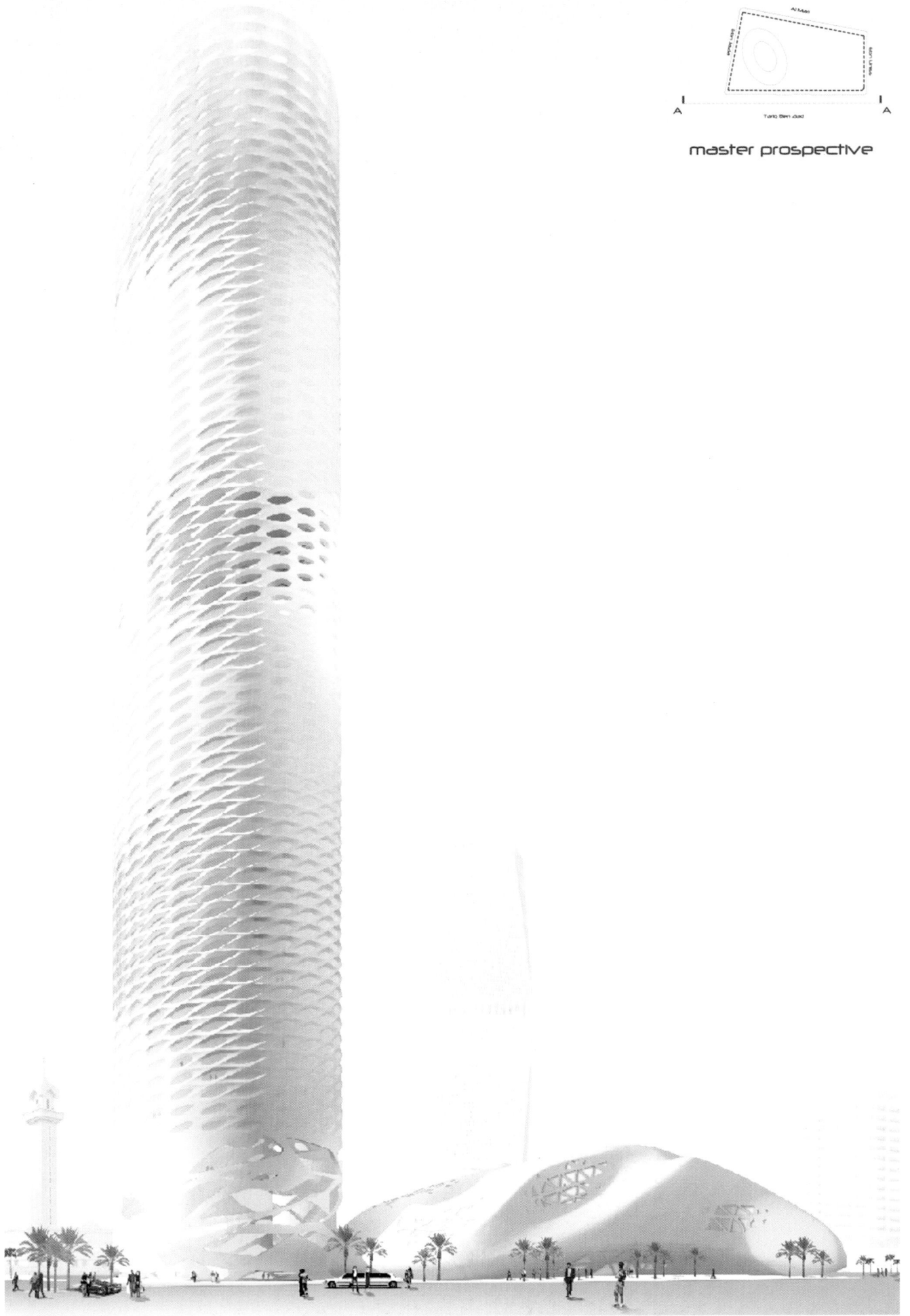

master prospective

Massimiliano Fuksas architetto, Rome

Tower at Suk Al Thalath Al Gadeem, Tripoli

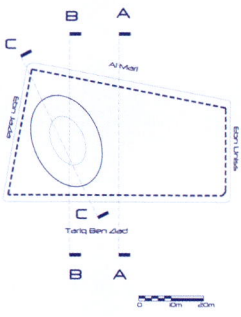

further elevations
scale 1:500

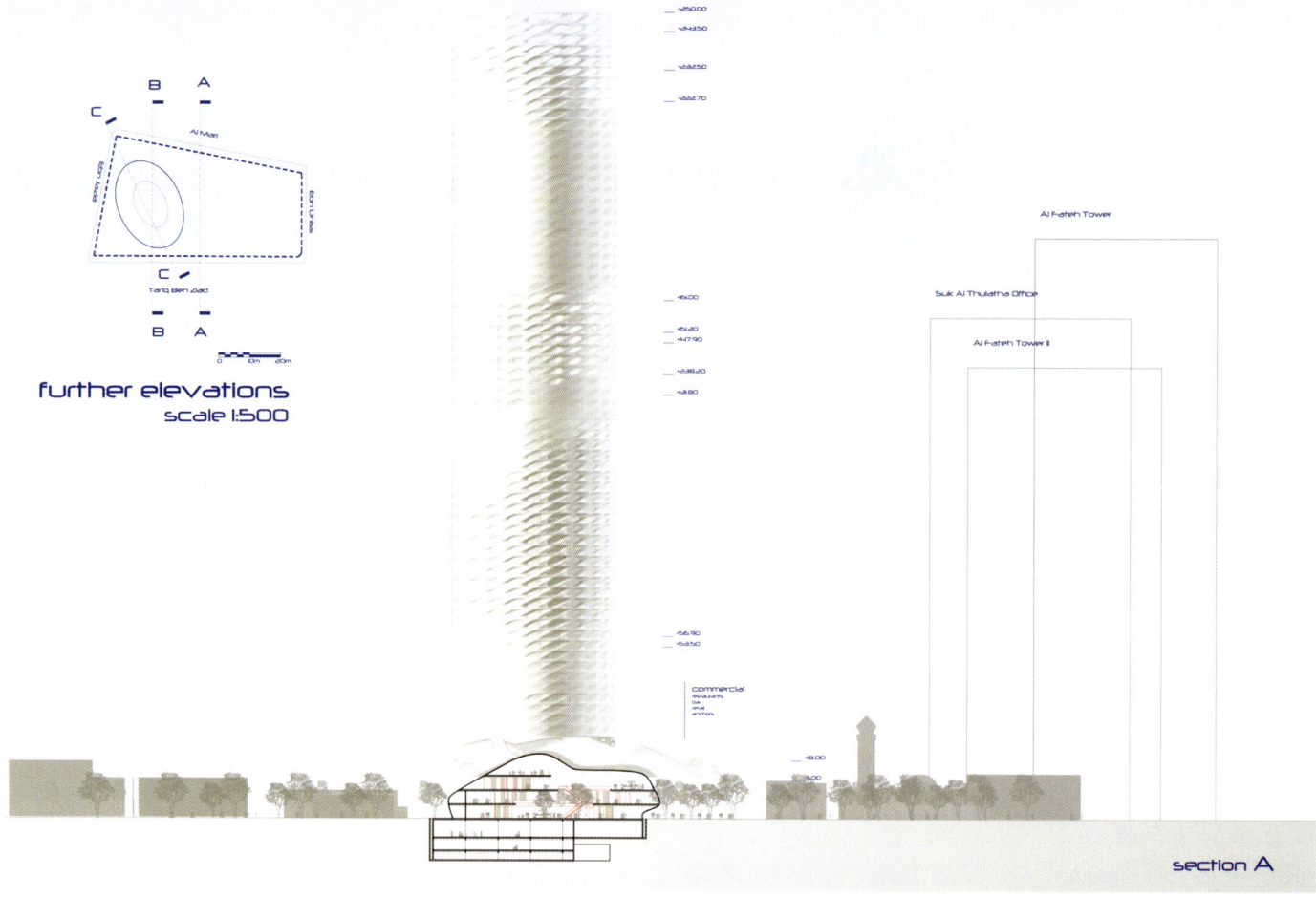

Al-Fateh Tower

Suk Al Thulatha Office

Al-Fateh Tower II

+50.00
+44.50
+43.50
+44.70

+6.00
+6.20
+47.90
+38.20
+4.80

commercial
restaurants
bar
retail
anchors

+56.90
+56.50

+8.00
+4.00

section A

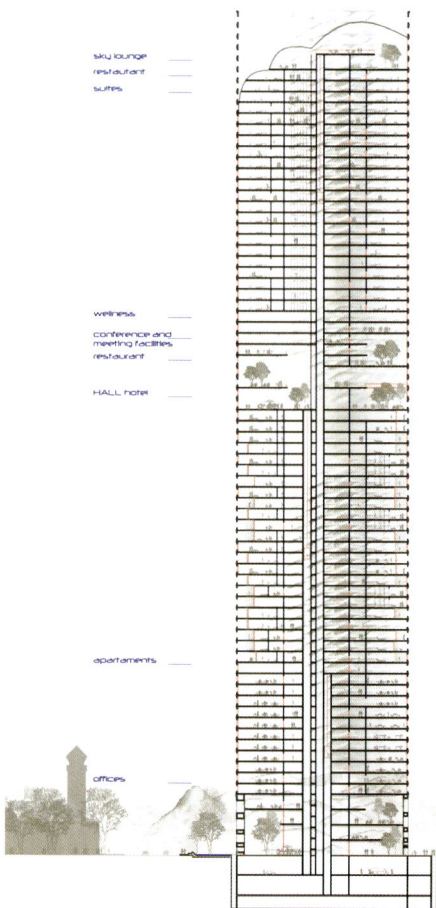

sky lounge
restaurant
suites

wellness

conference and
meeting facilities
restaurant

HALL hotel

apartaments

offices

+50.00
+44.50
+43.50
+44.70

+6.00
+6.20
+47.90
+38.20
+4.80

+56.90
+56.50

+8.00
+4.00

Al-Fateh Tower

section B

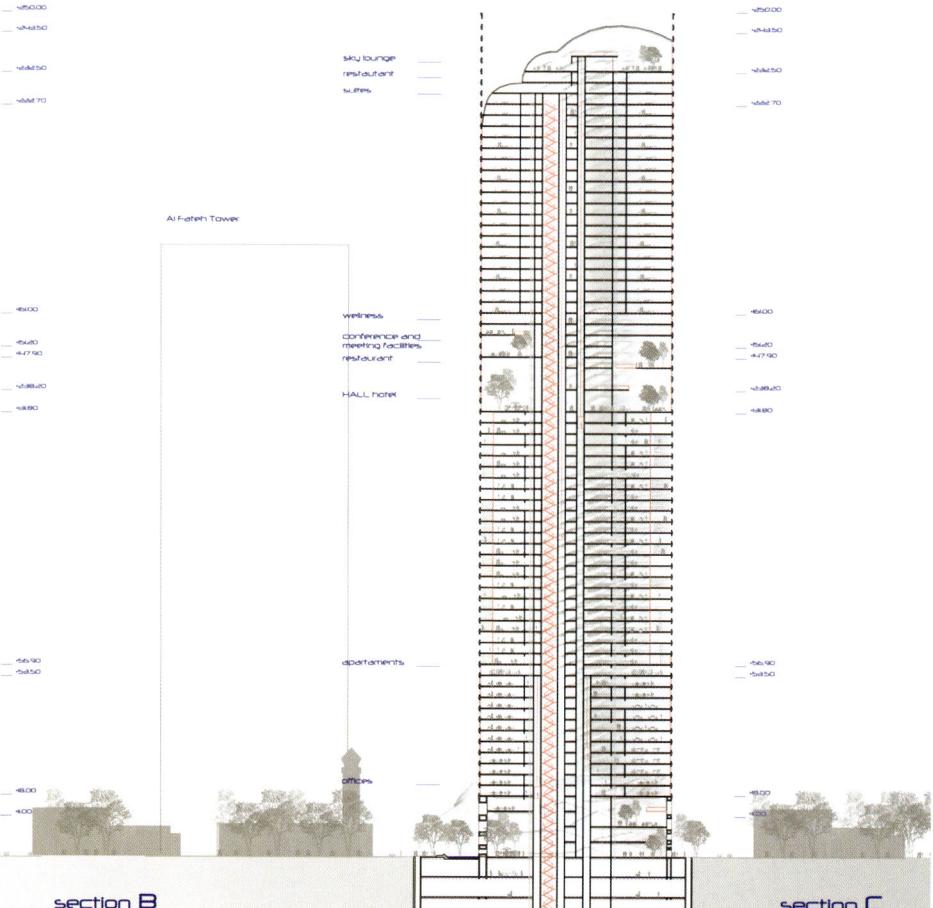

sky lounge
restaurant
suites

wellness

conference and
meeting facilities
restaurant

HALL hotel

apartaments

offices

+50.00
+44.50
+43.50
+44.70

+6.00
+6.20
+47.90
+38.20
+4.80

+56.90
+56.50

+8.00

section C

Al Mari

Tariq Ben 4iad

Ebn Ja√la

1 hotel public entrance
2 hotel vip entrance
3 office entrance
4 apartment entrance
5 cafè
6 winter garden
7 lounge
8 bistrot
9 kitchen
10 winter garden
11 vip lounge

groundfloor scale 1:200

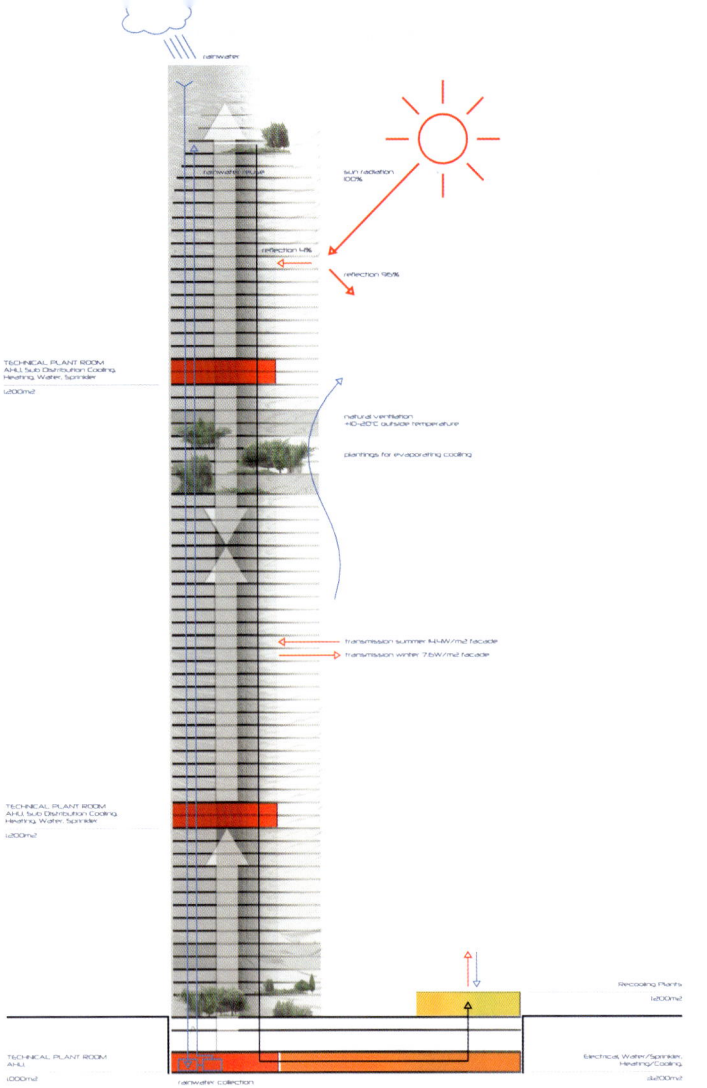

technical and ecological concept

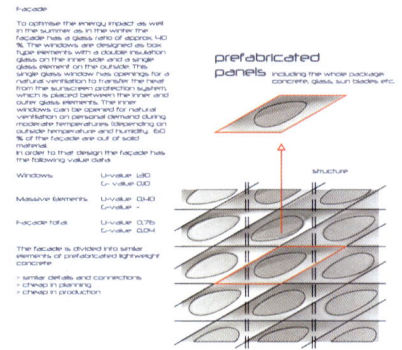

rainwater

rainwater reuse

sun radiation 100%

reflection 4%

reflection 96%

TECHNICAL PLANT ROOM
AHU, Sub Distribution Cooling,
Heating, Water, Sprinkler
≤200m2

natural ventilation
+10-20°C outside temperature

plantings for evaporating cooling

transmission summer 14 W/m2 facade
transmission winter 7.6W/m2 facade

TECHNICAL PLANT ROOM
AHU, Sub Distribution Cooling,
Heating, Water, Sprinkler
≤200m2

Recooling Plants
≤200m2

TECHNICAL PLANT ROOM
AHU
≤1000m2

Electrical, Water/Sprinkler,
Heating/Cooling
≤4700m2

rainwater collection

---

facade concept

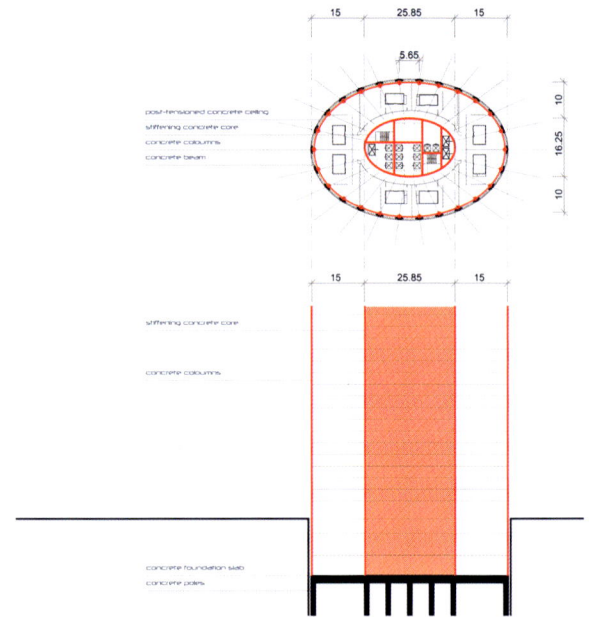

facade

To optimise the energy impact as well
in the summer as in the winter the
facade has a glass ratio of approx. 40
%. The windows are designed as box
type elements with a double insulation
glass on the inner side and a single
glass element on the outside. This
single glass window has openings for a
natural ventilation to transfer the heat
from the sunscreen protection system,
which is placed between the inner and
outer glass elements. The inner
windows can be opened for natural
ventilation on personal demand during
moderate temperatures (depending on
outside temperature and humidity. 60
% of the facade are out of solid
material.
In order to that design the facade has
the following value data.

Windows   U-value 1.80
               L-value 0.10
Massive Elements U-value 0.40
               L-value -
Facade total   U-value 0.78
               L-value 0.04

The facade is divided into similar
elements of prefabricated lightweight
concrete.

> similar details and connections
> cheap in planning
> cheap in production

prefabricated panels including the whole package
concrete, glass, sun blades etc.

structure

---

structural concept

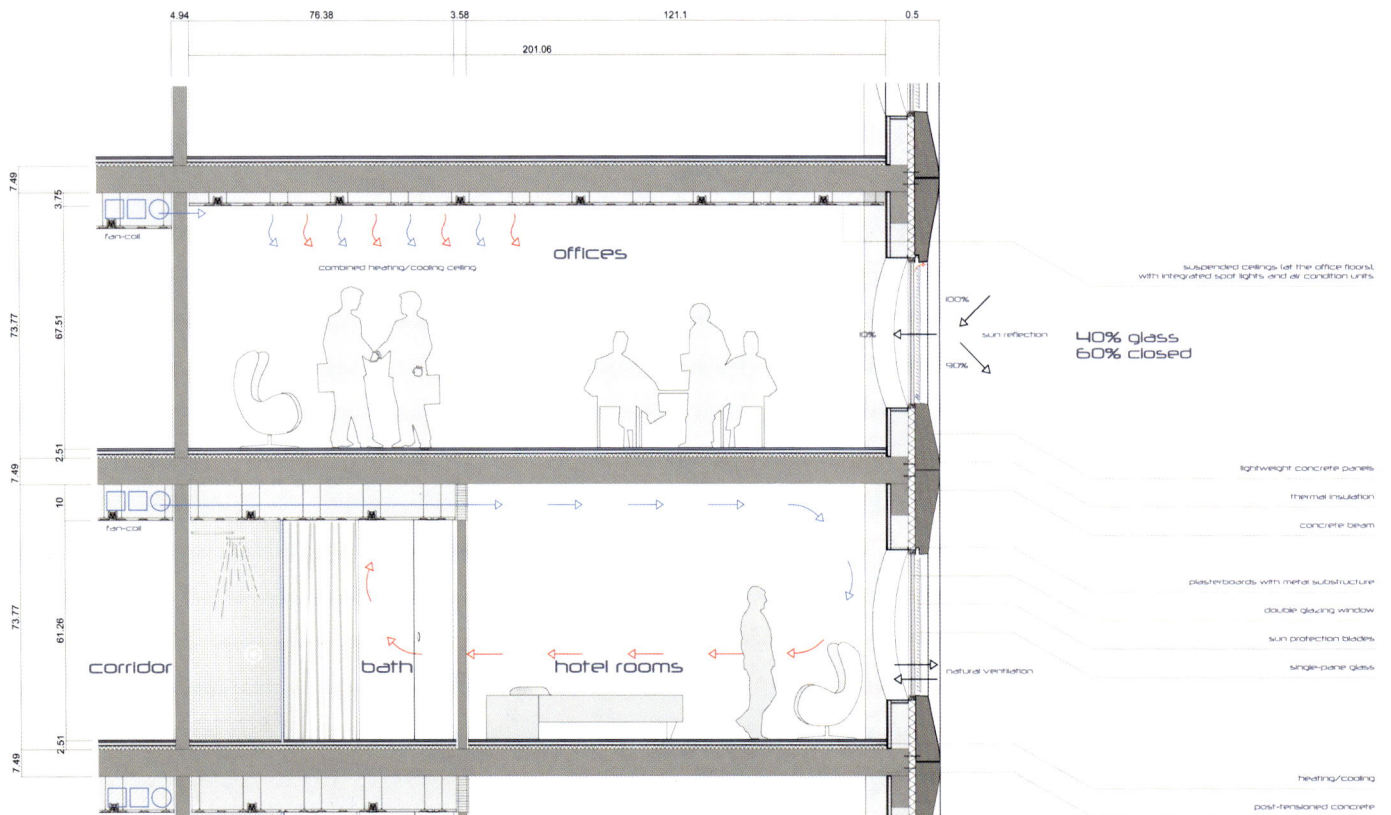

15   25.85   15
5.65
16.25   10
10

post-tensioned concrete ceiling
stiffening concrete core
concrete columns
concrete beam

15   25.85   15

stiffening concrete core

concrete columns

concrete foundation slab
concrete piles

---

section scale 1:200

4.94   76.38   3.58   121.1   0.5
201.06

7.49
3.75
73.77
67.51
fan-coil

offices

combined heating/cooling ceiling

7.49
2.51

7.49
10
fan-coil

73.77
61.26

corridor   bath   hotel rooms

7.49
2.51

suspended ceilings (at the office floors)
with integrated spot lights and air condition units

100%
sun reflection
0%
90%

**40% glass
60% closed**

lightweight concrete panels
thermal insulation
concrete beam
plasterboards with metal substructure
double glazing window
sun protection blades
single-pane glass

natural ventilation

heating/cooling
post-tensioned concrete

interior of the sky lounge

interior of the hotel foyer

Tower at Suk Al Thalath Al Gadeem, Tripoli

Harry Seidler and Associates Sydney
Harry Seidler and Associates建筑事务所 悉尼

"Hotel Concourse as a kind of Libyan urbanism which is characterised by a strict distinction between public and private use of space."

"酒店广场作为一种利比亚城市化的表现，特点是在公众和私人空间使用之间进行严格的区分。"

作者 Penelope Seidler, Peter Hirst, Hiromi Shiraishi, Henry Feiner   合作伙伴及建筑系在校生 John Curro, Dirk Meinecke, Katrin Schmidt-Dengler   专家 structural analysis: Birzulis Associates, Sydney; Mechanical, Electrical, Hydraulic Engineers, Sydney

View along Al"Mari Street

Harry Seidler and Associates, Sydney

475

Tower at Suk Al Thalath Al Gadeem, Tripoli

3

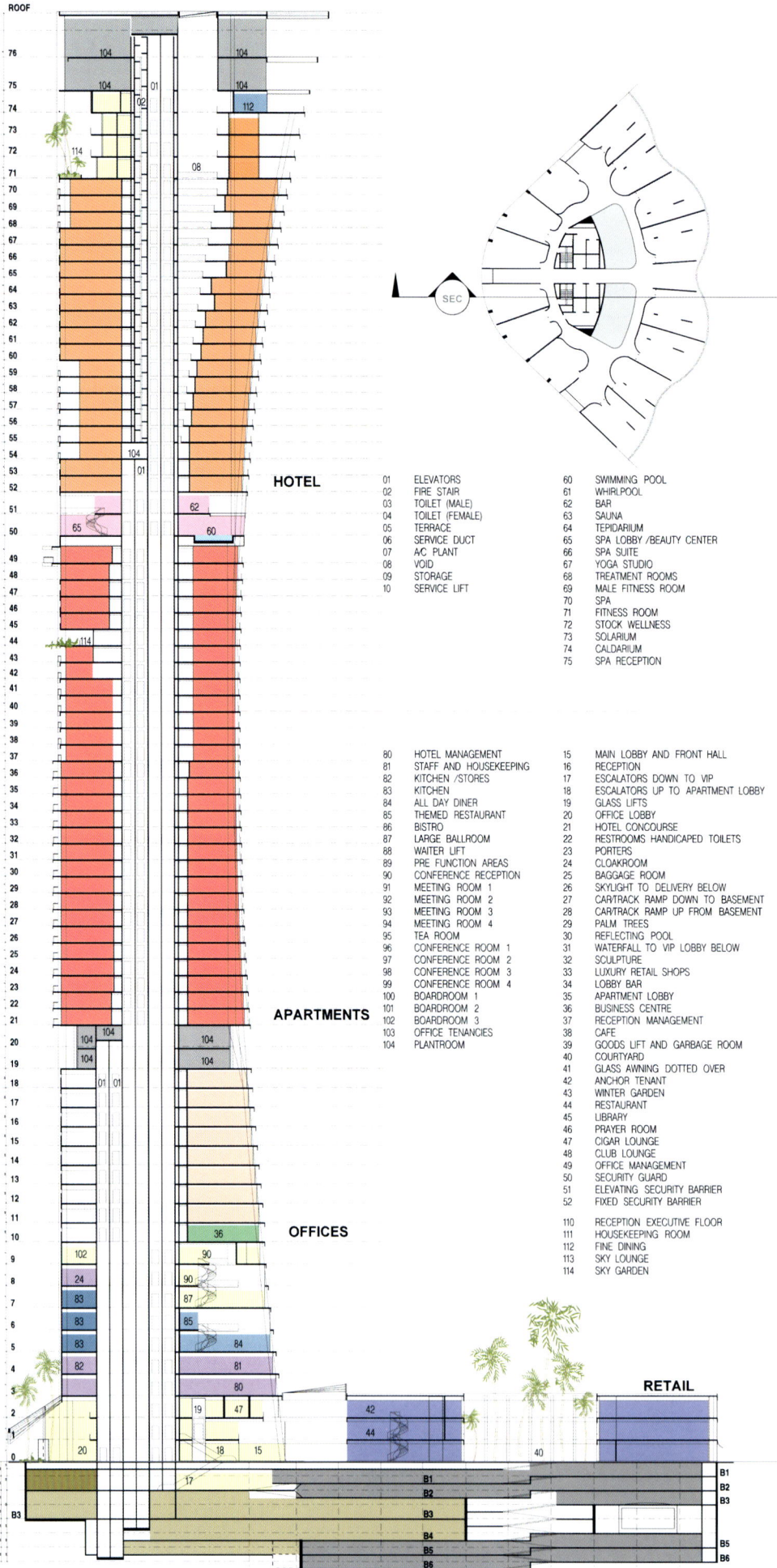

ROOF

76
75
74
73
72
71
70
69
68
67
66
65
64
63
62
61
60
59
58
57
56
55
54
53
52
51
50
49
48
47
46
45
44
43
42
41
40
39
38
37
36
35
34
33
32
31
30
29
28
27
26
25
24
23
22
21
20
19
18
17
16
15
14
13
12
11
10
9
8
7
6
5
4
3
2
1
0

B1
B2
B3
B4
B5
B6

HOTEL

APARTMENTS

OFFICES

RETAIL

01	ELEVATORS	60	SWIMMING POOL
02	FIRE STAIR	61	WHIRLPOOL
03	TOILET (MALE)	62	BAR
04	TOILET (FEMALE)	63	SAUNA
05	TERRACE	64	TEPIDARIUM
06	SERVICE DUCT	65	SPA LOBBY /BEAUTY CENTER
07	AC PLANT	66	SPA SUITE
08	VOID	67	YOGA STUDIO
09	STORAGE	68	TREATMENT ROOMS
10	SERVICE LIFT	69	MALE FITNESS ROOM
		70	SPA
		71	FITNESS ROOM
		72	STOCK WELLNESS
		73	SOLARIUM
		74	CALDARIUM
		75	SPA RECEPTION

80	HOTEL MANAGEMENT	15	MAIN LOBBY AND FRONT HALL
81	STAFF AND HOUSEKEEPING	16	RECEPTION
82	KITCHEN /STORES	17	ESCALATORS DOWN TO VIP
83	KITCHEN	18	ESCALATORS UP TO APARTMENT LOBBY
84	ALL DAY DINER	19	GLASS LIFTS
85	THEMED RESTAURANT	20	OFFICE LOBBY
86	BISTRO	21	HOTEL CONCOURSE
87	LARGE BALLROOM	22	RESTROOMS HANDICAPED TOILETS
88	WAITER LIFT	23	PORTERS
89	PRE FUNCTION AREAS	24	CLOAKROOM
90	CONFERENCE RECEPTION	25	BAGGAGE ROOM
91	MEETING ROOM 1	26	SKYLIGHT TO DELIVERY BELOW
92	MEETING ROOM 2	27	CAR/TRACK RAMP DOWN TO BASEMENT
93	MEETING ROOM 3	28	CAR/TRACK RAMP UP FROM BASEMENT
94	MEETING ROOM 4	29	PALM TREES
95	TEA ROOM	30	REFLECTING POOL
96	CONFERENCE ROOM 1	31	WATERFALL TO VIP LOBBY BELOW
97	CONFERENCE ROOM 2	32	SCULPTURE
98	CONFERENCE ROOM 3	33	LUXURY RETAIL SHOPS
99	CONFERENCE ROOM 4	34	LOBBY BAR
100	BOARDROOM 1	35	APARTMENT LOBBY
101	BOARDROOM 2	36	BUSINESS CENTRE
102	BOARDROOM 3	37	RECEPTION MANAGEMENT
103	OFFICE TENANCIES	38	CAFE
104	PLANTROOM	39	GOODS LIFT AND GARBAGE ROOM
		40	COURTYARD
		41	GLASS AWNING DOTTED OVER
		42	ANCHOR TENANT
		43	WINTER GARDEN
		44	RESTAURANT
		45	LIBRARY
		46	PRAYER ROOM
		47	CIGAR LOUNGE
		48	CLUB LOUNGE
		49	OFFICE MANAGEMENT
		50	SECURITY GUARD
		51	ELEVATING SECURITY BARRIER
		52	FIXED SECURITY BARRIER
		110	RECEPTION EXECUTIVE FLOOR
		111	HOUSEKEEPING ROOM
		112	FINE DINING
		113	SKY LOUNGE
		114	SKY GARDEN

SEC

Aerial Perspective

250623

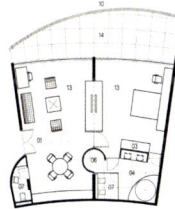

TYPICAL SUITE PLAN

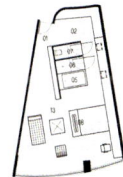

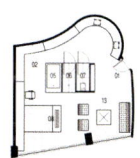

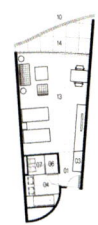

TYPICAL JUNIOR SUITE PLANS

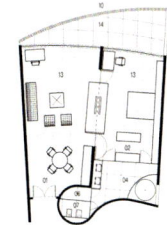

TYPICAL SUITE PLAN

TYPICAL HOTEL ROOM PLANS

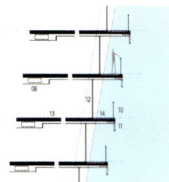

TYPICAL FACADE SECTION

LEGEND

01 ENTRANCE
02 DRESSING
03 WARDROBE
04 BATHROOM (POLISHED GRANITE FLOOR SLABS)
05 SUK BATH
06 SHOWER (GLASS DOOR & WALL)
07 WC (FROSTED GLASS DOOR)
08 BED BASE (WITH ELEVATING
    BACK-TO-BACK PLASMA TV SCREEN)
09 AC UNITS IN CEILING SPACE
10 CANTILEVERED TOUGHENED GLASS
    BALUSTRADE WITH CURVED S.S. HANDRAIL
11 S.S. ALUCABOND CURVED PANEL DOWN-TURNS
12 SOLAR GREY TINTED TOUGHENED FULL-HT GLASS
13 CARPET ON CONCRETE SLAB
14 FLAMED GRANITE PAVING TO TERRACES

TYPICAL HOTEL ROOMS S 1 : 50   0   2,5   5   10

COOP HIMMELB(L)AU, Prix/Dreibholz & Partner Vienna
COOP HIMMELB(L)AU, Prix/Dreibholz & Partner建筑事务所 维也纳

"... shaped by urban, cultural, climatic, and view considerations, as well as economies of structure, material, and inner functions, creating a memorable and unmistakable icon in the city-shape ..."

"通过城市、文化、气候和观点，以及结构经济性、材料和内部功能等方面的考虑，塑造出一个城市形状的令人难忘的显而易见的地标性建筑。"

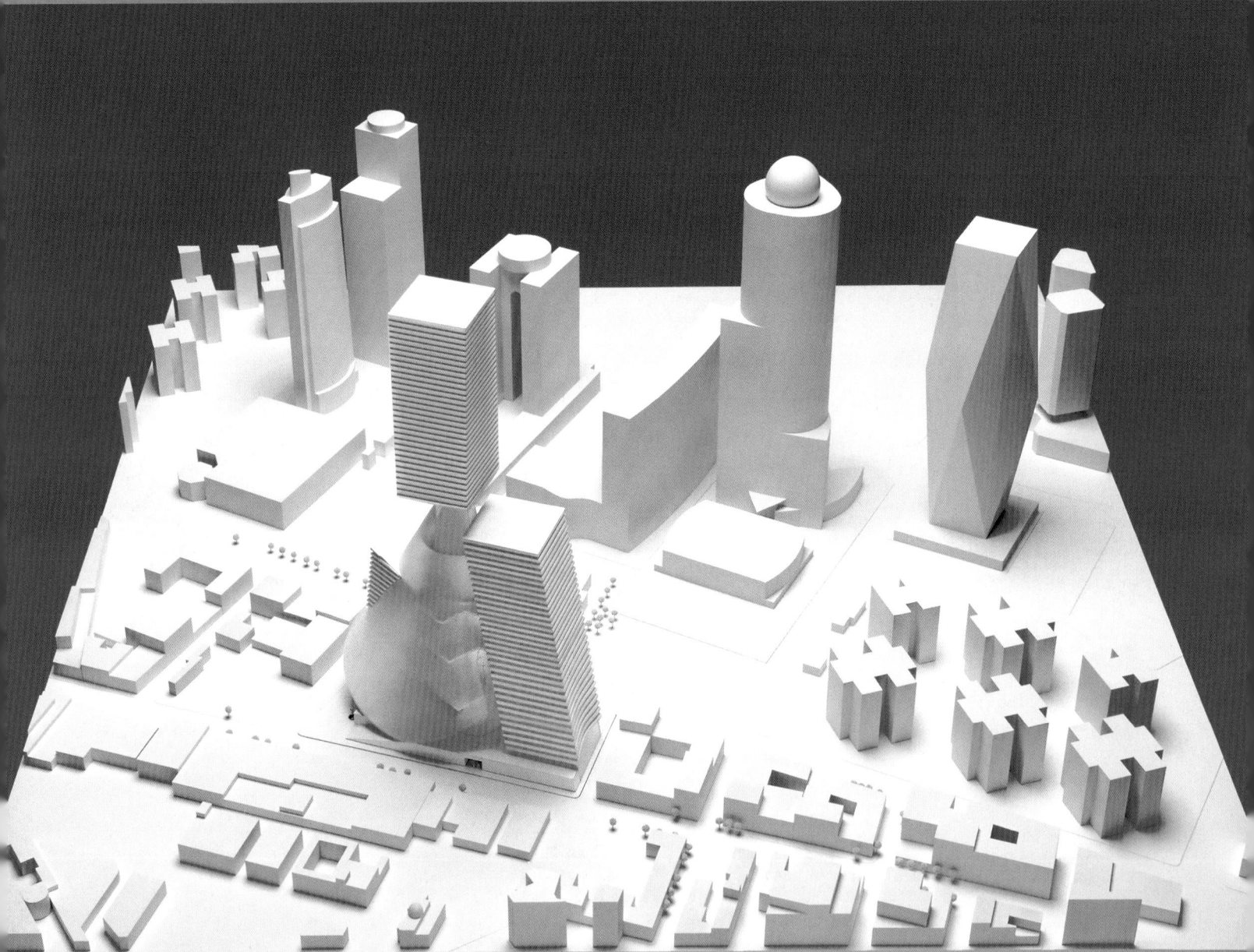

作者 Univ. Prof. Arch. DI Dr. Hc Wolf D. Prix   合作伙伴及建筑系在校生 Michael Volk, Andrea Graser, Pete Rose, Victoria Coaloa, Vinnento Possenti, Juhong Park, Anja Sorger, Guiseppe Zagaria, Elisabeth Swiss, Frank Hildebrandt, Daniela Comito, Chih-Bin Tseng, Paul Hoszowski, Markus Pillhofer   专家 structural analysis: B+G Ingenieure/Bollinger und Grohmann, Frankfurt/Main; building services: ARUP/Brian Cody, London/Frankfurt/Main

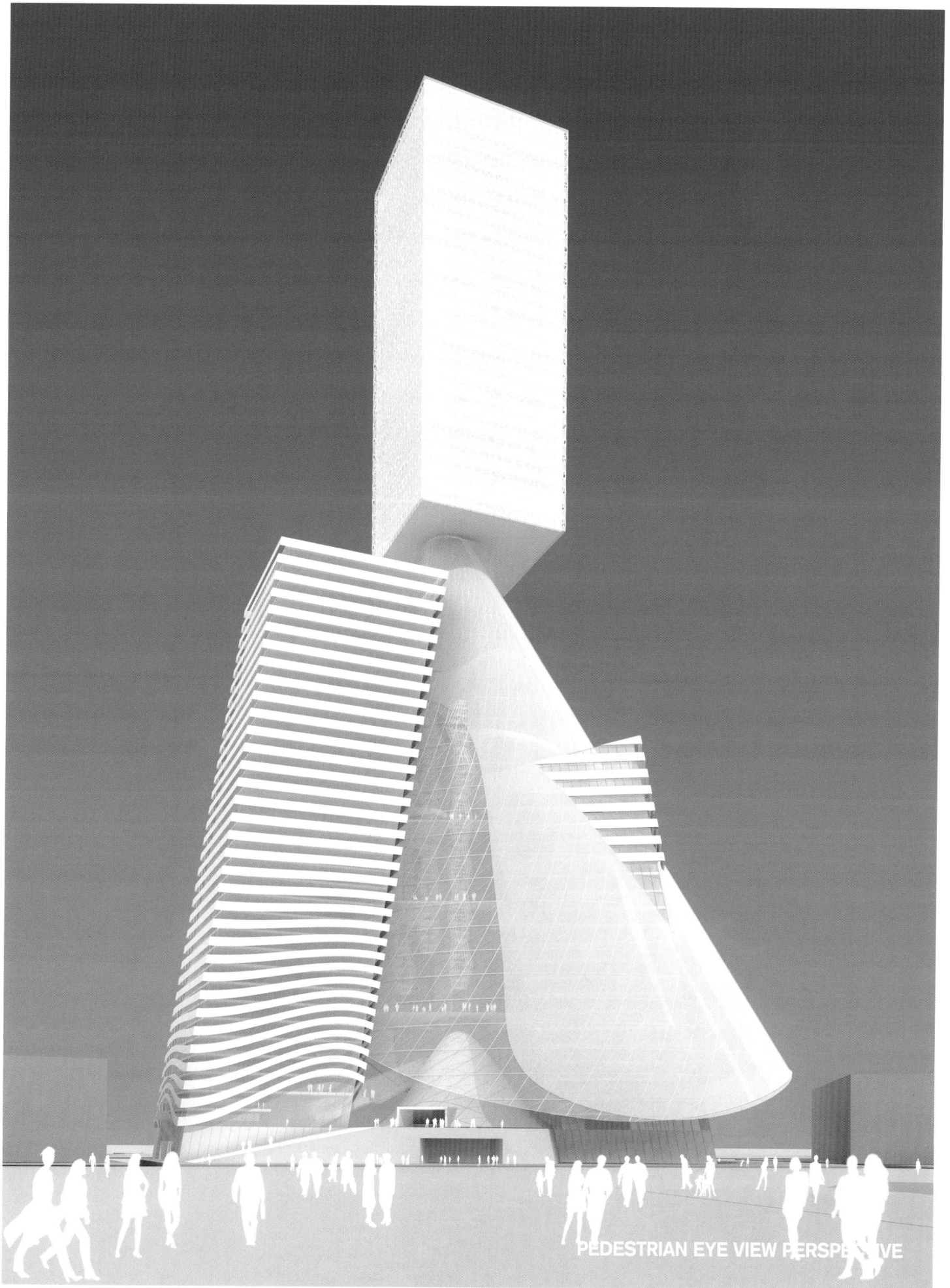

COOP HIMMELB(L)AU, Prix/Dreibholz & Partner, Vienna

479

Tower at Suk Al Thalath Al Gadeem, Tripoli

SHEET #6.1

SECTION A-A

0   4   8   12   16   20
SCALE | 1:200

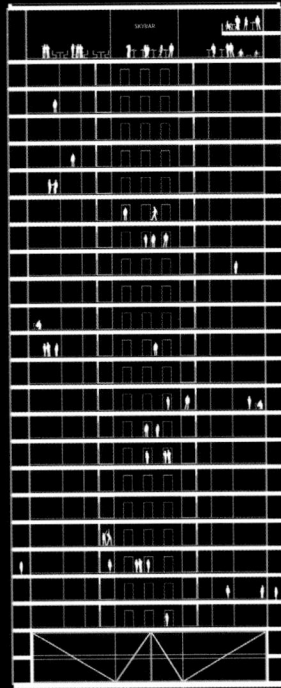

+225.00m ▼
+218.00m ▼ SKYBAR
+190.00m ▼ HOTEL
+169.00m ▼ HOTEL

SKYBAR

HOTEL TOWER

APARTMENT TOWER

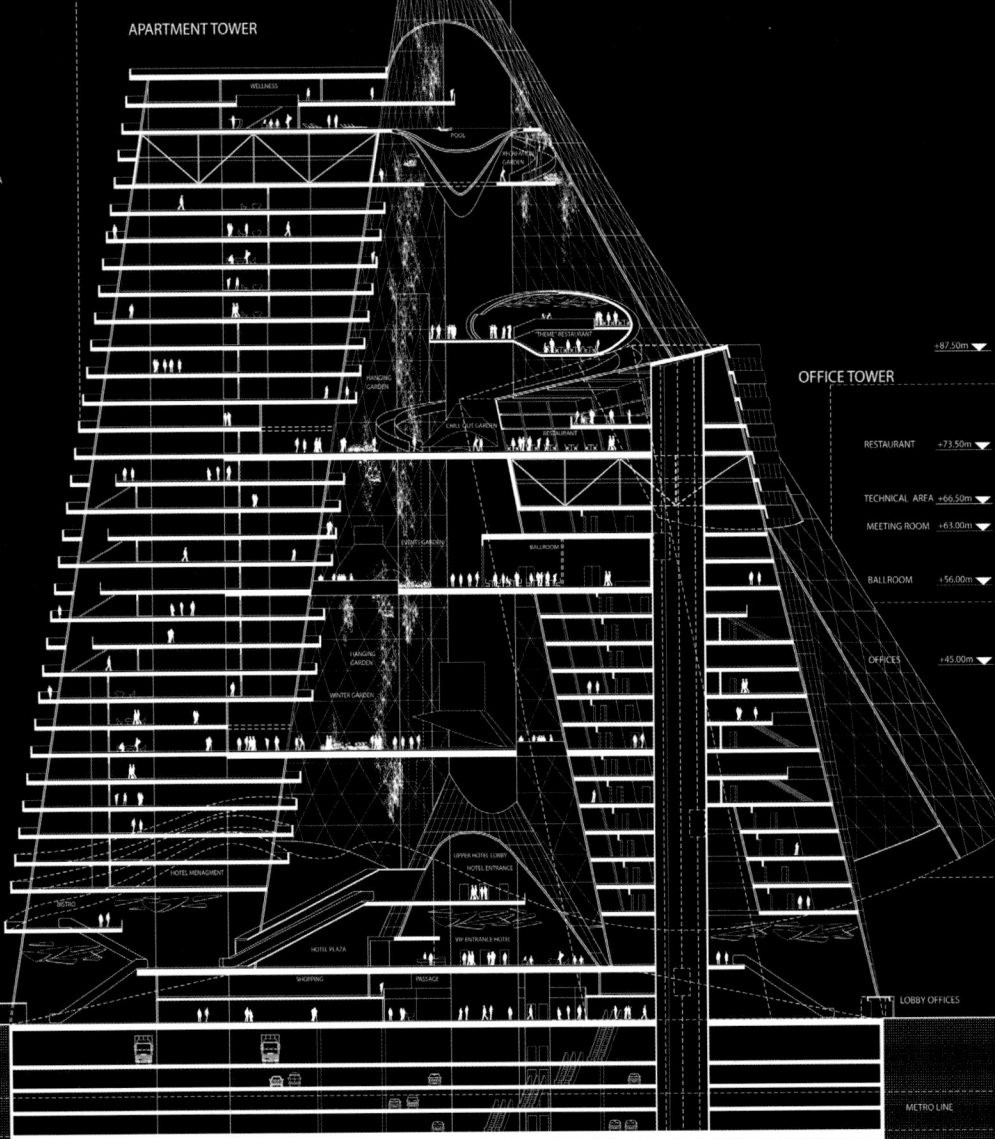

+37.00m ▼ TECHNICAL AREA
+23.00m ▼
+115.50m ▼ WELLNESS
+115.50m ▼ WELLNESS
+108.50m ▼ TECHNICAL AREA
+73.50m ▼ APARTMENTS
+63.00m ▼ APARTMENTS
+56.00m ▼ APARTMENTS
+45.50m ▼ APARTMENTS

WELLNESS

POOL

"THEME" RESTAURANT

HANGING GARDEN

CHILL OUT GARDEN

RESTAURANT

EVENTS GARDEN

BALLROOM

HANGING GARDEN

WINTER GARDEN

OFFICE TOWER

+87.50m ▼

RESTAURANT   +73.50m ▼
TECHNICAL AREA   +66.50m ▼
MEETING ROOM   +63.00m ▼
BALLROOM   +56.00m ▼
OFFICES   +45.00m ▼

HOTEL MANAGMENT

METRO

UPPER HOTEL LOBBY
HOTEL ENTRANCE

+7.50m ▼

HOTEL PLAZA

VIP ENTRANCE HOTEL

SHOPPING   PASSAGE

LOBBY OFFICES

+0.00m ▼ LOBBY HOTEL

SERVICE HOTEL

METRO LINE

+-225.00m ▼

ENVIRONMENTAL, ENERGY AND BUILDING SERVICES CONCEPTS

Strategies employing the form of the building to assist natural ventilation together with the use of renewable energy sources (wind and solar power) assure an energy efficient design and reduce energy consumption and reliance on fossil fuel energy sources.

Photovoltaic Shield
A transparent "shield" in the form of a second skin made from thin, semi-transparent photovoltaic film, wraps itself around the south facing office tower providing effective solar shading for the south facing offices, enclosing the atrium volume which in turns acts as a buffer zone using gardens to improve air quality, and generating electrical energy via the photovoltaic cells. The density of these varies according to orientation and tilt angle, thus creating a visually interesting pattern in the building skin. Tripoli enjoys some of the best climatic conditions in the world for the generation of electricity employing photovoltaic cells.

Wind energy
The building design also enables wind energy to be captured and employed via wind generation plant to generate renewable electrical energy. The building form is used to accelerate the predominantly north-westerly winds, increasing the local wind speed via the Venturi effect and the energy output of the generation plant.

Facade
The design of the building skin allows the use of natural ventilation throughout the entire winter period when temperatures are mild. Excessive wind pressures are reduced via an additional outer façade construction, which is also a solar screen formed in a pattern optimized to orientation.

Alternative energy sources
Biomass or possibly gas fueled combined heat and power generators provide the building with both heat and electrical power. This solution has both ecological and economic advantages compared to more conventional alternatives (c. 60% less CO2 emissions), and also provides a major advantage with regard to security of supply. In warm weather the heat is used to drive an absorption chiller that supplies chilled water to cool the building. All rainwater is collected and used for irrigation and toilet flushing.

Plant rooms
Plant rooms for technical equipment are located as shown in the diagrams and are connected via vertical shafts and risers to the individual floors to provide an efficient technical infrastructure. An integrated security system is provided including CCTV (closed circuit television) surveillance of public areas, full function access control at selected entrances and lifts and central monitoring equipment within a main lobby security/ reception desk. A complete Building Management System (BMS) is provided consisting of multiple Direct Digital Control (DDC) data processing outstations and a central management system.

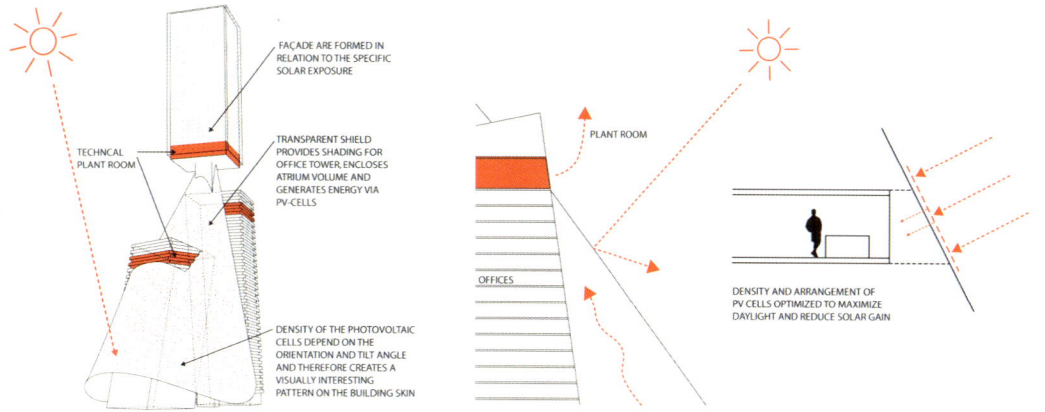

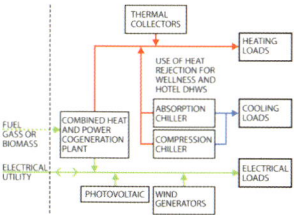

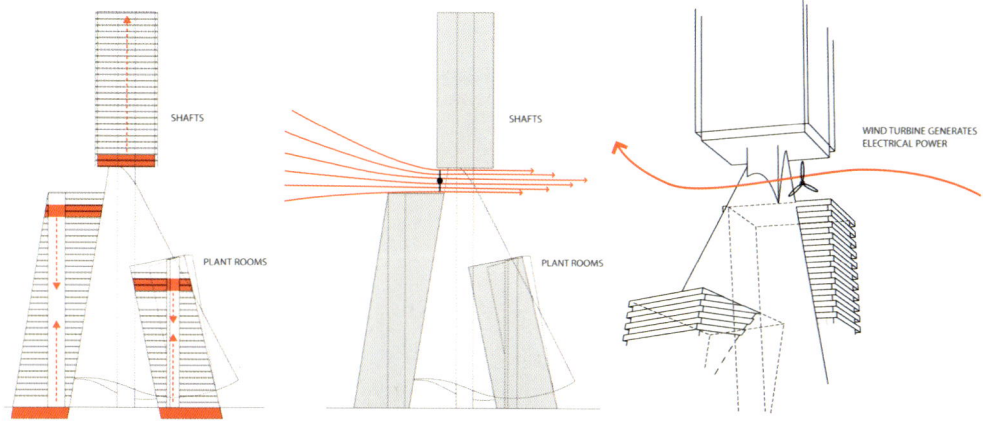

FAÇADE ARE FORMED IN RELATION TO THE SPECIFIC SOLAR EXPOSURE

TECHNICAL PLANT ROOM

TRANSPARENT SHIELD PROVIDES SHADING FOR OFFICE TOWER, ENCLOSES ATRIUM VOLUME AND GENERATES ENERGY VIA PV-CELLS

DENSITY OF THE PHOTOVOLTAIC CELLS DEPEND ON THE ORIENTATION AND TILT ANGLE AND THEREFORE CREATES A VISUALLY INTERESTING PATTERN ON THE BUILDING SKIN

PLANT ROOM

OFFICES

DENSITY AND ARRANGEMENT OF PV CELLS OPTIMIZED TO MAXIMIZE DAYLIGHT AND REDUCE SOLAR GAIN

SHAFTS

SHAFTS

WIND TURBINE GENERATES ELECTRICAL POWER

PLANT ROOMS

PLANT ROOMS

THERMAL COLLECTORS → HEATING LOADS

USE OF HEAT REJECTION FOR WELLNESS AND HOTEL DHWS

FUEL GAS OR BIOMASS → COMBINED HEAT AND POWER COGENERATION PLANT → ABSORPTION CHILLER → COOLING LOADS

COMPRESSION CHILLER

ELECTRICAL UTILITY → ELECTRICAL LOADS

PHOTOVOLTAIC | WIND GENERATORS

COOP HIMMELB(L)AU, Prix/Dreibholz & Partner, Vienna

STRUCTURAL DESIGN

The Suk Al Thalath building complex consists of three rectilinear high-rise towers and a cone-shaped atrium that interact with each other structurally to form a cohesive single structural system.

Earthquake
The building site belongs to a seismic zone with high requirements. As the seismic and wind loads are the governing parameters affecting the stability of the structure, these loads are mainly considered by the structural design of the building complex. Sufficient ductility is achieved by designing the structures as highly integrated and developed total systems, and by accurately detailing the individual elements with special respect to connections.

Apartment and Office Towers
The 33-storey Apartment tower and 22 Storey Office tower are both rectangular shaped concrete structures of 123 meters and 90 meters tall respectively.
The main structure of the Apartment and Office Towers are comprised of the inner core of concrete walls and outrigger systems situated at their top levels as service storeys. These outriggers provide necessary stiffness against horizontal forces for the Apartment and Office Towers, and are further utilized to make a robust connection to stabilize the upper Hotel Tower.
Vertical forces are carried through concrete columns in a grid of 8.75 by 4.60 m along the facades, and through the walls of the concrete core. The floor slabs are designed as flat slabs with a maximum span of 10.00m.

Hotel Tower
The hotel tower is a 24-storey building located atop the Cone structure of the Atrium. As the first floor is situated at a height of about 137m, the concrete core below supports the building vertically. To carry the vertical forces from façade and inner columns into the core, the two lower floors are designed as service storeys in a steel framework. Unlike the apartment and office towers, floors and columns are provided as a composite structure. This has advantages for construction sequencing and the reduced mass in the design is favourable for load bearing behaviour in relation to earthquake forces.

Cone
The cone is made of a triangulated steel grid with the concrete core of the hotel tower in its centre. As a lightweight structure it is stabilised vertically by the apartment and office tower as well as at its top by the core and horizontally by the platforms placed between both. The platforms are again vertically supported by the steel structure of the cone.

Subsurface Building Complex
The subsurface tract includes four storeys. The grid of 8.20 by 8.20 m allows that the floors be constructed as flat concrete slabs. The Horizontal stiffness of the lower floors is provided through concrete walls and concrete cores.

Foundation
A combined pile-raft-foundation is recommended. It is a geotechnical composite structure comprising piles, a thick foundation slab and the soil below. The building bears on the ground and transfers its vertical loads through compression underneath the foundation slab and by surface friction and base pressure of the piles. The slab thickness is adapted to the individual stress at each area and varies between 1.00m and 3.00m. Piles are arrayed below the cores and columns of the high-rise buildings.

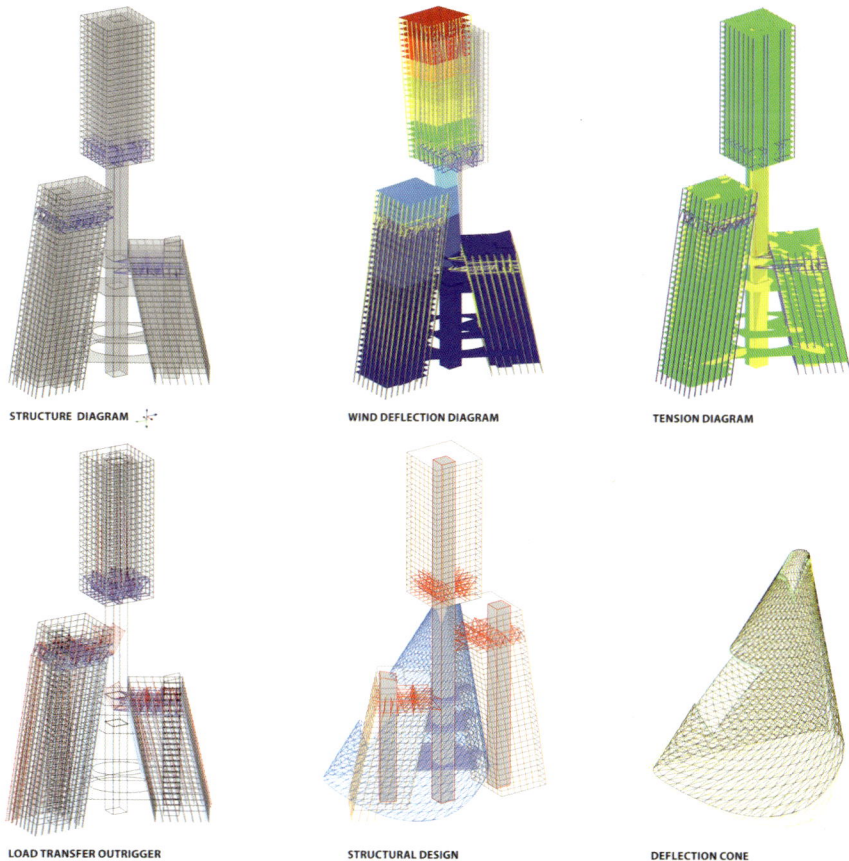

STRUCTURE DIAGRAM

WIND DEFLECTION DIAGRAM

TENSION DIAGRAM

LOAD TRANSFER OUTRIGGER

STRUCTURAL DESIGN

DEFLECTION CONE

Tower at Suk Al Thalath Al Gadeem, Tripoli

Aedas Dubai
Aedas建筑事务所 迪拜

"The 5 elements which graciously form 'the hand of Fatima' are transfered into a landmark to commemorate the people's past adversities, preserve their present way of life, and enhance future prosperity."

"优雅的形成'法帝玛之手'的五个元素被转化为一个地标性建筑，纪念人们苦难的过去，保留现在人们的生活方式，增强未来的繁荣。"

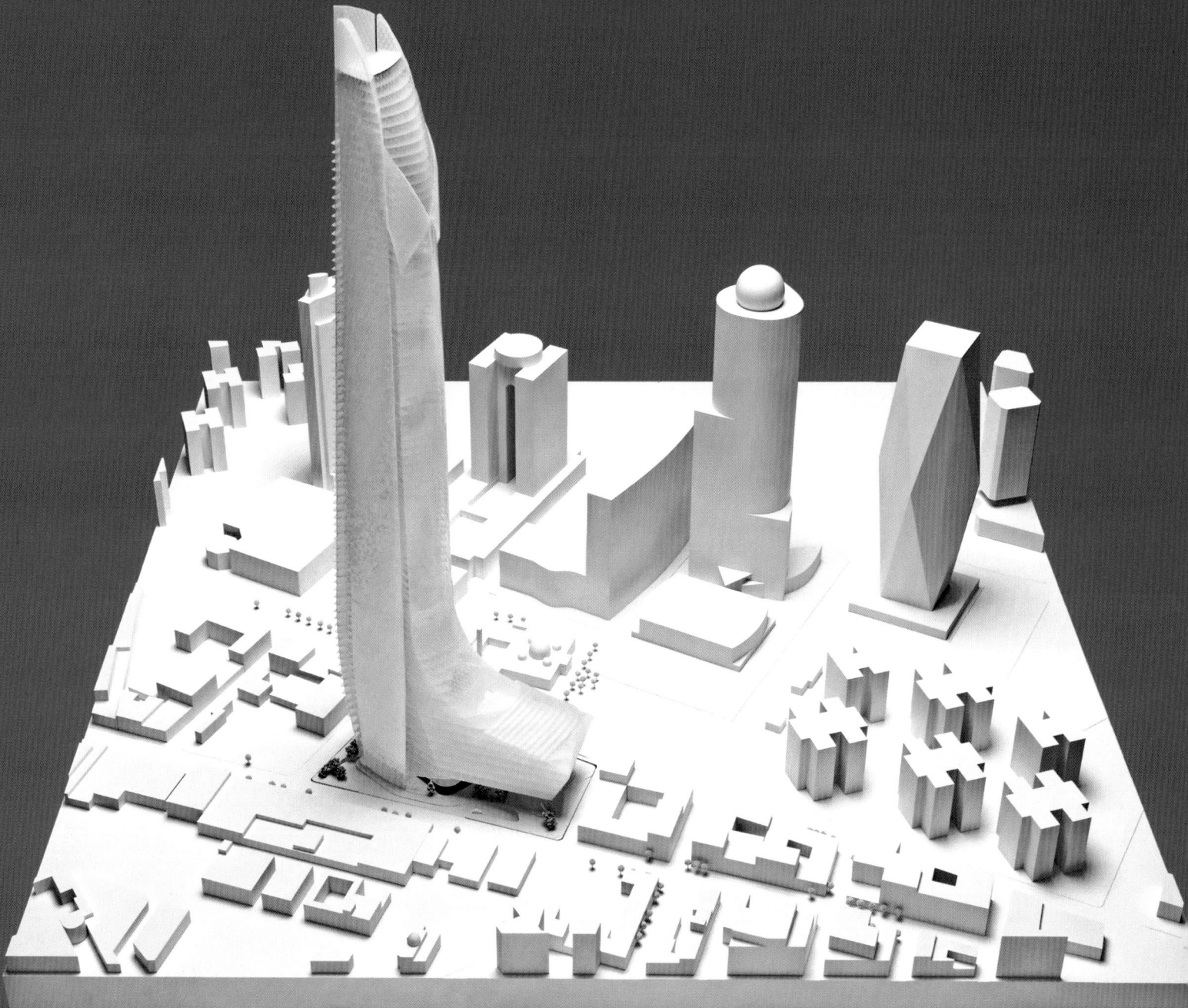

作者 Boran Agoston, Peter Engstrom   合作伙伴及建筑系在校生 Erwin Paul Lota, Byron Emmerson, Jonathan Wong, Rey Majadillas, Reynaldo Cleofe, Dan Albert Formalejo, June Marzan, Eric Chan, Raymond, RJ Models, Judith Kimpran (Aedas - Advanced Modeling Group)
专家 structural analysis: Predrag Eror, Meinhardt Pte Ltd., Singapore; sustainability: Sinsia Stankovic, BDSP Partnership, London; Alan Harles, BDPS Partnership, London; landscape planning: Sinsia Stankovic, BDSP Partnership, London; Alan Harles, BDPS Partnership, London

master perspective

3.0

tower at suk al thalath al gadeem

PHASE 1

PHASE 2

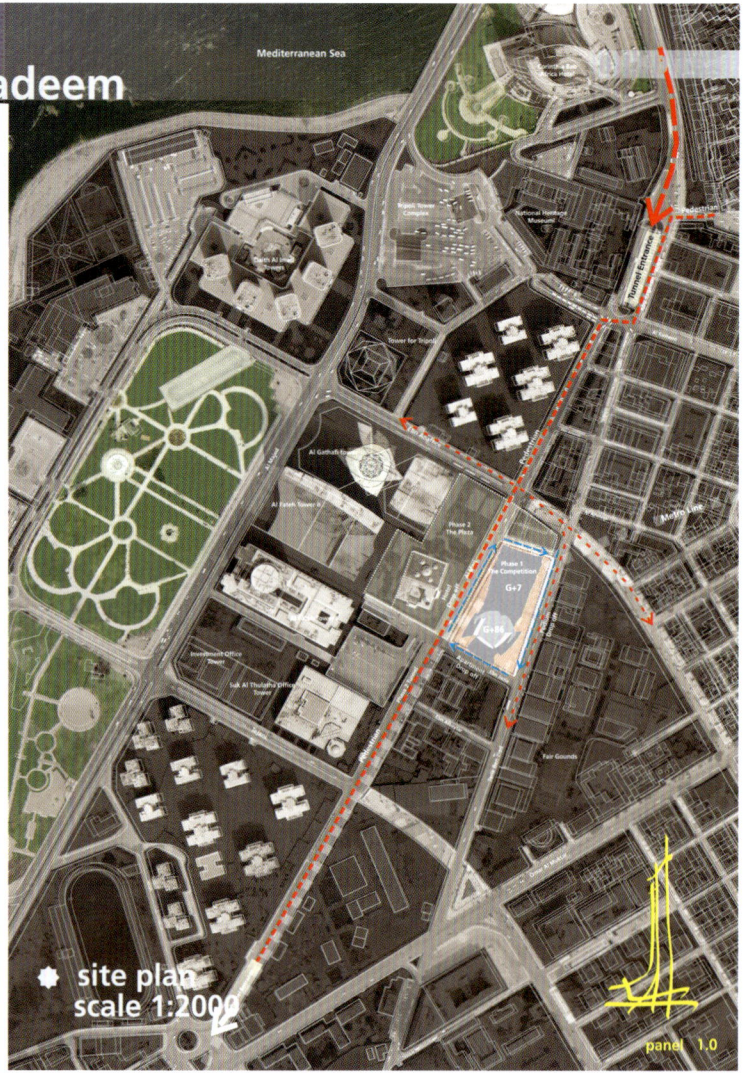

Mediterranean Sea

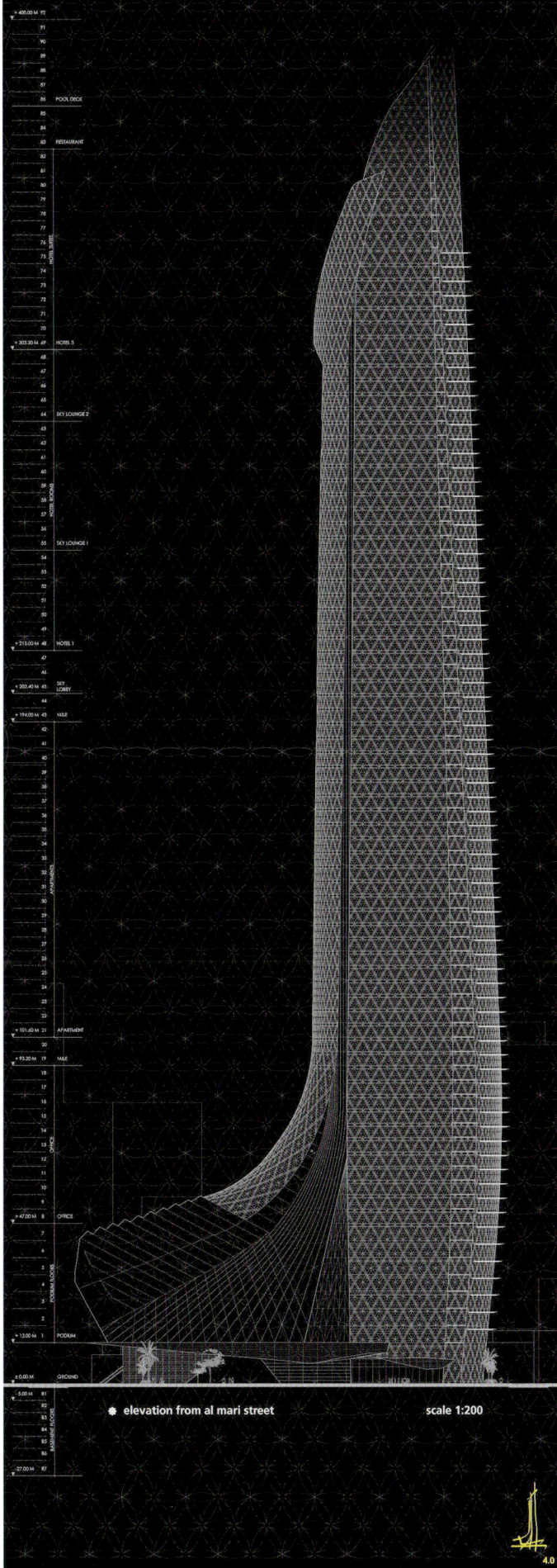

● elevation from al mari street · · · · · · · · · · · · · · scale 1:200

4.0

● longitudinal section · · · · · · · · · · · · · · scale 1:200

① ② ③ ④ ⑤ ⑥ ⑦ ⑧ ⑨ ⑩ ⑪ ⑫ ⑬ ⑭

6.1

Henning Lars

Kerry Hill Architects

Zaha

Snøhetta

DELUGAN MEISSL ASS

Atelier Christian de

n Architects

Hadid

OCIATED ARCHITECTS

Portzamparc

Darat King Abdullah II
in Amman

Darat King Abdullah Ⅱ
约旦安曼

Darat King Abdullah II for Culture and Arts Amman

Darat King Abdullah Ⅱ 文化和艺术中心，约旦安曼

Restricted project competition preceded by an application procedure

限制严格的项目竞赛，开始前有申请程序。

Wasfi Al-Tall Square

Ma'in Bin Za'edah

Al-Kulliyah Al-Ilmiyah Al-Islamiyah

Abdul Qader

9 Sha'ban

King Abdullah I Square

Al-Farabi

Al-Bayhaqi

Abu Baker Al Sidde

Zyad Bin Abeeh

Al-Buhtori

Zayd Bin Harethah

Abdul Mun'em Riyadh

Ja'far Al-Askari

Saleem Abdulhadi

Al-Buhtori

Princess Basma

Wadi Sir Bus Station

Omar Bin Al-Khattab

Omar Matar

Zyad Bin Abeeh

Ali Bin Abi Talib

Al Muhajereen

Khirfan

Omar Matar

Noor Al-Din Zanki

Atared

Adel Jaber

Ali Bin Abi Talib

Aljour

Sayf Al-Dawlah Al-Ham Adani

Beit Jala

Shukri Al-Quwatli

Sayf Al-Dawlah Al-Ham Adani

Al-Quds

Faleh Al-Daboubi

Abdullah Bin Al-Zubayr

Wasei

Noor Al-Din Zanki

Rushdie Al-Sham'ah

Hartha

Abdullah Baha Al-Din

Sayf Al-Dawlah Al-Ham Adani

Sa'd bin Ubadah

Alahal Madadhah

Ja'Farral Hassani

地点 Amman   时间 10/2007–04/2008   主办方 Greater Amman Municipality (GAM)   参赛者 6   面积 12,000 sq m
竞赛费用 110,000 EUR   专业评奖委员会 Prof. Dr. Gulzar Haider, chairman , Lahore; Prof. Klaus Kada, Graz; Prof. Hilde Léon, Berlin; Jafar Tukan, architect, Amman   其他专业评奖委员会 Dominique Lyon, architect, Paris   专家委员会 Omar Maani, mayor of Amman; Dr. Kifah Fakhoury, head of "National Music Conservatory", Amman; Michael Schindhelm, head of culture department, Dubai
其他专家委员会 Amer Bashir, architect and vice mayor of Amman; Lina Attel, general director "Performing Arts Center" (PAC), Amman

With a population of over 2 million, Amman is one of the most vibrant and modern cities in the Middle East. The task set for this competition was a design for Amman's new culture and art center, the "Darat King Abdullah II". The project's principal, the Greater Amman Municipality, or GAM, envisions the center as the hot spot of the regional art scene. It will accommodate the performing art center and host events like concerts, music, and theatre performed by foreign and Jordanian artists, among them the Amman Symphony Orchestra, the State Music Conservatory and various other sources. The "Darat King Abdullah II" is one of several projects meant to liven up the city center and bridge social barriers separating its residents. To this end, a number of public buildings will be constructed in what is called the "GAM Strip", a site at the heart of the city, and a point of contact between two hugely different neighborhoods. The "Darat King Abdullah II" aside, the city administration building, the Hussein Culture Center, and the National Museum (due to open in 2008) are already sited along a stretch of valley at the foot of Jabal Al Akhdar. The 19,000 square metres site of the new culture and art center is at the western end of the strip. The spatial program comprises 12,000 square metres of ancillary usable area, a concert theater with 1,600 seats, a smaller theater with 400 seats, rehearsal rooms and lecture halls, top end stage equipment, a sweeping foyer, a restaurant, a café and club combination, as well as administrative offices.

安曼人口过200万，是中东最有活力和最现代的城市之一。竞赛任务是为安曼设计一个的新文化和艺术中心 "Darat King Abdullah II"。项目的主办方，大安曼市政当局把这个中心看成是该区艺术界的热点。里边的表演艺术中心可以举办像音乐会、戏剧的活动，由外国和约旦艺术家与安曼交响乐队、国家音乐学院和其他组织一起表演。"Darat King Abdullah II" 与其他几个项目的目的一样，都是要增强城市中心的活力，消除离间居民的社会障碍。为了达到这一目的，很多公共建筑物将会在市中心的一个被称为 "GAM带" 的地点建起。除了 "Darat King Abdullah II" 以外，还有城市政府大楼、Hussein文化中心和国家博物馆。新文化和艺术中心，占地1.9万平方米，位于 "GAM带" 的西端。空间规划包括1.2万平方米的辅助使用面积、一个有1600个座位的大电影院、一个有400个座位的小电影院、彩排室、讲座厅、尖端舞台设备、宽敞的休息室、饭店、咖啡厅俱乐部以及行政办公楼。

Aerial view　鸟瞰图

Competition site　竞赛地点

Competition site　竞赛地点

Ancient amphitheater　古圆形剧场

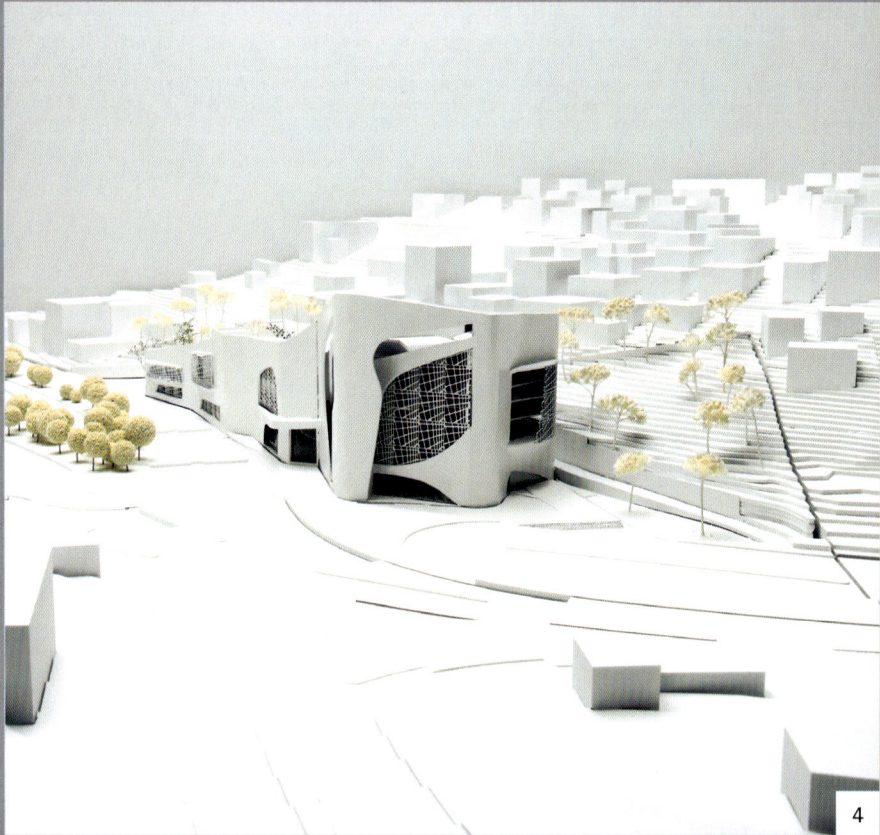

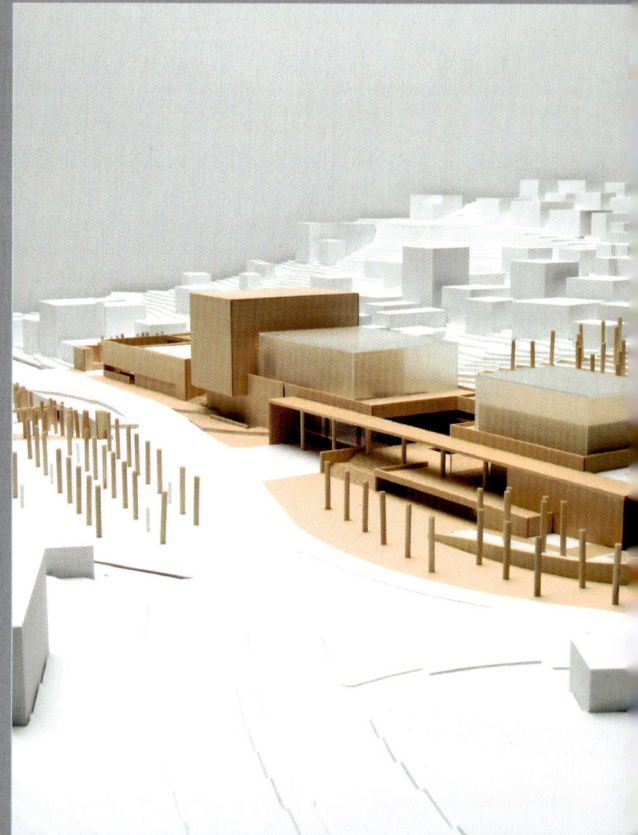

Qualified participants
合格的参赛者

1
1st prize　一等奖
Zaha Hadid Architects,
London

2
1st prize　一等奖
DELUGAN MEISSL ASSOCIATED
ARCHITECTS, Vienna

3
3rd prize　三等奖
Snøhetta, Oslo

4
Further participant　其他参赛者
Atelier Christian de Portzamparc,
Paris

5
Further participant　其他参赛者
Kerry Hill Architects,
Singapore

6
Further participant　其他参赛者
Henning Larsen Architects,
Copenhagen

Zaha Hadid Architects London
Zaha Hadid建筑事务所 伦敦

"... inspired by the uniquely beautiful monument of Petra [...] we are applying the principle of fluid erosion and carving to the mass of the building for the performing arts centre."

"受具有独特美感的Petra遗址的灵感激发，我们在建设表演艺术中心的时候运用流体冲蚀和雕刻的原理。"

作者 Zaha Hadid, Patrick Schumacher   合作伙伴 Christos Passas, Tariq Khayyat, Dominiki Dadatsi, Marya Araya, Sylvia Georgiadou, Bence Pap, Eleni Paviidou, Daniel Santos, Daniel Widrig, Sevil Yazipi

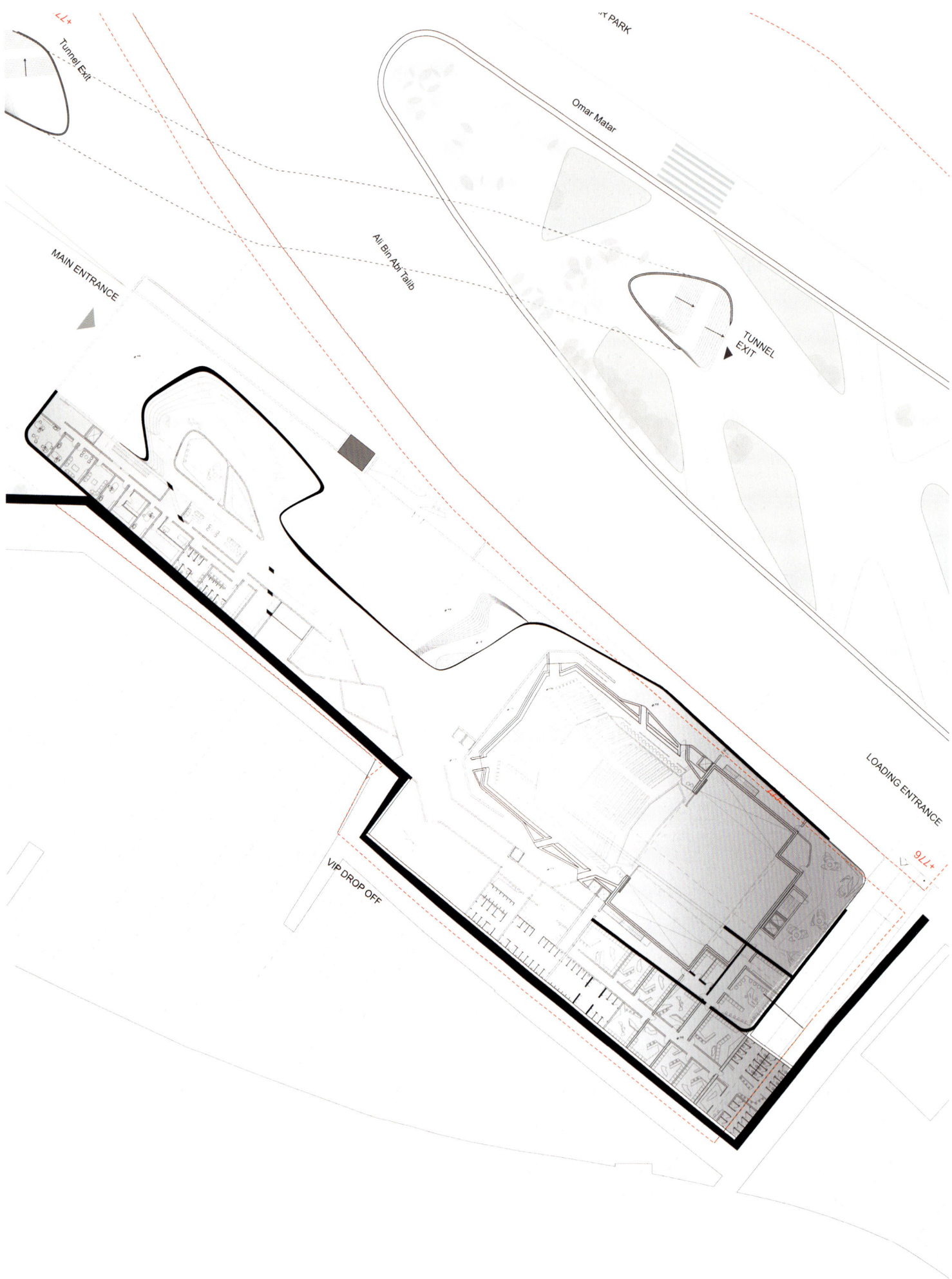

PARK

Omar Matar

Ali Bin Abi Talib

Tunnel Exit

TUNNEL EXIT

MAIN ENTRANCE

LOADING ENTRANCE

VIP DROP OFF

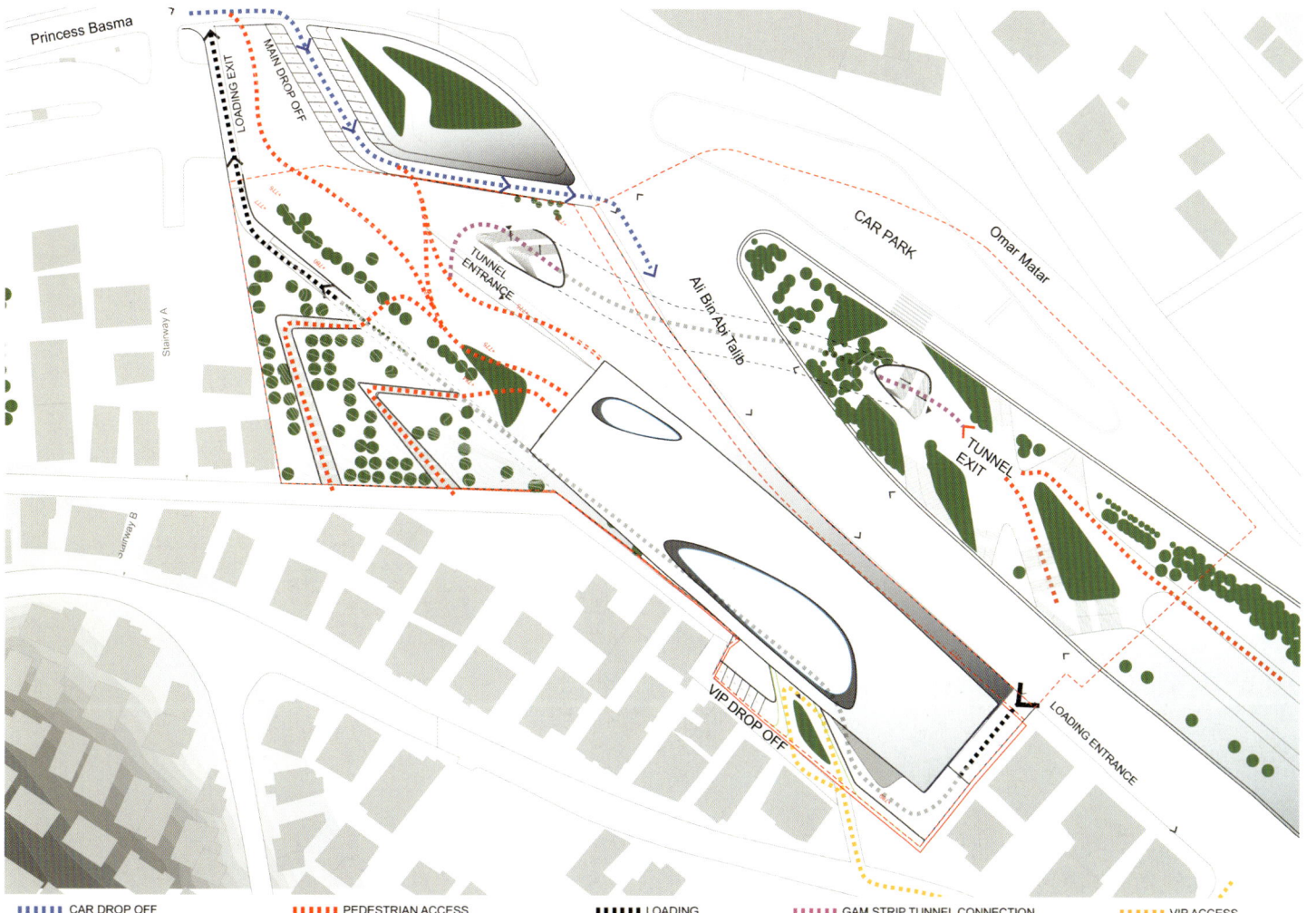

Princess Basma

MAIN DROP OFF

LOADING EXIT

Stairway A

Stairway B

TUNNEL ENTRANCE

Ali Bin Abi Talib

CAR PARK

Omar Matar

TUNNEL EXIT

VIP DROP OFF

LOADING ENTRANCE

| ▮▮▮▮▮▮ CAR DROP OFF | ▮▮▮▮▮ PEDESTRIAN ACCESS | ▮▮▮▮▮ LOADING | ▮▮▮▮▮ GAM STRIP TUNNEL CONNECTION | ▮▮▮▮▮ VIP ACCESS |

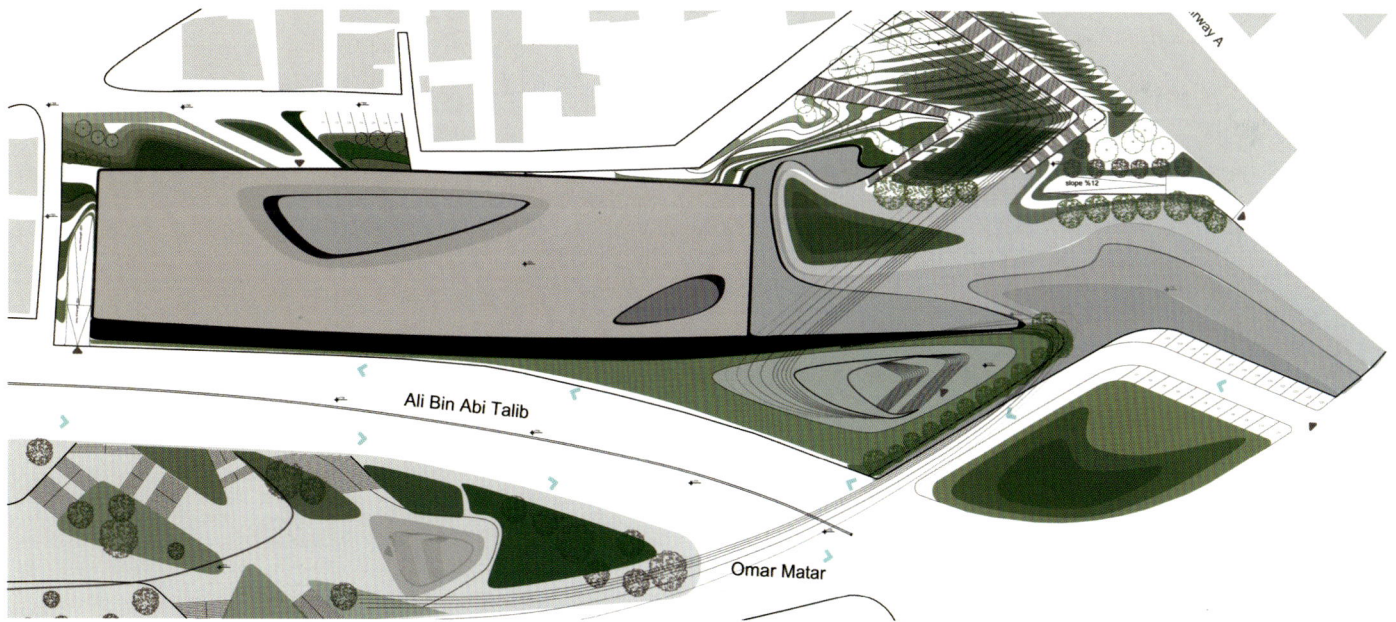

Stairway A

slope %12

Ali Bin Abi Talib

Omar Matar

VIEW FROM THE PARK AT ZONE A

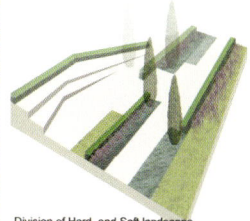

Division of Hard, and Soft landscape

-3.75_LEVEL_MAIN AUDITORIUM (VIP) / REHEARSAL ROOMS   1:500

10 m   50 m

+5m_LEVEL MAIN AUDITORIUM   1:500

10 m   50 m

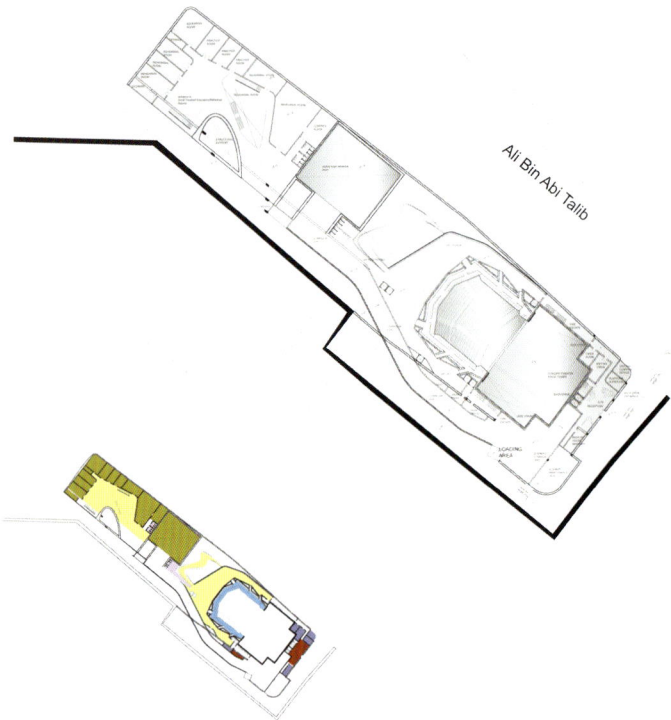

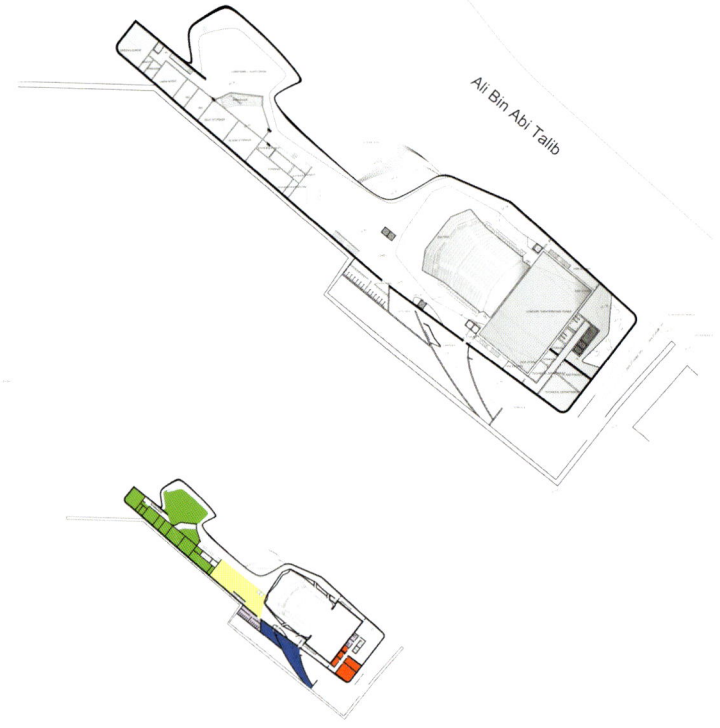

+18m_LEVEL_CAFE AND RESTAURANTS   1:500

10 m   50 m

+22_LEVEL   1:500

10 m   50 m

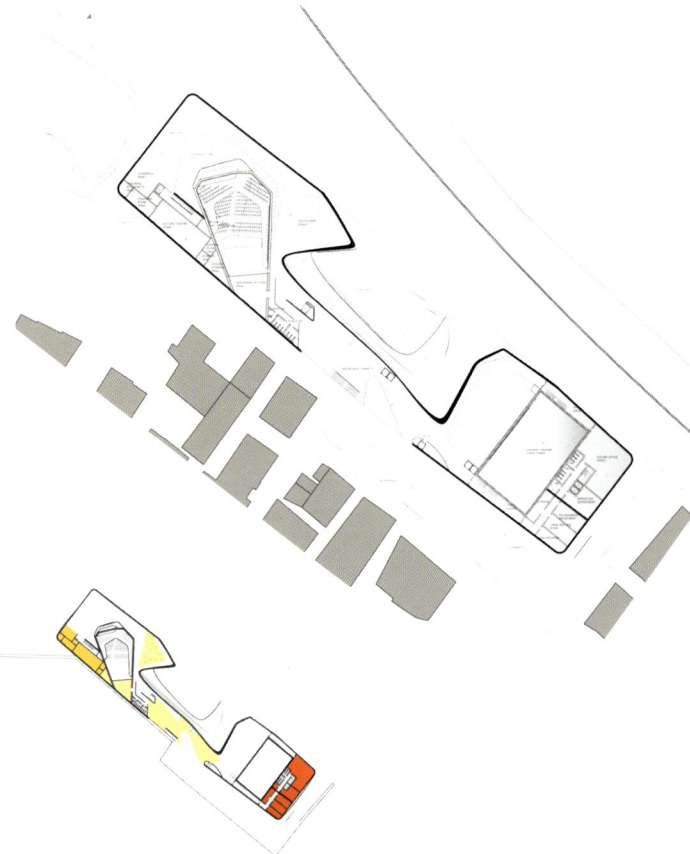

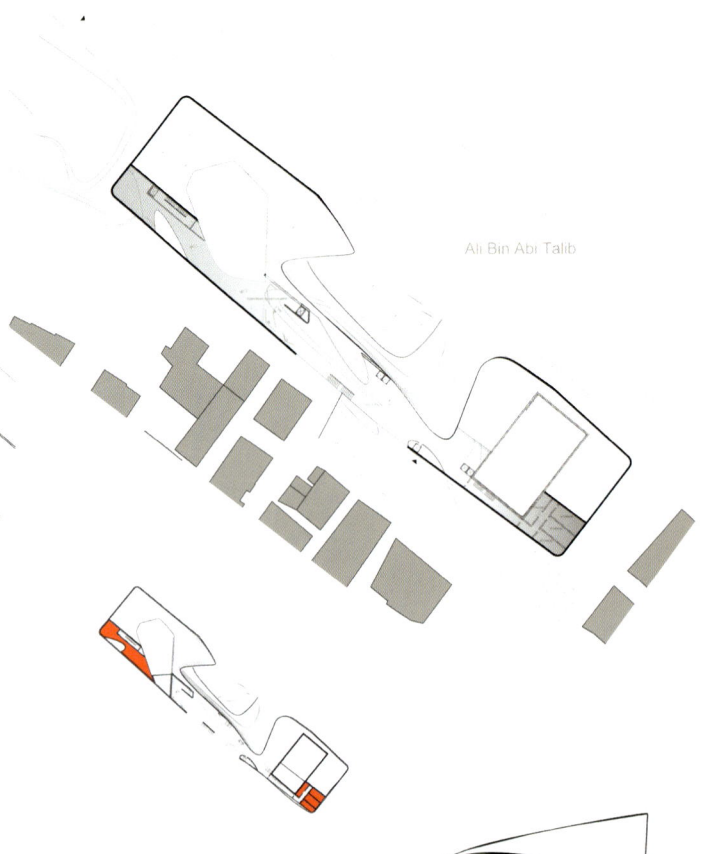

| 1 | 2 | 3 | 4 | 5 | 6 | 7 | 8 | 9 | 10 | 11 | 12 |

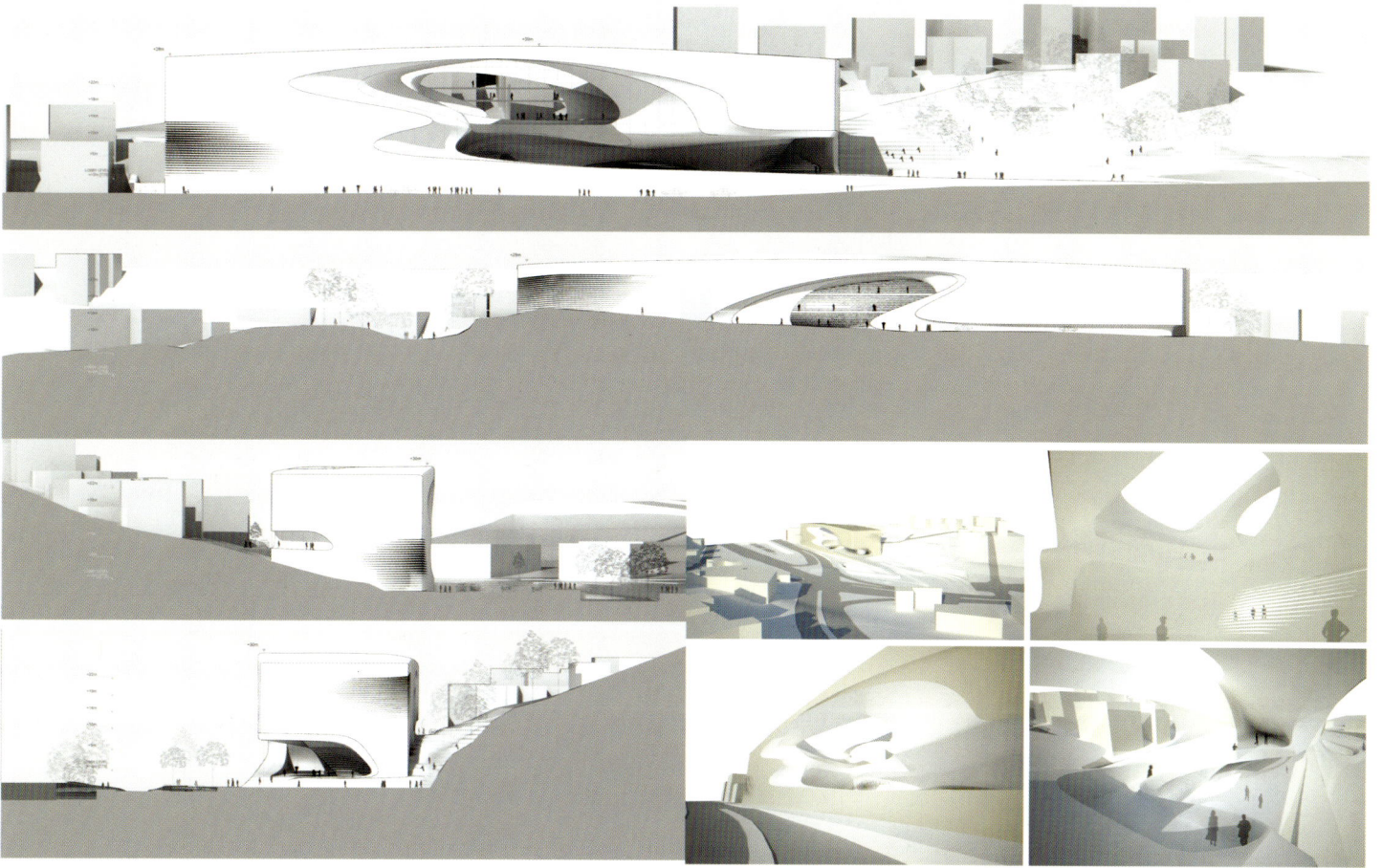

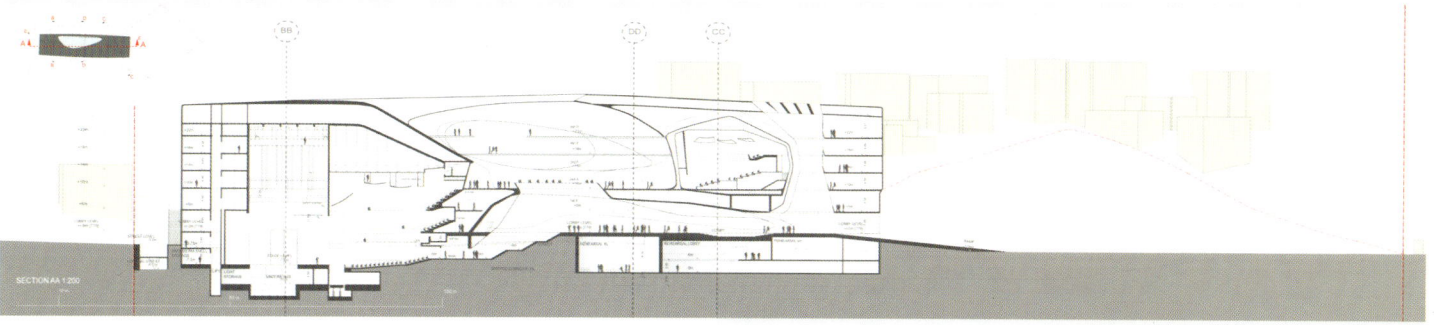

SECTION AA 1:200

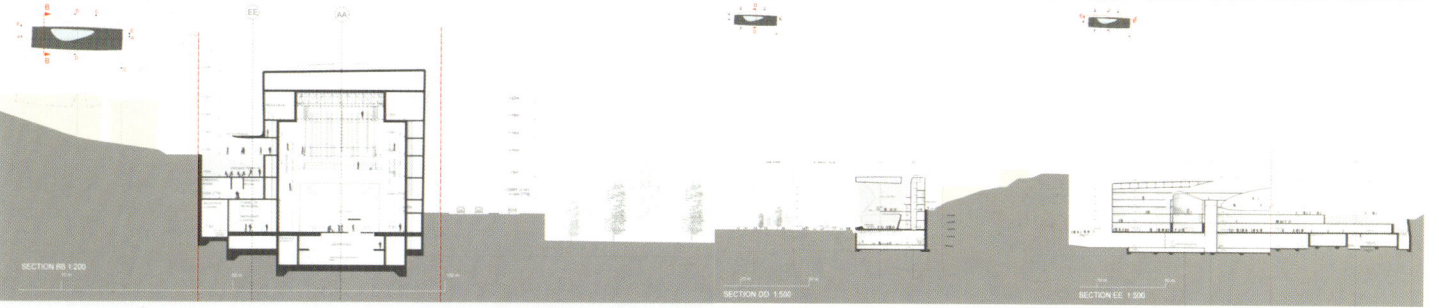

SECTION BB 1:200

SECTION DD 1:500

SECTION EE 1:200

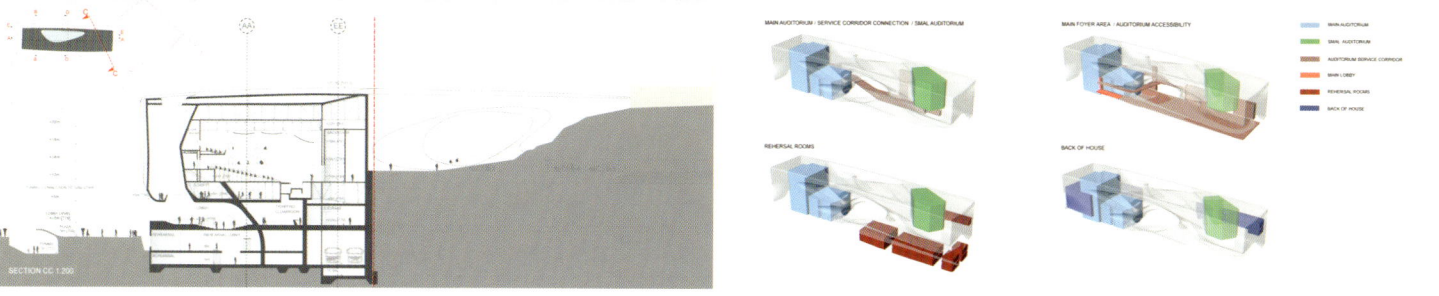

SECTION CC 1:200

MAIN AUDITORIUM - SERVICE CORRIDOR CONNECTION / SMAL AUDITORIUM

MAIN FOYER AREA - AUDITORIUM ACCESSIBILITY

REHERSAL ROOMS

BACK OF HOUSE

MAIN AUDITORIUM
SMALL AUDITORIUM
AUDITORIUM SERVICE CORRIDOR
MAIN LOBBY
REHERSAL ROOMS
BACK OF HOUSE

PEDESTRIAN- EYE VIEW

2   1 2 3 4 5 6 7 8 9 10 11 12

MAIN AUDITORIUM

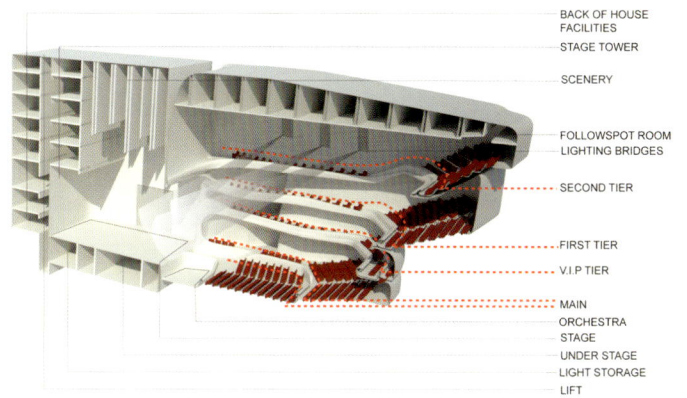

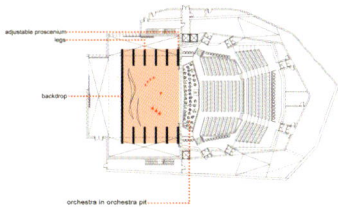

MAIN AUDITORIUM

BACK OF HOUSE
FACILITIES

STAGE TOWER

SCENERY

FOLLOWSPOT ROOM
LIGHTING BRIDGES

SECOND TIER

FIRST TIER

V.I.P TIER

MAIN
ORCHESTRA
STAGE
UNDER STAGE
LIGHT STORAGE
LIFT

adjustable proscenium
legs

backdrop

orchestra in orchestra pit

dance and opera configuration

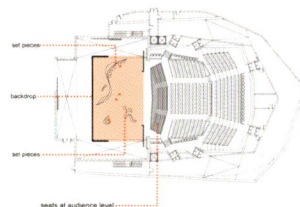

set pieces

backdrop

set pieces

seats at audience level

drama configuration

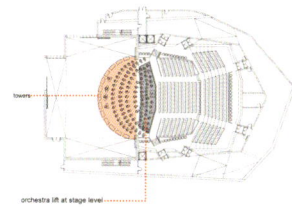

towers

orchestra lift at stage level

large orchestra configuration

DELUGAN MEISSL ASSOCIATED ARCHITECTS Vienna
DELUGAN MEISSL联合建筑事务所 维也纳

"The differentiated but interconnected spatial sequences of public spaces, foyers, and theater halls turn the Darat King Abdullah II into a lively platform for conversations, performances, and societal action ..."

"互不相同而又彼此连接的公共空间、休息室和剧院大厅等一系列的空间建筑使Darat King Abdullah II成为人们谈话、表演和进行社交活动的场所……"

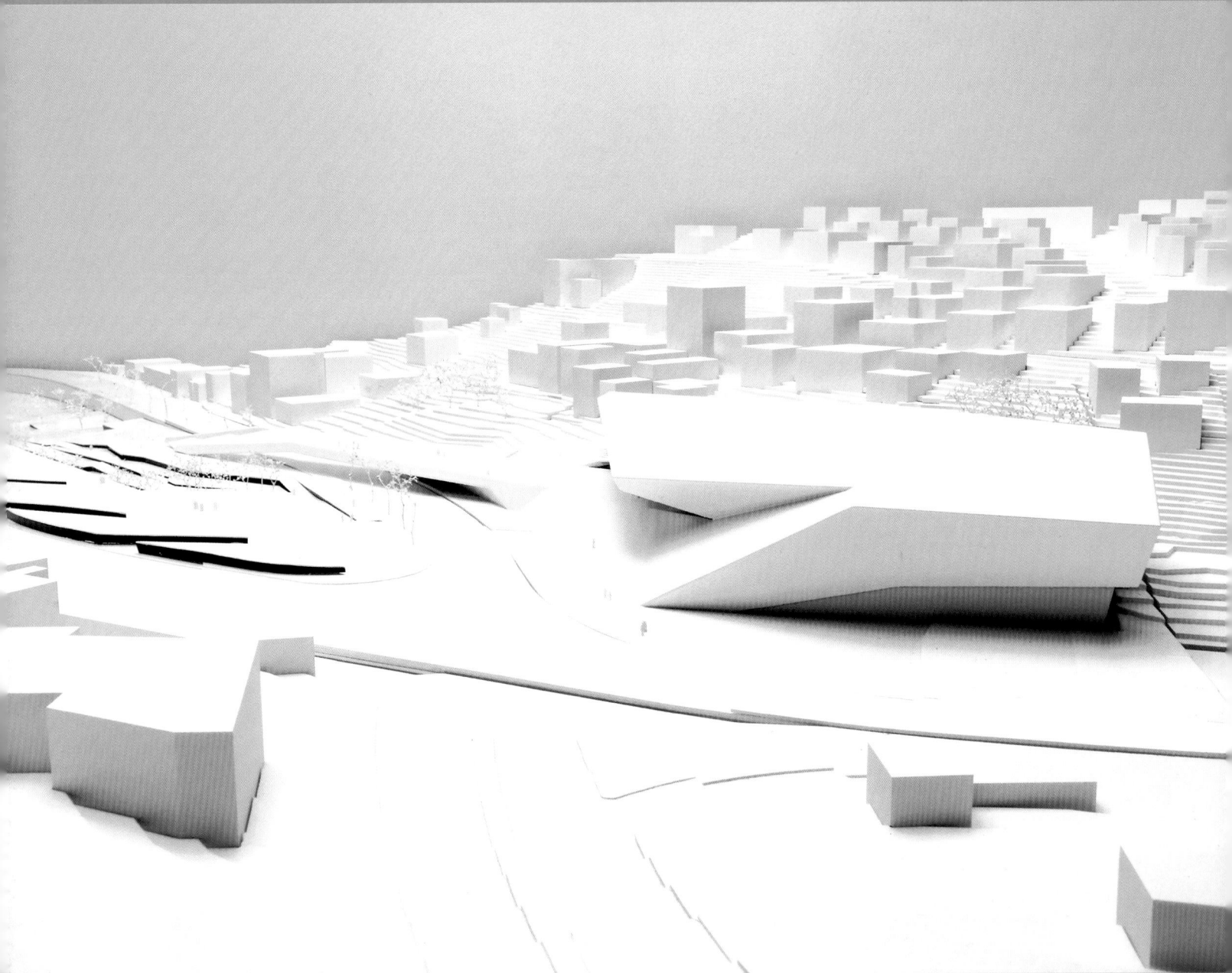

作者 Elke Delugan-Meissl, Roman Delugan   合作伙伴 Martin Josst, Sebastian Brunke, Jörg Rasmussen, Oana Maria Nituica, Claudiu Barsan-Pipu, Marina Kolloch, Thomas Theilig, Xiaozhen Zhu, Peter Pichler, Jan Saggau   专家 structural analysis: Werkraum Wien; open space planning: Rajek Borosch, Landschaftsarchitektur, Vienna; concert, theatre, and acoustics: Müller BBM Akustik, Munich; building services and air conditioning: Scholzegruppe, Vienna

DARAT KING ABDULLAH II

PUBLIC STAGE

RESTAREA

CONNECTING

BUFFER ZONE TO TRAFFIC

EXISTING FOREST + NEW FOREST

TERRACES

PUBLIC PLAZA

SITE PLAN 1:500

FOREST

GREEN BUFFER

PUBLIC STAGE

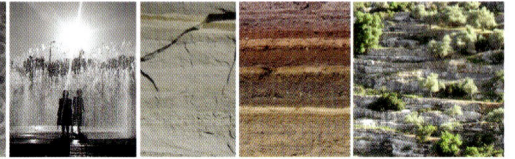

TERRACES

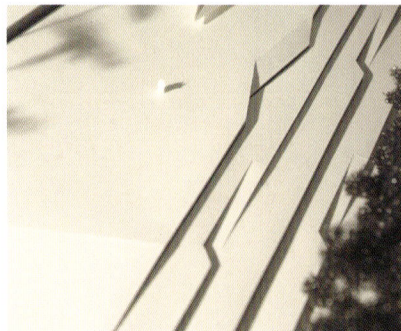

DESIGN OF OPEN-AIR SPACES:

The new landscape of the Darat King Abdullah II is characterized by two basic movements that develop from the existing geological layering and the morphology of the city. The trees are embedded into these horizontal and vertical movements that connect the city districts or are generously compensated by new plantings of trees. The entire landscape is made accessible to the public.

Horizontal layering
The first movement follows the sloping line of the hill chains and picks up the principle of the geological layering in the form of diverse terraces. The urban terrace landscape permeates the concert hall, staggers across the "bridge" and the "city stage" that is on the street level, in order to continue across several levels in the park of the GAM Stripe. The different levels are connected across ramps and stairs and create additional access and connection points in the surrounding cityscape.

With the aid of the park terraces, the level is raised in the direction of the bridge so that the height of the stairs can be reduced. Similar to the geological layering of the sandstone, the walls of the terraces consist of layers of differently colored concrete. The seating areas on these plateaus are in the shade of pergolas and provide small areas of respite. The upper landscape layer is formed by the expanded forest. As far as possible, the existing trees in the north-east are kept and the required reductions on the south-east slope are generously compensated. The entire area will be accessible to the residents.

City and landscape axis
The second movement is developed as an axis of green and open space from the GAM Stripe, flows into the street area of the Princess Basma Road as a tree structure, and continues to follow the traffic axis through the city. The new design of the city park here integrates the existing trees, which becomes a shaping element in that the terraces are adjusted to the trees in terms of levels and locations. The terrace system enables a new sequence of important urban open space functions and becomes a connecting space in the urban landscape. Starting at the Hussein Cultural Center, the park is organized into a sport, play, quiet and connection zone in order to open up as an urban location in the intersection area of the Ali Bin Abi Talib, Princess Basma and the new service road towards the concert hall. The urban "stage", easily visible from the concert hall and the pedestrian bridge, is enveloped by raised botanical terraces. This green buffer towards the Oma Mater Road is interrupted by broad aisles and thus provides interesting views of the square and the concert hall. The square is structured and enlivened with a play of water.

The horizontal and vertical movement melt into a new city and park landscape that connects the city areas, leads to the concert hall and allows the creation of new, adequate open-air stages.

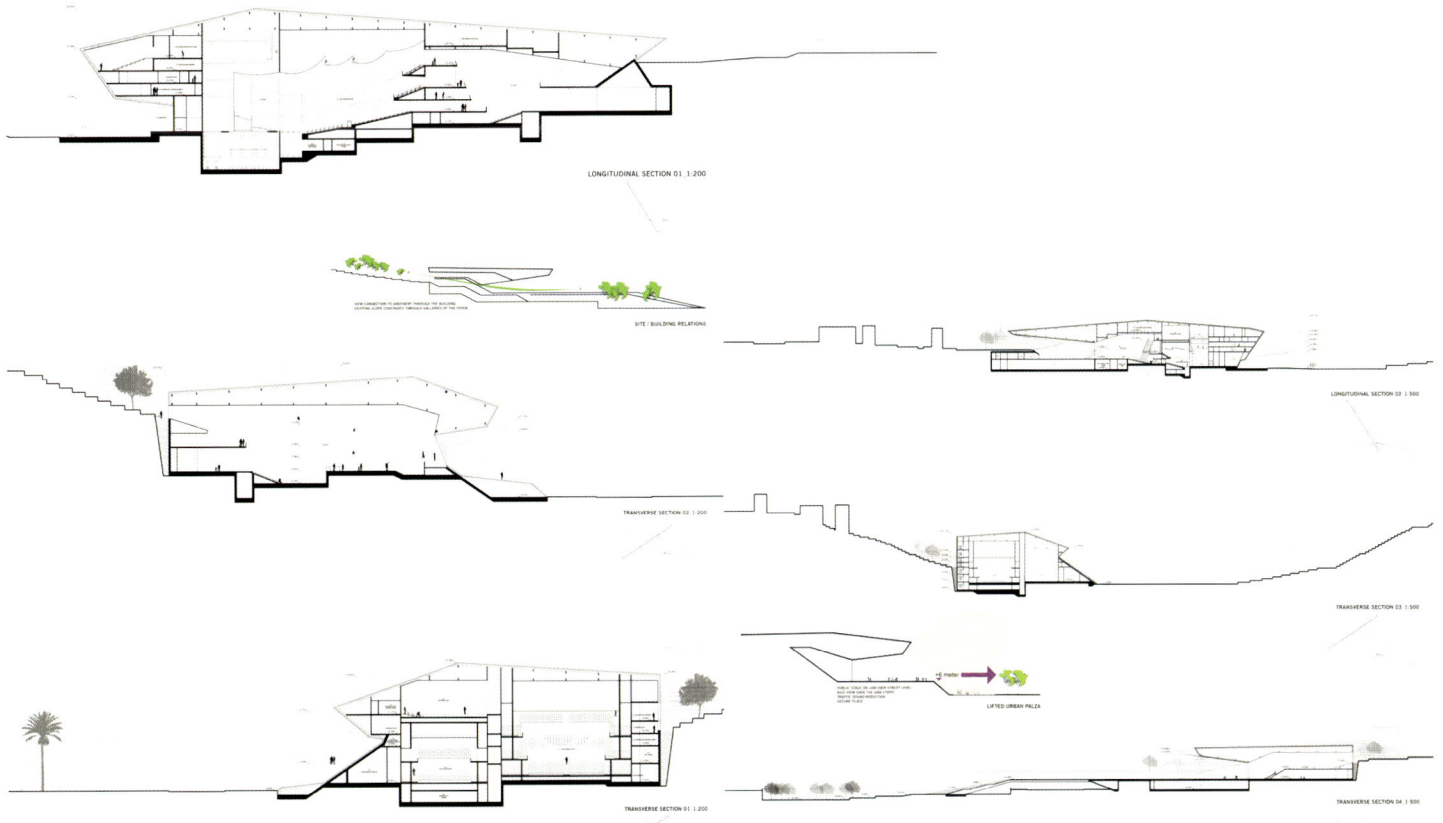

LONGITUDINAL SECTION 01 1:200

SITE : BUILDING RELATIONS

LONGITUDINAL SECTION 02 1:500

TRANSVERSE SECTION 02 1:200

TRANSVERSE SECTION 03 1:500

LIFTED URBAN PALZA

TRANSVERSE SECTION 01 1:200

TRANSVERSE SECTION 04 1:500

FUNCTIONAL OVERVIEW

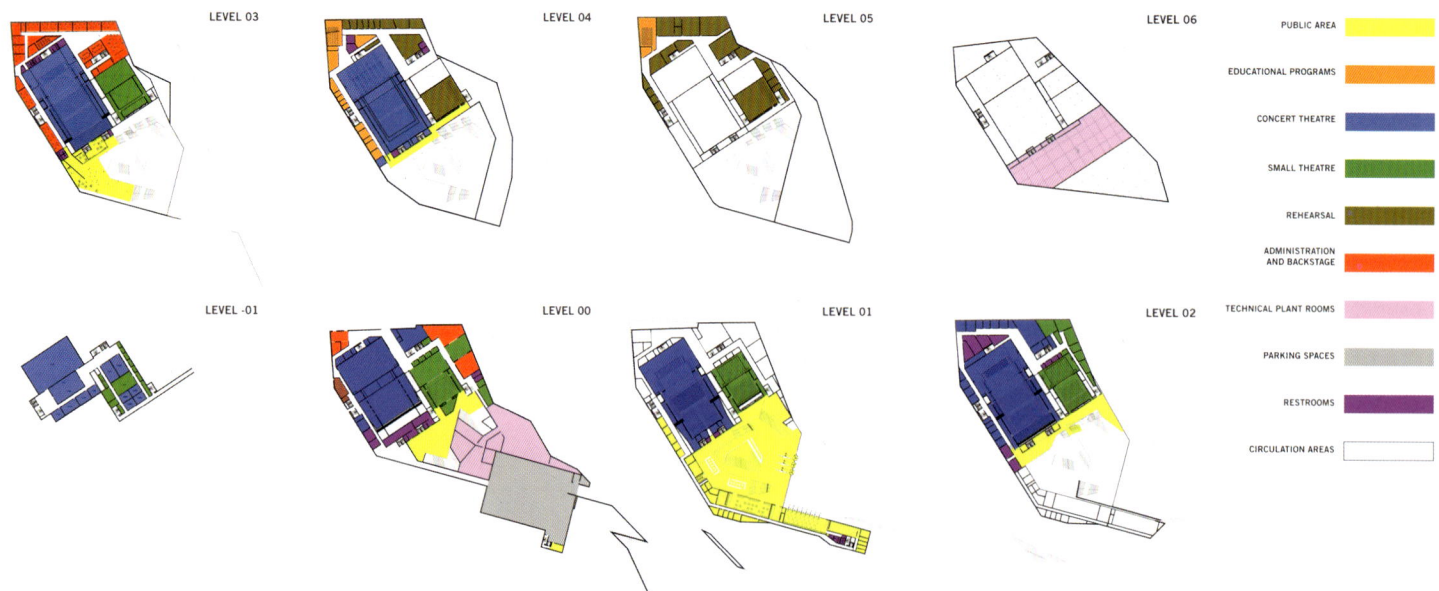

REHEARSAL AND PRACTICE ROOMS

EDUCATION AND SEMINARS

ADMINISTRATION

CHANGING AND DRESSING ROOMS

GENERAL BACKSTAGE / LOADING DOCK

CATERING AND RETAIL
CONCERT THEATER
SMALL THEATER
REHEARSAL ROOM
PUBLIC ACCESS
VIP ACCESS / HOSPITALITY

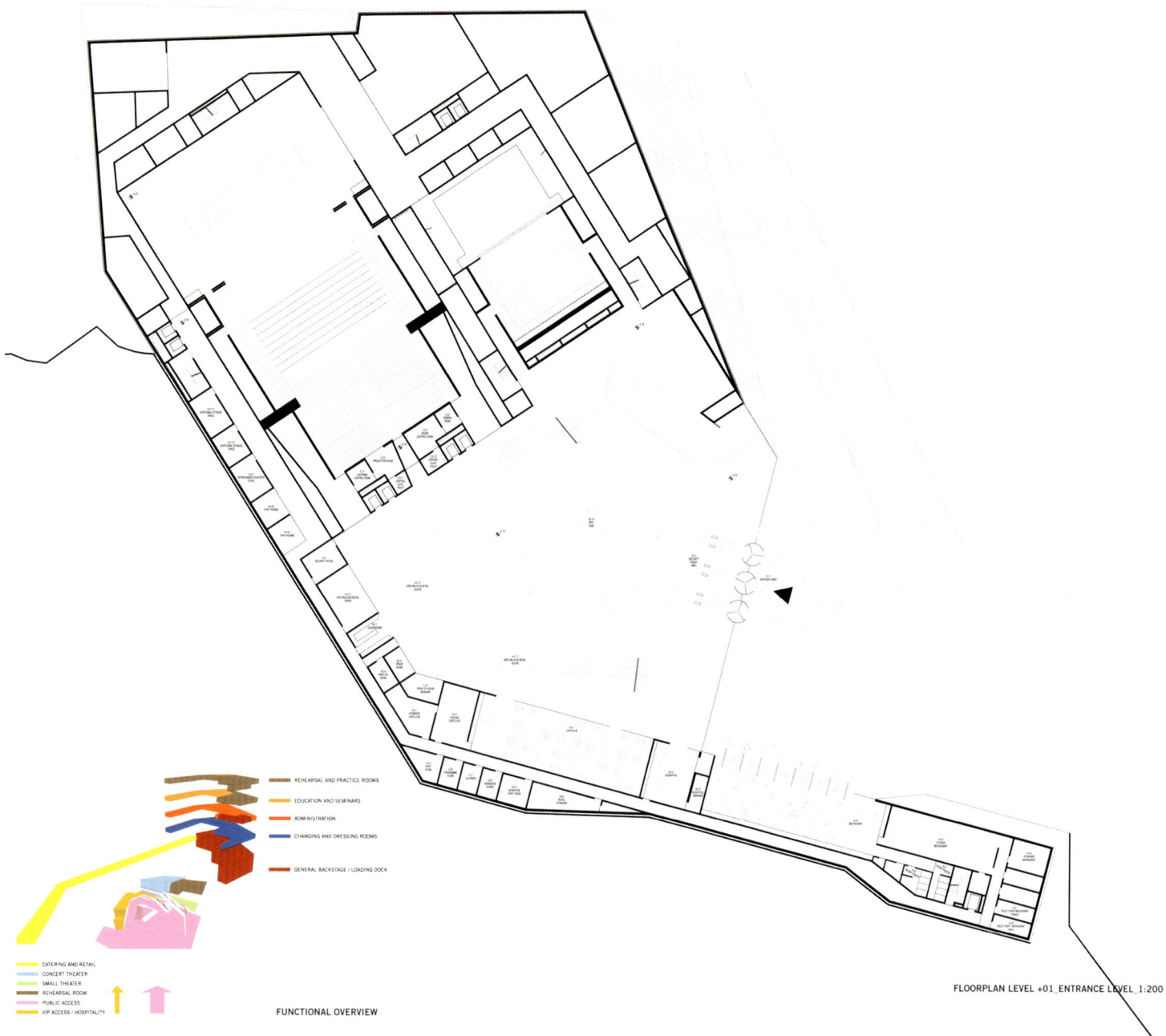

FLOORPLAN LEVEL +01_ENTRANCE LEVEL_1:200

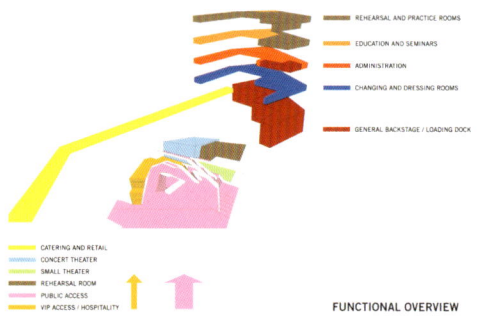

LEVEL 03

LEVEL 04

LEVEL 05

LEVEL 06

LEVEL -01

LEVEL 00

LEVEL 01

LEVEL 02

PUBLIC AREA

EDUCATIONAL PROGRAMS

CONCERT THEATRE

SMALL THEATRE

REHEARSAL

ADMINISTRATION
AND BACKSTAGE

TECHNICAL PLANT ROOMS

PARKING SPACES

RESTROOMS

CIRCULATION AREAS

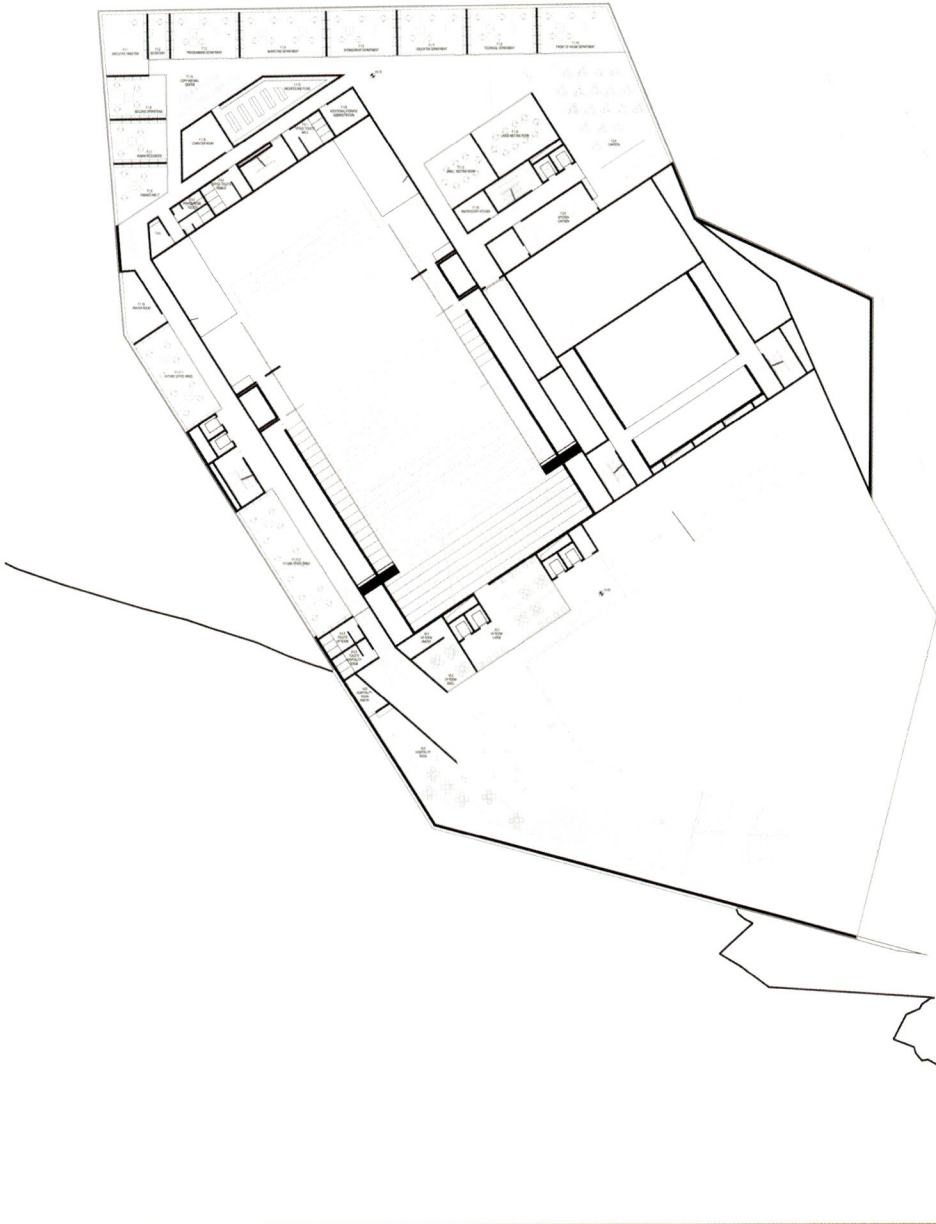

FLOORPLAN LEVEL +03_VIPLEVEL_ADMINISTRATION_1:200

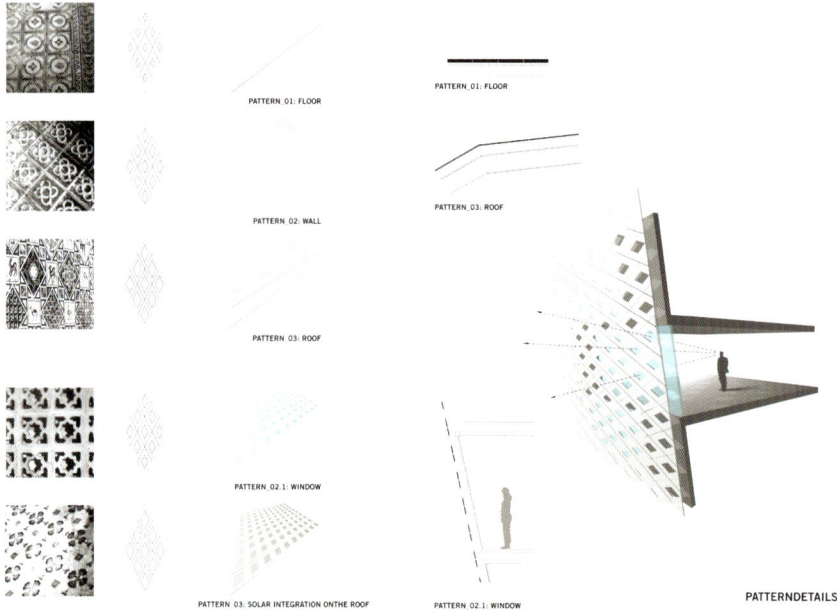

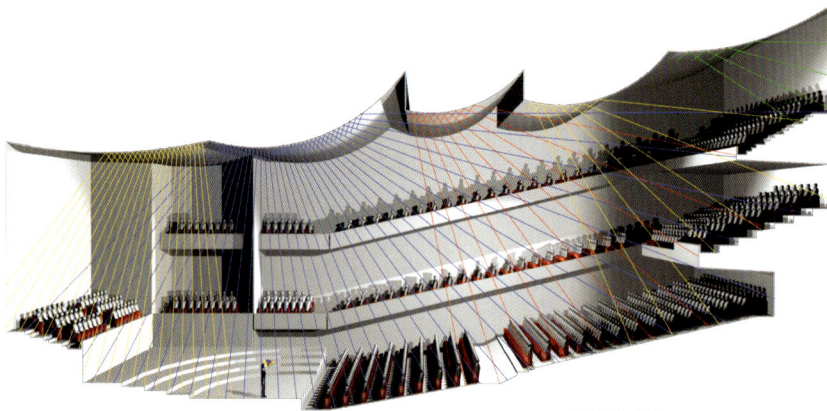

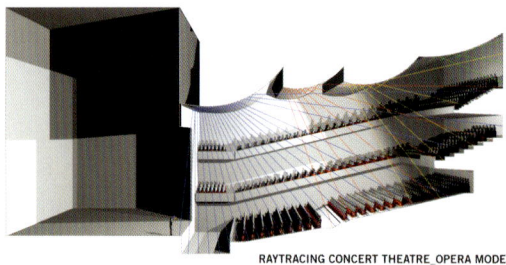

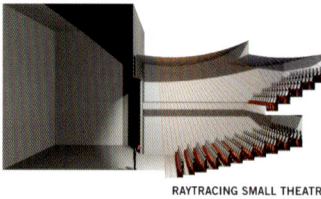

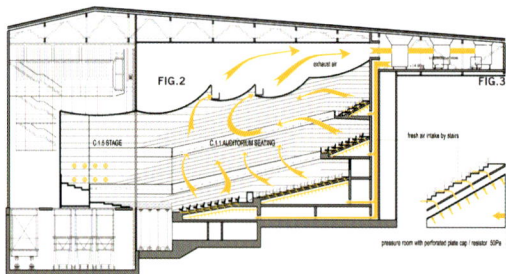

PATTERN_01: FLOOR

PATTERN_01: FLOOR

PATTERN_02: WALL

PATTERN_03: ROOF

PATTERN_03: ROOF

PATTERN_02.1: WINDOW

PATTERN_03: SOLAR INTEGRATION ONTHE ROOF

PATTERN_02.1: WINDOW

PATTERNDETAILS

RAYTRACING CONCERT THEATRE_CONCERT MODE

RAYTRACING CONCERT THEATRE_OPERA MODE

RAYTRACING SMALL THEATRE

FIG.2

FIG.3

exhaust air

C-1.6 STAGE

C-1.1 AUDITORIUM SEATING

fresh air intake by stairs

pressure room with perforated plate cap / resistor 50Pa

DISPLACEMENT FLOW FOR THE HALLS

THE MAIN PART OF THE SUPPLY AIR IS LED BENEATH THE FLOOR LEVELS OF THE HALLS AND WILL BE DISTRIBUTED VIA A DOUBLE FLOOR AND FINALLY VIA STEP - OR CHAIR - OUTLETS DIRECTLY TO THE PERSONS. A METHOD WELL ACCEPTED IN MANY OPERA HOUSES. FIG.2 SHOWS THE PRINCIPAL AIR FLOW WITH DISPLACEMENT FLOW PATTERN THAT HAS BEEN PROVED AS OPTIMAL. ON STAGE WE PROPOSE SWIRL NOZZLES IN SIDE WALLS.

FIG.3 IS AN ENLARGEMENT OF FIG.2 AND SHOWS THE DETAIL OF AIR ENTRANCE IN THE LOWER FLOOR PART. THE DISTRIBUTION IN THE WHOLE FLOOR AND THE BLOWING OUT VIA THE STEP-OUTLETS.

VENTILATION CONCEPT THEATRES

3D MODEL OF THE STRUCTURE DESCRIPTION

THE DESIGNED CONCERT HALL IS A BUILDING CONSTRUCTED IN STEEL AND CONCRETE, WITH THE LOCATION AMMAN / JORDANIA

DUE TO THE HIGH SPANS AND CANTILEVERS AND THE GEOMETRICAL REQUIREMENTS, THE ROOFCONSTRUCTION IS DESIGNED AS A ORTHOGONAL STEEL TRUSS, UP TO 10M HIGH, WHICH COVERS THE WHOLE BUILDING. THE TRUSS ITSELF STANDS IN ONE PART ON THE WALLS OF THE CONCRETE HALLS, IN THE OTHER PART ON COLUMNS AND CONCRETE WALLS ALONG THE ENTRANCE. FOR THE TWO COCERT HALLS AND THE SURROUNDED STOREYS, WE PROPOSE A CONCRETE CONSTRUCTION. THIS PART WILL WORK AS MASSIVE CORE AND SERVE FOR THE STIFFNESS OF THE BUILDING IN HORIZONTAL DIRECTION IN THE CASE OF WIND OR EARTHQUAKE IMPACT.

THE FOUNDATION OF THE BUILDING IS CONSIDERED TO BE A PILE FOUNDATION. PILES WILL BASICALLY BE NECASSARY BELOW ALL COLUMNS AND WALLS.

PRIMARY TRUSSWORK OF THE ROOF STRUCTURE (MARKED IN RED). THE ROOF CLADDING IS FIXED TO THIS TRUSS.

SECUNDARY TRUSSWORK OF THE ROOF STRUCTURE (MARKED IN RED) ALONG THE BORDER OF THE CANTILEVER AND THE ENTRANCE FACADE.

THE SUPPORTING STRUCTURE IN THE AREA OF THE ENTRANCE IS PROPOSED IN CONCRETE.

THE TWO CONCERT HALLS AND...

... AND THE SOURROUNDED WALLS ARE PROPOSED IN CONCRETE TOO. THEY SERVE THE STRUCTURE FOR THE STIFNESS AGAINST WIND LOADS AND EARTHQUAKE.

DISCUSSION OF THE RESULTS

DEFLECTION OF THE STRUCTURE UNDER DEAD LOADS. THE CANTILIVER HAVE A SPAN OF 40M RESPECTIVELY 75M.

DEFLECTION OF THE STRUCTURE UNDER WIND LOADS (ASSUMED WIND SPEED: 135KM/H)

BOTH RESULTS SHOW THE EXCELLENT LOAD BEARING BEHAVIOUR OF THE STRUCTURE, IN PARTICULAR THE ROOF STRUCTURE.

CAPACITY OF STEEL STRUCTURE. APART FROM SOME PEAKS THE STEEL STRUCTURE IS FAR FROM BEING AT ITS LIMITS, THERE IS STILL POTENTIAL FOR OPTIMIZATION.

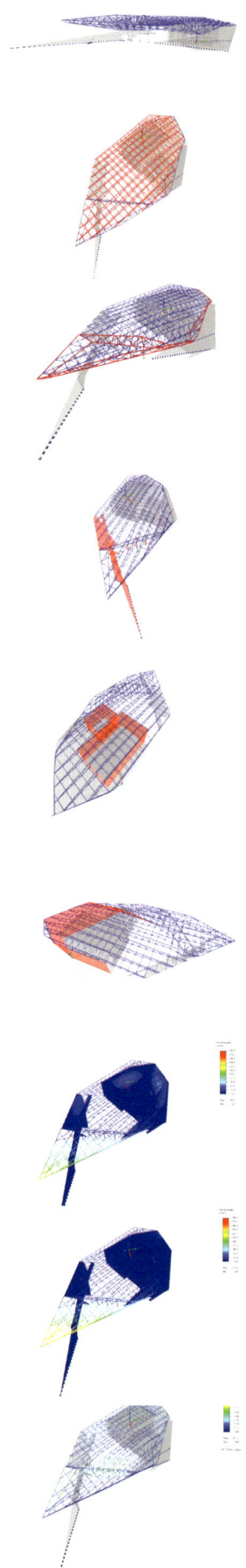

STRUCTURAL DESIGN CONCEPT

DELUGAN MEISSL ASSOCIATED ARCHITECTS, Vienna

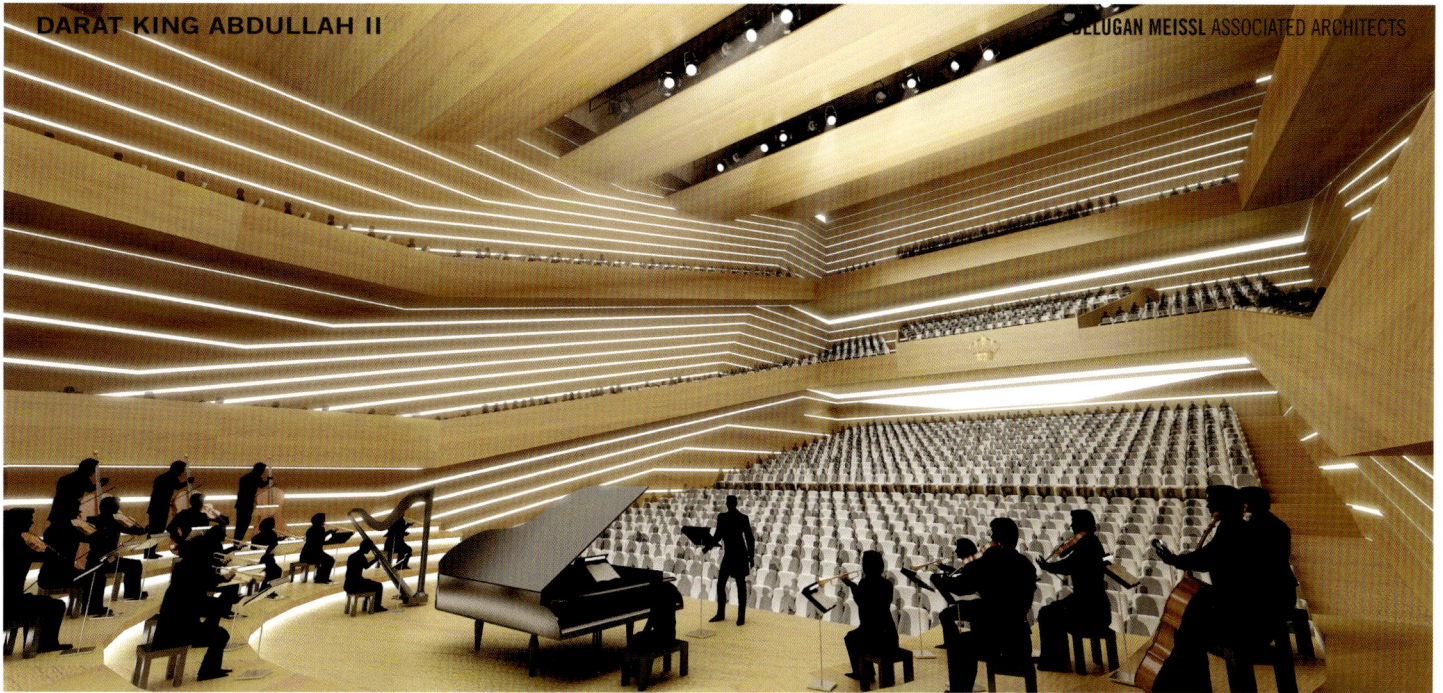

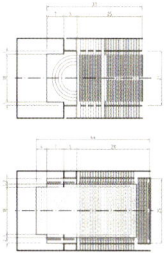

CONCERT THEATRE_CONCERT MODE

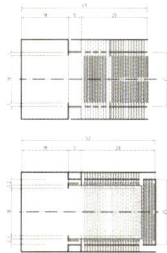

CONCERT THEATRE_OPERA MODE

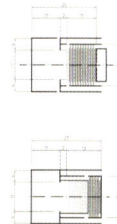

SMALL THEATRE

Snøhetta Oslo
Snohetta建筑事务所 奥斯陆

"A prestigious, contemporary, and stimulating complex for the performing arts, truly integrated into the landscape and urban context."

"一个有声望的、现代的和令人兴奋的表演艺术建筑群真正与周围景观和城市的文脉融合在一起。"

作者 Robert Greenwood, Oslo   合作伙伴 Kjerstin Bjerka, Peter Dang, Peter Girgis, Tine Hegli, Andreas Nypan, Julian Prizes, Erik Vitanza   专家 civil structural services and specialist engineers: Buro Happold Ltd., Glasgow; acoustical consultant: Arup Acoustics, Winchester, Hampshire

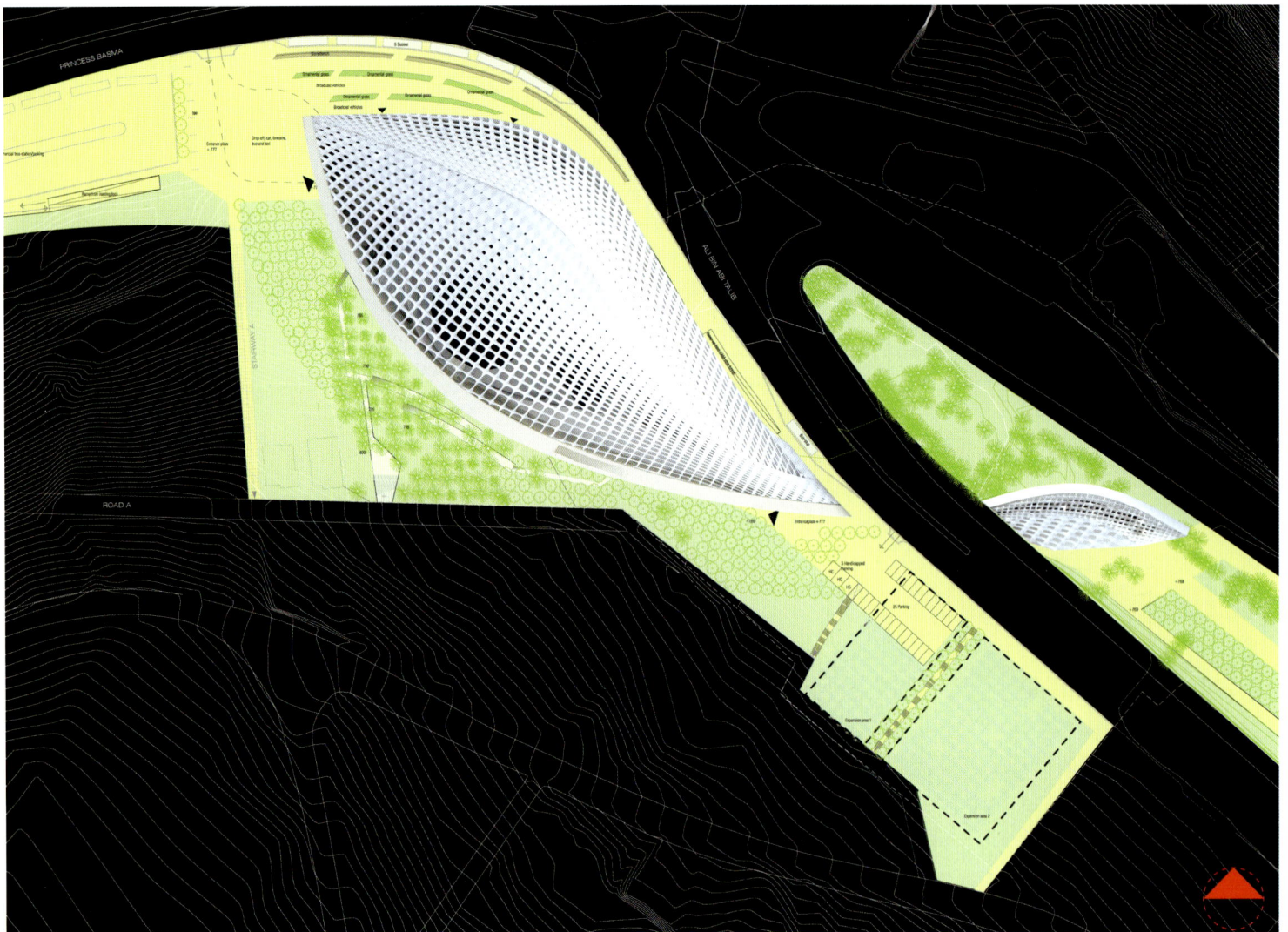

SITE PLAN
scale 1:500

0   50   100m

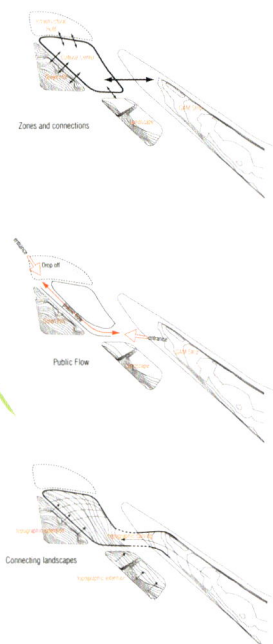

Zones and connections

Public Flow

Connecting landscapes

Perforated Canopy Structure

Landscape Connection

Park Entrance
Through tunnel under Ali Bin Abi Talib Street

Public Area / Foyer

Back of House / Services

Arched Back Wall
with Carved-in Balconies, stairs, program

Truck Delivery Access
Underground

A Concert Theater
B Small Theater
C Rehersal Rooms

Public Area / Foyer

Main Entry from West

Public Circulation

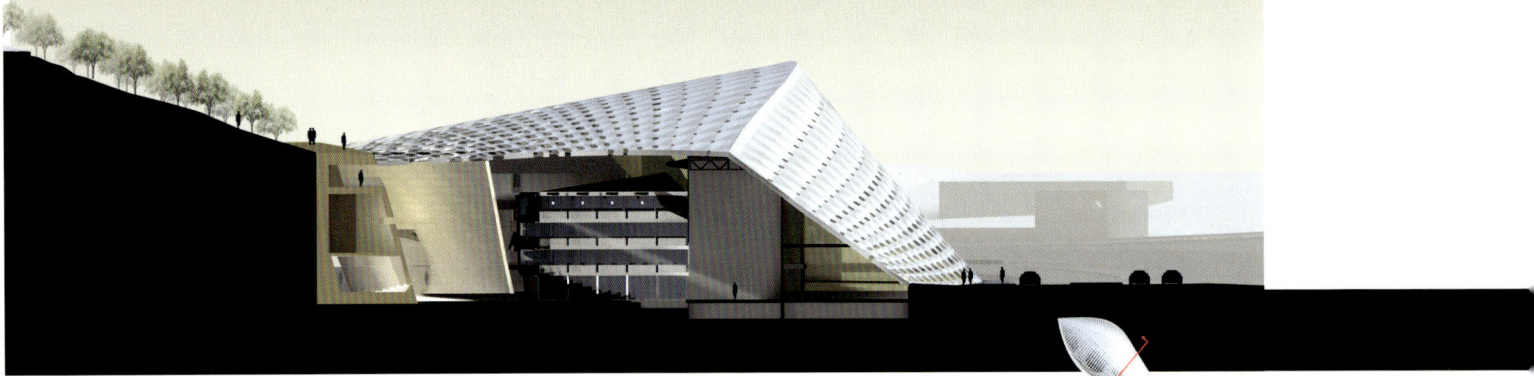

CTION THROUGH SMALL THEATER
e 1:200

NG SECTION THROUGH SMALL THEATER & CONCERT THEATER
e 1:200

NG SECTION - ARCHED WALL
e 1:200

N THROUGH CONCERT THEATER

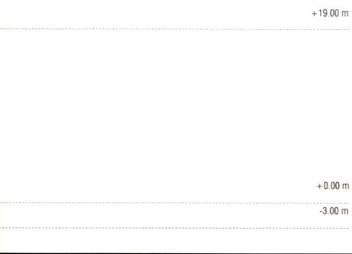

CONCEPT COLLAGE REPRESENTING COLOUR AND TEXTURE INSPIRATION REFERENCING NATURAL ENVI-
RONMENTS, EMPHASISING SPATIAL DIFFERENCES OF THE OPEN AND MAJESTIC   AS WELL AS INTIMATE
MEETING PLACES FOR THE PEOPLE OF AMMAN

+ 19.00 m

+ 0.00 m
– 3.00 m

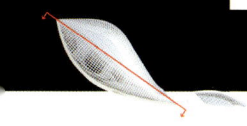

Darat King Abdullah II, Amman

Atelier Christian de Portzamparc Paris
Atelier Christian de Porzamparc建筑事务所 巴黎

"The main component is the desire for a lively building that can become a binding factor between the two halves of the city."

"设计的主要组成部分是想设计出一个充满活力的建筑物，把城市的两个部分加以结合。"

作者 Christian de Portzamparc   合作伙伴 Bertrand Beans, M-E Nicoleau, Andre Terzibachian, Duccio Gardelli, Burkhart Schiller, Bettina Reali, Hyun-Jung Song, Isabella Burck, Paul Chaulet, Fabiana Aravjo, Ricardo Marotta, Veronica Fiorini, Luisa Fonsela
专家 structural analysis: SIDF, Marseilles; acoustics: XU Acoustics, Paris; scenograph: Ducks Slend, Les Pleiades

NORTH-EAST ELEVATION

OPEN AIR-THEATER BY NIGHT

VIEW FROM THE NORTH BY NIGHT

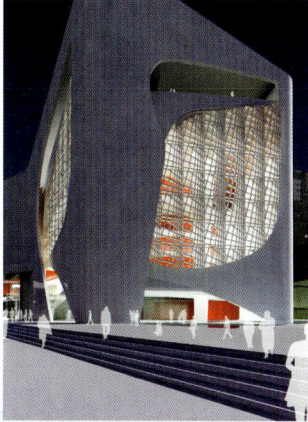

VIEW FROM THE NORTH

THE SPICE TERRACE AND THE RESTAURANT

THE PARFUM GARDEN

VIEW FROM THE WEST

VIEW FROM THE CONNECTING BRIDGE

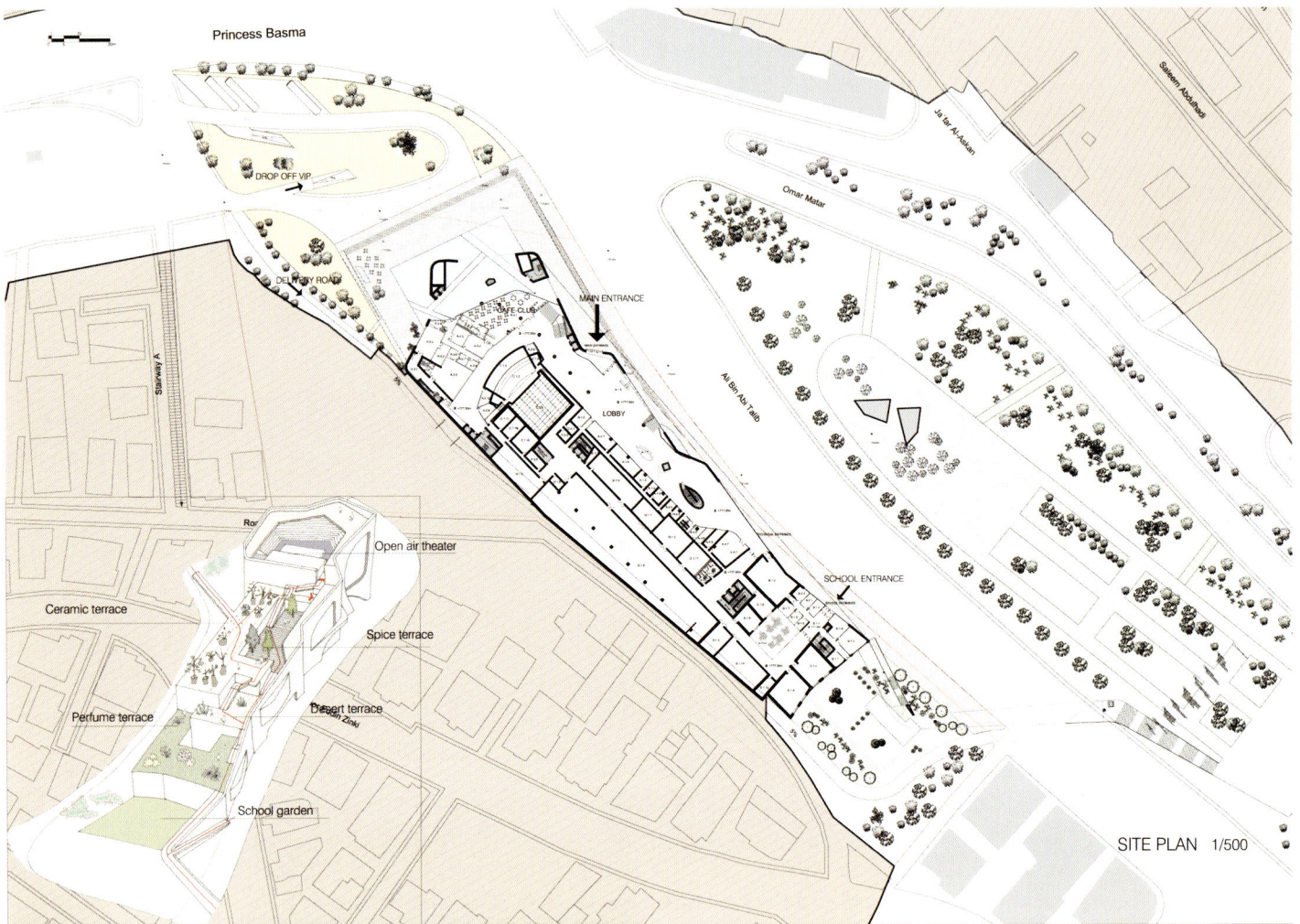

Princess Basma

DROP OFF VIP

DELIVERY ROAD

CAFE-CLUB

MAIN ENTRANCE

LOBBY

SCHOOL ENTRANCE

Ja far Al-Asfan

Omar Matar

Ali Bin Abi Talib

Saleem Abdelhadi

Stairway A

Open air theater

Ceramic terrace

Spice terrace

Perfume terrace

Desert terrace

School garden

SITE PLAN   1/500

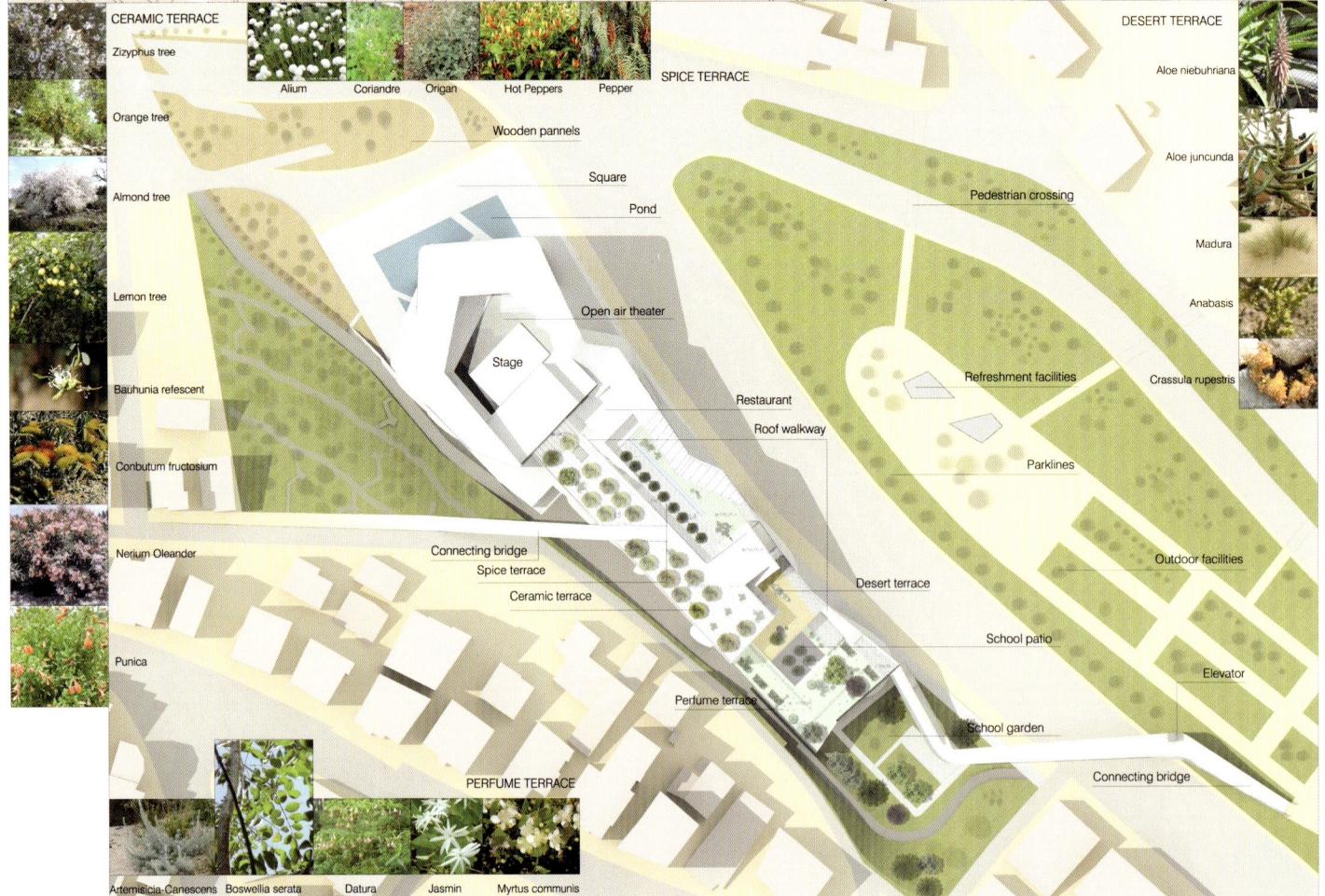

CERAMIC TERRACE

Zizyphus tree

Orange tree

Almond tree

Lemon tree

Bauhunia refescent

Conbutum fructosium

Nerium Oleander

Punica

SPICE TERRACE

Alium   Coriandre   Origan   Hot Peppers   Pepper

DESERT TERRACE

Aloe niebuhriana

Aloe juncunda

Madura

Anabasis

Crassula rupestris

Wooden pannels

Square

Pond

Open air theater

Stage

Restaurant

Roof walkway

Connecting bridge

Spice terrace

Ceramic terrace

Perfume terrace

Pedestrian crossing

Refreshment facilities

Parklines

Desert terrace

School patio

Outdoor facilities

Elevator

School garden

Connecting bridge

PERFUME TERRACE

Artemisicia-Canescens   Boswellia serata   Datura   Jasmin   Myrtus communis

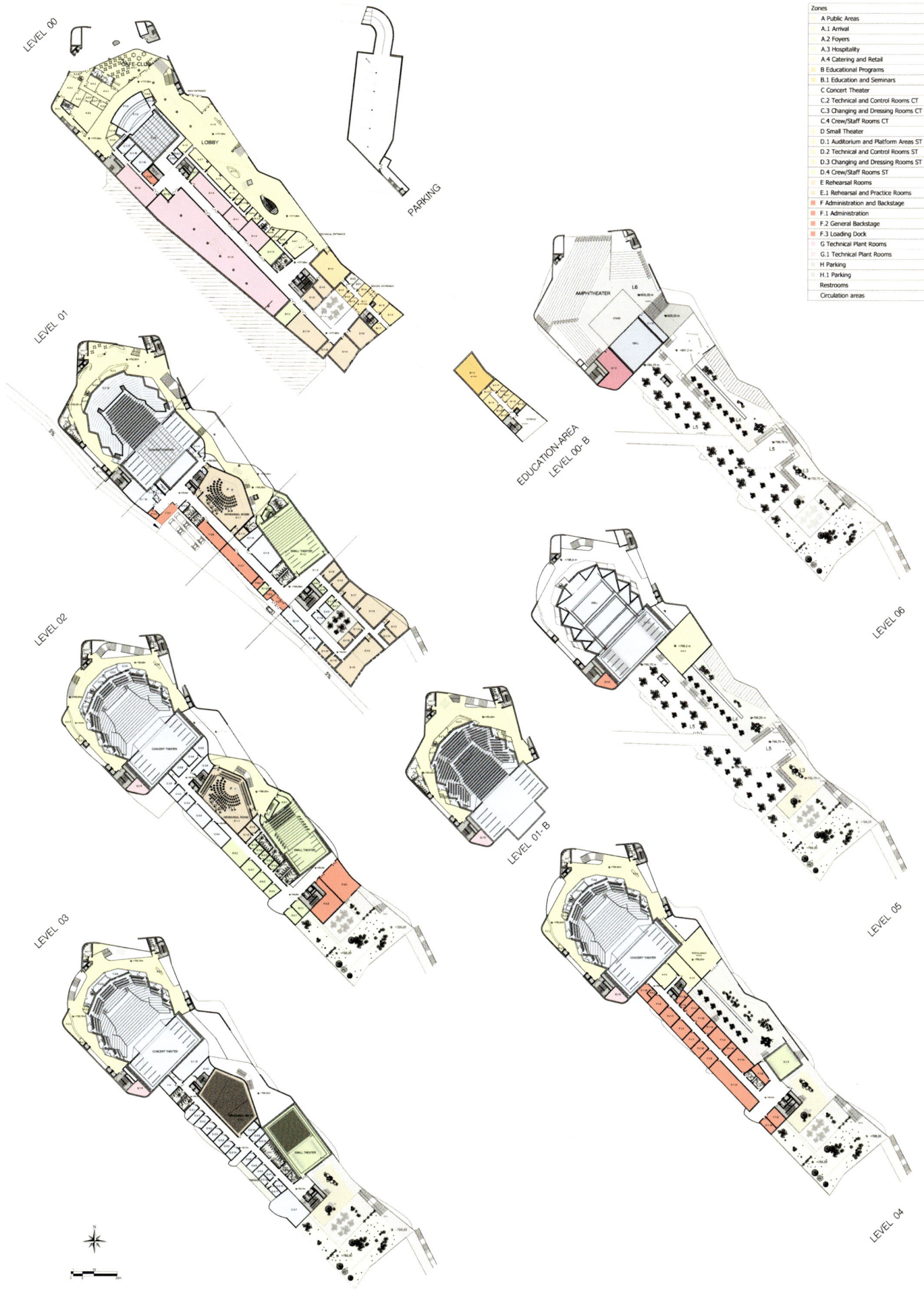

LEVEL 00

PARKING

LEVEL 01

LEVEL 02

LEVEL 03

EDUCATION-AREA
LEVEL 00-B

LEVEL 01-B

AMPHITHEATER

LEVEL 06

LEVEL 05

LEVEL 04

CAFE-CLUB

LOBBY

CONCERT THEATER

REHEARSAL ROOM

SMALL THEATER

CONCERT THEATER

Zones

A Public Areas
A.1 Arrival
A.2 Foyers
A.3 Hospitality
A.4 Catering and Retail
B Educational Programs
B.1 Education and Seminars
C Concert Theater
C.2 Technical and Control Rooms CT
C.3 Changing and Dressing Rooms CT
C.4 Crew/Staff Rooms CT
D Small Theater
D.1 Auditorium and Platform Areas ST
D.2 Technical and Control Rooms ST
D.3 Changing and Dressing Rooms ST
D.4 Crew/Staff Rooms ST
E Rehearsal Rooms
E.1 Rehearsal and Practice Rooms
F Administration and Backstage
F.1 Administration
F.2 General Backstage
F.3 Loading Dock
G Technical Plant Rooms
G.1 Technical Plant Rooms
H Parking
H.1 Parking
Restrooms
Circulation areas

ACOUSTIC STUDY OF THE CONCERT THEATRE

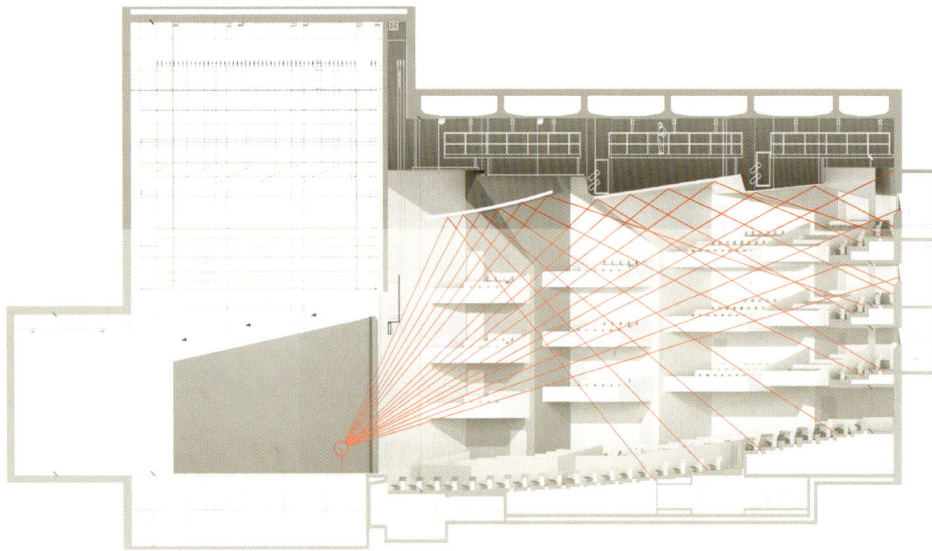

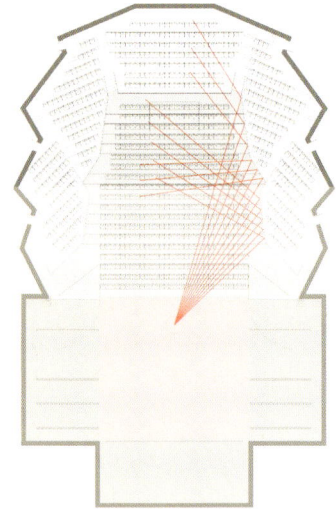

The wall structures of the "baskets" zone allows close proximity for ideal sound wave reflexions; the form that we propose assures excellent acoustic results.

VARIABILITY OF THE STAGE - CONCERT THEATRE

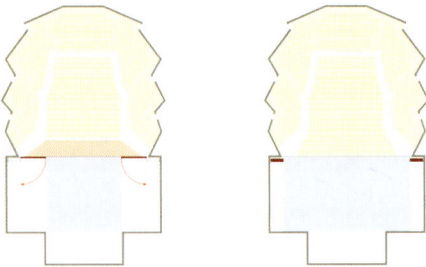

ACOUSTIC SLIDING PANNEL ALONG THE BALCONY TOWER

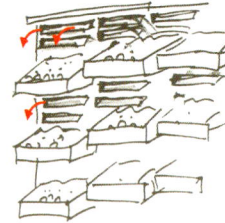

SLIDING PANNEL FOR WIDE FREQUENCY BAND ABSORPTION ADJUSTMENT

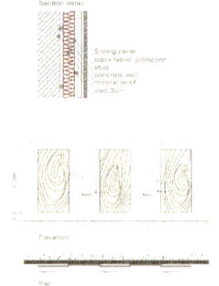

VARIABILITY OF THE STAGE - SMALL THEATRE

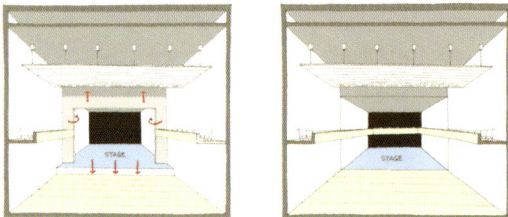

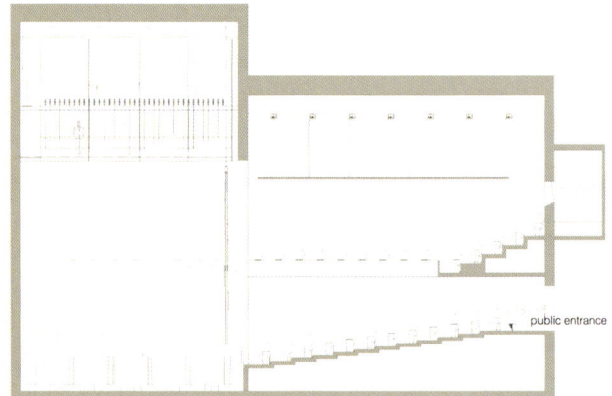

THE SMALL THEATRE

public entrance

STRUCTURAL CONCEPT

A simplicity of construction with vertical concrete walls
3 expansion joints at the separation between the large halls
Each blocks has its own structure
The concrete facade wall with large opening

One typical curved angle
in a gentle curve

Only one panel type
of prefabricated concrete
large scale lattices
(2m x 2m)

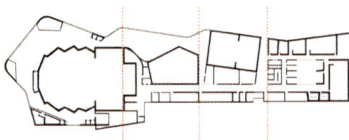

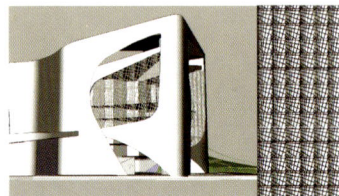

ECOLOGICAL CONCEPT

The garden on the roof protects the building from the heat and sun's rays

Photovoltaic pannels

Cooling air due to the chiminey effect

THE ENTRANCE - LOBBY

THE THEATER WITH TWO CONFIGURATIONS:

SUROUNDING AUDIENCE- CENTRAL STAGE

FRONT STAGE THEATER

THE LOBBY OF THE THEATER

THE CONCERT HALL

THE CONCERT HALL

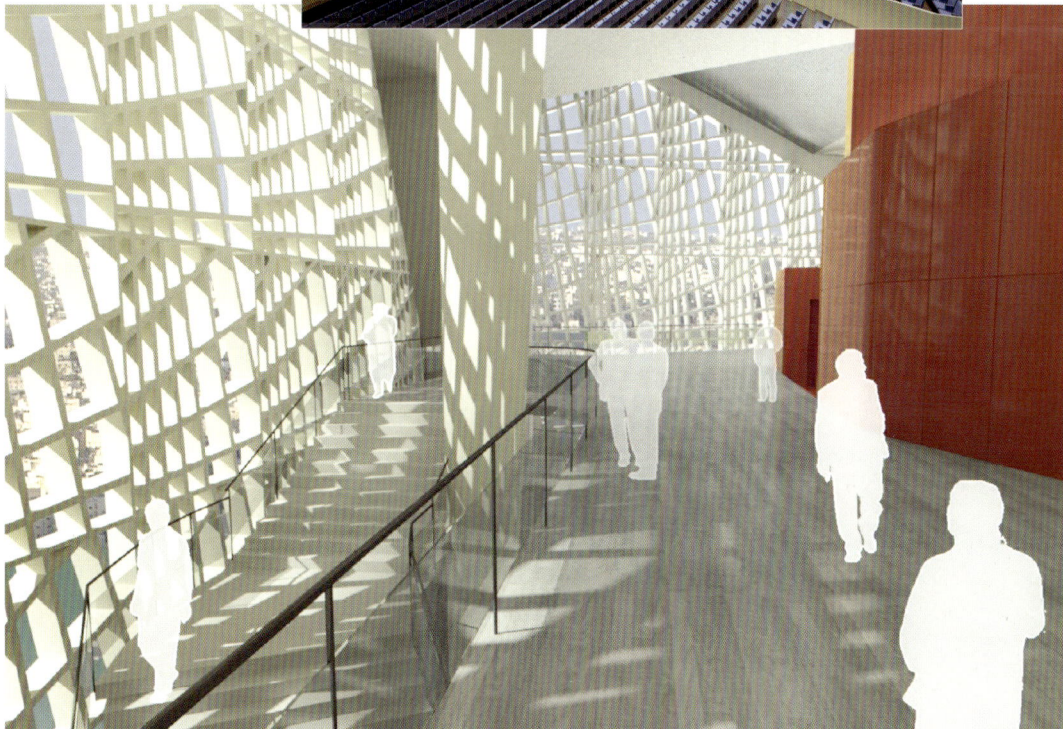
THE LOBBY OF THE CONCERT HALL

Atelier Christian de Portzamparc, Paris

Darat King Abdullah II, Amman

Kerry Hill Architects Singapore
Kerry Hill建筑事务所 新加坡

"The building is to be a stage for the performance, creation, and education of performing arts in Jordan – a building which welcomes people of all ages and backgrounds, that respects a rich architectural past, and aspires to innovation for the future."

"建筑物将成为约旦表演艺术的表演、创造和教育舞台——欢迎不同年龄和背景的人，尊重建筑丰富的历史，渴望未来的创新。"

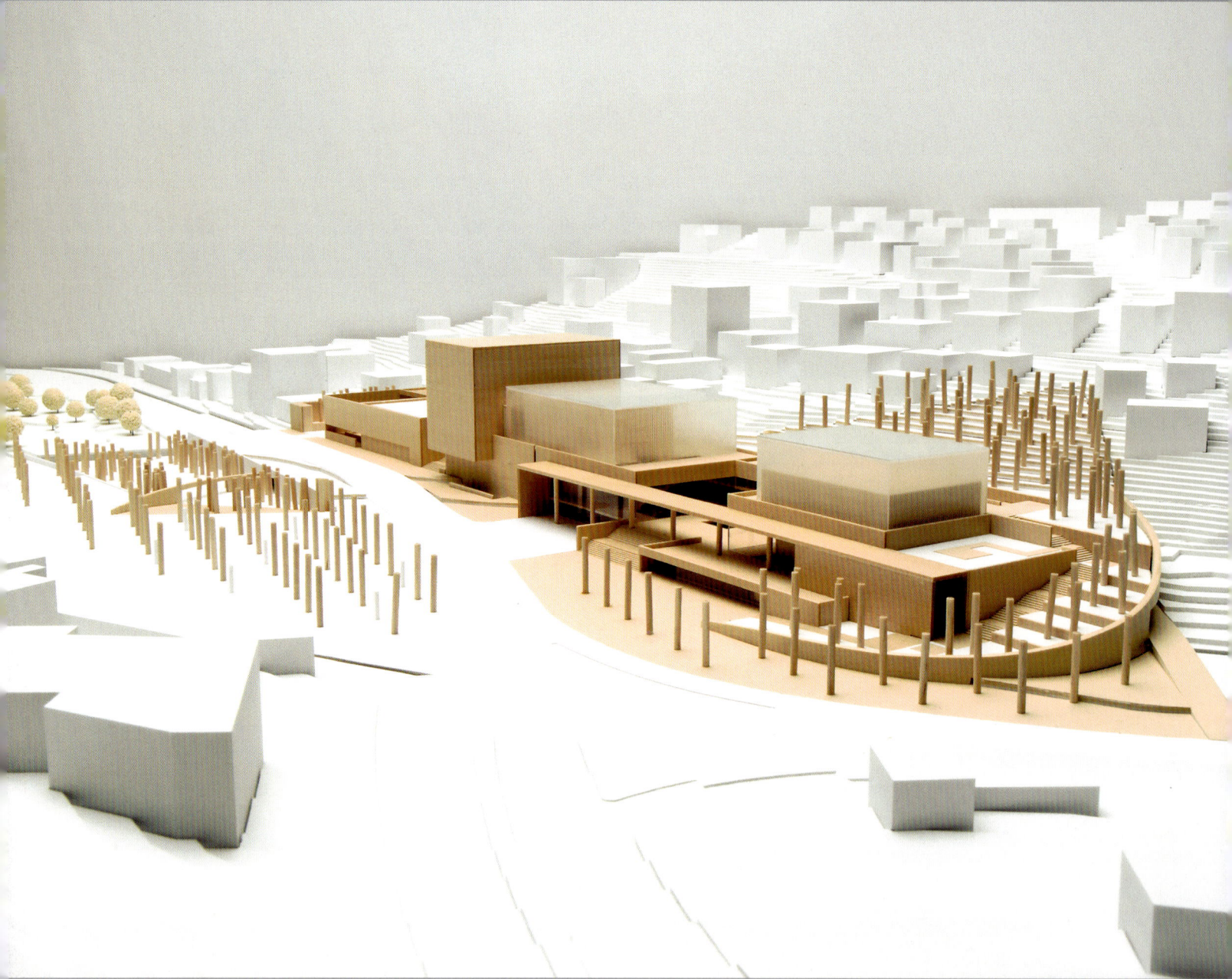

作者 Author A: Kerry Hill; Author B: Justin Hill   合作伙伴 Patrick Kosky, Cheng Ling Tan, Ken Lim, David Gowty, Bernard Lee, Chee Hong Lim, Duncan Payne, Alessandro Perinelli, Lidya Koes   专家 acoustics: Marshall Day Acoustics, Melbourne; landscape architecture: Tierra Arcsign, Netherlands; lighting: The Flaming Beacon, Melbourne; ESD consultant: Arup, Singapore

'The colour, the grace and levitation, the structural pattern in motion, the quick interplay of live beings, suspended like fitful lighting in a cloud, these things are the play'

TENNESSEE WILLIAMS

An important civic space generously accomodates the arrival and departure of large audiences

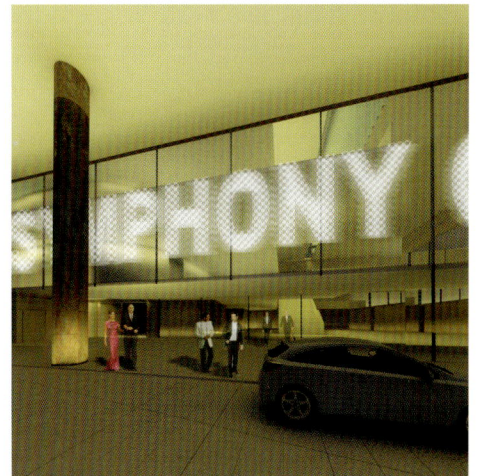

ILLUMINATED DIGITIZED SIGNAGE

Our vision is to proclaim "pride in the present without overwhelming the quiet history of the past". An architecture that clearly identifies itself through place, purpose and material. One that is contemporary, yet filtered through a sieve of traditional values.

Kerry Hill Architects   Darat King Abdullah II

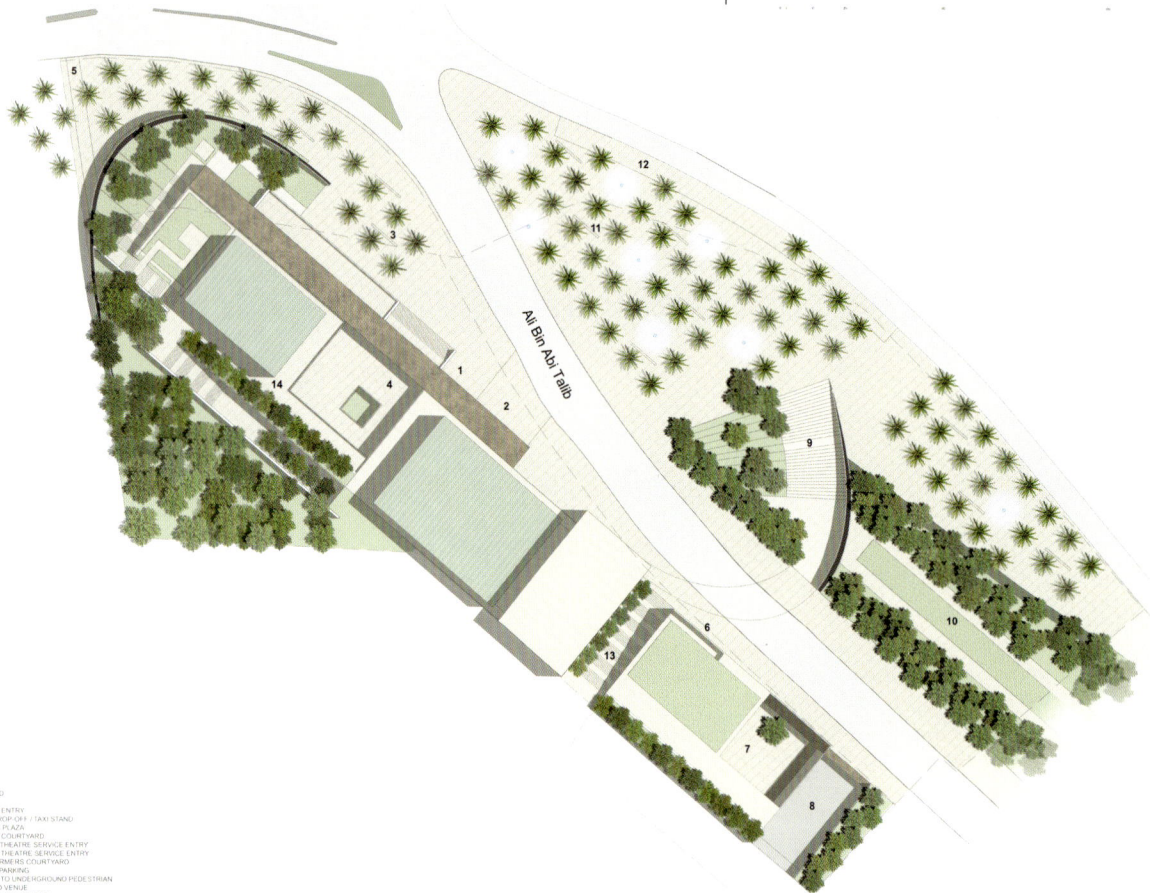

The stone podium merges with the landscape as a system of paths and stairways that evokes an urban experience distinct to Amman

APPROACH TO UNDERGROUND PEDESTRIAN LINK

STAIRS TO ROOF TERRACE

APPROACH FROM G.A.M PARK

LANDSCAPE ROOF

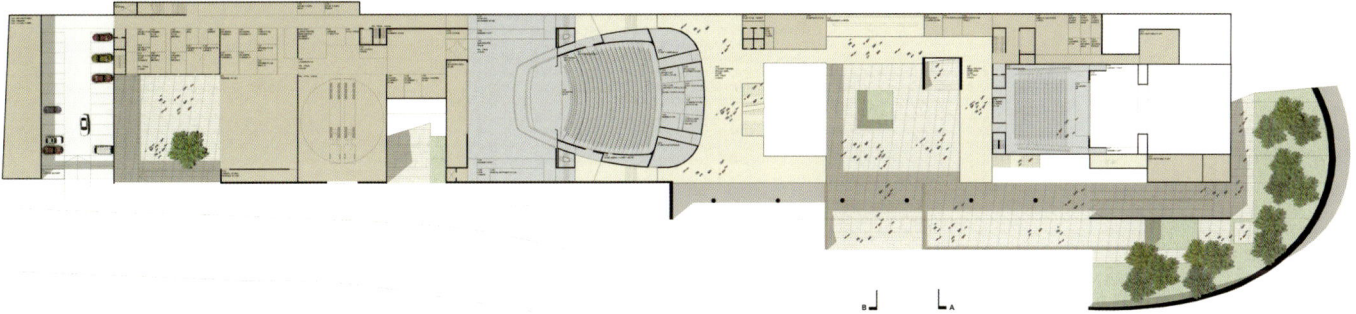

The education centre is given prime importance - located between the two theatres, in close
proximity to the small theatre, it is entered past a cafe that spills onto the public plaza.

Ali Bin Abi Talib

1. Pre-cooling of makeup air provision via the concrete structure of the thermal labyrinth significantly reducing active mechanical cooling energy consumption.

2. Rainwater / grey water harvest and a/c condensate water collection for purification and subsequent soft usages (irrigation, cooling tower, etc.)

3. Exterior facade in stone clad pre-cast concrete panels provides thermal mass and insulation to inner theatre spaces, low velocity displacement air conditioning through plenum and swirl diffusers.

4. Precast flooring system that allows cavities to be ventilated during the cooler night, purging the heat built up during the summer day.

5. Rooftop greenery to mediate micro-climate and provide insulation for building roofs.

6. A combination of photovoltaic panels and fritted glass to allow optimization of shading and natural illumination, while also generating electricity on site (building integrated photovoltaic panels), explore opportunities for a grid correct system to avoid large storage requirements. Circulation spaces designed to maximize daylight utilization and cross ventilation.

7. Hierarchy of air conditioned zones to maximize efficiency and minimize energy consumption.

PHOTOVOLTAIC PANEL

TYPICAL DETAIL THROUGH AUDITORIUM ROOF

NATURAL VENTILATION
MIXED MODE VENTILATION
FULLY AIR CONDITIONED

Intermission....

The pianist

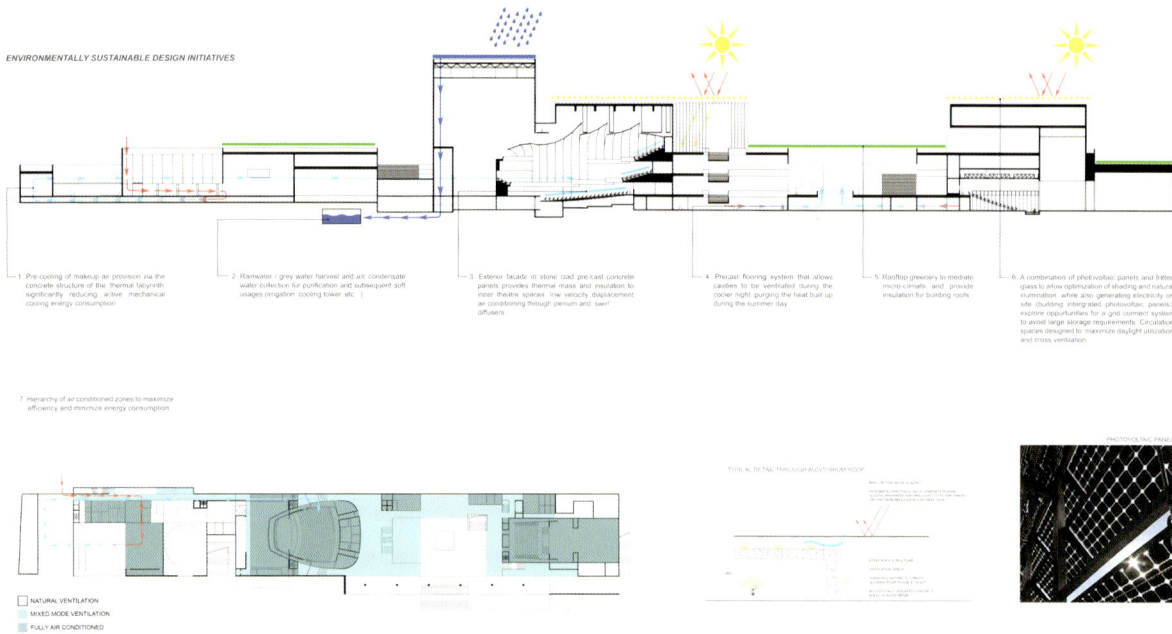

Theatre Acoustics

The Large Theatre

The functional brief gives highest priority to the symphonic use of the large theatre. The volumetric provision should reflect this. The design provides a gross symphonic volume of 19,250 m3 of which 2,624 m3 are on stage within the shell and 5,300 m3 are in the upper volume. That is a net volume per seat in the lower space of approximately 9 m³ per person.

Acoustical concept. In our realisation of the brief, the early reflected sound field beneath the visual ceiling provides the required clarity. Surfaces on the walls and soffits have been designed to ensure uniform high clarity of symphonic sound to all seats. Late sound is supplied by the upper volume outside these reflecting surfaces. Diffusion is graded throughout the space from a maximum intensity close the source to a minimum on remote surfaces.

The stage is to be equipped with an orchestra shell which provides an acoustical as well as a visual completion of the concert hall at the stage end. The shell is in two parts. An up-stage part is constructed as a single piece integral with its seating and top, and is able to be moved into storage as a unit at the rear of the stage. The sides and the remainder of the top are demountable elements, flown in a conventional manner, or formed by wheeled towers. Stage shells of several configurations can therefore be matched to the size of the performing group.

The many reflecting surfaces produce lateral reflections required for the symphonic function. More precise localisation of the voice is required for Opera. This is achieved by a forestage reflector flown in front of the proscenium.

Variation of the reverberation time is achieved by deploying purpose designed acoustic banners in the upper volume of the auditorium above the acoustically transparent ceiling.

The Small Theatre

Simplicity and adaptability are the two themes of this intimate theatre.

The principal and perhaps most demanding function is drama, and care has been taken to ensure excellent speech clarity The room has been designed to have a suitable reverberance to permit fullness in the sound. This volume at 7m³ per seat is rather higher than many conventional theatres. The high degree of clarity is assured by the provision of three early reflections to each seat. Even when in thrust stage mode, the same construction of reflected sound distribution has been applied. The provision for reverberance will be particularly valued during the briefed chamber music/recital function. This will be an engaging room for performers.

The main reflecting surfaces are the side walls which bound the seating plane. These surfaces are supplemented by the overhead reflectors which are incorporated into the design under the catwalks, lighting bridges, and particularly the forestage reflector. The principal mode of varying the acoustics is by way of deployable absorption mounted on the upper side wall. This would be in use for all ampli- fied functions and probably for Arabic and world music.

Symphony Mode

Theatre Mode

Theatre Mode Clarity

Reflector Coverage

Symphony Mode Reverberation Time

Theatre Mode Reverberation Time

Reverberation Time

Symphony Mode Clarity

Theatre Mode Reverberation Time

Theatre Mode Speech Intelligibility

Speech Intelligibility

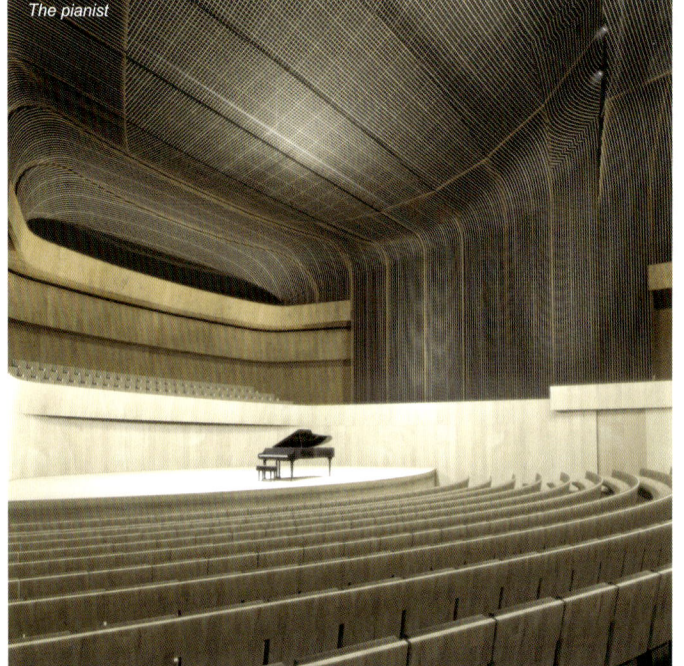

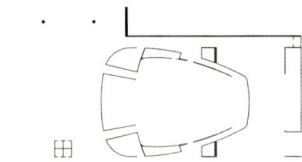

The Large Theatre - Symphonic configuration - Stalls

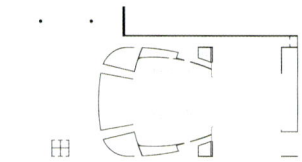

The Large Theatre - Lyric configuration - Stalls

SEATING: Stalls - 752, Removable seats (Orchestra lifts) - 152, Circle 1 - 365, Circle 2 - 337   TOTAL - 1606

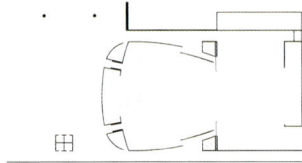

The Large Theatre - Circle 1

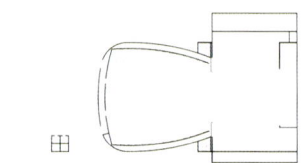

The Large Theatre - Circle 2

The fabric of the theatre architecture and the sound it sustains melds seemlessly into one entity

A building that celebrates activity by day and by night. Allowing spectators to momentarily become spectacle

'From the start it has been theatres job to entertain people... it needs no other passport than fun,

BERTOLT BRECHT

525

Henning Larsens Tegnestue Copenhagen
Henning Larsens Tegnestue建筑事务所 哥本哈根

"The heart of the building is the magnificent amphitheatric foyer. Here all the qualities of the project are brought together in a lively, social meeting point."

"建筑物的灵魂是华丽高贵的圆剧场的休息厅。所有项目的品质都被一起带到一个充满活力的社交场所。"

作者 Troels Troelsen 合作伙伴 Viggo Haremst, Nina La Cour Sell, Charlotte Soderhamn Nielsen 专家 Buro Happold; Ove Arup & Partners, London

Approaching the new Darat King Abdullah II
after a stroll in the park, looking forward to a cultural evening.

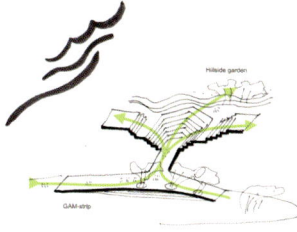

Hillside garden

GAM-strip

FLOW

The flow of the public starts in the GAM-strip and creates a crescendo culminating in the building complex thus creating a prominent landmark. Passers by are invited to take an informal stroll through the building to experience and partake in its life i.e. through the foyer amphitheatre to the hillside garden and further down to the park. Thus the flow does not terminate in the building, but divides further into several branches and creates circles and loops.

The continuous flow may include a wide underpass making the lower part of the park continue under the road to the site and further uphill to the garden at the top of the foyer.

The western end of the GAM-strip is progressed modelled to create a fluent integration between the valley landscape with the front plaza, overbuilding the road and extending into the 'stage' of the amphitheatric foyer. Further the landscape modelling may in the future extend to the areas to the west, creating extension of the park overbuilding the bus station.

THE HEART OF THE BUILDING

The heart is the magnificent amphitheatric foyer. Here all the qualities of this place are brought together in a lively, social meeting point.

The public is invited to pass through the building from the GAM-strip to the hillside garden. Thus the vivid heart is the shadowy crossing point between public landscape promenade and the multi-level foyer.

The transparent foyer visually opens to the gardens at the hillside and to the valley park, thus creating a link between the communities of the two sides of the valley.

THE MEETING PLACE

The shape of the foyer unifies the characteristics of the place. On one side the theatrical spaces of the building, on the other sides the amphitheatric character of the surrounding city.

The foyer is created as a terraced landscape giving access to the different levels of the auditoria via openings in the terraces.

From the terraces of the foyer you overview the GAM-strip below and the garden terraces at the hillside above.

The stage level of the foyer is extended across the street creating a front plaza unifying the amphitheatre with the GAM-strip.

LANDSCAPE AND BUILDING

The landscape and the building are unified. Indoor and outdoor areas are merged to a continuous and coherent sequence of spaces. Contrast and interaction are accentuated between the terrain and the prismatic elements of the building with the amphitheatric foyer. Furthermore a contrast between the monolithic base and the floating curved roof like an artificial cloud is created.

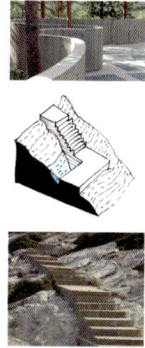

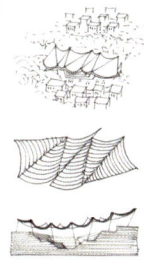

THE FIFTH FAÇADE

This roof structure is a modern transformation of native roof structures. It is something between inverted domes and solidified tents.

The curved roof seen from the hills, shining in the sun symbolizes the social meeting place linking the two sides of the city together. Thus the appearance from the southern hillside becomes appealing like from the GAM-strip.

THE ENTRANCE

The main access is from the GAM-strip with the underground parking structure. Here in the park, a drop-off can be reached from the left lane of the Ali Bin Abi Talib Street.

Additional / alternative drop off points can be created.

A street level entrance is created at Ali Bin Abi Talib Street. From here there is direct access to the education rooms and to the lower level of the small stage and the large rehearsal room as well as to the box office, lockers and toilets. Wide stairs lead up to the lower level of the amphitheatric foyer.

Accesses from the open spaces in the western and eastern ends of the site have built-in walkways at first floor along the street to the main foyer.

VIPs have direct access to the top level of the foyer via the hillside street (A-street). Busses can be parked during performances at the eastern end of the site.

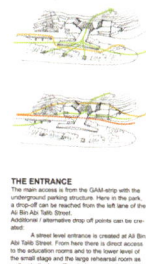

The main foyer amphitheatre.
The new meeting point for Amman citizens

restaurant and
view terrace

Road A

VIP and
hillside entrance

side entrance

pedestrian bridge

access to GAM Strip

Ali Bin Abi Talib Road

GAM Strip

+ 22.95m

+ 28.56m

0m

+ 4.86m

+ 5.8m

0m

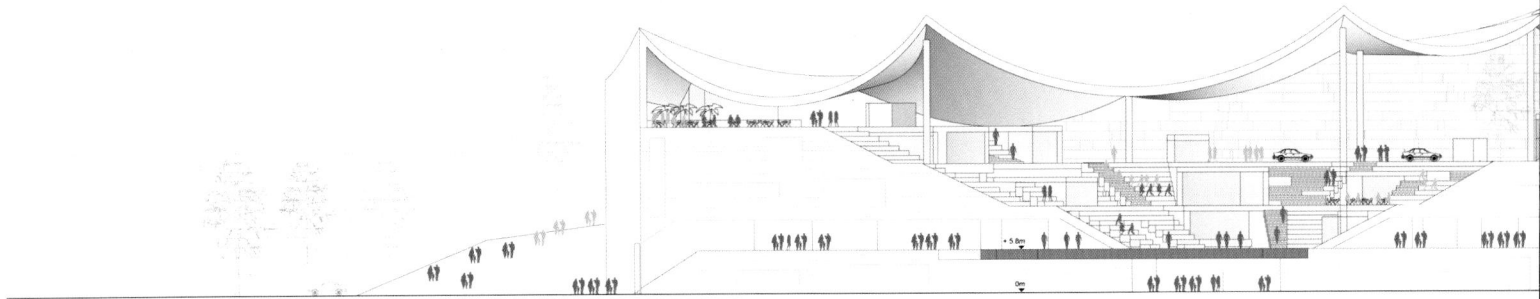

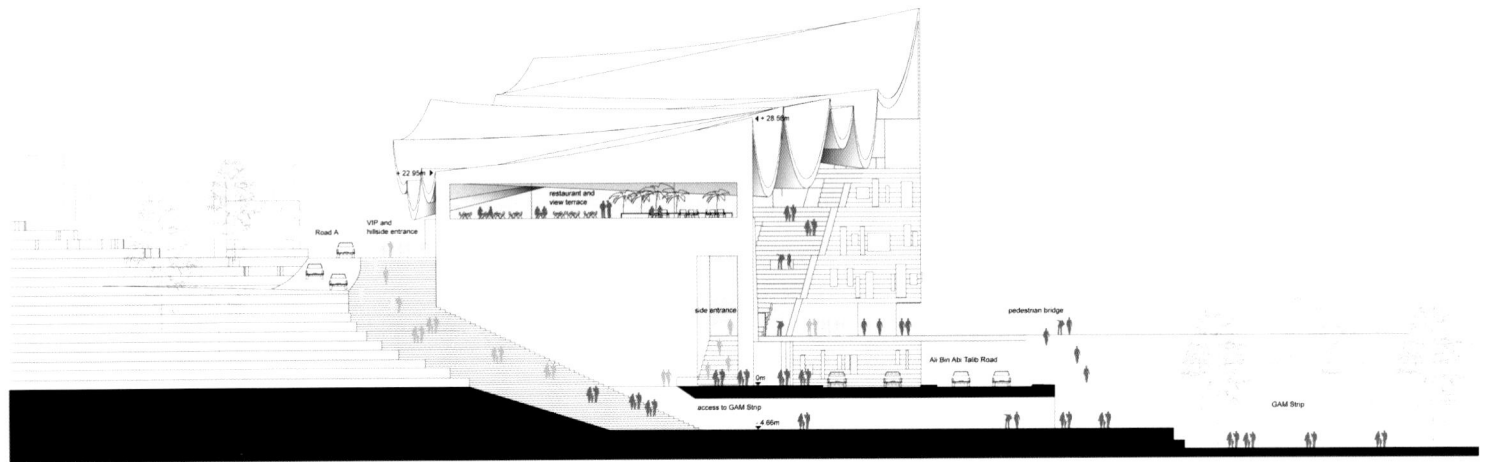

east elevation 1:200

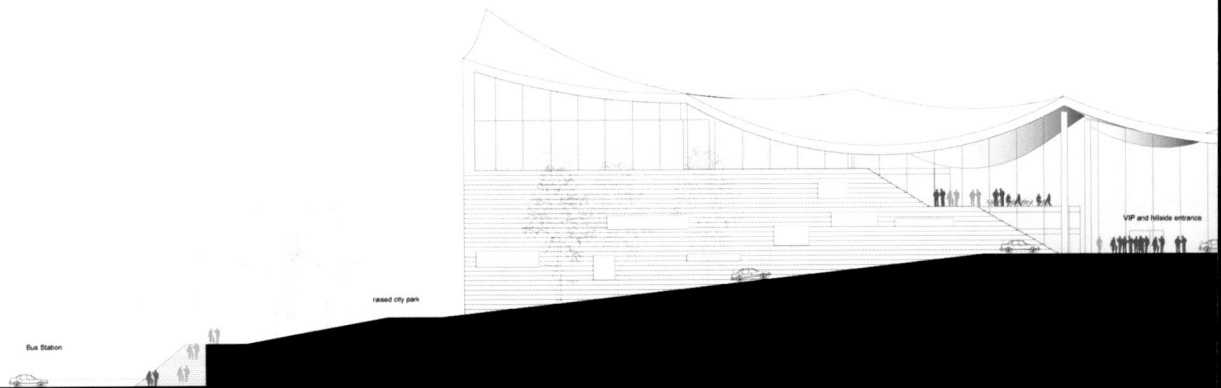

raised city park

VIP and hillside entrance

Bus Station

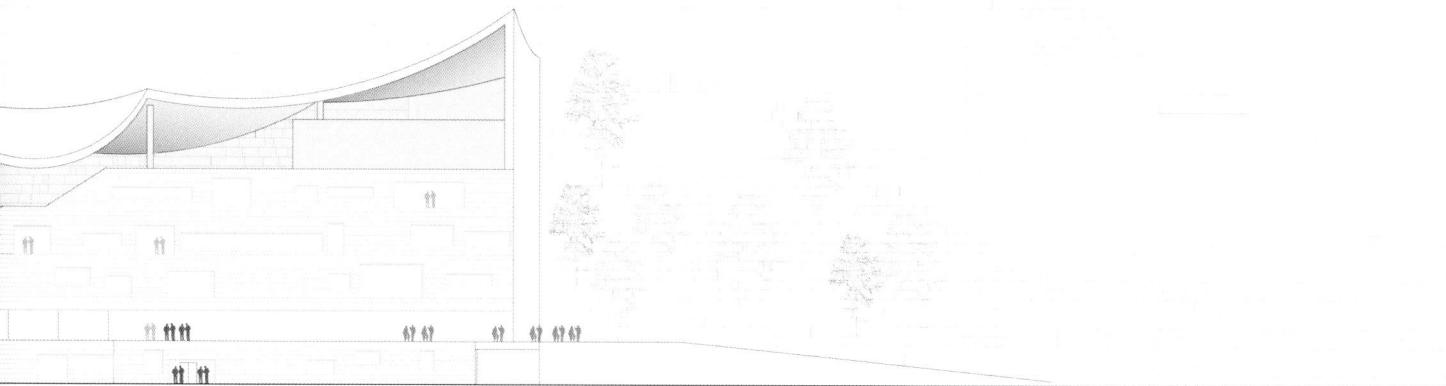

north elevation 1:200

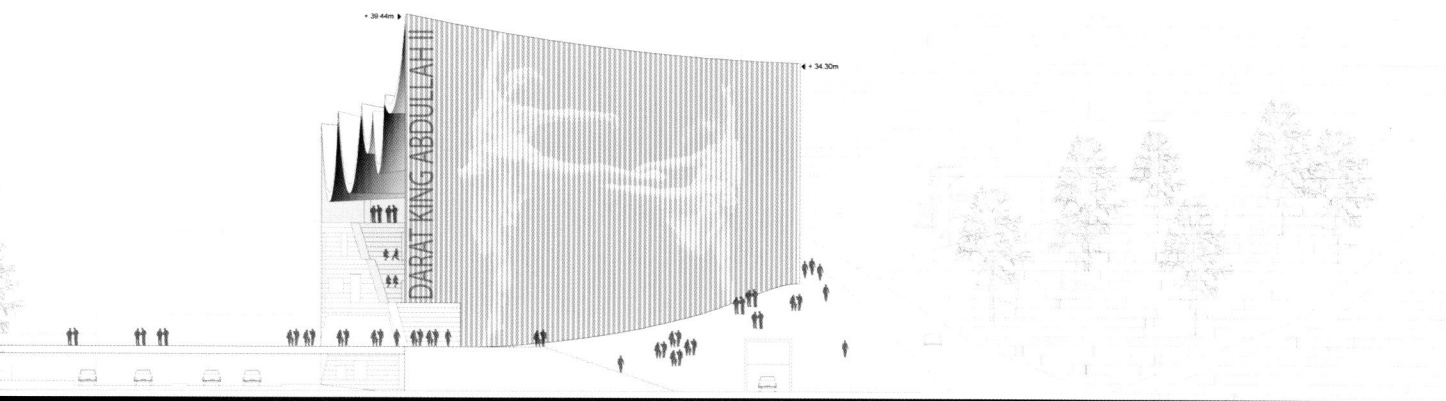

DARAT KING ABDULLAH II

+ 39.44m ▶

◀ + 34.30m

west elevation 1:200

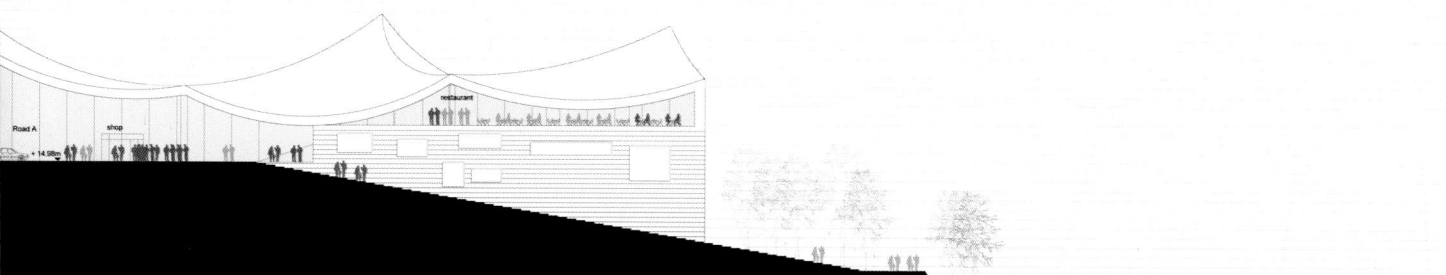

Road A

shop

+ 14.98m

restaurant

longitudinal section 1:200

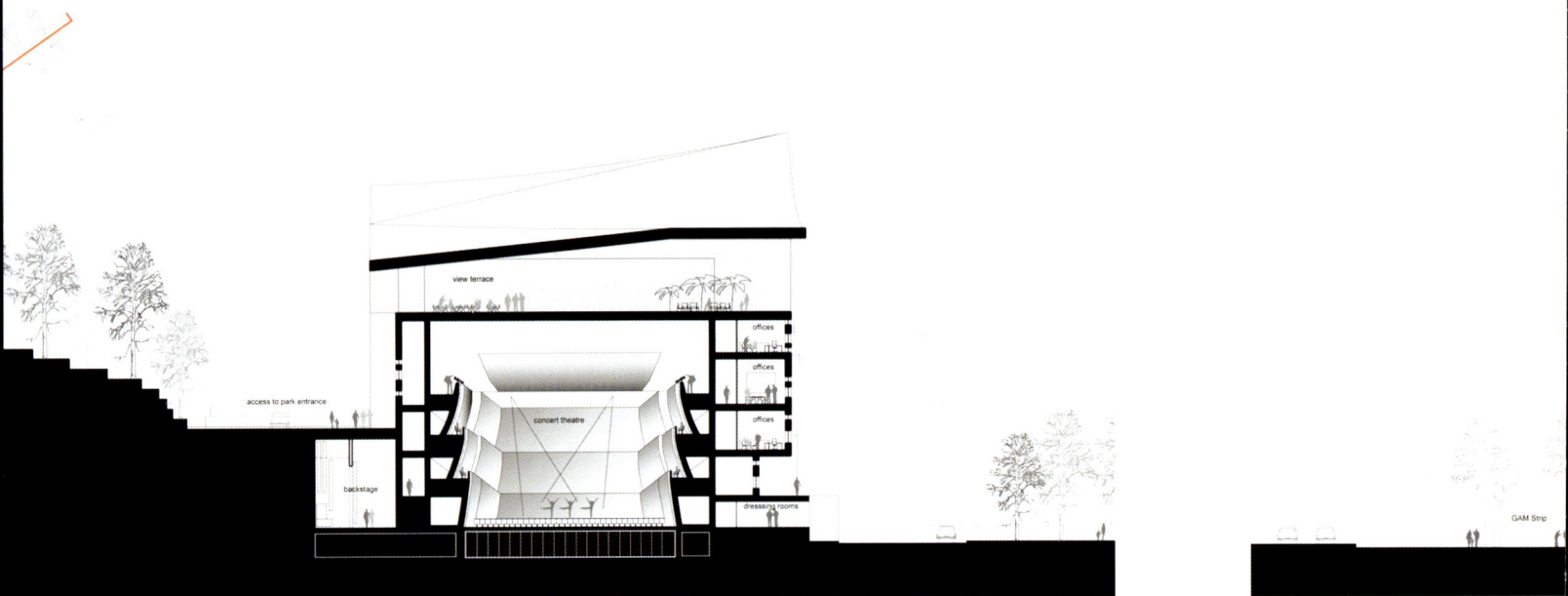

access to park entrance

view terrace

offices

offices

offices

concert theatre

backstage

dressing rooms

GAM Strip

section through concert theater 1:200

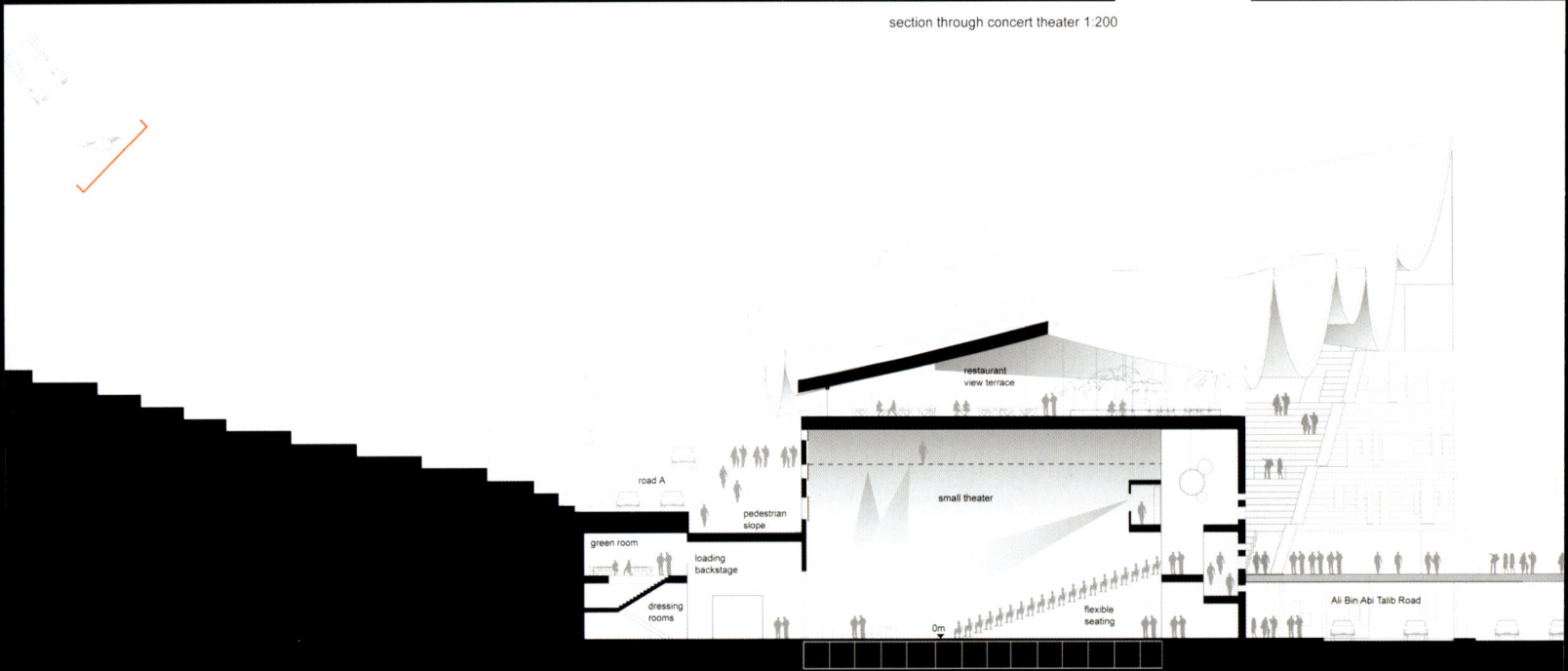

restaurant
view terrace

road A

pedestrian
slope

small theater

green room

loading
backstage

dressing
rooms

flexible
seating

Ali Bin Abi Talib Road

0m

The many informal stages in the foyer communicate the rich variety of cultural possibilities of Darat King Abdullah II

THE FOYER

The foyer becomes a focus for Amman's cultural life, a social, vivid meeting place with the audience as actors and also an informal amphitheatre for an occasional large event or for several small performances at different levels of the foyer.

The terraced amphitheatre landscape divides between the public areas above and the private or semi-private spaces below.

A variety of recesses and projections create additional terraces with bars as well as enclosed spaces for music-shop, VIP room, education spaces etc. overlooking the grand foyer.

Like the roman theatre the foyer has sitting-steps combined with zones of normal steps, but in a more fluid and informal way.

Restaurant, VIP-rooms and café are in the upper part of the amphitheatre overlooking the foyer and the GAM-strip parkland and facing the hillside gardens to the south.

POSSIBLE FOYER SECTIONS

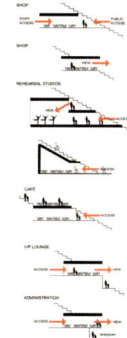

THE GROTTO

Under the stepped foyer a grotto-like space is created giving access to the theatre halls and to the education rooms, rehearsal spaces etc. exposed as separate volumes. Slots under the sitting-steps together with the larger openings and transparent volumes create dramatic light to the space.

THE MAGIC OF LIGHT

The interplay of light in the recesses and perforations of the stone material contrasts to the polished, shining surfaces of the roof structure and makes the building stand out like a sculpture or a piece of jewellery. At night, seen from the park, life and light in the lit, amphitheatric foyer will stand out as a grand festive stage.

During the day, the play of the sun will emphasise the monolithic character, whereas from the inside, openwork parts will appear transparent and weightless. At night when viewed from the outside, the density of the monolith will dissolve from the artificial light from the lively inner house.

FUNCTIONAL LAYOUT

A large, efficient main floor with stage related functions connects the two theatres as well as the large rehearsal theatre and the loading docks. Rehearsal rooms and educational spaces are directly accessible from the amphitheatric foyer as well as from the backstage side.

FLEXIBILITY

Multi-applicability and accessibility are keywords. Large or smaller informal performances can take place in the festive, amphitheatric foyer with the park as the backdrop.

Like the small hall is conceived as a multi-flexible space with either flat or sloping floor, the large theatre has the possibility of creating a horizontal parterre for conferences, exhibitions, a celebration or a feast.

One or more of the halls and education spaces can be directly accessed from the street level foyer, thus creating an independent conference centre.

The different levels of the amphitheatric foyer will have access to the levels of the halls as well as to different education, exhibition and rehearsal spaces. Due to their situation between a public and a private domain they have optional connection to either of the two sides and their use may shift according to the needs and desires. Also the VIP rooms can become a natural part of the public areas when not used for officials.

The park of the Cultural Centre has multiple functions: picnics, outdoor serving, music, theatre, lectures etc.

SMALL THEATRE FLEXIBILITY

CONCERT HALL FLEXIBILITY

THE LARGE HALL

A fluid visual coherence is created between hall and foyer via sliding gates, at stalls level opening the entire width of the hall. When closed these offer sound insulation to allow simultaneous use.

The interior of the hall resembles a faceted grotto. There is a festive and intense atmosphere with the audience distributed in the stalls, the balconies and the side balconies that lean into the volume like terraces, creating intimacy and involving everyone.

They are designed in a dynamic staggered way like cone scales to emphasize the intimate atmosphere but are carefully sculpted to augment the acoustics. When the hall is used as a theatre, the "scales" closest to the proscenium can be moved/turned to enhance acoustics and mould the space to an even more intimate framing of the stage opening.

At concerts the "scales" are moved back to create a unity with the orchestra shell completing the hall with slanting wall and ceiling "scales".

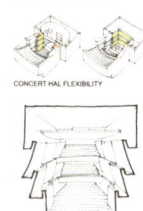

The concert theater embraces a wide variety of different events engaging all senses

Lightning, music and performance play together creating different worlds of experiences

[annex].
[附录].

Glossary
词汇表

Glossary
词汇表

A to Z of architectural competitions
建筑竞赛A——Z

Anonymity　匿名方式

A basic principle of architectural competitions, anonymity is to ensure that the jury decides solely on the basis of each entry's merit, i.e. without consideration of its author's reputation. Procedurally, during the otherwise anonymous process, anonymity may be lifted in favour of enhanced communication between sponsor and competitors in the cooperative phases.

Application procedure　申请程序

Restricted or partially restricted competitions are preceded by an application procedure where a selection committee picks a suitable number of competing candidates. Selection criteria must be unequivocal and non-biased.

Architectural Juror　建筑评奖委员会

The jury is composed of architectural jurors and technical jurors. The former are so designated because they must possess at least the same level of professional qualifications as the competitors. This is to ensure that the jury meets the standards required for its task. The distinction between architectural and technical jurors can be traced back to regulations of the German Public Contract Code, but in practice its importance has become less and less significant.

Author　作者

The project design is submitted by its author who is also the beneficiary of the promise of contract agreement. In this respect the author is the sponsor's contracting party. An author must meet the eligibility criteria set forth in the competition brief. In the case of a design team, all its members must meet these criteria and are considered as the project's authors.

Chamber of Architects　建筑师协会

National or regional professional association of architects and town planners within the relevant jurisdiction. The Chamber of Architects monitors architectural competitions for compliance with competition rules and, upon acceptance, issues a registration number. In Germany this is done by the State Competition Committees made up of members of the regional Chamber.

Competition Amount　竞赛费用

The overall monetary sum allotted by the sponsor to prize money, purchase awards, and payment for architectural services constitutes the competition amount. It is calculated on the basis of the fees normally charged, according to the official scale, for the services provided in the competition.

Competition Brief　竞赛任务说明

The competition brief unites, in the form of a brochure, all the textual information, plans, images, and tables required for undertaking the tasks of the competition. Its addenda include information drawings, working drawings, and other informational material as required.

Competition Cost Estimate　竞赛成本预算

Included in the offer that a competition manager makes to a (potential) sponsor for preparing and conducting a competition is an estimate of the overall cost of the procedure. Besides the remuneration due to the competition manager, this estimate includes the competition amount, the remuneration due to the jury, experts, and examiners and other costs incurred (third-party costs) such as printing cost, travel expenses, venue rental, etc. The cost estimate is continually updated as a basis for controlling costs during the course of the competition.

Competition Documents　竞赛文件

The documents made available to competitors as the basis for their design work comprise the brochure of the competition brief (with the programme of functions and spaces), the information drawings and the working drawings. Also included are a number of forms to be filled in and various tables, possibly the pilot plate for the environment model, the minutes of the online inquiry procedure, and the competitors' colloquium, if any.

Construction (GRW 1995), the rules for the
Award of Professional Services Contracts
(VOF), the Public Contract Code (VgV),
and the Law against Restraint of Trade (GWB).

Competition Homepage　竞赛主页

This webpage helps to disseminate information on the
competition. It forms the virtual space to distribute
and exchange data and other pieces of information
relevant to the project, to conduct an online forum,
and to document the outcome of the competition.
It facilitates the administration of major parts of the
competition (application procedure, response
to inquiries, document management) and thus
helps to keep down competition costs.

Competition Management　竞赛管理

Competition management (a.k.a. competition coordination) involves all the activities required for the preparation,
conduct, and documentation of an architectural competition.

Competitor　竞赛者

Participation in architectural competitions is open to
(a) all individuals or legal entities whose statutory purpose
includes pertinent planning services, and (b) design teams
consisting of such individuals or legal entities. Eligibility
may be restricted on the basis of regulations governing
the procedure in question (such as the rules for the
Award of Professional Services Contracts – VOF).

Competitors' Colloquium　参赛者专题座谈会

To provide a thorough understanding of the project task a
competitors' colloquium may be held before the midpoint
of the design period, normally at the project location so
that competitors can inspect the project site.
The jury, too, attends this colloquium. Its outcome is
recorded in the colloquium minutes and is considered part
of the competition task. The colloquium may be comple-
mented with or replaced by an online forum.

Contract Award Law　合同授予法

In Germany the legal basis for architectural competitions
and for awarding planning contracts are the EU Service
Directive, the Principles and Guidelines for Competitions in
Regional Planning, Town Planning and

Cooperative Competition　合作型竞赛

A competition where an immediate dialogue takes place
between the sponsor, the jury, the experts and the
examiners on one side and the competitors on the other.
This dialogue provides an opportunity for optimising one's
approach to problems arising from the designs proposed,
at stage of the competition most conducive to solutions.
A special feature of cooperative competitions is waiving
anonymity during at least parts of the procedure.

Data Bank　数据库

Once an applicant is approved to take part in a competition managed by [phase eins]., his data (name of
practice, address, authors' data, reference project data,
etc.) are stored in an online data bank. The planner
can access his data record at any time for updating
or reuse in case of a later application. This practice
ensures full and coherent presentation of the elements
of an application; in addition, it simplifies the (often very
time-consuming) application procedure and speeds up
subsequent processing to a high degree of precision.

Declaration of Authorship　原著声明

The anonymously submitted competition entry is
accompanied by a sealed envelope containing a declara-
tion that reveals the author's name and the names of
his collaborators and consulting experts. The sealed
envelopes are opened only at the end of the jury
meeting, i.e. after the short-listed designs have been
ranked and prizes have been awarded. The reading
out of each author's name concludes the meeting.

Design Elaboration, Suggestions as to...　设计详细说明

The jury's recommendations to the sponsor, prior to
revealing the identities of the designers, should include
suggestions for elaborating the short-listed entries. If
required, this document should also include suggestions as to
any necessary change to the project task and what conclusions the sponsor should draw from the competition and
its outcome.

Documentation　文档资料

Upon conclusion of the competition, all the parties
involved are provided with an account of the
procedure and of the design proposals submitted.

This documentation adds to the transparency of the procedure and to the plausibility of the decision taken.

Downloads 下载

The Internet constitutes an ideal platform for the distribution of information, to as many recipients as required. Where competitions are concerned the applicants, competitors, and all interested parties can download various data files (forms, CAD files, illustrations, movie clips, text) from the competition homepage.

Eligibility for Participation 参赛资格

The eligibility for participation depends on the pertinent regulations of the public contract-award code and on the content of the task. Eligible to participate in an architectural competition are generally individuals who are entitled to designate themselves as an "architect", under the law of their country of origin. Analogous regulations govern which town planners and graduated engineers may participate in urban-planning competitions, engineering competitions, and integrated competitions.

Evaluation Round 评估会议

During the jury meeting, the informational round is followed by several evaluation rounds when jurors
(a) discuss which entries meet to a higher or lesser degree the requirements set forth in the competition brief and
(b) decide which entries are to remain for further consideration and which entries to eliminate from the selection procedure.
During the first evaluation round, one vote in favour suffices for the entry in question to be kept in the contest. In other words, rejection in round one requires a unanimous vote. In subsequent rounds decisions are taken simply by a majority. If required, between evaluation rounds discussions can be held that help to form opinions but where no actual decisions are taken.

Examination Drawings 测试设计图纸

In addition to presentation drawings, calculations, and other explanatory texts, the examination drawings form a part of the material to be submitted by each competitor. The examination drawings must contain all pieces of information (dimensions, distribution of uses, etc.) required to make the project intelligible and examinable in quantitative ways. To complement or replace the examination drawings, CAD data files as the basis for examination play an ever more important role.

Examiner 主考人

The examiners are professionals, i.e. generally architects or engineers from the fields covered by the competition. Upon receiving the entries, they appraise the designs for compliance with the competition requirements. The examiners are appointed by the sponsor, they attend to his interests and provide advice to the jury in their role as advocates of the project authors. The examiners should be involved with the competition throughout its course. Normally the competition manager coordinates their activity, and in particular during the quantitative preliminary examination the competition manager's team supports the examiners.

Exhibition 展览

In the interest of public acceptance of the project it is recommended that the submitted proposals be exhibited for at least two weeks, upon conclusion of the competition. The exhibition serves the double function of being a forum for discussion and an event celebrating the contributions made by each competitor.

Experts 专家

These are specialists from various fields that may be called up for setting up the competition and for contributing to the preliminary examination. Experts do not have a vote at the meeting of the jury.

Feasibility Study 可行性研究

A feasibility study may, for example, determine a site's potential for building development. It may be carried out in preparation of a competition or it may be included among the services to be rendered within the framework of project development.

Ideas Competition 观点竞赛

An ideas competition, as opposed to a project competition, seeks to obtain a variety of solutions to a given problem without the immediate intention of implementing the project under consideration. Consequently no promise of contract award is made, but in return a higher amount of money is allotted to prizes. An ideas competition may become the basis for a subsequent project competition.

Information Drawings 施工图

All information required (site plan, built environment, topography, existing vegetation, utility lines, programme of functions and spaces, etc.) is made available to all competitors on CD-ROM, by means of download, or through illustrations in the competition brochure, or, if required, as hardcopy.

Information drawings (as opposed to working drawings) are not a graphic model to be followed for the representation of each competitor's design.

Informational Round 资料会议
The jury begins design appraisal with an informational round where the examiners present a detailed, unbiased explanation of each entry. In this round all the designs are inspected, but judgement is withheld for the time being.

Inquiry 咨询
At the start of the period allotted to the task, competitors may ask for clarifications by sending written inquiries by mail, fax, or e-mail and, if a competitors' colloquium is held, by asking questions at that event. The sponsor will answer these questions in the shortest time possible, after consulting with jurors or experts. Prepared and moderated by the competition manager, this inquiry process is often conducted as an online forum on the competition webpage and concluded before the midpoint of the design period by distributing the forum minutes to all interested parties. The questions asked and the answers given define requirements that are considered as parts of the competition task.

Integrated Competition 综合竞赛
In an integrated competition (a.k.a. multidisciplinary competition) the concern is to integrate issues arising from various different disciplines. Consequently, the entries present design approaches and planning services that transcend the borders of any single area of expertise. An integrated competition is the best option in cases where the complexity of the task calls for the collaboration of experts from various fields.

International Competition 国际竞赛
Originally adopted as recommendations by UNESCO in 1956, the "Standard Regulations for International Competitions in Architecture and Town Planning" were last revised in November 1978. This document defines the basic rules for international competitions, i.e. any competition in which architects or town planners of more than one country are invited to participate. Among the issues settled are the various types of competitions, the composition of the jury, the commissioning process subsequent to the competition, the announcement and publication of competition results, the language or languages to be used in the brief and in deliverables, the anonymity of the competition procedure, and the obligation to appoint a professional adviser and supervisor. The body UNESCO entrusted with the preparation of a detailed set of rules and supervision of international competitions is the "union internationale des architectes" (uia). Thus, the "uia Guide for International Competitions in

Architecture and Town Planning" constitutes the recognised basis for competitions organised on an international level.

Issuing of Competition Documents 竞赛文件的发布
Describes the formal procedure for issuing competition documents to competitors or making the data files required for the task available for download from the competition homepage. This takes place on a given date.

Juror 评奖委员会
Jurors exercise their duties independently, in person, and are guided solely by their professional expertise. Due to their professional qualifications, architectural jurors must match the requirements placed on competitors at a particular high level. Technical jurors should be familiar with the content of the competition task and with local conditions in particular. In addition, architectural and technical alternates with voting capacity should be appointed so that they can replace regular jurors who may be absent.

Jurors' Colloquium 评奖委员会的专题座谈会
Jurors, alternate jurors, experts, and examiners should participate in the formulation of the competition brief. For this purpose a jury colloquium (a.k.a. jury "kick-off") is held where the project content and the competition task are discussed, possibly reformulated and ultimately adopted.

Jury 评奖委员会
The jury's remit is to decide which submitted proposals are to be admitted to the competition, to appraise the admitted entries, to pick those proposals that best comply with the competition requirements, and to suggest the direction of the process. The jury is appointed by the sponsor. It makes its decisions solely on the basis of the criteria listed in the competition brief. Its voting members are architectural jurors on the one hand and technical jurors on the other. They are supported in their decision-making by non-voting alternate jurors, experts, and examiners.

Jury Meeting 评奖委员会会议
Chaired by an experienced architectural juror appointed by consensus amongst its members, the jury meets in a closed session to select the competition entry that best responds to the task. Jury meetings are jointly moderated by the jury chair and the competition manager. Their proceedings and outcome are recorded in the jury meeting minutes. Meetings are attended by jurors, experts, examiners, representatives of the sponsor, by the competition

manager, and his agents. Guests too may be invited. All attendees must declare not to have discussed the competition task with any of the competitors and must undertake to keep deliberations confidential.

Life-cycle Cost Analysis 生命周期成本分析

This cost analysis relates planning and construction costs to those incurred over the lifetime of a (proposed) building: management costs of the real estate, the costs of consumables, of conversion and adaptation, of renovation and restoration plus costs associated with demolition and disposal. This provides the basis for appraising (initial) investment costs against consequential costs and for assessing the impact of planning decisions on costs arising during the overall (expected) lifetime of a building.

Minutes 会议记录

All meetings held in the course of the competition and during its preparation are documented in the form of minutes. The minutes of the colloquia and the jury meeting are automatically distributed to all the parties involved.

Official Scale of Fees for Services by 支付建筑师和 Architects and Engineers 工程师服务酬金的官方标准

In short referred to as HOAI, this by-law (a) defines the official profile of services rendered by architects, interior architects, landscape architects, town planners, and graduated engineers from various fields and (b) regulates their remuneration. The HOAI as a price-regulatory regime has binding force upon the contracting parties. However, remuneration for the consulting services provided by a competition manager is not covered by the HOAI.

Online-Forum 在线论坛

Clarification of issues raised may take the form of an Internet-based online forum. Questions about the project task can be posted anonymously in a password-protected area of the competition homepage and coordinated answers are made available, so as to ensure a uniform level of information on the part of all the competitors. Online forums improve the level of dialogue with competitors as they allow exchanges to take place over a longer period of time. They also contribute to cost saving as they may replace colloquia, or make colloquia more efficient.

Open Competition 公开竞赛

A participant in an open competition may be any individual with the required professional qualification. The number of competitors is not restricted by a candidature procedure or any similar scheme. In order to reduce the burden on

the parties involved it is customary to conduct an open competition in two stages.

Ordnung SIA 142 SIA142条令

This standard established by the Swiss Institute of Architects and Engineers (SIA) lists the rules governing Swiss planning and comprehensive-services competitions in architecture, engineering, and similar fields. The more detailed annexes ("guidelines") complementing it deal with determining the amounts of prize money, the selection of competitors, and procedural options. In 2007, SIA 143 was issued as a separate body of rules governing the commissioning of studies.

Period Allotted to the Task 任务分配阶段

The period allowed to the competitors for preparing their design proposals, i.e. the space of time between issuing of competition documents and projects due-date. Customarily competitors are allowed about eleven weeks or, in the instance of a two-stage competition, five weeks in stage one and seven weeks in stage two.

Preliminary Examination 初试

This term covers two different notions, as it designates
(a) the examiners as a group entity, together with the competition manager's support team, and
(b) the activity exercised by these individuals.
The preliminary examination – as a group – is responsible for appraisal of the competition entries and processing the relevant data and pieces of information, prior to the meeting of the jury.
The outcome of the preliminary examination – as an activity – is recorded in the Report on the Preliminary Examination. At the meeting of the jury it is up to the examiners to point out to the jury the essential functional and economic characteristics of the competition entries and to make the jury aware of any aspect that in their opinion the jury should not disregard.

Preparation of a Competition 竞赛的准备工作

The tasks in the run-up to a competition include formulating the task, drafting the competition materials, preparing, conducting, moderating, and documenting all preparatory meetings and colloquia. Next are scheduling and preparing all meetings and colloquia to be held during the competition proper, advice as to the appointment of the jury, and coordinating the procedure with the competent permit authorities and the Chamber of Architects. This includes monitoring costs and controlling schedule during this stage and, in the case of a restricted open competition, carrying out an application procedure resulting in the selection of a suitable number of competitors.

Presentation Drawings 示意图

The materials to be submitted by the competitors normally include presentation drawings, examination drawings, calculations, and documents such as explanatory texts and building specifications. The presentation drawings are the original drawings that are to be presented to the jury and are ultimately exhibited.

Press Conference 记者招待会

Upon conclusion of a competition, the outcome is normally announced at a press conference that may coincide with an exhibition opening for all the designs submitted.

Principles and Guidelines for Competitions in Regional Planning, Town Planning, and Architecture 地区规划、城市规划和建筑竞赛的原则和指导方针

This refers to the German Public Code governing the preparation and conduct of competitions in the fields specified (short reference: GRW 95). Its preamble states that it "should form the basis [...] for a fair cooperation, in a spirit of partnership, of all the parties involved in a competition and help to develop architectural culture [...] in pursuit of societal, economic, ecologic, and technologic progress." Other regulations that apply to public authorities in Germany are those of the Public Code for Contracting Services from the Liberal Professions (VOF).

Prizes 奖金

The competition amount is split into prize money, money for purchase awards, and (possibly) remuneration for services rendered. The amounts allotted to prizes are mentioned in the public announcement of the competition. The first prize is to be awarded to the entry that best satisfies the problem described by the sponsor. If a prize winner is commissioned with the elaboration of his proposal, remuneration for services rendered during the competition will be reduced by the amount of prize money received, provided that essential parts of the submitted design form the basis for the detailed plan.

Programme of Functions and Spaces 功能与空间的规划

This is one of the task-defining elements of a competition. When not provided by the sponsor, it is up to the competition manager to prepare, in cooperation with the sponsor, the programme of functions and spaces.

Project Competition 项目竞赛

A project competition is intended to demonstrate, on the basis of a clearly defined programme and unambiguous service requirements, the feasibility of a project in terms of planning and to help select the party to be awarded the planning contract.

Project Cost Estimate 项目成本预算

A prerequisite for calculating competition costs is the project cost estimate based on the available characteristic data. This cost estimate may also be a part of the problem posed to the competitors or it may be included in the preliminary examination.

Project Development 项目开发

This term designates all the activities intended to integrate the location elements, project idea, and capital in a way that ensures the creation and sustained profitable utilisation of a competitive, socially acceptable, and environmentally compatible built object. In a more restricted sense project development comprises the stage from project kick-off (or conception of the project idea) to the award of planning commissions, while in a wider sense it covers the whole life cycle of the project.

Project Due-date 项目到期日

The moment when the time period allowed for the task ends and all competitors must hand in their completed proposals in full compliance with competition requirements. As a rule, it is a given point in Central European Time by which a competitor must deliver his entry to a courier service or to the postal service. This practice ensures that each competitor is accorded the same amount of time regardless of geographical location. On demand, the competitors must provide proof of timely delivery.

Project Management 项目管理

This term describes the unbiased, independent exercise of (delegable) technical, economic, and legal activties otherwise exercised by the sponsor.

Promise of Contract Award 合同判授承诺

A competition constitutes a contractual relationship between the sponsor and each competitor. An essential element of this relationship is the pledge to award one of the prize-winners, in return for the mostly unremunerated services rendered by the competitors, the planning commission for the project. The promise of contract award must state accurately and in a legally sound manner the scope of services offered.

Purchase Award 购买奖金

In open or (partially) restricted competitions, the jury customarily reserves part of the prize money (usually 20 percent) for purchase awards. These are given to entries that, while not realisable in full, are distinguished by outstanding partial solutions. The purchase award implies that the sponsor may make use of that partial solution.

RAW 2004 RAW2004
Derived from the GRW 1995, the Rules for the Conduct of Competitions in Regional Planning, Town Planning and the Construction Industry (RAW 2004) are applicable law in only three German states (Bremen, Lower Saxony, and North Rhine-Westphalia). The discrepancies reflected by this state of affairs, outcome of many years of dispute over a reform of GRW, are to be resolved with the upcoming revision of GRW. In contrast to GRW, RAW rules seek to introduce less stringent requirements, an attempt that partly succeeds and partly results in ambiguities or, to give it a positive spin, more leeway.

Remuneration for Services Rendered 支付提供服务的酬金
In the case of competitions by invitation and in stage two of some two-stage competitions, the competitors are paid a portion of the competition amount as remuneration for the services rendered. The remainder of the competition amount is distributed, as in any ordinary competition, in the form of prizes and purchase awards.

Report on the Preliminary Examination 初试报告
Jury sessions begin with the delivery of the report on the preliminary examination. In the first round, the procedure to be followed during preliminary examination is explained and the outcome of the formal inspection of entries is announced. During the subsequent informational round, the examiners present the designs in detail and in an unbiased manner. They explain the results of the preliminary examination for each entry so as to acquaint the jury, within a short span of time, with the range of problems posed and the set of solutions provided.

Service Phase 使用阶段
The official scale of fees for services by architects and engineers (HOAI) divides the performance profiles of German architects and engineers into service phases, to each of which is assigned a portion of the remuneration due. The service phases describe a regular planning and construction process with reference to the different activities performed by an architect or engineer. Service phase one, establishing the basis of the project, normally pertains to the competition manager, while service phase two, preliminary design, pertains to the competitors.

Specialist Consultant 专家咨询人员
The competitor submits his design to the competition, and in return receives sponsor's promise of contract award for one of the entries submitted. The design may call for input from specialist consultants, experts from various disciplines such as landscaping and engineering. The sponsor may even suggest the participation of specialists with specific expertise. These experts are not required to furnish the same evidence or level of eligibility as that demanded of applicants to the competition, in the procedural prerequisites.

Sponsor 主办方
As a rule, a competition's sponsor is the party that subsequently will award the planning contract for the project emerging from the competition. Concerning the promise of contract award, the sponsor is the contracting party for each competitor.

Suggestions for Work 设计作品建议
Essential elements of the competition brief are its precision and inclusiveness. To ensure uniformity, the brief should comprise clear statements as to what items are to be included in the design proposal and how they are to be presented.

Technical Juror 专家评奖委员会
The jury comprises architectural jurors and technical jurors. What is required of the latter is particular familiarity with the content of the competition task and with local conditions. Generally, representatives of the sponsor and of the local community act as technical jurors.

Two-Stage Competition 两个阶段的竞赛
To limit overall cost and improve the competition results, a competition may be conducted in two stages, thus allowing for a more intensive dialogue with the competitors and for phased elaboration of designs. At the end of phase one, the jury picks, from a comparatively large number of not highly developed entries, the ones that respond best to the requirements and that are to be elaborated upon during stage two, at the end of which the jury takes its ultimate decision.

Urban-planning Competition 城市规划竞赛
An urban-planning competition, as opposed to a building project competition, aims at the solution of town-planning tasks. Its outcome may form the basis for an urban land-use planning procedure or a building project competition.

Working Drawings 施工图

As a basis for their design work, the competitors are provided with all the required planning documents, digitally formatted. Working drawings (as opposed to information drawings) are a graphic model to be followed for the representation of each competitor's design. They contain such elements (e.g. neighbouring buildings and borders of the competition site) as are required to ensure that the plans submitted are at a uniform scale and depict identical elements.

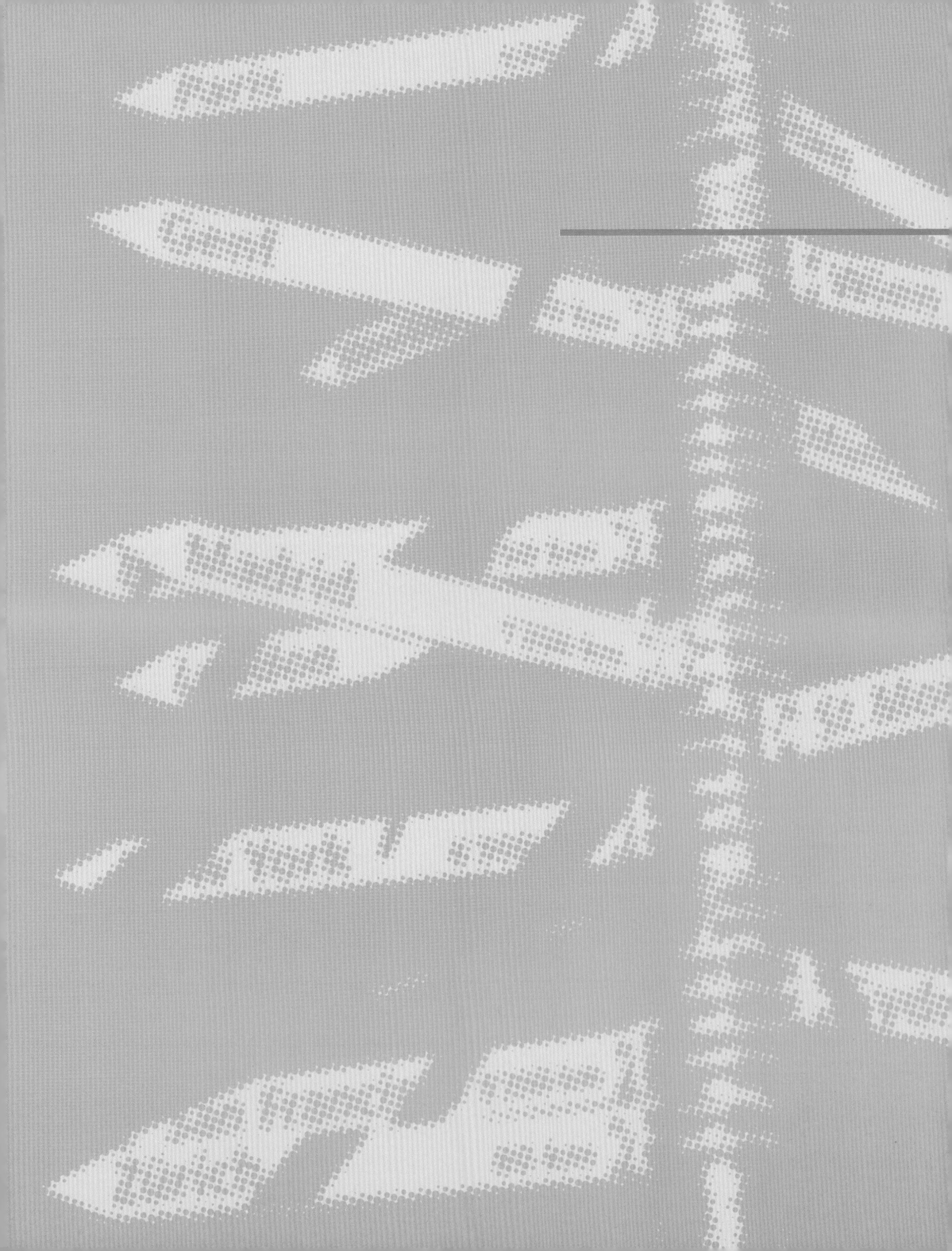

Links to Platforms, Organisations, and Legal Regulations

与论坛、机构和法律条文的相关链接

Links to Platforms, Organisations, and Legal Regulations

与论坛、机构和法律条文的相关链接

Topical links to websites covering competitions, the building industry, and pertinent regulations

与竞赛、建筑业和有关法律条文的代表性链接

Architectural competitions 建筑竞赛

- www.BauNetz.de/arch/_wettbewerbe
  The competition section of Baunetz
- www.archi.fr/EUROPAN
  Europan
- www.archinform.de
  Literature, works of reference, biographies
- www.newitalianblood.com
  Interactive platform for the announcement of competitions and of their outcome
- www.europaconcorsi.com
  Competitions in Europe (I)
- www.nextroom.at
  Architecture and competitions in Austria (AT)

Platforms for the building industry 建筑业平台

- www.baunetz.de
  Internet portal of Bertelsmann AG
- www.archi.fr
  French portal to topics architectural
- www.archguide.com
  Architecture network in Belgium
- www.irish-architecture.com
  Architecture network in Ireland
- www.archicool.com
  Architecture network in France
- www.archined.nl
  Architecture network in The Netherlands
- www.architettura.it
  Architecture network in Italy
- www.architectenwerk.nl
  Architecture network in The Netherlands
- www.architektura.info
  Building portal in Poland
- www.vitruvio.ch
  Architecture network in Switzerland
- www.europaconcorsi.com
  Competition portal from Italy
- http://archnet.org/lobby.tcl
  Architects' portal focusing on Islamic architecture
- www.archijob.co.il
  Israeli architecture portal
- www.abbs.com.cn
  Chinese architecture portal

Trade journals 贸易杂志

- www.wettbewerbe-aktuell.de
  Current competitions (Germany)
- www.BauNetz.de/arch/bauwelt
  Bauwelt (Germany)
- www.baumeister.de
  Baumeister (Germany)
- www.dbz.de
  DBZ (Germany)
- www.tema.de/mass/news/cl4.htm
  AIT (Germany)
- www.detail.de
  Detail (Germany)
- www.sia.ch
  Tec21 (Switzerland)
- www.hochparterre.ch
  Hochparterre (Switzerland)
- www.penrose-press.com/IDD/pub/cards/J11726.html
  International directory of design (Switzerland)
- www.archis.org
  Archis (Netherlands)
- www.arplus.com
  Architectural Review (UK)
- www.elcroquis.es
  El Croquis (Spain)
- www.competitions.org
  Competition (USA)

Chambers of architects and federations 建筑师协会和建筑师联盟

- www.bak.de
  Bundesarchitektenkammer Germany
- www.architekt.de/bdia
  Bund Deutscher Innenarchitekten (BDIA)
- www.ak-berlin.de
  Architektenkammer in Berlin
- www.aknw.de
  Architektenkammer Norh Rhine-Westphalia
- www.architekten-thueringen.org
  Architektenkammer Thuringia
- www.akh.de
  Hessische Architektenkammer (Hesse)
- www.byak.de
  Bayerische Architektenkammer (Bavaria)
- www.akrp.de
  Architektenkammer Rhineland-Palatinate
- www.akbw.de
  Architektenkammer Baden-Württemberg

Organisations 组织机构

- www.bda-architekten.de
  Bund Deutscher Architekten
- www.architekten-ueber-grenzen.de
  Architekten über Grenzen (D)
- www.planned-in-germany.de
  Planned in Germany
- www.vbi.de
  Verband Beratender Ingenieure

- www.ak-hh.de
  Hamburgische Architektenkammer
- www.architektenkammer-bremen.de
  Architekten- und Ingenieurkammer Bremen
- www.ak-brandenburg.de
  Brandenburgische Architektenkammer Potsdam
- www.aksachsen.org
  Architektenkammer Saxony
- www.architektenkammer-mv.de
  Architektenkammer Mecklenburg-Western-Pomerania
- www.aknds.de/htm/start.htm
  Architektenkammer Lower Saxony
- www.ak-lsa.de
  Architektenkammer Saxony-Anhalt
- www.archi.fr/UIA
  Union internationale des architectes
- www.sia.ch
  Schweizer Ingenieurs- und Architektenverein (Switzerland)
- www.bna.nl
  Bond van Nederlandse Architecten (Netherlands)
- www.architecture.com
  Royal Insitute of British Architects (UK)
- www.aik-sh.de/
  Architekten- und Ingenieurkammer Schleswig-Holstein
- www.oai.lu
  Luxemburgische Architektenkammer (Luxembourg)
- www.rias.org.uk
  Royal Institute of Architects in Scotland
- www.riai.ie
  Royal Institute of Architects in Ireland
- www.aia.org
  American Institute of Architects
- www.architecture.com.au
  The Royal Australian Institute of Architects
- www.comarchitect.org
  The Commonwealth Association of Architects
- www.aua-architects.com
  Africa Union of Architects

Legal regulations 法律条文

- "Grundsätze und Richtlinien für Wettbewerbe auf den Gebieten der Raumplanung, des Städtebaus und des Bauwesens", novel version of the 22nd December 2003 – GRW 1995 (Principles and Guidelines for Competitions in Regional Planning, Town Planning, and Construction)
- "Verdingungsordnung für freiberufliche Leistungen" of the 26th August 2002 – VOF (Rules for the Award of Professional Services Contracts)
- "Gesetz gegen Wettbewerbsbeschränkungen" of the 26th August 1998 – GWB (Law against Restraint of Trade)
- Council Directive 92/50/EEC of the 18th June 1992 relating to the Coordination of Procedures for the Award of Public Service Contracts
- "Verordnung über die Vergabe öffentlicher Aufträge" of the 9th January 2001 (novel version of the 11th February 2003) – VgV (Ordinance on the Award of Public Contracts)

http://www.phase1.de/forum_links.htm
This page of the [phase eins]. website provides active links to the above legal regulations and other topical web pages.

Partners and Collaborators 2006-2008

合作人与共事人
2006－2008

Partners and Collaborators 2006–2008
合作人与共事人2006 – 2008

The people behind [phase eins].
[phase eins].公司背后的工作人员

The partners 合作伙伴
Benjamin Hossbach (co-founder, since 1998),
Christian Lehmhaus (since 2001)

Christine Eichelmann (since 2008), Martin Linz (since 2008)

The permanent staff 正式员工
Alexander Bulgrin, Uwe "Matt" Dahms, Barbara Frei,
Julia Grahl, Stefan Haase, Raschid Hafiz, Marc Havekost,
Michaela "Svea" Heinemann, Sebastian Illig, Maja Kastaun,
Ronny Kutter, Susanne Mocka, Brigitte Panek,
Birgit Pfisterer, Angela Salzburg, Björn Steinhagen,
Harald Theiss, Silke Wischhusen

The freelancers, students, and trainees 自由建筑师、建筑系在校生和培训生
Mogdeh Ali, Katrin Bade, Sameh Balo, Uwe Barsch,
Mario Bär, Anna-Luisa Bories, Martin Bütow, Annika Bleckat,
Jewgeniy Borshchevskiy, Max Dölling, Marc Dufour-Feronce,
Philippe Dufour-Feronce, Lana Eichelmann, Nicole Erbe,
Oliver Gassner, Philipp Haas, Michael Kandel, Jens Kärcher,
Patrick Kutterolf, Kornelia Klimmeck, Paul Maiwald,
Jürgen Middelberg, Paul-Merlin Müller, Viet Dung Nguyen,
Florian Pacher, Michael Pawelzick, Anne Peters,
Andreas Reeg, Ceva Sahinarslan, Ina Schoof, Ben Tullin,
Yakup Vardar, Lisamarie Villegas Ambia, Anina Wagner,
Alexander G. Williams

The independent examiners 独立的主考官
Annette Bresinsky, Heinrich Burchard, Friedhelm Gülink,
Helmut Hanle, Roland Kuhn, Birgit Petersen

The third-party specialists 第三方专家
- Michael Rädler (exhibition building, display panels), Berlin, www.mraedler.de
- Klaus Rupprecht + Bernard D. Wilmot (translations), Berlin
- Olaf Schreiber, Fa. Raecke & Schreiber (Internet and data bank solutions), Berlin, www.raecke-schreiber.de
- Sirko Sparing, Fa. dBusiness (repro and printing), Berlin, www.dbusiness.de
- Olaf Thiede, Fa. Jack-in-the-Box (hardware and server administration), Berlin, www.jack-in-the-box.de
- Hans-Joachim Wuthenow, (photography), Berlin, www.wuthenow-foto.de

Acknowledgement   致谢

We extend our thanks to all our fellow-architects who with
their designs contributed to this book, and our particular
gratitude is due to our clients, including those whose projects
did not fit into the limited space of this book. Their commitment to architectural excellence and their aspiration to make
contributions of their own to a high-quality built environment
are at the very basis of the work performed by [phase eins].

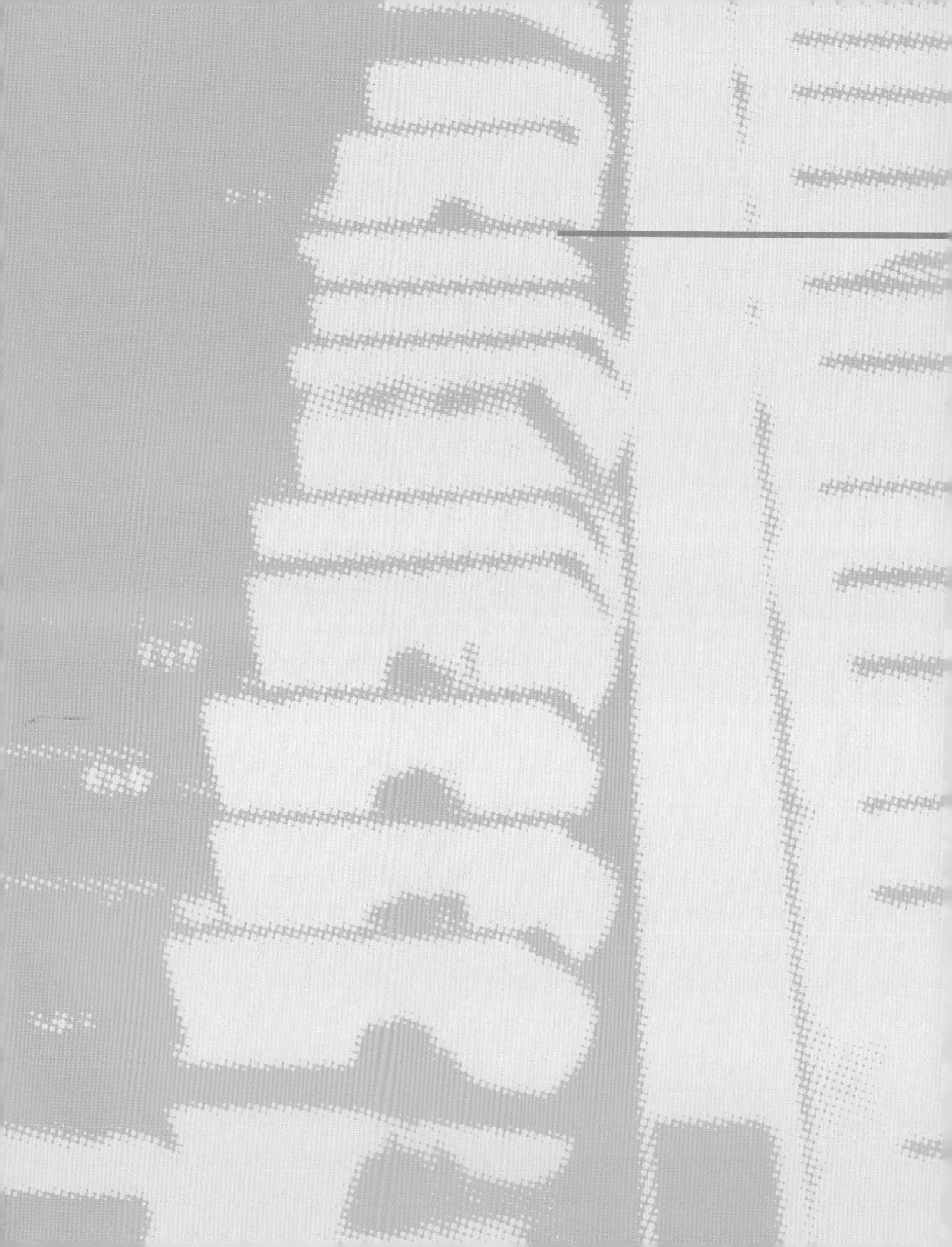

Index

索引

Index
索引

Complete index of authors and offices
作者及建筑事务所的全部索引

V

W

Y

Z

图书在版编目（ＣＩＰ）数据

建筑竞赛／（德）本杰明·胡斯巴赫，（德）克里斯蒂安·雷姆豪斯编；
王晨晖译. —— 沈阳：辽宁科学技术出版社，2009.5
ISBN 978-7-5381-5923-3

I. 建… II.①胡…②雷…③王… III. 建筑设计－竞赛－
简介－世界 IV. TU206

中国版本图书馆CIP数据核字（2009）第055326号

- -

出版发行：辽宁科学技术出版社
　　　　　（地址：沈阳市和平区十一纬路29号　邮编：110003）
印 刷 者：利丰雅高印刷（深圳）有限公司
经 销 者：各地新华书店
幅面尺寸：225mm×285mm
印　　张：35
插　　页：8
字　　数：200千字
印　　数：1~3000
出版时间：2009年5月第1版
印刷时间：2009年5月第1次印刷
责任编辑：陈慈良
封面设计：张　迪
版式设计：张　迪
责任校对：周　文

书　　号：ISBN 978-7-5381-5923-3
定　　价：498.00元（1、2册）

联系电话：024-23284360
邮购热线：024-23284502
E-mail: lkzzb@mail.lnpgc.com.cn
http://www.lnkj.com.cn